NATO ASI Series

Advanced Science Institutes Series

A series presenting the results of activities sponsored by the NATO Science Committee, which aims at the dissemination of advanced scientific and technological knowledge, with a view to strengthening links between scientific communities.

The Series is published by an international board of publishers in conjunction with the NATO Scientific Affairs Division

A Life Sciences B Physics	Plenum Publishing Corporation London and New York
C Mathematical and Physical Sciences D Behavioural and Social Sciences E Applied Sciences	Kluwer Academic Publishers Dordrecht, Boston and London
F Computer and Systems Sciences G Ecological Sciences H Cell Biology I Global Environmental Change	Springer-Verlag Berlin Heidelberg New York London Paris Tokyo Hong Kong Barcelona Budapest

NATO-PCO DATABASE

The electronic index to the NATO ASI Series provides full bibliographical references (with keywords and/or abstracts) to more than 30000 contributions from international scientists published in all sections of the NATO ASI Series. Access to the NATO-PCO DATABASE compiled by the NATO Publication Coordination Office is possible in two ways:

- via online FILE 128 (NATO-PCO DATABASE) hosted by ESRIN, Via Galileo Galilei, I-00044 Frascati, Italy.
- via CD-ROM „NATO Science & Technology Disk" with user-friendly retrieval software in English, French and German (© WTV GmbH and DATAWARE Technologies Inc. 1992).

The CD-ROM can be ordered through any member of the Board of Publishers or through NATO-PCO, Overijse, Belgium.

Series I: Global Environmental Change, Vol. 15

The ASI Series Books Published as a Result of Activities of the Special Programme on Global Environmental Change

This book contains the proceedings of a NATO Advanced Research Workshop held within the activities of the NATO Special Programme on Global Environmental Change, which started in 1991 under the auspices of the NATO Science Committee.

The volumes published as a result of the activities of the Special Programme are:

Vol. 1: **Global Environmental Change.** Edited by R. W. Corell and P. A. Anderson. 1991.
Vol. 2: **The Last Deglaciation: Absolute and Radiocarbon Chronologies.** Edited by E. Bard and W. S. Broecker. 1992.
Vol. 3: **Start of a Glacial.** Edited by G. J. Kukla and E. Went. 1992.
Vol. 4: **Interactions of C, N, P and S Biogeochemical Cycles and Global Change.** Edited by R. Wollast, F. T. Mackenzie and L. Chou. 1993.
Vol. 5: **Energy and Water Cycles in the Climate System.** Edited by E. Raschke and D. Jacob. 1993.
Vol. 6: **Prediction of Interannual Climate Variations.** Edited by J. Shukla. 1993.
Vol. 7: **The Tropospheric Chemistry of Ozone in the Polar Regions.** Edited by H. Niki and K. H. Becker. 1993.
Vol. 8: **The Role of the Stratosphere in Global Change.** Edited by M.-L. Chanin. 1993.
Vol. 9: **High Spectral Resolution Infrared Remote Sensing for Earth's Weather and Climate Studies.** Edited by A. Chedin, M.T. Chahine and N.A. Scott. 1993.
Vol. 10: **Towards a Model of Ocean Biogeochemical Processes.** Edited by G. T. Evans and M. J. R. Fasham. 1993.
Vol. 11: **Modelling Oceanic Climate Interactions.** Edited by J. Willebrand and D.L.T. Anderson. 1993.
Vol. 12: **Ice in the Climate System.** Edited by W. Richard Peltier. 1993.
Vol. 13: **Atmospheric Methane: Sources, Sinks, and Role in Global Change.** Edited by M. A. K. Khalil. 1993.
Vol. 14: **The Role of Regional Organizations in the Context of Climate Change.** Edited by M. H. Glantz. 1993.
Vol. 15: **The Global Carbon Cycle.** Edited by M. Heimann. 1993.
Vol. 16: **Interacting Stresses on Plants in a Changing Climate.** Edited by M. B. Jackson and C. R. Black. 1993.

The Global Carbon Cycle

Edited by

Martin Heimann

Max-Planck Institut für Meteorologie
Bundesstrasse 55
D-20146 Hamburg, Germany

Springer-Verlag
Berlin Heidelberg New York London Paris Tokyo
Hong Kong Barcelona Budapest
Published in cooperation with NATO Scientific Affairs Division

Proceedings of the NATO Advanced Study Institute on the Contemporary Global Carbon Cycle, held in Il Cioccio, Italy, September 8–20, 1991

ISBN-13:978-3-642-84610-6 e-ISBN-13:978-3-642-84608-3
DOI: 10.1007/978-3-642-84608-3

Library of Congress Cataloging-in-Publication Data
The Global carbon cycle / edited by Martin Heimann. p. cm. – (NATO ASI series. Series I, Global environmental change ; vol. 15)
"Published in cooperation with NATO Scientific Affairs Division."
"Proceedings of the NATO Advanced Study Institute on Contemporary Global Carbon Cycle, held in Il Ciocco, Italy, September 8-20, 1991" – T.p. verso.
ISBN-13:978-3-642-84610-6
1. Atmospheric carbon dioxide–Environmental aspects–Congresses. 2. Carbon cycle (Biochemistry)–Environmental aspects–Congresses. I. Heimann, Martin, 1949- . II. North American Treaty Association. Scientific Affairs Division. III. NATO Advanced Study Institute on the Contemporary Global Carbon Cycle (1991 : Il Ciocco, Italy) IV. Series. QC879.8.G54 1993 574.5'222–dc20 93-31766

Softcover reprint of the hardcover 1st edition 1993

Typesetting: Camera ready by authors
31/3145 - 5 4 3 2 1 0 - Printed on acid-free paper

PREFACE

This book is the outcome of a NATO Advanced Study Institute on the contemporary global carbon cycle, held in Il Ciocco, Italy, September 8-20, 1991. The motivation for this ASI originated from recent controversial findings regarding the relative roles of the ocean and the land biota in the current global balance of atmospheric carbon dioxide. Consequently, the purpose of this institute was to review, among leading experts in the field, the multitude of known constraints on the present day global carbon cycle as identified by the fields of meteorology, physical and biological oceanography, geology and terrestrial biosphere sciences. At the same time the form of an Advanced Study Institute was chosen, thus providing the opportunity to convey the information in tutorial form across disciplines and to young researchers entering the field.

The first three sections of this book contain the lectures held in Il Ciocco. The first section reviews the atmospheric, large-scale global constraints on the present day carbon cycle including the emissions of carbon dioxide from fossil fuel use and it provides a brief look into the past. The second section discusses the role of the terrestrial biosphere and the third the role of the ocean in the contemporary global carbon cycle. The final section contains the summaries of the workshop discussions where the participants of the ASI tried to identify within crucial areas what we do know, what we do not know and which research efforts should be pursued in order to improve our knowledge on the global carbon cycle.

The science of global change is a rapidly evolving field and meetings, such as the ASI of Il Ciocco, are essential to maintain the coherence in an interdisciplinary field like the study of the global carbon cycle. It is hoped, that the outcome of the ASI, as witnessed by this document, will be remembered as another milestone documenting the progress in this field.

ACKNOWLEDGMENTS

I sincerely thank Graeme Pearman, whose initiative made the ASI possible in the first place, and Mike Fasham and Colin Prentice for their guidance and assistance as members of the scientific committee of the ASI. I also thank all the lecturers for devoting their time to attend the ASI and for submitting their written contributions. I also appreciate the support of the numerous reviewers. Furthermore I thank Bruno Giannasi and his staff of the International Center Il Ciocco who made our stay at Il Ciocco remarkable even beyond the scientific discussions. I also thank Karin Niedl for her diligent help in the thousand tasks that have to be looked after in planning and conducting the ASI. I also thank Marion Grunert for her support in preparing this book.

The Advanced Study Institute was sponsored by NATO under the Special Programme on the Science of Global Environmental Change, the U.S. Department of Energy, the U.S. National Science Foundation, the Australia/New Zealand/UK Tripartite Initiative and the Commission of the European Communities.

Martin Heimann

25 February 1993

TABLE OF CONTENTS

LECTURE 1: GLOBAL OBSERVATIONS OF ATMOSPHERIC CO_2

Dr. Charles D. Keeling
Scripps Institution of Oceanography
University of California, San Diego
La Jolla, California 92093-0220
U.S.A.

Introduction. The global carbon cycle consists of the geochemical reservoirs that store carbon on the earth and the pathways that transport carbon between them. With respect to annual through decadal time scales to be discussed here, the principal reservoirs of interest are the atmosphere, the oceans, and three land compartments consisting of land plants, their detritus, and soils, and called collectively the terrestrial biosphere. Rivers, lakes, and animals are of only marginal importance to the global carbon inventory, but they provide some of the pathways linking the terrestrial biosphere to the oceans and the atmosphere.

A fourth reservoir, the lithosphere, stores very large amounts of carbon in limestone and shale, but this carbon interacts so slowly with carbon in the three above named reservoirs that we need not be concerned with it here except to keep in mind that coal, petroleum and natural gas are extracted from it. These fossil fuels, when they are burned, create carbon dioxide (CO_2) as a by-product. This unnaturally produced gas is presently entering the atmosphere globally at a rate that far exceeds the natural rate of evolution of CO_2 from the lithosphere, and even exceeds the net rates of carbon transfer between the atmosphere, the oceans, and the terrestrial biosphere on time scales longer than annual.

The oceanic carbon reservoir contains a biospheric component comprising living marine plants and animals and their detritus, but it also stores a very considerable quantity of inorganic carbon as dissolved bicarbonate and carbonate salts. The latter, although products of the slow weathering of carbonaceous rocks, are important on short time scales as well, because they mediate the release and absorption of CO_2 exchanged with the atmosphere.

The atmosphere serves mainly as a passive reservoir for carbon. It stores carbon containing gases that have escaped from the other two reservoirs owing to their volatility.

Human activities today influence the carbon cycle far more than in the past. Industrial society, accompanied by a rapidly rising world population and increasing mechanization of agriculture and forestry, has modified extensive areas of previously undisturbed landscape over the past two centuries, and thus altered the terrestrial biosphere. Above all, however, industry has impacted all of the major carbon reservoirs through the combustion of fossil fuels. Ever since the adoption of the coal-fired steam engine, these fuels have been the major source of mechanical energy for human society. As a consequence, the concentration of atmospheric CO_2 has been rising steadily for more than a hundred years. A well publicized consequence of this rise is an attending greenhouse effect, caused by the ability of atmospheric CO_2 to turn back heat upwelled from the earth's surface, and is expected to cause noticeable warming on a global scale when the

NATO ASI Series, Vol. I 15
The Global Carbon Cycle
Edited by M. Heimann

CO_2 concentration has increased substantially above preindustrial levels.

Rapid conversion of native forest and grasslands to agriculture and the harvesting of timber at unprecedented rates have also produced changes in the carbon cycle, but the alterations are still too subtle to make obvious the extent of the changes to the carbon cycle and their likely global consequences. A major unknown is how greatly deforestation and land clearing for agriculture have aggravated the build-up of atmospheric CO_2 from fossil fuel combustion.

The oceanic carbon reservoir, because of its vast size and relative remoteness from the centers of human activity, has so far largely escaped the impact of modern industrial civilization, but not entirely. It is a sink for excess atmospheric CO_2 accumulating in the atmosphere, because of the ability of sea water to absorb CO_2 gas. The sea water retains much of the excess gas that it absorbs by reducing its volatility through a chemical reaction with dissolved carbonate salt to form additional dissolved bicarbonate salt. The consumed carbonate is then partially replenished by transport of additional carbonate to the sea surface from the deep oceans.

The strength of this sink for excess atmospheric CO_2 is limited by the vigor of the ocean's vertical transport process. Because we do not know precisely the strength of this process, the strength of the sink for excess CO_2 cannot be precisely determined. The few available direct measurements of oceanic carbon are inadequate to establish the sink strength directly.

The terrestrial biosphere may also be a sink for excess atmospheric CO_2 because plants are sensitive to the amount of CO_2 in the air. When plants secure adequate light and nutrients in controlled experiments, they grow faster in ambient air containing elevated concentrations of CO_2. Whether such additional growth is prevalent in the natural environment is not well established, however, nor how such growth may be inhibited when the biosphere is also disturbed directly by human activities.

Although human disturbances to the terrestrial biosphere and oceans are suspected to be occurring, the extent of these disturbances is not well known. The rise in atmospheric CO_2 presently furnishes the strongest direct evidence of human impact on the carbon cycle, but this information doesn't indicate whether the rise is almost entirely due to fossil fuel combustion or is also due to disturbances to the biosphere and oceans. Essentially, the strengths of the global sinks of fossil fuel derived CO_2 are precisely known only in combination, the sum being equal to the difference between the rate of industrial CO_2 emissions and the observed rise in atmospheric CO_2, a difference that is known to be about 45% of the emissions rate. Although there is great interest in predicting the future rise of atmospheric carbon dioxide for a prescribed scenario of fossil fuel use or change in agriculture and forestry practice, it is impossible to decide reliably how the various reservoirs of the carbon cycle will separately respond to further human interventions.

It should also be kept in mind that other carbon compounds are involved in the natural and perturbed carbon cycle, including carbon monoxide and methane gases, which contribute to the greenhouse effect along with CO_2. A large suite of organically derived compounds also play a role in the terrestrial and oceanic carbon cycles. Though not further discussed here, these gases and other compounds also deserve attention.

To improve knowledge of the carbon cycle, both for the sake of advancing environmental science in general as well as for assessing possible threats to the local and global environment, a strategy of multiple approaches is desirable. Two such approaches are, firstly, in a local setting to carry out specialized studies at judiciously chosen experimental sites to learn more about the pathways of carbon, but, in addition, in a global setting to measure repeatedly a set of key parameters that monitor more or less continuously the general state of the carbon cycle over time. In this lecture, I will focus on the second approach. This approach is particularly rewarding when applied to measurements of atmospheric gases because the turbulent mixing of the air by the atmospheric circulation insures that such chemical measurements, even though made at only a few strategic sites, portray the chemical state of extended regions of the atmosphere, and even the global average state.

Oceanic measurements of carbon parameters are more difficult to make than measurement in air, and typically represent the state of the carbon cycle only in restricted regions of the oceans. It is not surprising that oceanic time series measurements have been given less attention than they deserve.

Time series measurements of carbon in the terrestrial biosphere are also worthwhile, although the representativeness of any particular site is difficult to establish and the carbon reservoirs are profoundly inhomogeneous. An asset to terrestrial studies is the global availability of remote sensing data from satellites, for example of the greenness of the land surfaces. An index of greenness from satellite data was employed in the modeling of the carbon cycle that is discussed below.

The first of these two lectures address atmospheric CO_2 measurements made at the Scripps Institution of Oceanography. These measurements, together with a large body of data gathered by other investigators, provide evidence bearing on the relative importance of the oceans and the terrestrial biosphere in reducing the rise in atmospheric CO_2.

The second lecture describes two case studies of the oceanic carbon cycle, based on near-surface data gathered as time series in the Sargasso Sea near the Atlantic island of Bermuda, and in the South Equatorial Current of the Pacific Ocean between Tahiti and the Marquesas Islands. Some data near Hawaii are also discussed. Although time series data of the oceanic carbon cycle are still very limited, it is likely that a greater effort to gather such data will be made in the future. This second lecture, therefore, draws attention to an evolving approach to studying the oceanic carbon cycle.

Many additional aspects of the carbon cycle could be addressed. A substantial number of these are the subject of additional lectures of this NATO Institute.

The topics of the first lecture are arranged as follows. After a brief description of measurements carried out, the seasonal cycle of atmospheric CO_2 will be described. Although this short-term cycle is a natural feature of the carbon cycle, it evidently has been altered recently so that the seasonal oscillation has increased in amplitude. Next, interannual variations will be examined. The analysis of these variations will draw upon measurements of the $^{13}C/^{12}C$ isotopic ratio of atmospheric CO_2 that supplement measurements of CO_2 concentration since 1978. By examining the two kinds of data simultaneously it is possible to estimate the relative contributions of the oceans and the terrestrial biosphere to these interannual variations. In particular, El Niño events in the tropics appear to influence atmospheric CO_2 globally by extensively perturbing both the oceanic and terrestrial biospheric carbon cycles, the latter more strongly than the former. Finally, decadal time scale changes in the north-south distribution of CO_2 from 1980 to 1990 will be described. Isotopic data will again be employed in determining the causes of the observed changes. A substantial increase in the CO_2 in the northern hemisphere relative to the southern hemisphere is evidently due not only to an increased release of CO_2 from fossil fuel combustion predominantly in the northern hemisphere, but also to an increased net annual release of CO_2 there by the terrestrial biosphere.

Data base. Modern measurements of atmospheric CO_2 concentration began in 1957 at the South Pole and at La Jolla, California, and were gradually extended during the 1960's and 1970's to comprise sampling at an array of stations from the Arctic to Antarctica. The main focus has been on time series at fixed locations, such as those shown in Figure 1, sampled by the Scripps Institution of Oceanography. Near the years 1962, 1968, and 1980 in addition, however, "snapshots" were produced of the global patterns to which data from ships contributed (Bolin and Keeling, 1963; Keeling et al., 1989a). Except for continuous measurements made *in situ* at a few stations of which

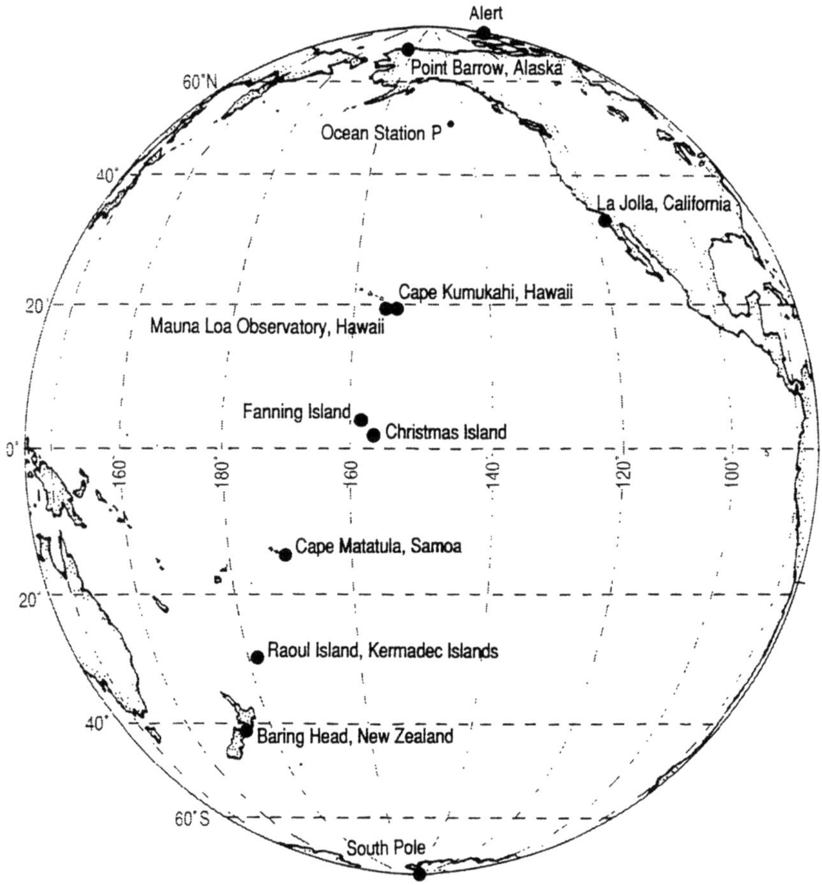

Figure 1. Locations for long term sampling. (Not all stations operated throughout the period of the study from 1957 to 1991).

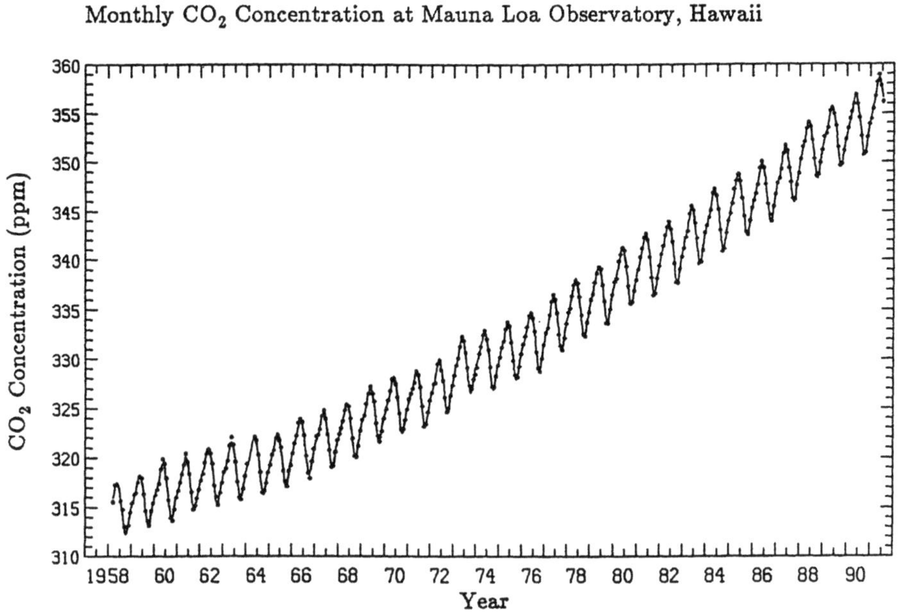

Figure 2. The concentration of atmospheric CO_2, in parts per million of dry air by volume (ppm), at Mauna Loa Observatory, Hawaii versus time. The dots denote monthly mean concentrations obtained from a continuously recording gas analyzer. The oscillating curve is a fit to a spline function combined with four harmonic terms.

Mauna Loa Observatory, Hawaii is the best known example (Figure 2), the data have been derived mainly from samples collected in flasks and returned to central laboratories for analysis. Isotopic measurements on flask samples, begun in 1978, have gradually been expanded to include most of the stations where the CO_2 concentration is measured. Plots of data since 1972, at the stations indicated in Figure 1, are shown in Appendix A, Figures A1-A4. North-south profiles for concentration and isotopic data, including 1990 as well as the years mentioned above, are shown in Appendix A, Figures A5-A10.

Some of the findings from atmospheric CO_2 measurements are highlighted below. Three time domains will be considered: seasonal, interannual, and decadal.

Changing amplitude of the seasonal cycle. At Mauna Loa Observatory the concentration of atmospheric CO_2 oscillates seasonally (see Figure 2). The pattern is similar from year to year, but the amplitude has increased appreciably since 1958 as shown in Figure 3. The rise, 0.54% per year on average, has not been steady, occurring mainly from 1973 to 1982. The South Pole CO_2 record shows an even larger increase in amplitude (Figure 4), but the change is more evenly distributed over time. A long, but incomplete record, for Point Barrow, Alaska, and shorter but unbroken records at three other stations, also show increases, as summarized in Table 1.

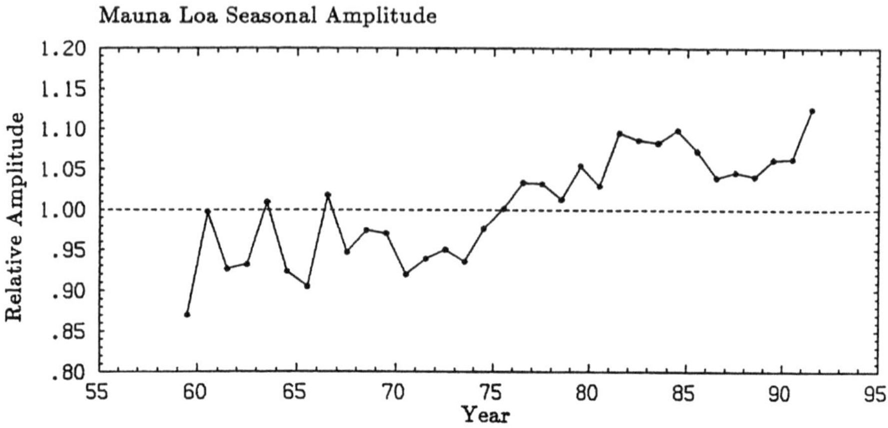

Figure 3. Amplitudes of the seasonal cycle in atmospheric CO_2 concentration at Mauna Loa Observatory, Hawaii. The amplitudes are expressed relative to the amplitude for the middle year of the record. The dots are connected by line segments to show their trend. Each year begins at a tick mark. Amplitudes are for annual intervals beginning in January.

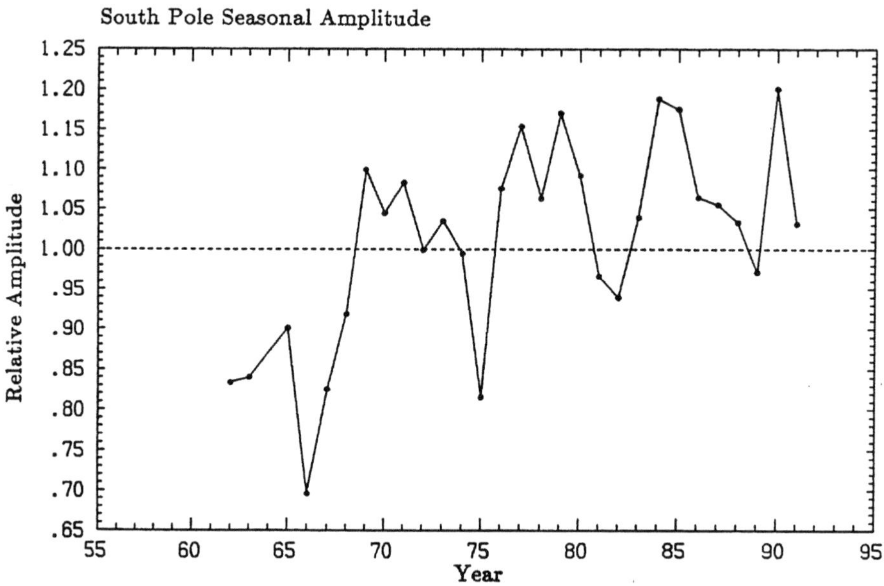

Figure 4. Same as Figure 3 for South Pole. Data before 1962 were too infrequent to establish a seasonal amplitude. No data were obtained in 1964. Amplitudes are for annual intervals beginning in July.

Three-dimensional tracer transport models predict regional differences in the character of the seasonal cycle in CO_2. These differences need to be understood before attempting to explain the observed changes in seasonable amplitude. Let us compare the atmospheric CO_2 data to predictions with a model that uses actual wind data (Heimann and Keeling, 1989). This model substantiates that the seasonal cycle at Mauna Loa and at more northerly stations, such as Point Barrow, Alaska (Figure 5), is mainly owing to land plant activity. Near the equator, however, where data show no significant increase in seasonal amplitude, the same model predicts that CO_2 from fossil fuel combustion, and CO_2 released by ocean water, contribute large, nearly opposing signals to the composite cycle (Figure 6). These signals arise mainly from seasonal changes in the north-south and vertical transports of CO_2 emitted to the air by relatively steady sources and sinks.

For the southern hemisphere the model predictions conform less well to the observations. It is difficult to correct for this shortcoming, because the seasonality in oceanic CO_2 exchange of the southern oceans is not known and cannot yet be reliably modeled, as noted by Fung (1986, pp. 464-468). Qualitatively, the large increase in amplitude of the seasonal cycle at the South Pole (see Figure 3) suggests that terrestrial and oceanic contributions to the seasonal cycle were nearly equal and opposite near the beginning of the record, but that the terrestrial contribution has increased in amplitude since, giving rise to a more and more dominant terrestrial signal in the record.

Table 1. Annual Rate of Increase in the Amplitude of the Seasonal Oscillation of Atmospheric CO_2 at Various Locations

Location	Approximate Latitude	Inclusive Dates of Observations	Rate of Increase* (in percent)
Point Barrow, Alaska	71°N	July, 1961 - Sep., 1967, Jan., 1974 - Apr., 1991	1.044 ± 0.070
Ocean Station 'P'	50°N	May, 1969 - June, 1981	0.738 ± 0.259
La Jolla, California	33°N	Jan., 1970 - June, 1991	1.247 ± 0.200
Mauna Loa Observatory	20°N	Mar., 1958 - June, 1991	0.535 ± 0.042
South Pole	90°S	June, 1961 - Feb., 1991	0.887 ± 0.199

* With reference to 1 January, 1980. The entries after the ± sign are standard deviations.

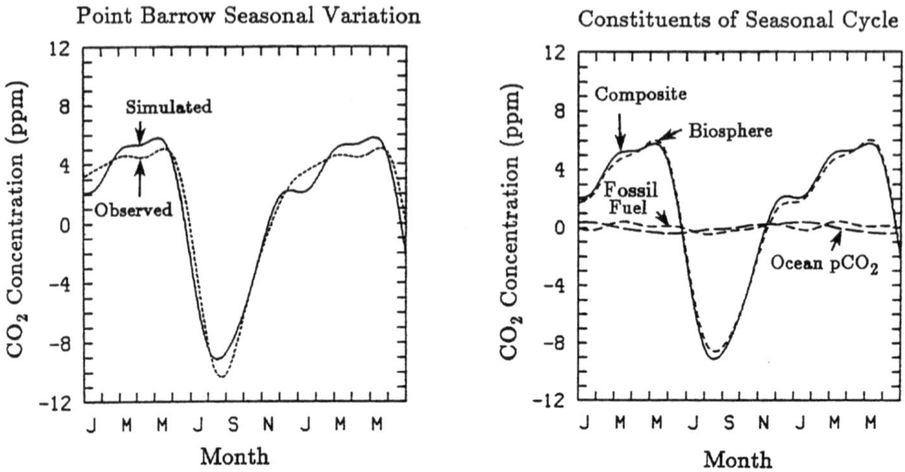

Figure 5.
Left Panel: Simulation with a three-dimensional transport model of the seasonal variation of atmospheric CO_2 in ppm (solid line) compared to observations (dashed line) at Point Barrow, Alaska. Simulated and observational concentrations are each represented by a sum of four harmonics fitted to the daily model prediction and to the individual observational data, respectively. The first six months are displayed twice in order to reveal the annual cycle more clearly.

Right Panel: Composite seasonal concentration variation (solid line, same as simulated curve of left panel) for Point Barrow, Alaska together with its constituents: variation due to the terrestrial biosphere (short dashes), variation resulting from the industrial CO_2 source (medium dashes), and variation from the oceanic pCO_2 source/sinks (long dashes). Industrial CO_2 refers to CO_2 released by combustion of fossil fuels and cement manufacture (Rotty, 1987).

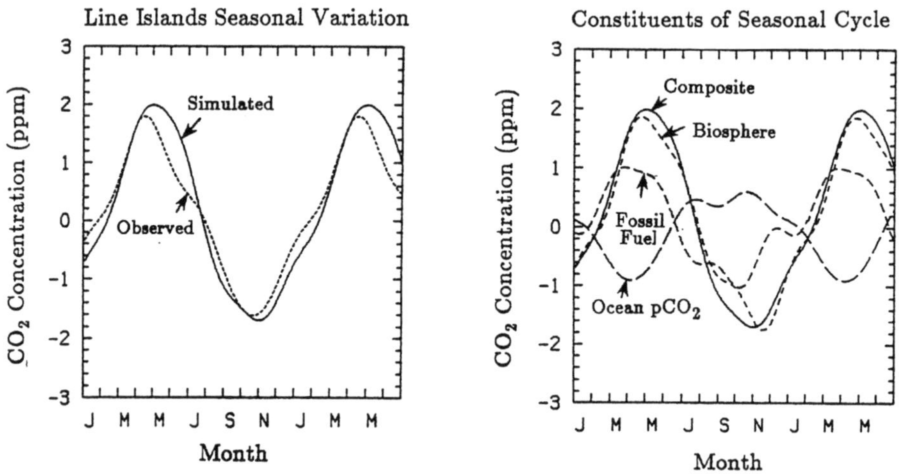

Figure 6. Same as Figure 5 for Line Islands (Fanning and Christmas Islands combined).

In the northern hemisphere, it is probably correct to infer that the increase in amplitude of CO_2 is predominantly owing to interannual changes in plant growth, since seasonality in growth in these regions is the principal cause of the seasonal CO_2 signal. If higher CO_2 concentrations were to stimulate plant growth proportionally (a so-called "growth factor" of unity), the net primary productivity (NPP) world-wide should have increased by 0.39% per year, the average percentage increase in CO_2 concentration per year since 1958. The resulting CO_2 seasonal signal, however, might have been amplified, because it reflects the difference between NPP and biospheric respiration, and the latter probably increased less rapidly than NPP. On the basis of model predictions of NPP and of respiration, Kohlmaier et al. (1989) estimated that the growth factor should have been about one tenth larger than otherwise on this account, implying for Mauna Loa an increase in amplitude of 0.43% per year. However, as they pointed out, botanical evidence suggests an upper limit to the growth factor of about 0.5, rather than unity. If so, an increase of 0.39% in ambient CO_2 should cause only a 0.215% increase in seasonal CO_2 signal.

What additional factors may be involved to account for an observed increase in amplitude of 0.54% per year at Mauna Loa Observatory? Keeling et al. (1989a, pp. 213-214) have hypothesized that changes in climate, especially the recent increase in surface air temperature, may have contributed. To prove this, however, will require model simulations of the regional impact of climate on the atmospheric CO_2 distribution to determine whether year to year variations in CO_2 amplitude are influenced regionally by climatic factors such as temperature and rainfall.

Also, there are complications in attempting a detailed analysis of the seasonal cycle of atmospheric CO_2. The exact peaks and troughs at observing stations are difficult to establish because weather systems introduce synoptic-scale variations into the CO_2 records. The three-dimensional model of Heimann et al. (1989, Figures 26 and 27) predicts synoptic variations about half of those observed at Mauna Loa and Ocean Station 'P' during 1979, the year of the actual wind data used to drive their model. If this model were improved by using contemporary wind, cloud, temperature and remote sensed data for NPP on a daily basis for a succession of years, if the model were to include a more realistic boundary layer, and if the spatial resolution of the model were increased, all of which are feasible improvements, it should be possible to establish more precisely the influence of NPP and respiration on the peaks and troughs of the seasonal cycle, and the extent to which the seasonality of plant growth affects the amplitudes seen at Mauna Loa Observatory and elsewhere.

Interannual variations. Interannual fluctuations in atmospheric CO_2 correlate world wide with large scale oscillations in the tropical atmospheric circulation associated with El Niño events. To explain this correlation, attention has hitherto been mainly focused on mechanisms involving air-sea exchange (Newell et al., 1978; Bacastow and Keeling, 1979; Newell and Hsiung, 1984), although Bacastow et al. (1980) and Pearman and Hyson (1986) pointed out difficulties in predicting either the magnitude or the phase of the associated CO_2 oscillations unless the terrestrial biosphere is also involved. Confirmation of such biospheric involvement has recently been provided by isotopic measurements of atmospheric CO_2 (Keeling et al., 1989a, p. 202-208, summarized by Siegenthaler, 1990).

Because the persistent rise in CO_2 from fossil fuel combustion tends to mask interannual changes in concentration from El Niño events and other causes, it is useful to prepare anomaly plots in which, from each CO_2 record, a constant fraction of the cumulative emissions of industrial CO_2 is subtracted (Keeling et al., 1989a, p. 168). The set of values of annual emissions, for this purpose, are replaced by a continuous spline function forced to agree for each calendar year with the annual emissions value. This procedure is useful to detrend the CO_2 records, even though, strictly speaking, the fraction of industrial emissions retained in the air is not exactly constant from year to year.

The anomaly plot for Mauna Loa Observatory, Hawaii (Figure 7, upper panel) shows evidence of significant fluctuations in the flows of carbon in and out of the atmosphere, fluctuations which are too large to arise from year to year variations in fossil fuel combustion. Corresponding anomaly plots for other stations show similar interannual fluctuations in CO_2. Thus it is evident that the interannual patterns seen at Mauna Loa occur globally. The average of the anomaly for Mauna Loa with that for the South Pole, correlates well with every El Niño event since 1958, as shown in Figure 8.

Isotopic data increase our ability to understand the cause of this CO_2 response to El Niño. Let us express variations in the $^{13}C/^{12}C$ of CO_2 in per mil (‰) by the reduced isotopic ratio, $^{13}\delta$, defined by the equation

$$^{13}\delta = r/r_s - 1$$

Here r denotes the $^{13}C/^{12}C$ ratio of the sample and r_s that of a reference standard, having the value 0.0112372. Since the observed variations in $^{13}\delta$ are quite small, they are conveniently expressed in per mil. Thus, for example, $^{13}\delta = 0.001$ is written $^{13}\delta = 1$‰. As with CO_2 concentration data, the variations in $^{13}\delta$, after seasonal adjustment, are similar in both hemispheres, so that the global trend is well represented again by averaging the records for Mauna Loa, Hawaii and the South Pole. The plot of $^{13}\delta$ in Figure 9, with the vertical scale inverted to aid in comparison with Figure 8, shows a drift towards more negative $^{13}\delta$ that can be shown to be in response to industrial CO_2 emissions. The plot also shows distinct anomalous fluctuations with peak values of negative $^{13}\delta$ in 1980, 1984, and 1988 similar to those of CO_2 concentration and, thus, in correlation with El Niño events. After adjustment for the long term trend towards more negative $^{13}\delta$, the peak in 1984 differs by 0.11‰ relative to troughs in negative $^{13}\delta$ in 1982 and 1986. This difference, if all biospheric in origin, is equivalent to a CO_2 rise of 2.2 ppm (parts per million of dry air by volume), and hence a CO_2 release of 4.7×10^{12} kg of carbon (4.7 GtC) by land plants and soil during the period from June 1982 to May 1984. This release is considerably greater than the 0.8 GtC indicated by the concentration data and implies that the oceans, on net, absorbed 3.9 GtC (see Keeling et al., 1989a, p. 204, Table 13). Such an absorption is qualitatively consistent with evidence that during the El Niño event of 1982 and 1983 the oceanic source of CO_2 normally present in the tropics almost disappeared, while, presumably, oceanic sinks continued to operate elsewhere (Keeling and Revelle, 1985; Feely et al., 1987).

Carbon isotopic data are not available to interpret El Niño events before 1979. Is CO_2 released by the terrestrial biosphere almost simultaneous with a reduced release by the tropical oceans during almost *every* El Niño event? Possibly. Keeling et al. (1989a, p. 200) hypothesized that a reduction in plant growth typically occurs shortly before an El Niño event because the southeast Asian monsoon rains are usually weak at this time. Reduced tropical rainfall may occur simultaneously in Africa, South America, and

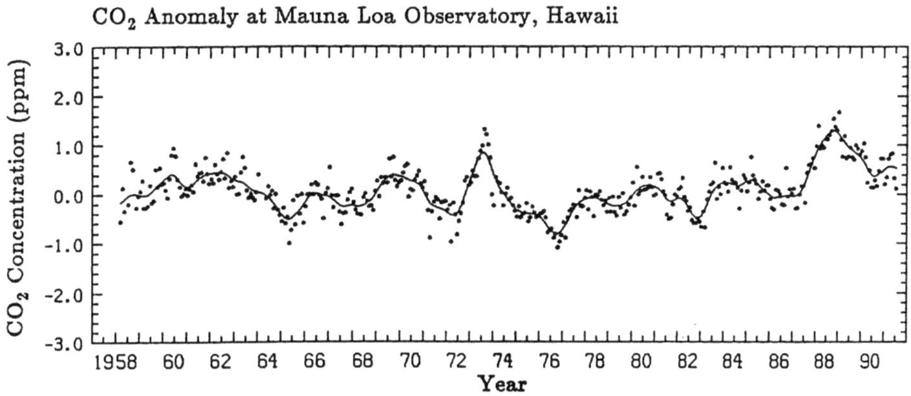

Figure 7. Anomaly in the concentration of atmospheric CO_2, in ppm, at Mauna Loa Observatory. The dots depict monthly mean concentrations after seasonal adjustment, and after subtraction of a constant fraction of the CO_2 release by combustion of fossil fuels and cement manufacture. The smooth curve is a spline fit of the monthly data.

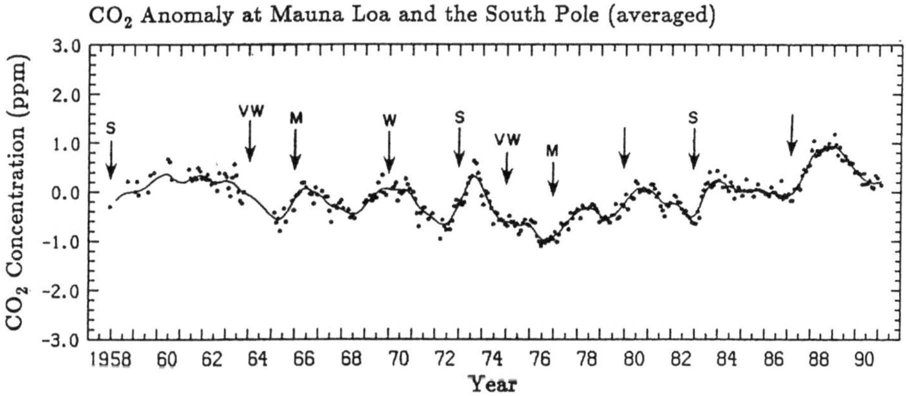

Figure 8. Average of the anomaly records for Mauna Loa Observatory (shown in Figure 7) and the South Pole. The arrows denote the times of El Niño events, each shown for December of the year of onset. The strengths are indicated as S, strong; M, moderate; W, weak; VW, very weak as designated by Rasmussen (1984). The event in 1979 (see Donguy et al., 1982) is not ranked by Rasmussen; that in 1986 occurred subsequent to his article. The CO_2 data are inadequate to establish a possible relationship to the 1963 El Niño event.

Monthly Carbon Isotopic Ratio at Mauna Loa and the South Pole (averaged)

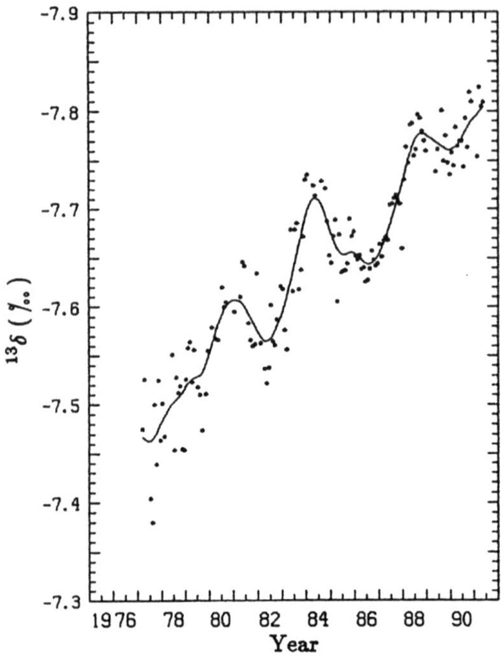

Figure 9. Seasonally adjusted record of the reduced isotopic ratio, $^{13}\delta$, of atmospheric CO_2 in per mil (‰), obtained by averaging monthly mean data of Mauna Loa Observatory and the South Pole. The curve is a spline fit of the monthly values, shown as dots.

Australia as well (Stoeckenius, 1981). Reduced rainfall, by inhibiting plant growth, is expected to lessen the uptake of atmospheric CO_2, whereas the release of CO_2 by respiration, especially of the soil, and especially in a drought which endures for only one season, will be less reduced. Reduced rainfall may also promote wildfires which emit CO_2. The individual recorded CO_2 anomalies persisted on average for about one and a half years, a duration consistent with a single event in one hemisphere, possibly followed by a similar event in the opposite hemisphere a half year later, as for example occurred near 1983.

Can this mechanism be confirmed? An obvious need is for even more precise isotopic data, because the total signal is less than 0.1‰ per event, and is thus within the uncertainty of present calibrations of isotopic standards. Records have improved considerably since measurements began in the 1970's, but still better sampling, and more precise mass spectrametric data would be helpful. Also, several more El Niño events need to be documented isotopically to determine how persistent the $^{13}\delta$ signal is.

Decadal variations. In addition to scrutinizing individual records of atmospheric CO_2, as just described, one can employ records in combination to produce seasonally adjusted "snapshots" of the spatial distribution of CO_2 concentration and of $^{13}\delta$, so as to reflect the slowly varying sources and sinks of the carbon cycle. From the results of the array of stations shown in Figure 1, profiles roughly north to south near the center of the Pacific ocean basin were prepared to reveal non-seasonal features of the carbon cycle relative to the South Pole. The inclusion of $^{13}\delta$ data allows discrimination between oceanic and terrestrial biospheric components after 1978. For three periods, near 1962, 1968, and 1980, additional available data from shipboard sampling increase the density of observations, as noted above. North-south profiles of these data, shown in the lower three curves of Figure 10, reveal a striking increase in the north-south gradient of CO_2 over 18 years. This increase has continued since, as shown by the upper curve for 1990, prepared from recent data. (Actual data points are shown in Appendix A, Figures A.5-A.8).

The global distribution of atmospheric CO_2 for 1962, 1968, and 1980 has been predicted using the three-dimensional transport model of Heimann and Keeling (1989) and the results compared with observations at this array of stations as reported by Keeling et al. (1989b). The salient features of the predicted sources and sinks, compressed from the original 8 by 10° model grid to four zones, are summarized in Figure 11. The main cause of the changing profiles is a near doubling of CO_2 emissions from industrial sources. It is not possible to prove conclusively how other changes in fluxes were shared between the oceanic or biospheric sources and sinks, because the first two profiles, for 1962 and 1968, lack isotopic data. Confronted with this handicap, Keeling et al. (1989b) assumed that CO_2 emissions from the terrestrial biosphere brought about by deforestation and expanding agricultural practices had increased with time according to the global estimates of Houghton et al. (1987), while atmospheric CO_2 had been simultaneously sequestered at increasing rates as a result of stimulated plant growth.

Because the identified biospheric releases were mainly tropical, while stimulation of growth in the model was assumed to be proportional to net primary production everywhere, the effect of changing biospheric CO_2 exchange was to counteract the predicted increasing north-south gradient produced by the combustion of fossil fuels sufficiently to account fully for the observed change in north-south gradient. Consequently, the model predicted that the natural oceanic sources and sinks of atmospheric CO_2 remained approximately constant from 1962 to 1980, as indicated by Figure 11. The model predictions for 1980, although not in any way based on isotopic data, turn out to be consistent with

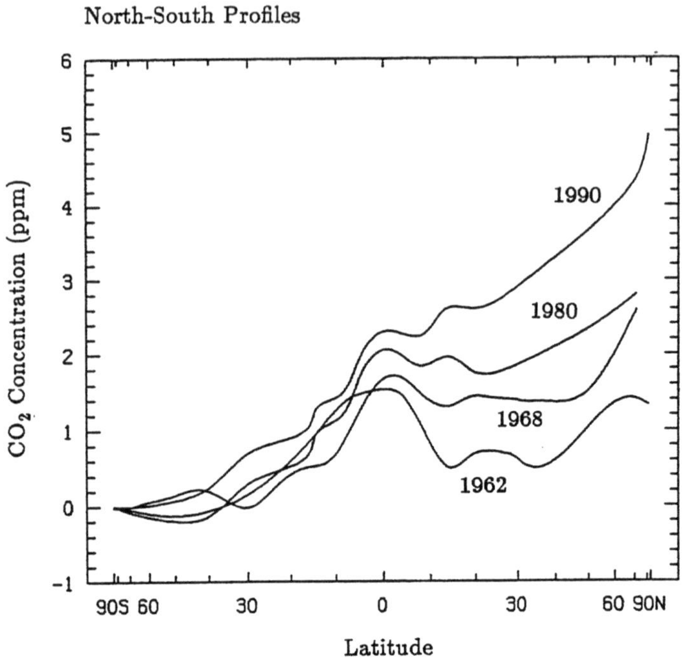

Figure 10. North-south profiles of the seasonally adjusted concentration of atmospheric CO_2 near the earth's surface, in ppm, plotted versus the sine of the latitude for January 1 of the indicated years. The profiles express concentration differences from the South Pole. Curves for 1962, 1968, and 1980 are equivalent to Figures 28, 29, and 30 of Keeling et al. (1989a), except that the 1980 curve is revised by a reappraisal of data at La Jolla near 30°N. The curve for 1990 is based on a new analysis, including more recent data, as shown below in the upper left panel of Figure 12, where the revised 1980 curve also appears.

Annual Fluxes of the Global Carbon Cycle
(10^{12} kgC yr^{-1})

1962

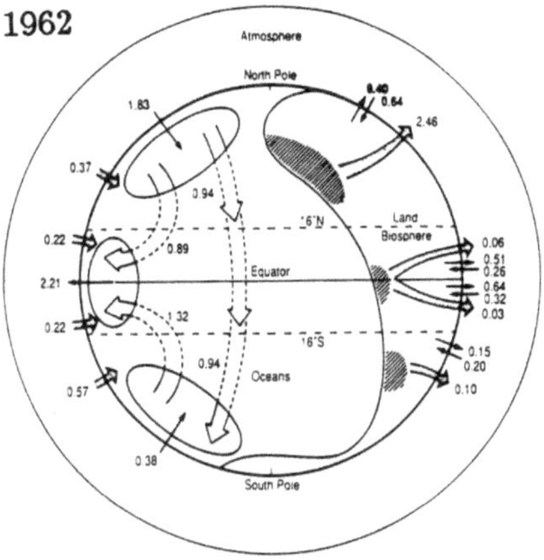

1980

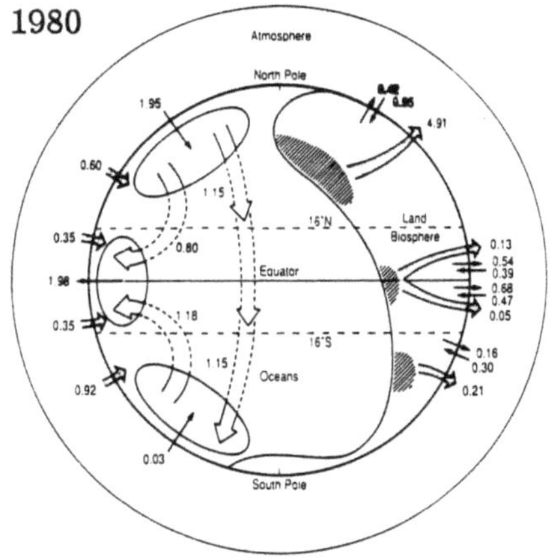

1968

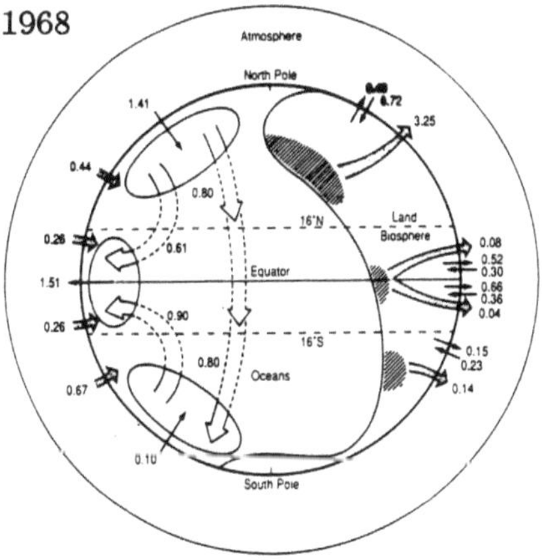

Figure 11. Schematic representations of the major sources and sinks of atmospheric CO_2 arising from perturbations of the carbon cycle, as predicted for the three years indicated. The figures are taken from Keeling et al. (1989b, pp. 355-356). The CO_2 transfers, in units of 10^{12} kgC yr^{-1}, are averages over one year centered on January 1, and are for four zonal compartments of the earth's surface divided by the equator and the 16°N and 16°S parallels. The transfers are divided into categories as follows: open dashed arrows, subsurface oceanic transfers with their sums for each region shown as line arrows; shaded arrows, CO_2 entering the atmosphere from fossil fuel combustion (shown on the right) and absorption of a fraction of this CO_2 by the oceans (shown on the left), line arrows on the right, CO_2 exchange with the land (terrestrial) biosphere.

the limited isotopic data available for that year, except in the far north where a biospheric source, not in the model, is needed to achieve agreement with isotopic observations. Thus it seems likely that, after fossil fuel emissions are accounted for, the main cause of changes in atmospheric CO_2 distribution over the 18-year period was changes in the terrestrial biospheric carbon cycle.

Since this published analysis was carried out, enough isotopic data have been acquired to construct profiles of the north-south gradients of isotopic ratio, as well as CO_2 concentration, spanning the decade from 1980 to 1990, as shown in the upper panels of Figure 12. Also plotted (lower panels) are the direct differences between profiles and these differences with the effect of fuel emissions removed. Evidently only about half of the shifts in north-south gradient from 1980 to 1990 are explained by fossil fuel combustion. The adjusted difference profiles for isotopic ratio and concentration are very similar. They scale at approximately -0.05‰ per ppm as expected for terrestrial biospheric CO_2 sources and sinks (Keeling et al., 1989a, p. 171).

Changes in profiles from 1980 to 1990 are significant in the northern hemisphere, but not in the southern hemisphere, or near the equator. Since the equatorial region is known to be an increasing source of CO_2 from deforestation (Myers, 1988), one would expect increased gradients in concentration and isotopic ratio from the equator to the South Pole. The adjusted profiles suggest, however, that a compensating vegetative sink has also operated near the equator so as to counteract an increased release of CO_2 by deforestation. South of the equator, a possibly significant peak in negative $^{13}\delta$ hints at increasing deforestation there, perhaps enhanced by biomass burning. An increase in biospheric release is far more clearly indicated north of 30°N, where massive deforestation before 1980 did not occur (Houghton et al., 1987), and where the earlier analysis of Keeling et al. (1989b) gave evidence of an increasing biospheric sink from 1962 to 1980 (see Figure 11). On the basis of the gradient differences shown in Figure 12, this implied northern source appears to be substantial; it is evidently of the same magnitude as the increase in fossil fuel CO_2 emissions in the northern hemisphere (about 0.7 GtC yr^{-1}).

From global average atmospheric CO_2 data, Keeling et al. (1989a) deduced that the trend in net terrestrial biospheric emissions changed sign in the 1970's. From 1959 to 1975 (approximately the first half of the CO_2 records) the terrestrial biosphere evidently became increasingly a sink for atmospheric CO_2 (by 0.7 GtC per year), while from 1975 to 1988, it became more of a source (by 0.9 GtC per year) (*loc cit.*, Table 14). The latter increase approached the increase from accelerating combustion of fossil fuels (1.3 GtC per year). The change in net flux over the entire period from 1959 to 1988 correlates positively with variations in global air temperature (*loc cit.*, Figure 67) as though cooling during the first half of the period of CO_2 observations caused the terrestrial biosphere to be increasingly a sink, while warming afterwards caused it to be increasingly a source. Newer data suggest that the biosphere, especially in the northern latitudes of the northern hemisphere, is increasingly a source, and hence that climatic change, especially warming, may be a contributing factor. Perhaps this change is sufficiently important to be affecting the global carbon budget in the manner just described.

A recent three-dimensional modeling study by Tans et al. (1990) appears to contradict the possibility of an increasing biospheric source in the north by suggesting that the northern terrestrial biosphere is a strong net sink for atmospheric CO_2. Their study was constrained by concentration data, as was the earlier study of Keeling et al. (1989a,b), but was alternatively based on a strategy in which vegetative, as well as oceanic, sources and

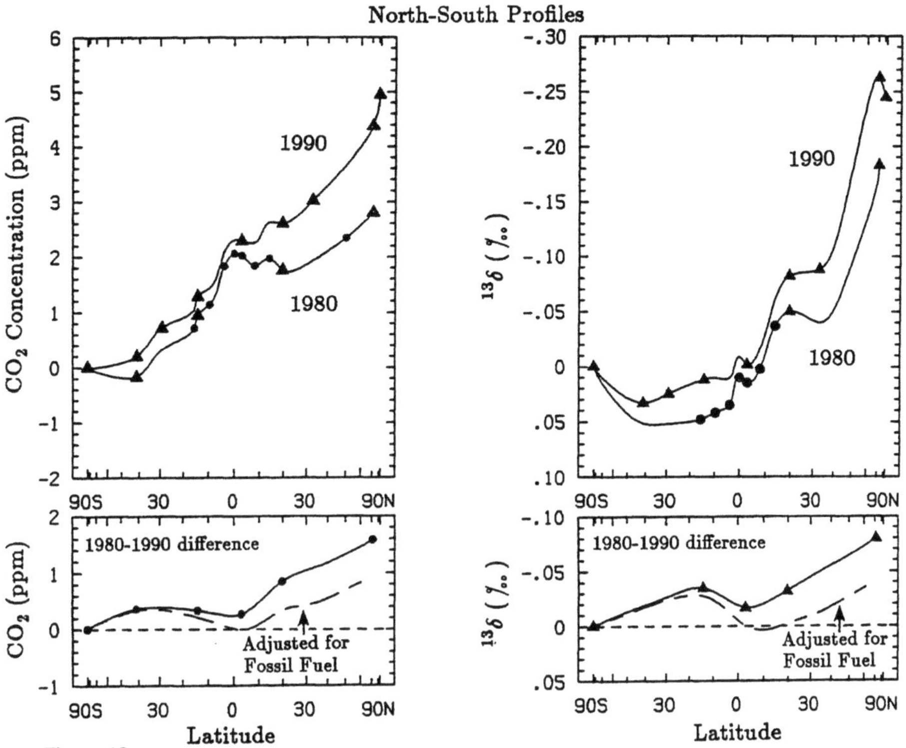

Figure 12.

Upper left panel. North-south profiles of the atmospheric CO_2 concentration, in ppm, for 1980 and 1990, plotted as differences from the CO_2 concentration at the South Pole. Observed data, shown as triangles, indicate seasonally adjusted values at land stations. (For more detailed information see Keeling et al. (1989a), p. 222). Circles indicate observed data over the Pacific Ocean from samples of air collected on the FGGE Shuttle Expedition in the central Pacific ocean near 150°W. The two curves are taut splines that pass through the contemporary observations. The fit for 1990 also passes through additional points in the tropics (not shown), established by adding to points on the 1980 profile the 1980-1990 differences, as prescribed by the solid curve of the lower left panel.

Lower left panel. Difference from 1980 to 1990 in the north-south profiles of atmospheric CO_2 concentration, shown separately in the upper left panel. The dashed curve denotes the difference in profiles after subtracting the influence of fossil fuel combustion, as calculated by the three-dimensional model of Heimann and Keeling (1989, p. 249).

Upper right panel. Same as upper left panel, but for the reduced isotopic ratio, $^{13}\delta$, of atmospheric CO_2 in per mil.

Lower right panel. Difference from 1980 to 1990 of north-south profiles of $^{13}\delta$, in per mil, shown separately in the upper right panel. As in the lower left panel, the dashed line denotes the difference in profiles, adjusted to remove the influence of fossil fuel combustion. The similarity of the two adjusted profile differences indicates that north-south changes in CO_2 between 1980 and 1990, to the extent that they are not attributed to fossil fuel combustion, are mainly a result of perturbations of the terrestrial biosphere.

sinks were adjusted to fit observations. Tans and his co-workers concluded, in contrast to Keeling et al. (1989a,b), that the northern terrestrial biosphere between 1981 and 1987 was a sink of approximately 1.7 GtC per year. A comparison of model predictions, with and without such a northern biospheric sink, shows that their distribution of sources and sinks is difficult to reconcile with isotopic evidence. As shown in the two left panels of Figure 13, it is possible to account for the north-south profile of CO_2 concentration with either an oceanic sink (upper panel) or a terrestrial biospheric sink (lower panel). The model predictions of the north-south profile of $^{13}\delta$ corresponding to these two possible sinks shown in the two right panels, are, however, distinctly different. The gradient is substantially less for the case of a biospheric sink because this sink strongly shifts the isotopic ratio to more positive values in the north relative to the south, whereas an oceanic sink does not. North of 30°N, where the postulated sinks should have had their major impact, the predicted profile with an oceanic sink (upper panel) considerably better fits the observations.

As just illustrated, CO_2 observations are capable of monitoring regional sources and sinks of the carbon cycle to within a few tenths of one GtC. The four profiles illustrated in Figure 13, and used in this analysis, were for times when no strong El Niño events were occurring. An obvious future goal is to carry out model predictions over periods of several years that include such events. It is clear, however, that the data need to be internally consistent to the order of 0.2 ppm to prevent substantial errors in the predictions and must be able to distinguish terrestrial from oceanic sources and sinks. With a continuation of precise measurements of CO_2 concentration and $^{13}\delta$ it should eventually be possible to distinguish all major oceanic and biospheric sources and sinks zonally to the resolution expressed in Figure 11, perhaps even better.

Epilogue. When atmospheric CO_2 data are scrutinized in detail, a complex suite of sources and sinks of the global carbon cycle are revealed. In this lecture an attempt has been made to distinguish natural sources and sinks in the presence of a large unnatural source produced by fossil fuel combustion. Also taken into account were the responses of the oceans and terrestrial biosphere to this industrial source, and to another possible unnatural source of atmospheric CO_2 produced by human disturbances to the terrestrial biosphere.

Observations reveal broad scale patterns in the concentration and isotopic ratio of atmospheric CO_2 from which one can infer the locations and strengths of these sources and sinks. These data also provide evidence of how these sources and sinks have changed over the past several decades in which the cycle has been under unprecedented stress from human activities. Even greater use is likely to be made of atmospheric data in the next few decades, as the build-up of atmospheric CO_2 from fossil fuel combustion continues and its consequences to climate become potentially more serious.

Acknowledgments. The research of the Scripps Institution of Oceanography reported here was supported by the Division of Atmospheric Sciences of the U.S. National Science Foundation, Grants ATM88-19398 and ATM91-21986, with co-support from the Chemical Oceanography Program of NSF and the U.S. Environmental Protection Agency. The research was also supported by the U.S. Department of Energy Grant, DE-FGO3-90-ER60940, and by the Electric Power Research Institute, Contract RP-8011-20. Computer time was provided by the San Diego Supercomputer Center. The author is indebted to Peter Guenther, Stephen Piper, Betsy Stewart, and Timothy Whorf for help in analyzing the data and for many useful suggestions regarding the topics under consideration.

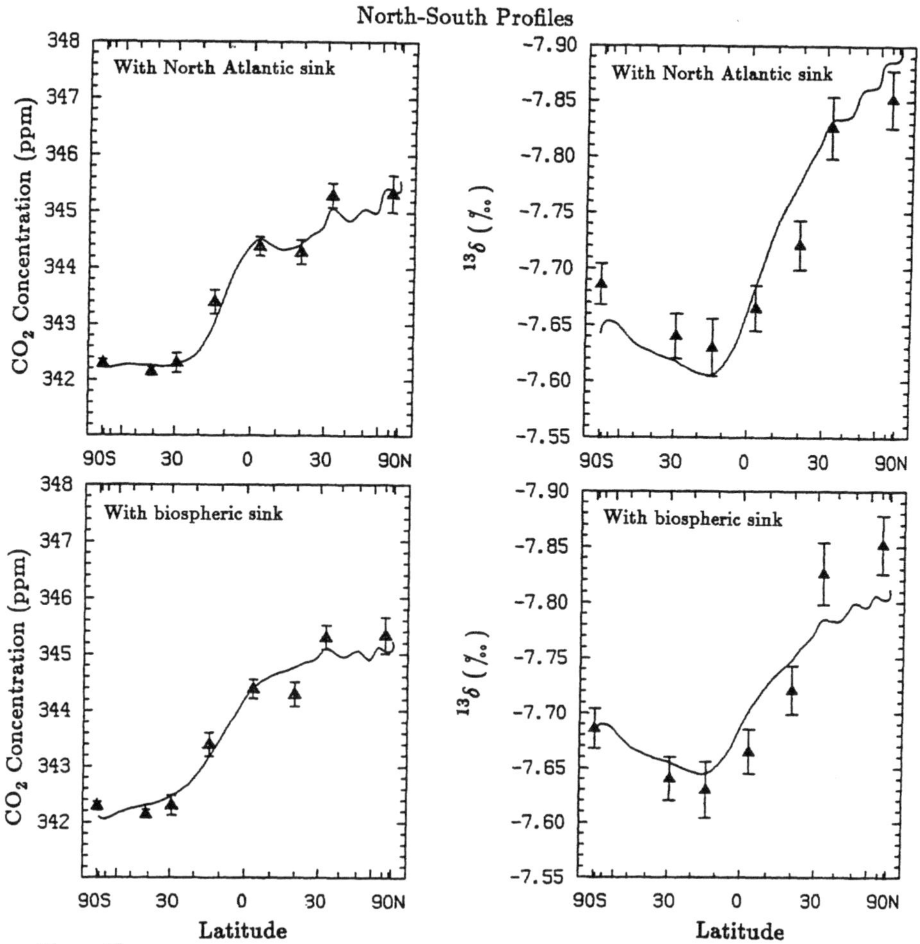

Figure 13.

Upper left panel. Mean annual atmospheric CO_2 concentration in ppm as a function of latitude in 1984, near 160°W. As in the upper panels of Figure 12, triangles denote observations from land stations. Bars indicate ± 2 standard errors of the estimated concentrations for each location. The curve indicates the profile of CO_2 along a transect near the stations as simulated by the three-dimensional model of Keeling et al. (1989b) with standard parameters that include a North Atlantic oceanic sink of atmospheric CO_2.

Upper right panel. Same as upper left panel but for the reduced isotopic ratio of atmospheric CO_2, $^{13}\delta$, in per mil.

Lower left panel. The same simulation for CO_2 concentration as shown above, except that a terrestrial biospheric sink has been substituted for the North Atlantic sink.

Lower right panel. Same as the lower left panel but for the reduced isotopic ratio, $^{13}\delta$, of atmospheric CO_2 in per mil.

The panels are adapted from Figures 46, 48, 53, and 54 of Keeling et al. (1989b).

APPENDIX A
SUMMARY OF ATMOSPHERIC CO_2 MEASUREMENTS

Of the following figures, the first four (A.1 to A.4) summarize CO_2 data obtained by the Scripps Institution of Oceanography with assistance of several national government programs, especially the United States National Oceanic and Atmospheric Administration (NOAA). The data are shown as time plots for individual stations over land. The scale of the reduced isotopic ratio, $^{13}\delta$, is inverted so that input of carbon of biospheric origin to the atmosphere produces higher values on both the concentration and $^{13}\delta$ plots. The isotopic ratios are corrected for the presence of N_2O as explained by Mook and van der Hoek (1983). The seasonal cycle at each station is represented by an annual fundamental and three higher harmonics. This cycle is superimposed on interannual variations, expressed by spline functions described by Bacastow et al. (1985).

To reveal the essential spatial variations in the concentration and $^{13}C/^{12}C$ ratio of atmospheric CO_2, north-south profiles have been prepared (Figure A.5 to A.10). Profiles for 1962, 1968, and 1980 were selected for years during which data from the array of stations over land were supplemented by data from sampling on ships. The seasonal cycle at each location was subtracted and the data detrended to remove the influence of the long term rise in atmospheric CO_2, as described by Keeling et al. (1989a). Profiles have also been prepared for 1990 based on an expanded array of data from land stations, but without supplementary data from ships. All profiles from 1962 to 1990 refer to periods without extreme El Niño conditions, and thus represent an approach to quasi-stationary conditions.

The previously unpublished profiles for 1990 were derived from the data in the same manner as the earlier profiles, and thus differ slightly from the profiles for 1990 shown in the main text in Figure 12, which were derived from linear fits of the seasonally adjusted data for each station. These linear-based profiles require some additional comment. The linear fits to the isotopic data were over the full records for each station, regardless of their length. The fits to the concentration data were over the same intervals as for the corresponding isotopic records. This approach reduced as much as possible the influence of El Niño events in producing the difference plots shown in the lower panels of Figure 12. For 1980 the profile for concentration, produced from the linear fits, agreed closely with the profile shown in Figure A.7. Therefore, because the latter was better documented, the published version of the data for 1980 has been substituted in producing the left panels of Figure 12. Likewise, the published version of the isotopic data for 1980, shown in Figure A.9, is substituted for the linear isotopic profile for 1980 in producing the right panels of Figure 12. Finally, the data for La Jolla at 33°N in 1980 have been disregarded in producing the solid curves in all the profiles shown, because too few air samples were collected and analyzed near 1980 for this station to produce results that are as reliable as the rest of the record.

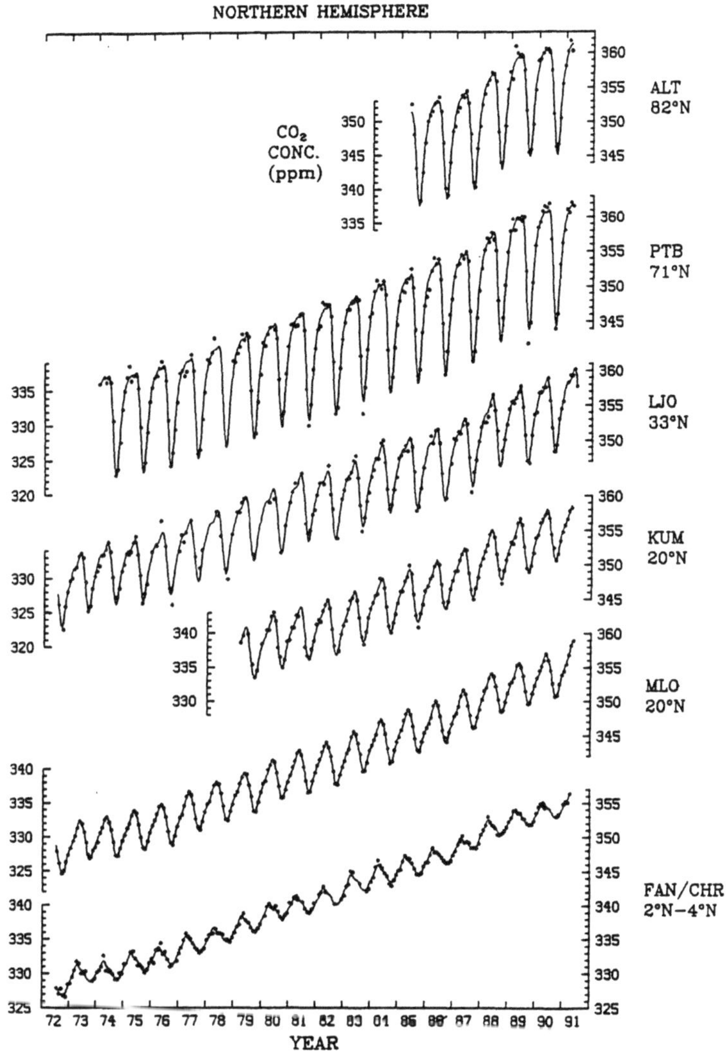

Figure A.1. The concentration of atmospheric CO_2, in ppm by volume, at northern hemisphere stations since 1972 shown from north to south. Station symbols are explained in the caption to Figure 1 of the proposal. Data for FAN, Fanning Island, are combined with data for CHR, Christmas Island. The oscillating curves are fits of monthly mean concentration to a spline function and four harmonic terms. The dots denote monthly mean concentrations.

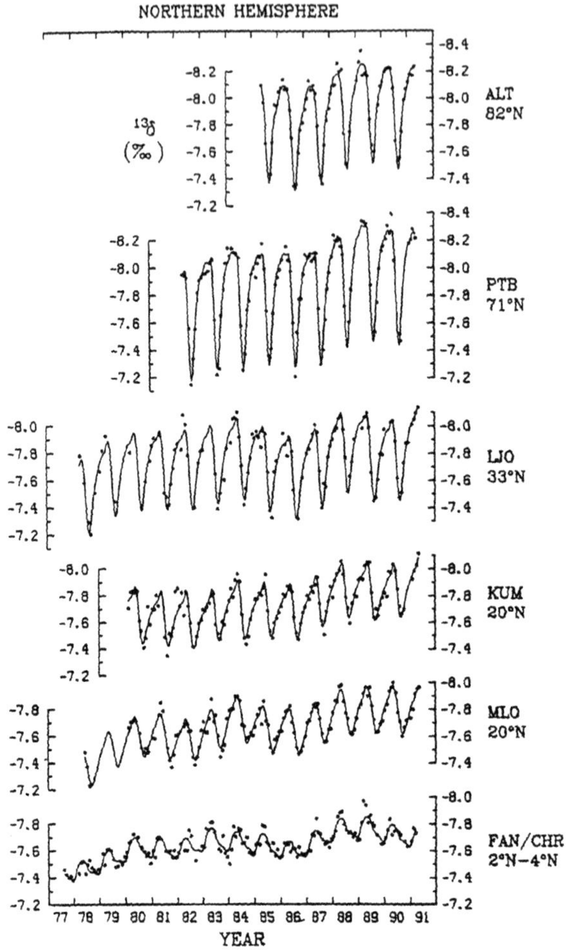

Figure A.2. The reduced isotopic ratio, $^{13}\delta$, in per mil from standard PDB, for the same stations as in Figure A.1. The dots indicate monthly averages. The oscillating curves are fits similar to those in Figure A.1.

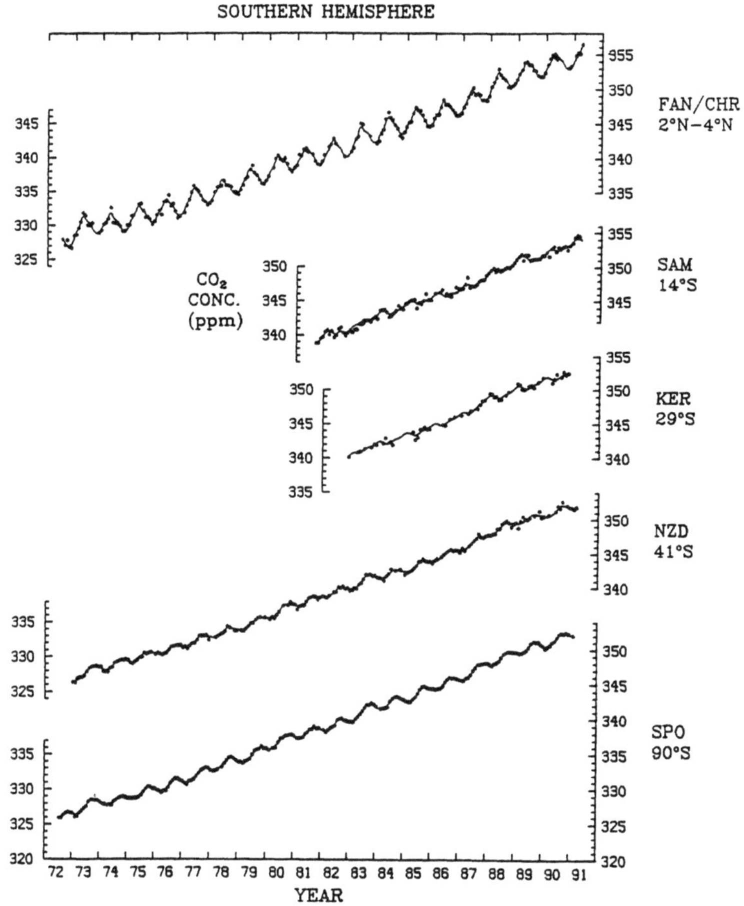

Figure A.3. Same as Figure A.1 for the southern hemisphere. Fanning and Christmas Islands combined (FAN/CHR) is repeated from Figure A.1 for comparison.

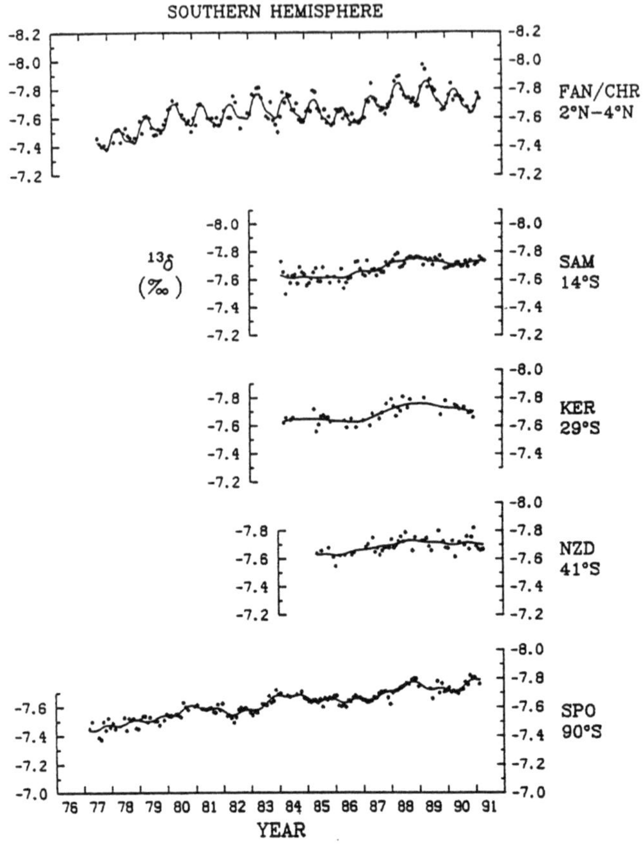

Figure A.4. Same as Figure A.2 for the southern hemisphere.

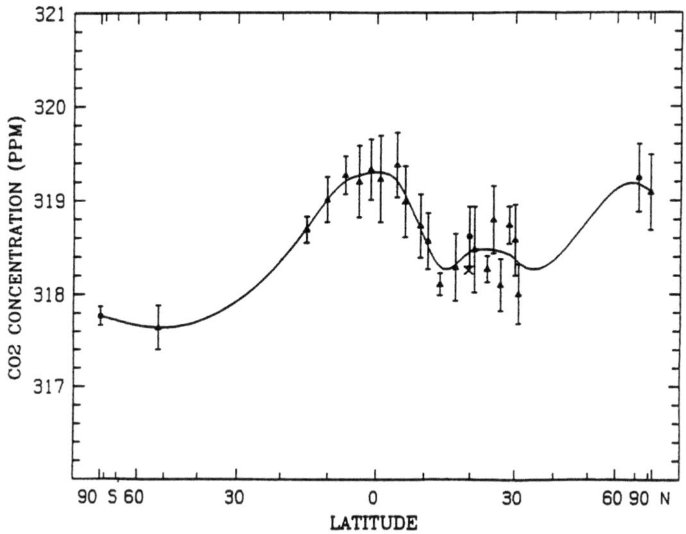

Figure A.5. North-south profile of the mean annual CO_2 concentration in ppm in 1962, near 160°W. Solid triangles denote observations from samples collected over ocean water on ships and dots samples collected over land at fixed stations. Bars indicate twice the standard error of the means for each location [Keeling et al., 1989a]. The solid curve is a weighted fit to the data, using a cubic spline function forced to agree with data for the South Pole and the northernmost station. The concentration at Mauna Loa Observatory is indicated by a cross, and is not included in the spline fit, owing to the high elevation of the station.

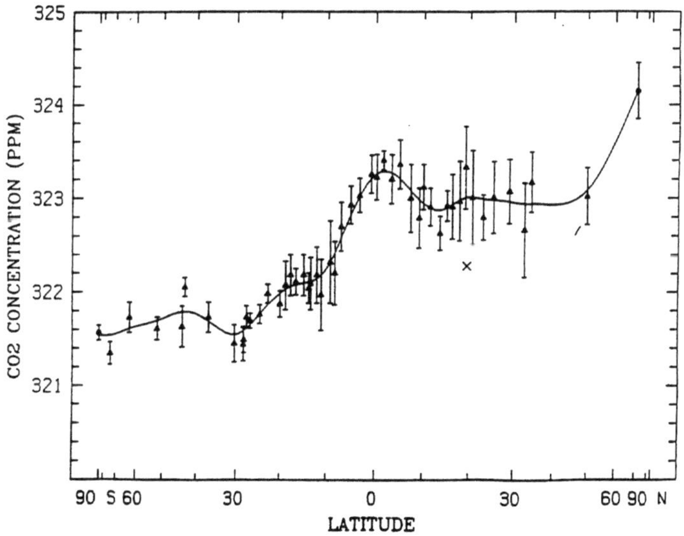

Figure A.6. Same as Figure A.5, but for 1968.

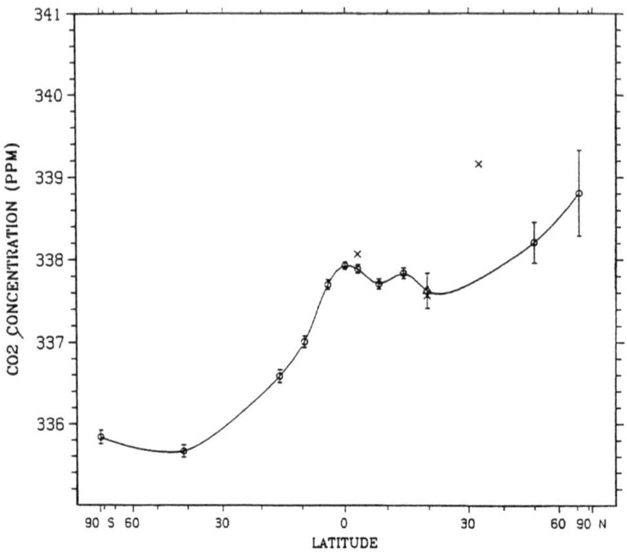

Figure A.7. Same as Figure A.5, but for 1980. The solid curve is an interpolating taut spline which passes through the fitted points. The point for La Jolla, at 33°N, shown as a cross, is not included in the fit owing to too few measurements in 1979 and 1980. Additional crosses denote data for Fanning and Christmas Islands combined, and for Mauna Loa Observatory (left and right, respectively). Both data were omitted from the spline fit.

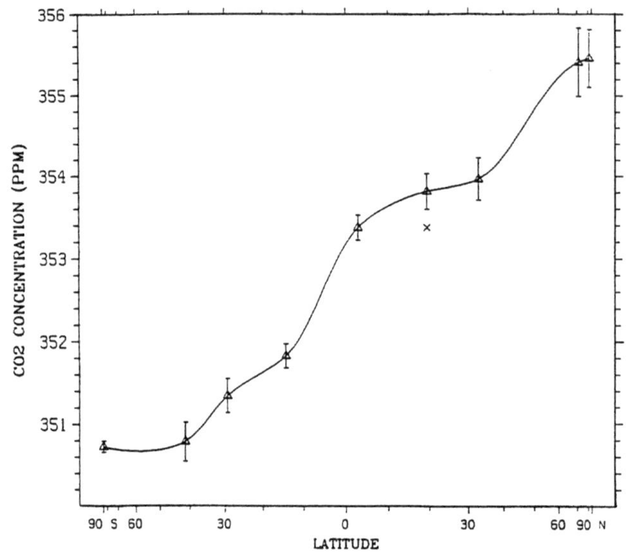

Figure A.8. Same as Figure A.7, but for 1990. Datum for Christmas Island is included in the fit of the solid curve.

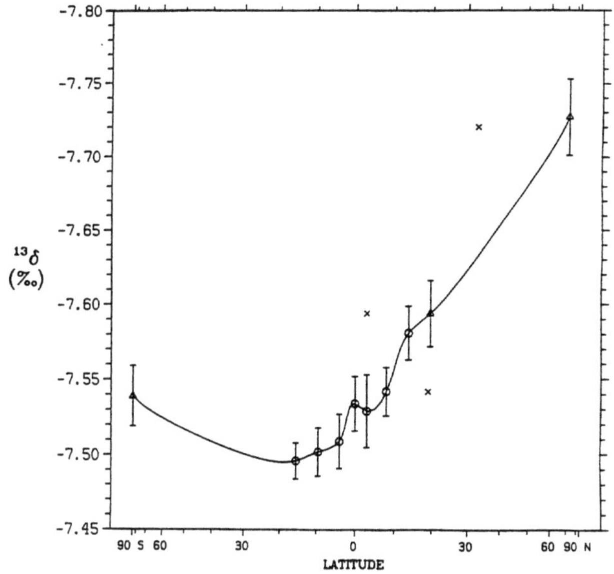

Figure A.9. North-south profile of the annual mean of $^{13}\delta$ of atmospheric CO_2, in per mil, in 1980 near 160°W. The symbols have the same meaning as in Figure A.7.

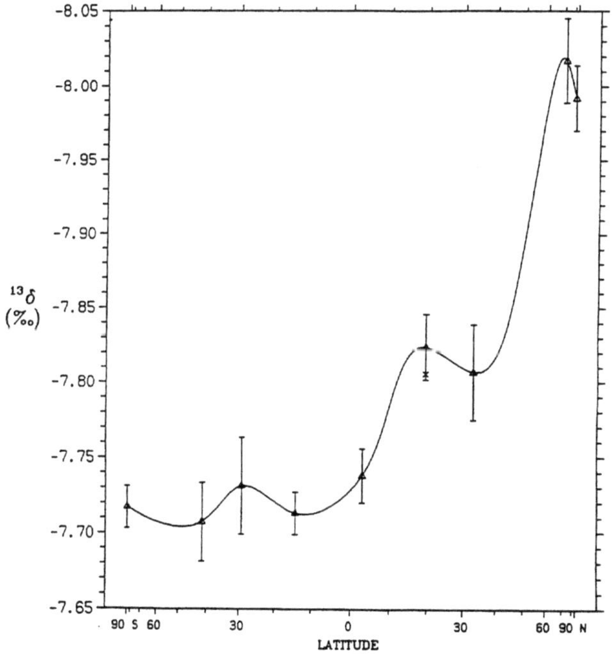

Figure A.10. Same as Figure A.9, but for 1990.

REFERENCES

Bacastow RB, Keeling CD (1979) Models to predict future atmospheric CO_2 concentrations. In: Elliott WP, Machta L (eds) *Workshop on the Global Effects of Carbon Dioxide from Fossil Fuels.* United States Department of Energy, 72-90

Bacastow RB, Adams JA, Keeling CD, Moss DJ, Whorf TP, Wong CS (1980) Atmospheric carbon dioxide, the Southern Oscillation, and the weak 1975 El Niño. *Science 210*:66-68

Bacastow RB, Keeling CD, Whorf TP (1985) Seasonal amplitude increase in atmospheric CO_2 concentration at Mauna Loa, Hawaii, 1959-1982. *J Geophys Res 90*:10529-10540

Bolin B, Keeling CD (1963) Large-scale atmospheric mixing as deduced from the seasonal and meridional variations of carbon dioxide. *J Geophys Res 68*:3899-3920

Donguy JR, Henin C, Morliere A, Rebert JP (1982) Appearances in the Western Pacific of phenomena induced by El Niño in 1979-1980. *Tropical Ocean-Atmosphere Newsletter 10*:1-2

Feely RA, Gammon RH, Taft BA, Pullen PE, Waterman LS, Conway TJ, Gendron JF, Wisegarver DP (1987) Distribution of chemical tracers in the Eastern Equatorial Pacific during and after the 1982-1983 El Niño/Southern Oscillation event. *J Geophys Res 92*:6545-6558

Fung IY (1986) Analysis of the seasonal and geographical patterns of atmospheric CO_2 distributions with a three-dimensional tracer model, Chapter 22. In: Trabalka JR, Reichle DE (eds) *The Changing Carbon Cycle, A Global Analysis.* Springer-Verlag, New York, 459-473

Heimann M, Keeling CD (1989) A three dimensional model of atmospheric CO_2 transport based on observed winds: 2. Model description and simulated tracer experiments. In: Peterson DH (ed) *Aspects of Climate Variability in the Pacific and the Western Americas.* American Geophysical Union, Washington, DC, 237-275

Heimann M, Keeling CD Tucker CJ (1989) A three dimensional model of atmospheric CO_2 transport based on observed winds: 3. Analysis of the seasonal cycle of CO_2. In: Peterson DH (ed) *Aspects of Climate Variability in the Pacific and the Western Americas.* American Geophysical Union, Washington, DC, 277-303

Houghton RA, Boone RD, Fruci JR, Hobbie JE, Melillo JM, Palm CA, Peterson BJ, Shaver GR, Woodwell GM (1987) The flux of carbon from terrestrial ecosystems to the atmosphere in 1980 due to changes in land use: geographic distribution of the global flux. *Tellus 39B*:122-139

Keeling CD, Revelle R (1985) Effects of El Niño/Southern Oscillation on the atmospheric content of carbon dioxide. *Meteoritics 20*:437-450

Keeling CD, Bacastow RB, Carter AF, Piper SC, Whorf TP, Heimann M, Mook WG, Roeloffzen H (1989a) A three dimensional model of atmospheric CO_2 transport based on observed winds: 1. Analysis of observational data. In: Peterson DH (ed) *Aspects of Climate Variability in the Pacific and the Western Americas.* American Geophysical Union, Washington, DC, 165-236

Keeling CD, Piper SC, Heimann M (1989b) A three dimensional model of atmospheric CO_2 transport based on observed winds: 4. Analysis of the mean annual gradients of CO_2. In: Peterson DH (ed) *Aspects of Climate Variability in the Pacific and the Western Americas*. American Geophysical Union, Washington, DC, 305-363

Kohlmaier GH, Sire E, Janecek A, Keeling CD, Piper SC, Revelle R (1989) Modelling of the seasonal contribution of a CO_2 fertilization effect of the terrestrial vegetation to the amplitude increase in atmospheric CO_2 at Mauna Loa Observatory. *Tellus 41B*:487-510

Mook WG, van der Hoek S (1983) The N_2O correction in the carbon and oxygen isotopic analysis of atmospheric CO_2. *Isotope Geoscience 1*:237-424

Myers N (1988) Tropical deforestation and climatic change. *Environmental Conservation 15*:293-298

Newell RE, Navato AR, Hsiung J (1978) Long-term global sea surface temperature fluctuations and their possible influence on atmospheric CO_2 concentrations. *Pure Appl Geophys 116*:351-371

Newell RE, Hsiung J (1984) Sea surface temperature, atmospheric CO_2 and the global energy budget: Some comparisons between the past and present. In: Mörner N-A, Karlén W (eds) *Climatic Changes on a Yearly to Millennial Basis*. D Reidel Publishing Company, 533-561

Pearman, GI Hyson P (1986) Global transport and inter-reservoir exchange of carbon dioxide with particular reference to stable isotopic distributions. *J Atm Chem 4*:81-124

Rasmusson EM (1984) El Niño: The ocean/atmosphere connection. *Oceanus 27*:5-12

Rotty RM (1987) A look at 1983 CO_2 emissions from fossil fuels (with preliminary data for 1984). *Tellus 38B*:203-208

Siegenthaler U (1990) El Niño and atmospheric CO_2. *Nature 345*:295-296

Stoeckenius T (1981) Interannual variations of tropical precipitation patterns. *Monthly Weather Review 109*:1233-1247

Tans PP, Fung IY, Takahashi T (1990) Observational constraints on the global atmospheric CO_2 budget. *Science 247*:1431-1438

AVERAGE GLOBAL DISTRIBUTIONS OF CO_2

I.G. Enting and G.I. Pearman
CSIRO, Division of Atmospheric Research
Private Bag 1, Mordialloc
Victoria, 3195
Australia

1. Introduction

It is now more than 3 decades since the first high-precision measurements of atmospheric carbon dioxide (CO_2) were commenced by C.D. Keeling in Hawaii and Antarctica (Keeling et al., 1976a, b). Those measurements, commenced primarily to assess whether the global atmospheric inventory of CO_2 was increasing, were the forerunners of what now is a somewhat *ad hoc* global network of observatories (see Figure 1). Given the relatively rapid mixing of the global atmosphere, one cannot justify such a network on the basis of observing the secular change in CO_2. The network has been established to describe the temporal and spatial distribution of CO_2 variations which can be used to estimate surface exchanges (inverse method) of CO_2 or for the validation of models of the global carbon cycle which generate estimates of such variations as a result of surface exchanges (forward method).

Such applications of the data will be described elsewhere in this volume (Keeling, this volume; Heimann, this volume). In this chapter, we outline what observational data are available, the gross features of those data, and observational issues to be considered in their application.

In assessing global networks, Pearman (1980) estimated the requirements as:

- *About 3 stations:* Determine global inventories and trends;
- *Additional 5 to 10 stations:* Determine meridional transport of CO_2;
- *Additional 10 to 20 stations:* Determine air-surface exchange between large-scale areas;
- *Over 100 stations:* Determine air-surface exchange within regions with significant anthropogenic influence.

It is an extremely difficult matter to quantify these estimates more precisely in order to define the resolving capability of the existing network or to suggest improvements in network coverage. However, in addition to the points listed by Pearman (1980), recent research is revealing the complexity of *natural* sources and sinks, implying the need for even denser spatial coverage by the observational network. In this presentation we give a preliminary

NATO ASI Series, Vol. I 15
The Global Carbon Cycle
Edited by M. Heimann

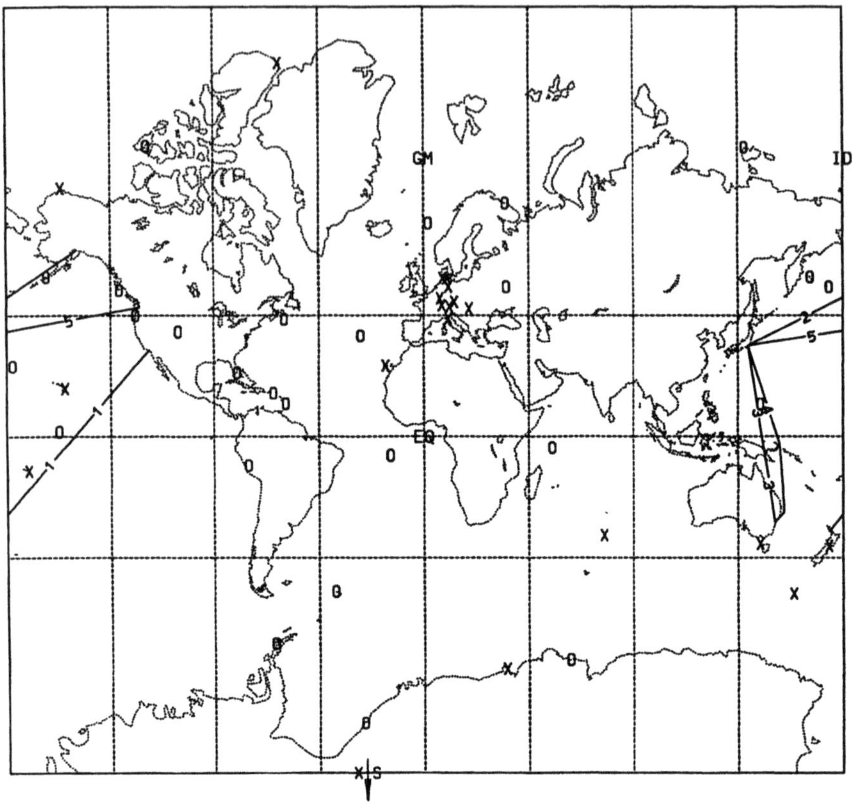

Figure 1. The CO_2 observing network. × denotes sites with continuous analysers (often with supplementary flask sampling programs). 0 denotes sites with only flask sampling. Lines 4 and 5 show the approximate track of the Tohoku University ship-based flask sampling programs. Line 1 approximates the NOAA ship-based sampling program. Lines 2 and 3 show the 1984–5 Tohoku aircraft-based program. Additional, more localised, aircraft programs operate over Japan and South-east Australia. Sites with only short records are not shown.

discussion of the problem in terms of the *signal-to-noise ratio* as a framework for focussing on those aspects of the network that limit our ability to estimate source-sink strengths.

The outline of the remainder of this chapter is as follows. Section 2 analyses the problem of deducing the space-time distribution of sources and sinks in terms of the signal-to-noise ratio and the effect of the 'noise-amplifying' characteristics of the source-deduction process. Section 3 describes various observational networks reporting CO_2 data. Section 4 discusses the selection of 'baseline' data, intended as representative of large-scale air-masses. Section 5 reviews observations at high altitudes. Section 6 describes the decompostion of the space-time variations into trends, cycles and spatial gradients, with emphasis on the spatial gradients. Seasonal cycles are reviewed by Manning (this volume). Section 7 discusses the interannual variability, primarily in its role as a complicating factor in estimating representative spatial distributions. Section 8 suggests possible future requirements for CO_2 observations. The discussion also makes use of some preliminary three-dimensional model results describing aspects of the small-scale variability in the data. The transport model was the GISS tracer model described by Fung et al. (1983), run at 8° by 10° resolution. In order to isolate the effects of transport, we included only broad-scale sources, taking zonally-uniform sources as deduced by Enting and Mansbridge (1991). An appendix gives further discussion of the statistical aspects of estimating surface sources from concentration data. A final section lists various abbreviations and acronyms used in this chapter.

2. Deducing Sources and Sinks

As noted in the Introduction, one of the main reasons for measuring the spatial distribution of concentrations of CO_2 (or any other trace gas) is to deduce the spatial distribution of sources and sinks. This deduction process requires the use of an atmospheric transport model. A limitation of such work is that none of the source-deduction studies to date have looked at the statistical characteristics of the source estimates, arising from observational errors. In this section we present some of the relevant concepts underlying such error analysis. The more technical details are given in the appendix to this chapter. We emphasise the *signal-to-noise ratio* as determining the limits of resolution and note the consequent need to characterise the 'noise' in the observations. In later sections we use this to review the observational data in terms of the requirements for deducing sources through atmospheric transport modelling.

There is a formal distinction between *forward modelling* and *inverse modelling*. In forward modelling one works from a specification of surface sources and calculates atmospheric concentrations for comparison with observations. In inverse modelling one uses a specification of the surface concentrations as a function of space and time. The model is forced to follow this surface CO_2 distribution — the non-surface distribution is calculated as in forward modelling and the surface sources and sinks are deduced by mass balance. In mathematical terms, the distinction between forward and inverse modelling is between specifying the lower boundary in terms of a flux or a concentration. In practice, the distinction between forward and inverse modelling can become blurred because it is possible to run a model in forward mode and iteratively adjust the sources until the calculated concentrations agree with observations (e.g. Pearman and Hyson, 1980, 1986; Pearman et al., 1983b; Fung et al., 1983; Heimann and Keeling, 1989; see also Heimann, this volume).

While the process of determining surface sources from surface concentrations formally has a unique solution, there are some significant practical difficulties:

i. The sources deduced by the inversion process are very sensitive to errors in either the observational data or in the transport model being used for the inversion. This is true, regardless of whether the deduction is performed by a direct inversion calculation or by iterative adjustment of the sources in a model run in forward mode. This extreme sensitivity to errors is a characteristic of a whole range of so-called *inverse problems* involving interpretation of indirect information and a variety of mathematical techniques have been developed to deal with the difficulties (see the Appendix and also Enting and Newsam, 1990b and references therein).

ii. For calculations in inverse mode, the concentration must be specified at all points on the surface, necessitating some form of (non-unique) interpolation of the observational data. Similarly, in forward mode the incompleteness of surface data will lead to ambiguities in the sources deduced.

iii. It is not possible to incorporate additional information such as non-surface data or constraints on the sources in the direct inversion calculations.

Newsam and Enting (1988) studied noise amplification in the inversion problem using a simple semi-analytic model in which atmospheric transport was modelled as a purely diffusive process (see also Enting and Newsam, 1990a). They worked with a set of basis functions $P_n^m(\cos\theta)e^{im\phi}e^{2\pi kt/T}$ (where θ and ϕ are the spherical coordinates, t is the time, T is one year and the P_n^m are spherical harmonics). These functions were used to expand concentrations in terms of coefficients c_{kmn} and the surface sources using coefficients s_{kmn}.

They found that source components s_{kmn} will produce surface concentration components c_{kmn} whose sizes are attentuated by a factor that goes to zero with a k^{-1} dependence on the meridional wave-number, k. This k^{-1} factor relating surface concentrations to surface sources was confirmed numerically in a two-dimensional (advective-diffusive) atmospheric transport model by Enting and Mansbridge (1989). The attenuation of the source contributions results in the more rapidly varying components being reduced to below the noise level. Consequently, the inverse operation, estimating the sources, involves an effective multiplication by factors of order k, acting on both the attentuated 'signal' and the dominant 'noise'. Thus without some form of filtering, the source estimates would be dominated by amplified noise.

In mathematical terms we can formulate the inversion problem as that of deducing a set of sources and sinks, s_j, from a discrete set of concentration data, c_i. Even if the amount of data matched the number of degrees of freedom in the source/sink distribution, this would be a poorly-conditioned inverse problem and so is subject to error amplification (Enting and Newsam, 1990b). In fact the sparseness of the sampling network exacerbates the difficulties in determining source distributions from concentration data. This leads us to ask the general question: *What can we learn about the sources, s_j, from the concentrations, c_i?*

Part of the answer to this question must be that we can only deduce some low-dimensional 'projection' of the source/sink distribution since the number of independently determined source components cannot exceed the number of data items. Studies with both an approximate, purely diffusive, analytic model and with a zonally-averaged numerical model show that what can be deduced is information about the long-wavelength (i.e. large-scale) variations in sources and sinks. Information about small-scale variations is lost due to the smoothing effect of atmospheric mixing. (See Bolin and Keeling, 1963; Enting and Newsam, 1988; Enting and Mansbridge, 1989). The effect of this progressive loss of information is that the 'signal' from the more variable components will be drowned out by the 'noise' in the observations. Unless very high precision can be obtained, the number of detectable source components will be very much less than the number of observational records.

In the Appendix we give a more detailed analysis of the 'trade-off' between suppressing noise and improving the resolution. The analysis is presented by analogy with the classic problem of digital filtering in time series analysis. Both the analogy and the specific analysis of the source estimation problem show the a statistically optimum estimation of source strengths requires a knowledge of the statistical characteristics of both the 'signal' and the 'noise'. In the Appendix we identify the possible 'noise' terms as coming from *transport model inadequacy* (due to limitations such inadequate resolution, lack of small-scale variability, lack of interannual variability and zonal averaging), *neglect of source components* and *actual observational errors*. A full statistical characterisation of all these influences is a major undertaking.

The first attempt to deduce the spatial distribution of sources and sinks of CO_2 from atmospheric data was by Bolin and Keeling (1963) using a vertically-averaged latitude-only diffusive model. They noted that the k^{-2} dependence of concentrations on sources implied a k^2 amplification of errors in the data and concluded that *'no details in the distribution of sources and sinks are reliable'*. In fact, part of the problem encountered by Bolin and Keeling was due to treating surface concentrations as if they were vertical averages. (Recently, Enting and Newsam, 1991, have described an *ad hoc* correction for incorporating surface data into vertically-averaged models to avoid this problem.)

The main applications of inversion calculations of CO_2 sources have been those of Enting and Mansbridge (1989, 1991) and Tans et al. (1989a). Each of these calculations used a two-dimensional transport model to deduce zonally-averaged sources. In each of the calculations, the inversion was stabilised by inverting a smoothed representation of the space-time distribution of atmospheric CO_2, with minor differences between the form of smoothing used in each of the three cases. In each case, the transport coefficients were obtained from the work of Plumb and Mahlman (1987) who obtained these transport fields as a parameterisation of the effective transport of the GFDL atmospheric general circulation model. The work of Enting and Mansbridge (1991) includes the role of CO in the atmospheric carbon budget, recognising that there is a small but significant source of CO_2 in the free atmosphere due to the oxidation of CO. Even in undisturbed ecosystems the surface CO_2 fluxes will not be in balance because some of the carbon taken up as CO_2 in photosynthesis will be returned to the atmosphere as CO, CH_4 and a range of other organic compounds. Incomplete oxidation of fossil carbon gives a further CO contribution to the atmospheric carbon budget.

One of the limitations of the direct inversion calculations is the difficulty of including additional information. It would be desirable to incorporate aircraft data into the inversions since such data should be dominated by the large-scale features of the source distribution. The simple model used by Newsam and Enting (1988) indicates that upper tropospheric data should reflect only the very large-scale source contributions.

The two-dimensional models used by Enting and Mansbridge (1989, 1991) and Tans et al. (1989a) have their transport coefficients varying smoothly in time, by virtue of being zonally averaged. In contrast three-dimensional models aim to reflect real small-scale synoptic variability in the atmospheric transport. Unless the transport matches the actual variation in the concentrations (i.e. by using observed winds from the same year as the concentration data) there will be a 'mis-match' in the small-scale variation. Forcing the model concentrations to follow either a smoothed concentration or a concentration record with synoptic variability that is not matched to the model will lead to spurious small-scale contributions to sources deduced by the model. The model calculations presented here determine the small-scale variation (in time) that results from small-scale variations in transport acting on broad-scale sources. Preliminary studies by the group at Melbourne University suggest that in spite of this 'transport-induced variability' it is possible to obtain meaningful time averages of the estimated sources (I. Simmonds and co-workers, personal communication). However it may be preferable to deduce sources such that the smoothed model concentrations track a smoothed observational record. This could be achieved by applying a one-sided (possibly recursive) filter to the model data.

3. Surface Observational Networks

Tables 1 to 6 summarize the CO_2 observational network currently in existence (together with some programs now concluded). The map in Figure 1 shows the coverage. One of the key distinctions is between the two types of observational site, those which maintain continuous or nearly continuous measurement, and those where regular or quasi-regular samples of air are collected and returned to a central laboratory for analysis. Typically continuous analyser data are obtained on a time-scale of minutes or less while flasks are sampled at weekly or greater intervals.

Site	Agency	Latitude	Longitude
Alert	AES	82°31′ N	62°18′ W
Amsterdam I.	CFR	37°48′ S	77°32′ E
Baring Head	DSIR	41°25′ S	174°52′ E
Bering I. *	Main Geophys. Obs.	55°12′ N	165°59′ E
Brotjackriegel	Umweltbundesamt	48°49′ N	13°13′ E
Cape Grim	CSIRO	40°41′ S	144°41′ E
Cape Matatula	NOAA	14°15′ S	170°34′ W
Cape St. James *	AES	51°56′ N	131°01′ W
Deuselbach	Umweltbundesamt	49°46′ N	7°03′ E
Izaña	INM	28°17′ N	16°29′ W
Kotelny I. *	Main Geophys. Obs.	76°06′ N	137°54′ E
K-Puszta	Hung. Met. Service	46°58′ N	19°35′ E
Langebruge	Umweltbundesamt	52°48′ N	10°46′ E
Mauna Loa	SIO	19°32′ N	155°35′ W
Mt. Cimone	Ital. Met. Service	44°11′ N	10°42′ E
Pt. Barrow	NOAA	71°19′ N	156°36′ W
Sable I. *	AES	43°56′ N	60°01′ W
Schauinsland	Unweltbundesamt	47°55′ N	7°55′ E
South Pole	NOAA	89°52′ S	24°48′ W
Station 'C'	Main Geophys. Obs.	52°45′ N	35°30′ W
Teriberka	Main Geophys. Obs.	69°12′ N	35°07′ E
Wank	Fraunhofer Inst.	47°31′ N	11°09′ E
Westerland	Umweltbundesamt	54°56′ N	8°19′ E
Zugspitze	Fraunhofer Inst.	47°25′ N	10°59′ E

Table 1. Stations reporting to WMO for 1989 (see WMO, 1990). Stations with only flask data are denoted '*'. In addition, the NOAA flask data (see Table 2) are published in BAPMoN reports. In many cases, programs operating continuous analysers also trap flask samples for various comparison purposes and often provide flask samples for the programs of other agencies.

The data from these observational programs are reported in a variety of ways. For stations in the WMO BAPMoN network, provisional selected daily data are published annually (most recently WMO, 1990). Table 1 list the sites in this network. In addition, many programs publish regular reports (often annually) of their operations. A large body of CO_2 concentration data was assembled by CDIAC (1990). (A list of abbreviations and acronyms is given at the end of this chapter).

The most extensive spatial coverage of CO_2 observations is that of the NOAA CMDL (formerly GMCC) program. This is based on regular (generally twice weekly) flask sampling from several dozen sites as listed in Table 2. Some of these sites are observing stations for other national programs. The NOAA program also operates continuous analysers at Barrow, Mauna Loa, Samoa and the South Pole. Details of the NOAA program are reported in a series of annual summaries (most recently GMCC, 1989; CMDL, 1990) and in journal articles (Komhyr et al., 1985; Conway et al., 1988; Tans et al., 1989a).

C.D. Keeling and co-workers at SIO have maintained a global sampling network for many years. Analysis of the data is presented by Keeling et al. (1989a). Data was obtained from the SIO continuous analyser at Mauna Loa, and (in cooperation with DSIR) from the continuous analyser at Baring Head, New Zealand. Long time-series of flask samples were obtained from Barrow, Ocean Station P (until 1981), La Jolla (California), Cape Kumakahi (Hawaii), Samoa, Fanning and Christmas Islands, the Kermadec Islands and the South Pole, together with shorter records from other sites.

The Canadian Atmospheric Environment Service maintains a number of flask monitoring sites, currently at Alert (in the high Arctic on Ellesmere Island), on Sable Island (off the east coast of Nova Scotia) and on Queen Charlotte Island (off the west coast) (Trivett et al., 1989). An earlier Canadian program (in co-operation with SIO) operated at Ocean Station P (Wong et al., 1984). More recently a continuous analyser has been installed at Alert.

The Division of Atmospheric Research of CSIRO (Australia) maintains a number of CO_2 monitoring programs (Enting, 1989a). The aircraft programs described below have operated since 1972. The main surface observatory is at Cape Grim, Tasmania where a baseline station monitors a wide range of atmospheric constituents. The observational techniques have evolved during the course of the program — most notably air has been sampled at the successive heights of 3 m, 10 m and 70 m above the top of the cliff on which the station is situated. Details of the Cape Grim programs are given in the annual reports (most recently Wilson and Ayers, 1990). CSIRO also operates a continuous analyser periodically on Macquarie Island and regularly obtains flask samples from the Australian station Mawson on the Antarctic coast. Flask samples are also obtained on a regular basis from Alert, Mauna Loa, Barrow, Samoa and the South Pole through co-operative arrangements with AES and NOAA. Descriptions of the CSIRO surface programs are given by Beardsmore and Pearman (1987).

Coverage over a wide range of latitudes was achieved by a ship-based flask sampling program

Code	Site	Latitude	Longitude	1st year
ALT+	Alert, NWT	82.50 N	62.33 W	1986
AMS+	Amsterdam I.	37.95 S	77.53 E	1979
ASC	Ascension I.	7.92 S	14.42 W	1979
AVI	St. Croix	17.75 N	64.75 W	1979
AZR	Azores	38.75 N	27.08 W	1980
BME	Bermuda (E)	32.36 N	64.65 W	I
BMW	Bermuda (W)	32.26 N	65.88 W	I
BRW†	Barrow, Alaska	71.32 N	156.60 W	1971
CBA	Cold Bay, Alaska	55.20 N	162.72 W	1979
CGO+	Cape Grim, Tas	40.68 S	144.73 E	1984
CHR	Christmas I.	2.00 N	157.30 W	1984
CMO	Cape Meares, Oregan	45.00 N	124.00 W	1982
GMI	Guam	13.43 N	144.78 E	1979
HBA	Halley Bay	75.67 S	25.50 W	1983
KEY	Key Biscayne, Florida	25.67 N	80.17 W	1973
KUM	Cp. Kumakahi, Hawaii	19.52 N	154.82 W	1976
MBC	Mould Bay Canada	76.23 N	119.33 W	1980
MID	Sand I., Midway	28.20 N	177.38 W	1986
MLO†+	Mauna Loa, Hawaii	19.53 N	155.58 W	1976
NWR	Niwot Ridge, Colorado	40.05 N	105.63 W	1968
PSA	Palmer Station	64.92 S	64.00 W	1978
RPB	Ragged Pt., Barbados	13.16 N	59.43 W	1988
SEY	Seychelles	4.67 S	55.17 E	1980
SGI	South Georgia I.	54.00 S	38.00 W	I
SHM	Shemya, I.	52.75 N	174.08 E	1986
SMO†	American Samoa	14.25 S	170.57 W	1973
SPO†	South Pole	89.98 S	24.80 W	1975
STM	Station 'M'	66.00 N	2.00 E	1981
SYO+	Syowa station	69.00 S	39.58 E	1989

Table 2. NOAA flask sampling sites for 1989. For the first year, 'I' denotes no complete years reported through to 1989. Sites denoted † have a NOAA continuous monitoring program. Sites denoted + have continuous monitoring by other agencies. Sampling by NOAA has ceased at a number of sites such as Cosmos (Peru) and Falkland Islands.

Site	Agency	Latitude	Longitude
Macquarie I.	CSIRO	54°29′ S	158°58′ E
Mawson *	CSIRO	67°37′ S	62°52′ E
Barrow, Alaska	SIO	71°19′ N	156°36′W
La Jolla, California *	SIO	32.9°N	117.3°W
Cape Kumakahi *	SIO	19°31′ N	154°49′W
Fanning I./Christmas I. *	SIO	3.9/2.0°N	159.3/157.3°W
Cape Matatula, Samoa *	SIO	14°15′ S	170°34′W
Raoul I., Kermadecs *	SIO	29.2°S	177.9°W
South Pole *	SIO	89°52′ S	24°48′W
Syowa	Tohoku	69°00′ S	39°35′ E

Table 3. Other surface sampling sites (neither NOAA nor BAPMoN). (For details see Beardsmore and Pearman, 1987; Keeling et al., 1989a; Tanaka et al., 1987c). Stations with only flask data are denoted '*'. The SIO program at the South Pole used a continuous analyser for the first few years of operation. In many cases, programs operating continuous analysers also trap flask samples for various comparison purposes and often provide flask samples for programs of other agencies.

operated by the group at Tohoku University. As reported by Tanaka et al. (1987b) they obtained samples at 5° latitude intervals between Yokohama, Japan (33°N) and Melbourne, Australia (39°S).

Other programs are operated in New Zealand (Lowe et al., 1979), Germany (Reiter et al., 1986; Levin et al., 1992), Italy (see Ciattaglia et al., 1987), at Amsterdam Island (Monfray et al., 1987; Gaudry et al., 1991), the USSR, Hungary, the Canary Islands (see Navascués and Rus, 1991) and the Japanese Antarctic station, Syowa (see Tanaka et al., 1987c; Nakazawa et al., 1991b). Additional details of the BAPMoN programs are given in WMO (1990).

Region	Agency	Latitudes	Longitudes	Years
Central Pacific	SIO	30N–20S	120W–180W	1967–68
Tropical Pacific	SIO	20N–17S	155W	1979–80
North Pacific	Tohoku U.	36N–50N	140E–125W	?
West Pacific	Tohoku U.	36N–38S	140E–155E	1982–.
Pacific	NOAA	33N–36S	120W–175E	1986–.

Table 4. Ocean transects. If no last year is specified, program is believed to be continuing. Shorter-term projects, oriented to single cruises have been undertaken by NOAA, CSIRO and other agencies. Further details are given by Keeling et al., 1989a; GMCC, 1989; Tanaka et al., 1987b.

The Japanese ship-based program has been noted above with respect to its regular sampling and wide range of latitudes covered. NOAA has established a flask sampling program using a ship travelling regularly between California and New Zealand (GMCC, 1989). In addition to these programs a number of short-term observational studies have been mounted. Those of most interest for establishing spatial distributions involved transects taken on oceanographic cruises. Keeling et al. (1984) report data from north-south transects during 1979–1980. Keeling et al. (1989a) report data from a number of other cruises. NOAA have obtained samples from a number of cruises in the Pacific (GMCC, 1983, 1986a, b) and the South Atlantic (GMCC 1986a). Beardsmore et al. (1984) report data from summer cruises between Australia and the Antarctic.

One important aspect of these observational programs is the range of intercomparisons that are undertaken. This is achieved by having two or more programs obtain regular samples from single sites as noted above and also by special purpose inter-comparison studies. In terms of the analysis in Section 2, such intercomparisons give a partial measure of the error or noise level in the observational records.

4. Baseline Measurements

The discussion in Section 2 above indicates that only the large-scale features of the CO_2 distribution can be interpreted in terms of sources and sinks. Information on small-scale variation will be incomplete and so largely uninterpretable. Thus any small-scale component in observational data must be treated as un-analysable noise. Therefore the most satisfactory data sets are those that reflect only large-scale variations. The term *baseline data* is given to those values that are believed to be representative of large-scale air masses.

In choosing sites for measuring such baseline data the key criterion is to avoid strong vertical gradients so that the measured concentrations are representative of at least the boundary layer. This requires both small surface sources and good vertical mixing with the greatest mixing generally occurring around mid-day. The small sources can be achieved if the surface is water, ice or rock and baseline data are collected from each of these types of site. This question is discussed further by Fraser et al. (1983). Overall, there are a number of criteria favouring the use of marine or coastal sites for baseline observations:

(i). surface fluxes are usually smaller than over land;
(ii). vertical mixing is less restricted by vertical stability than over land;
(iii). the diurnal cycle of atmospheric stability is much smaller than over land;
(iv). there is much less day-to-day variation in atmospheric stability than over land.

In addition to these environmental characteristics, a number of practical considerations impinge on the selection of sites for baseline monitoring. Some of the most important of these are:

(i). ownership of land;
(ii). long-term control of the site environment;
(iii). accessibility by personnel;
(iv). availability of electric power and water.

The raw data measured at each site are generally subjected to a number of transformations before being used in interpreting atmospheric carbon budgets. Typically, the main processes are:

Editing: To remove faulty data such as that arising from known instrument malfunction;

Calibration: To convert the data to an international standard concentration scale. NDIR instruments compare a sample to a reference gas. To produce concentrations it is first neccessary to establish the relation between signal ratio and concentration ratio (there is significant non-linearity, particularly in some older instruments) and then link the concentration of the reference gas back to an international standard. What is required is to define standards to an accuracy of 0.1 ppmv or better in concentrations of over 300 ppmv. This standardisation has to be maintained across a series of local calibration gases, through to local standards and ultimately tied back to international standards. This degree of precision needs to be maintained over periods of many decades. For details of the historical development and difficulties in achieving this see Pearman (1980), Keeling et al. (1986) and WMO (1987).

Selection: To remove data which are believed not to represent baseline conditions;

Averaging: To reduce the effects of 'noise' and to reduce the data volume for publication.

Even in the best of surface sites, it is necessary to perform some degree of data selection in order to remove local influences on the concentrations. The criteria for selection need to be based on extensive studies of the variability at the site. Some of the main criteria that are used (or could be used) are:

Wind direction: At many sites, particularly coastal sites, a 'baseline sector' is defined so that only data with winds from that sector are taken as baseline. For example, Baring Head uses a southern sector, Cape Grim uses a south-westerly sector.

Data variability: The degree of variability within short runs of continuous analyser data is used as a selection criterion at several sites. For example at Izaña, variability within each half-hour period is used. The Cape Grim selection criteria are applied to hourly means. An hourly mean is selected as 'baseline' if it is part of a run of 5 hours during which the none of the other 4 hourly means differs from it by more than 0.3 ppmv. The NOAA continuous analyser programs use both within-hour variability and (except at the South Pole) between-hour variations as selection criteria.

Wind speed: In order to reduce local contamination, a minimum wind-speed can be specified. For Amsterdam Island, the minimum used depends on the sector. Early reports of provisional data from Cape Grim used wind-speed and direction as the baseline selection criteria.

Wind consistency: The persistence of the wind direction is another criterion that can indicate large-scale air-masses.

Particle count: Particulate counts are used at Cape Grim as part of the sampling criteria for flask programs.

Flask consistency: A common practice in flask sampling programs is to take pairs of samples in the same sampling session and reject data when concentrations from the flasks disagree, assuming this to be due to either highly variable local contamination and/or operator or handling error.

Other constituents: Anomalous concentrations of other atmospheric constituents in flasks can be indicative of contamination. In the CSIRO programs it has been more common to use this criterion the other way round, with anomalous CO_2 data being used as a criterion for rejecting other data from the same flask.

^{14}C: Levin (1987) has described the use of ^{14}C data to remove data affected by local fossil fuel sources in the CO_2 record from Schauinsland (a mountain above the Rhine valley).

Airmass trajectories: Computed airmass trajectories are seldom available for routine use in data selection but have been used in a number of exploratory studies.

Radon: Radon-222 (with a half-life of 3.8 days) provides a way of distinguishing air masses that have recently passed over land since the radon sources from land are much greater than from the ocean. For example, exploratory studies using radon data have been undertaken for CO_2 records from Amsterdam Island (Gaudry et al., 1991), Schauinsland (Levin, 1987) and Cape Grim.

The determination of appropriate selection criteria for a particular site requires comprehensive studies of the variability in CO_2 at the site. While some of these studies can be undertaken in assessing the *a priori* suitability of the site, a full evaluation requires the data from an operational period of several years. Examples of sites for which studies of the baseline selection criteria (or more generally of the relation between CO_2 and other data) have been published are:

Cape Grim: Beardsmore and Pearman (1987) have reviewed the reasons for the choice of the current operational definition of Cape Grim baseline data. More recent studies (e.g. Enting, 1989b and references therein) have suggested that a narrower sector of wind directions would reduce the variability in the data.

Izaña Schmitt et al. (1988) have analysed the role of long-range transport in determining CO_2 concentrations at Izaña.

Mauna Loa: For Mauna Loa, Keeling et al. (1976a) have discussed local influences on the SIO record. Thoning (1989) describes the selection criteria for the NOAA continuous analyser program.

Mt. Cimone: Ciattaglia (1983) has analysed the first three years' data in terms of wind direction.

Flask Sampling: With flask sampling, the limited amount of data restricts the possibilities for detailed studies of the selection criteria. Tans et al. (1989b, 1990b) addressed this problem by looking at the statistics of flasks samples from the NOAA sites where continuous analysers are operated. They obtained estimates of the statistical errors expected from the limited sampling rate by studying the statistics of sets of pseudo-flask data, sampled from the continuous analyser record. They then estimated the experimental errors by comparing the real flask data to the continuous data. They concluded that while long-term averages were in good agreement, interannual differences could be as large as 0.4 ppmv.

Apart from the greater frequency of continuous analyser data as compared to flask data, an important advantage of the continuous data is the ability to go back over the record retrospectively and change the selection criteria. With flask sampling, most of the selection takes place in the definition of sampling conditions — once the sample has been taken, these can not be changed retrospectively.

Figure 2 shows the record of hourly mean selected baseline CO_2 data from Cape Grim for the year 1989. For comparison, Figure 3 shows concentrations (at four-hour time steps), selected for wind direction, obtained from a three-dimensional model run using zonally uniform sources as deduced by Enting and Mansbridge (1991) (but without correcting for the role of CO). We included only these broad-scale sources in order to isolate the effects of transport. The transport model was the GISS tracer model described by Fung et al. (1983), run at 8° by 10° resolution. Convective mixing is treated using monthly averages from the original GISS general circulation model. The model also includes an enhanced horizontal diffusion. The variability in the concentrations calculated in this way will reflect the influence of advective transport but will exclude the effects of:
(i). Small-scale variations in the sources; *(ii).* Day-to-day (and hour-to-hour) variations in convective mixing; *(iii).* Advective transport on scales less than 10° by 8°.

The comparison implies that a significant fraction of the small-scale variability in the Cape Grim data is due to the effect of transport acting on broad-scale sources but the majority of the variability arises from more local small-scale or regional effects.

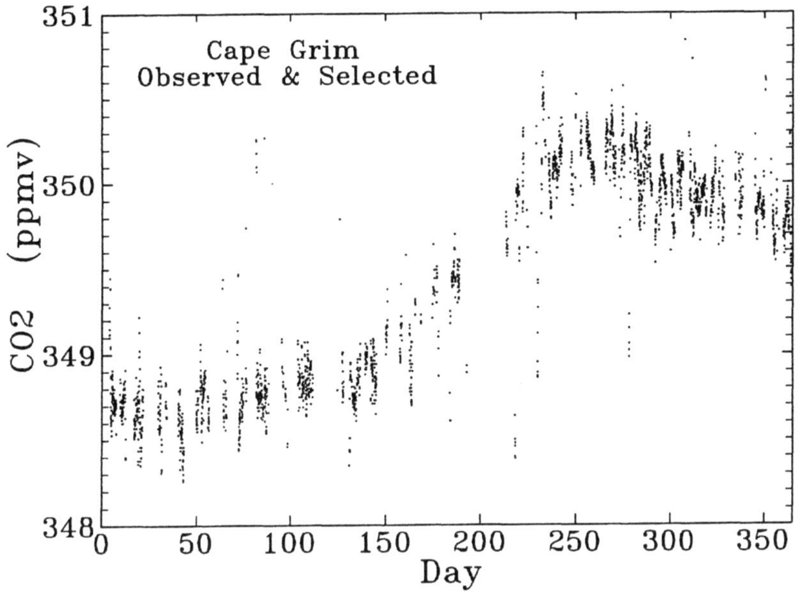

Figure 2. Hourly mean baseline selected CO_2 concentrations from Cape Grim for 1989.

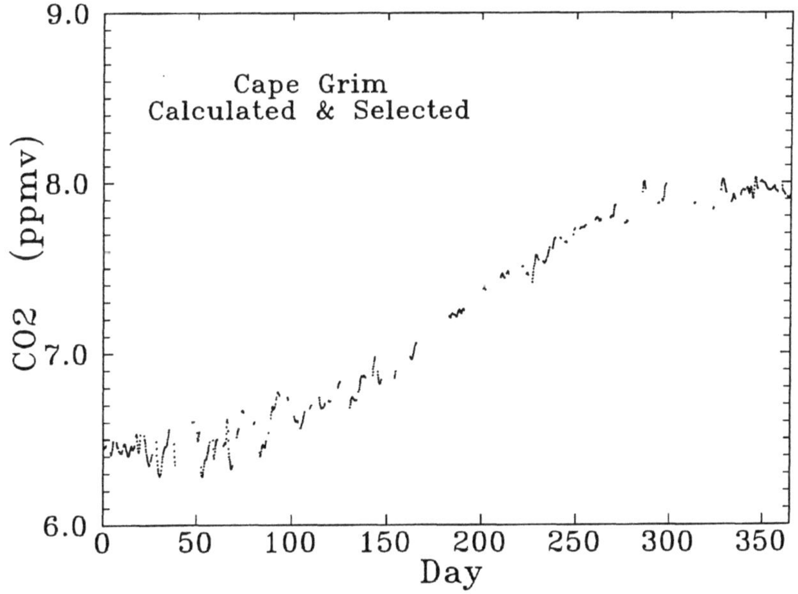

Figure 3. Model CO_2 concentrations, at 4-hour intervals, selected for wind direction, calculated using zonally uniform sources.

The observational record shows an increasing trend and a well-defined seasonal cycle. In practice, the definition of a mean seasonal cycle is limited not by the small-scale variability shown in Figure 2 but rather by the interannual variability. Spectral analysis (Thompson et al., 1986) found that at most sites, only the 12-month and 6-month components of the cycle could be resolved (Barrow being the exception). Further discussion of the cycle is outside the scope of this chapter, but a similar situation exists for defining mean gradients in that the interannual variability places more restrictions on the definition of representative gradients than does the small-scale variation in time.

As noted above, the criteria for obtaining baseline data strongly favour coastal (and often island) sites. There are few CO_2 data from continental regions. Some of the difficulties involved in interpreting data from continental sites can be appreciated by studying the data from the German CO_2 program as reported in WMO (1990). The data are not selected for baseline conditions, generally reflecting an inability to define such conditions. The standard deviations in the monthly means can be as much as 8 ppmv, generally decreasing with the altitude of the site.

The work of Levin (1987) shows the possibility of extracting a 'continental clean air' record from such data (albeit from one of the higher sites). Data for ^{222}Rn and ^{14}C were used in the interpretation of the CO_2 record. The seasonal cycle that was extracted shows significant differences (mainly in phase) from other data from the same latitude (see also Manning, this volume). However it is not clear that the 'clean-air' analysis can be made sufficiently precise to extract quantitative information about longitudinal variations in the annual mean.

5. High Altitude

A serious limitation to the current global data base is the lack of information concerning the vertical distribution of concentration through the troposphere and stratosphere. The bulk of the observational data are from surface sites at low altitude. Of course, some indication of the vertical distribution of concentration is obtained by virtue of the fact that several of the ground-level observatories are in fact well above sea level (those at over 1000 m are listed in Table 5). For example, the South Pole Observatory is at 2810 m altitude, Mauna Loa is at 3397 m, and the Mt. Cimone observatory is at 2165 m. In the case of the first two of these, comparison can be made with observations collected at sea level in the vicinity (Mawson or Syowa and Cape Kumakahi respectively).

In addition, a few aircraft-based programs have been established to provide additional high-altitude data. Some of the reasons for seeking such data are:

i. Remoteness from contamination by local sources, thus reducing or eliminating the need for data selection;

Site	Agency	Latitude	Longitude	Alt
Brotjacklriegel	Umweltbundesamt	43°49′ N	13°13′ E	1016m
Izaña	INM	28°17′ N	16°29′W	2367m
Mauna Loa	SIO	19°32′ N	155°35′W	3397m
Mauna Loa	NOAA	19°32′ N	155°35′W	3397m
Mt. Cimone	Ital. Met. Service	44°11′ N	10°42′ E	2165m
Niwot Ridge *	NOAA	40°03′ N	105°38′W	3749m
Schauinsland	Umweltbundesamt	47°55′ N	7°55′ E	1205m
South Pole *	SIO	89°52′ S	24°48′W	2810m
South Pole	NOAA	89°52′ S	24°48′W	2810m
Wank	Fraunhofer Inst.	47°31′ N	11°09′ E	1780m
Zugspitze	Fraunhofer Inst.	47°25′ N	10°59′ E	2960m

Table 5. High altitude sites (over 1000 m). Stations with only flask data are denoted '*'. NOAA previously obtained samples from Cosmos, Peru (4600 m) but was unable to obtain satisfactory data quality. NOAA also operated a short-term sampling program at Kitt Pk., Arizona, primarily for comparison with archived solar spectra from Kitt Pk. observatory.

ii. To obtain vertical gradients from which to relate surface source strengths to vertical mixing rates;

iii. To use latitudinal gradients at high altitudes to determine the broad-scale features of the surface sources with a minimum of 'aliasing' effects from small-scale components (see the analysis by Newsam and Enting, 1988 noted in Section 2, above).

During the 1960s and early 1970s, a Swedish (University of Stockholm) sampling program collected data using aircraft flying over the north Atlantic and polar regions of the northern hemisphere (Bischof 1965, 1970, 1971, 1973, 1981; Bolin and Bischof, 1970). The program

Region	Agency	Latitudes	Longitudes	Altitude	Years
Nth Atlantic	U.Stockholm	≈70N–35N	15E–120W	various	1963–?
S.E. Australia	CSIRO	≈40S	≈145E	3.5–4.5 km	1972–.
Tasman Sea	CSIRO	35S–45S	145E–170E	>9 km	1972–9
Gt. Aust. Bight	CSIRO	35S–32S	115E–145E	> 8.5 km	1974–.
Japan	Tohoku U.	≈36N	≈140E	0–8 km	1979–.
West Pacific	Tohoku U.	36N–30S	140E–155E	10–12 km	1984–5
North Pacific	Tohoku U.	40N–60N	140E–150W	10–12 km	1984–5

Table 6. Aircraft programs. If no last year is specified, program is believed to be continuing. (For details, see Bischof, 1965, 1970, 1971, 1973; Pearman and Beardsmore, 1984; Tanaka et al., 1987a; Nakazawa et al., 1991a).

was based on flask samples of air collected through the main air-conditioning system of DC8 aircraft in which there was no cabin recirculation system. Consequently the air entering the cabin was uncontaminated and truly representative of the outside air.

A similar program was commenced in 1972 in Australia (Pearman et al., 1983a) by the CSIRO Division of Atmospheric Physics. It had three components. First, flask samples were collected from Qantas B707 aircraft flying between Australia and New Zealand at cruise altitude (upper troposphere and lower stratosphere) by the flight engineers. This program ran for 10 years and provided a detailed description of concentration variations at that level and latitude (Pearman and Beardsmore 1984). Second, samples were collected during quasi-regular mid-tropospheric flights of government aircraft (Fokker F27 and F28) flying over the south-eastern part of Australia. This program has continued until present with, in recent years, a greater dependence on hired small aircraft to ensure at least one flight per month. The samples are collected manually and a range of different aircraft have been used for the purpose. Finally, through the cooperation of Australian Airlines (formally Trans Australian Airlines), automatically collected samples have been taken at cruise altitude of B727 aircraft flying over the Great Australian Bight.

In Japan, an aircraft monitoring program is operated by the Upper Atmospheric Physics Department of the University of Tohoku, Sendai. This program utilizes DC8 aircraft and concentrates on the Japan region. It has established the mean vertical gradient and the vertical attenuation of the seasonal cycle above Japan (Tanaka et al., 1987a). In addition during 1984–1985, the Tohoku group operated aircraft sampling programs between Tokyo, Japan and Sydney, Australia and between Tokyo and Anchorage, Alaska. The cruise height for the Japan–Australia flights is in the upper troposphere — the Japan–Alaska flights are in the lower stratosphere, but additional tropospheric samples were obtained at the end-points (Nakazawa et al., 1991a).

Measurements higher in the stratosphere are much less plentiful. Bischof et al. (1985) made measurements from samples collected by balloons to an altitude of 25 km. More recently, preliminary observations have been made on samples collected as part of the Jülich stratospheric atmospheric chemistry program (Schmidt and Khedim, 1991).

Observations at altitude represent concentrations and isotopic compositions characteristic of air previously at the earth's surface. In principle, stratospheric samples give the opportunity to retrospectively examine the relationships between CO_2 concentration changes and those of isotopic composition or the concentration of other conserved gases.

As noted above, vertical measurements can be used to validate the mixing characteristics of transport models in the vertical. Bolin and Bischof (1970) used the seasonal cycle to characterise vertical mixing in terms of a purely diffusive approximation. In the context of two-dimensional advective-diffusive models, the use of the vertical attenuation of the amplitude of the seasonal cycle to estimate vertical mixing rates has been noted by Pearman and Hyson (1980) and Tans et al. (1989a).

Transport times from the surface to the upper troposphere vary from days in the case of strong tropical convective mixing to a month or more in the more stratified higher latitudes. In the northern hemisphere, the annually averaged concentration in the lower layers of the atmosphere are generally higher than those of the upper atmosphere and lower stratosphere, as might be expected given the net source of fossil fuel CO_2 released into the lower layers. In the southern hemisphere, on the other hand, in the mid and higher latitudes surface concentrations are generally lower than at altitude, reflecting the net surface sink for CO_2 and the mode of transport of low-latitude air enriched with CO_2 of northern hemisphere origin in the middle and upper troposphere through the Hadley circulation.

We have performed some preliminary three-dimensional model calculations, analogous to those presented in Figure 3 for mid-tropospheric data without any selection for wind direction. A somewhat surprising result was that the mid-troposphere concentrations showed more variability than the surface concentrations. It has not been possible to test this result against our observational data since the surface data show much additional variabilty, presumably due to local effects and the aircraft data are inherently more noisy, due to being obtained from flask samples.

6. Spatial Distributions

6.1 Decomposition of the distribution

The full space-time distribution of atmospheric CO_2 is a four-dimensional function of latitude, longitude, altitude and time. However for the purposes of discussion it is appropriate to express it in terms of a number of components. These are:

Long-term trends: As noted by Pearman (1980) the rapidity of atmospheric mixing means that the trend in the global inventory can be established from a small number of stations. Indeed much analysis of long-term trends has been based on the single SIO record from Mauna Loa.

Horizontal gradients: The distribution of annual mean surface concentrations provides the most important information for deducing the annual mean sources and sinks as a function of position.

Vertical gradients: As noted above, there are relatively few records from sites well above sea-level. It is expected that as the height increases, the features of the horizontal distribution will be progressively attenuated.

Average seasonal cycles: The amplitude and phase of the cycle vary as a function of both latitude and height and to a lesser extent as a function of longitude (Manning, this volume). For comparisons of the cycle at different sites, consistency of calibration is much less important than for the spatial gradients.

Other variation is regarded as interannual variation in one of the above components (see the following section) or as small-scale 'noise' that is not of interest to global-scale studies.

Figure 4 shows the annual means from the NOAA flask network, for several years plotted as a function of latitude. These show a broad-scale interhemispheric gradient, reflecting the major input of fossil carbon into the northern hemisphere. The curves also show a high degree of interannual variability and a considerable degree of scatter around the mean line.

In this chapter we consider the spatial variation — the seasonal variation is described by Manning (this volume). In terms of the discussion above, we seek to express the horizontal variation at the surface as:

Zonal mean: The direct inversion calculations carried out to date (Enting and Mansbridge, 1989, 1991; Tans et al., 1989a) have only used the zonal means, thus deducing sources as a function of latitude.

Longitudinally varying component: There are almost certainly significant land-sea differences in annual mean concentrations, but few data to quantify these because of the difficulty (see Section 4 above) of defining representative data over land.

Systematic interannual variations in the above: Both the estimation of the latitudinal variation and detection of systematic longitudinal variations are complicated by the high degree of interannual variability; this is discussed further in Section 7 below.

'Un-interpretable' noise: The types of variation contibuting to this component will depend on the sophistication of the model used to interpret the data. Most obviously, if the data are interpreted using a two-dimensional (zonally-averaged) model then any longitudinal variation becomes part of the 'noise'. Similarly the interannual variation must be regarded as part of the noise unless it is possible to model interannual variations in sources and transport. Small-scale synoptic variations must usually be regarded as noise, but the study by Heimann et al. (1989, Figure 26, p299) shows that some of these variations can be represented using a transport model based on observed winds for the same time as the concentration data.

6.2 Zonal mean

In analysing the zonal mean we ask the question of how much information is contained in the latitudinal dependence. One of the simplest ways in which this question can be addressed is

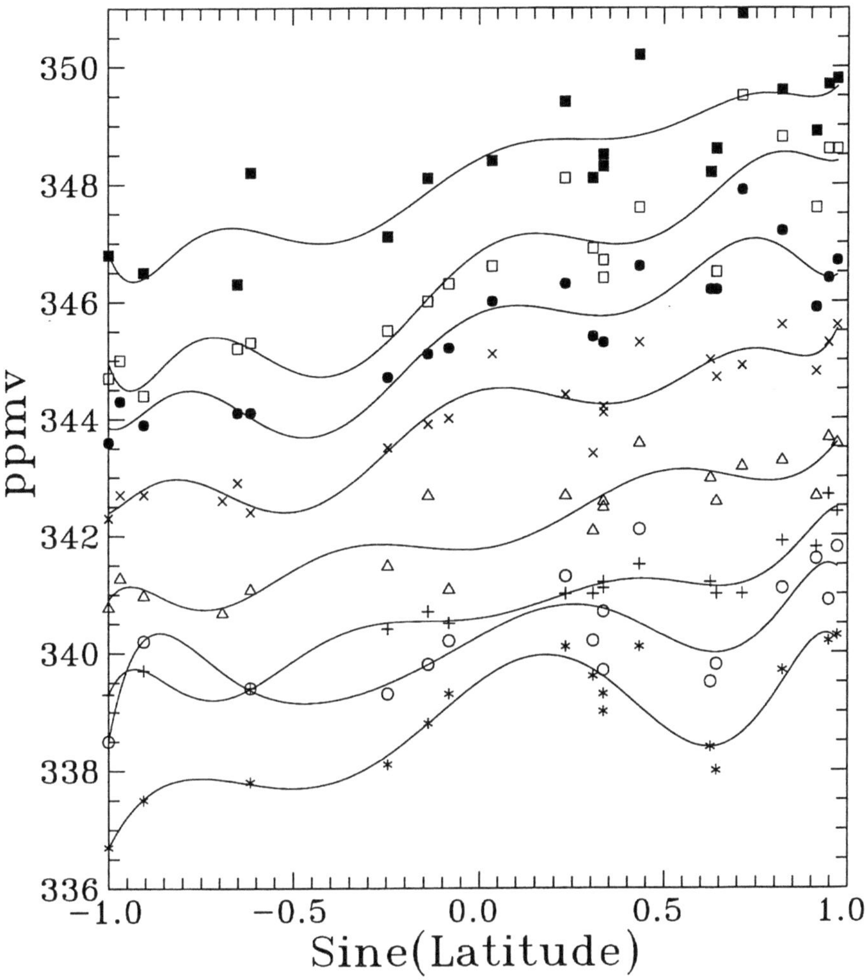

Figure 4. Latitudinal distribution of annual mean CO_2 concentrations. Data from the NOAA flask sampling program as reported in CDIAC (1990). Curves are 8th degree polynomial fits to each year's data. The symbols are '*' 1980; 'o' 1981; '+' 1982; '△' 1983; '×' 1984; '•' 1985; '□' 1986; '■' 1987.

to ask how many regression components are required to represent all statistically significant detail in the curve. The answer seem to be about 5, in terms of the polynomial fit (Pearman and Hyson, 1986) or spectral fits (Enting and Mansbridge, 1989). The degree of smoothing used in the spline fitting by Tans et al. (1989a) and Enting and Mansbridge (1991) also seems to imply about 5 independent degrees of freedom in the data fit.

These preliminary estimates of the information content of the annual means are subject to two related qualifications:

i. The estimates of statistical significance in a standard regression analysis assume that the data represent a polynomial plus independent errors. If some of the variability about the zonal mean represents a coherent longitudinal variation that can be modelled then the effective noise level in estimation of the zonal mean would be reduced and more components of the zonal mean may become statistically significant.

ii. Conversely, if the data are subject to coherent but unexplainable variations then significance levels based on the assumption of independent errors will be overestimates and fewer than 5 components may be significant.

The NOAA flask data in Figure 4 show the extent to which interannual variability complicates the interpretation of spatial gradients. The data give a possible indication that, compared to earlier and later periods, the years from 1980 to 1985 had an anomalously low southern sink and an anomalously high northern sink. In other words, the results of Enting and Mansbridge (1989, 1991) and Tans et al. (1989a, 1990a), which appear to be in conflict with ocean models based in ^{14}C, may represent a short-term anomalous situation. In this regard, Enting and Mansbridge (1991) noted that the study by Pearman and Hyson (1986) used larger interhemispheric differences (based on pre-1982 data) than the later inversion studies (Enting and Mansbridge, 1989, 1991; Tans et al., 1989a) and consequently found a larger southern hemisphere ocean uptake. Unfortunately, this difference can not be definitively identifed as showing a secular change in gradient because the inter-calibration accuracy of the earlier data is uncertain. Indeed, data from Keeling et al. (1989a) suggest a progressive evolution of the interhemispheric gradient, rather than indicating an anomalous period in the early 1980's.

6.3 Longitudinal dependence

As noted above, if some of the variability in the data could be attributed to coherent longitudinal variations then a more precise specification of the zonal mean could be obtained. The simple diffusive modelling described above predicts that the effects of longitudinal variations in the sources will be highly attenuated in the concentration record. Three-dimensional modelling suggests that the effects of longitudinal variations in sources are not detectable outside a latitude range of about ±10° around the source region. Preliminary results from our modelling studies of CO_2 indicate that the largest contributions to longitudinal variations in the annual mean concentration may come from the effects of seasonally varying transport acting on seasonally varying sources rather than from longitudinal variations in the annual mean sources and sinks.

A number of studies have noted a longitudinal variation in the seasonal cycle in the northern hemisphere, primarily as a phase difference (e.g. Wong et al., 1984; Levin, 1987; Tans et al., 1989a; Navascués and Rus, 1991). Detection of zonal variations in the annual means is much less clear-cut. The study by Tans et al. (1989a) used a two-dimensional transport model to deduce annual mean sources using two distinct data sets, each regarded as zonally representative. This issue is considered further in Section 7 where differences between three southern hemisphere stations are examined.

6.4 Vertical profiles

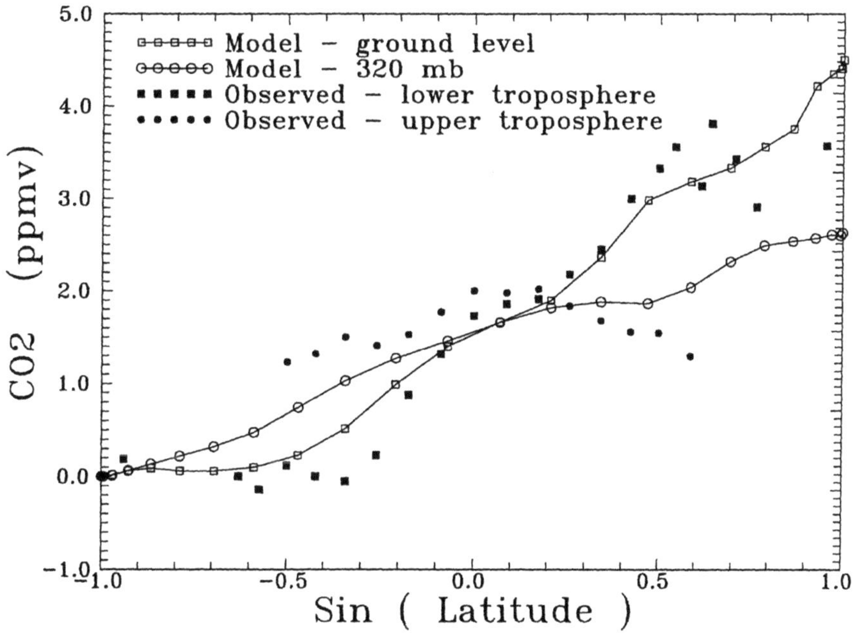

Figure 5. Latitudinal CO_2 variations, 1984–5 average, for the west Pacific from the Tohoku University ship-based (■) and aircraft (•) programs. Data from Nakazawa et al. (1991a). Solid lines are CO_2 concentrations at the surface (□) and 320 mb (o) for longitude 145E, as calculated using the 3-d model with zonally uniform sources.

As noted above, there are relatively few data sets defining the distribution of CO_2 above the earth's surface. The aircraft programs over Japan and SE Australia give useful information about the shift in the seasonal cycle with altitude. More specific profile data over Australia were given by Garratt and Pearman (1973). The altitude dependence of the seasonal cycle in the northern hemisphere was also described by Bolin and Bischof (1970). However, for establishing mean spatial distributions above the surface, the most important record is from the 1984–5 Tohoku aircraft program (Nakazawa et al., 1991a). The data are reproduced in Figure 5, along with the surface data for essentially the same longitude, taken from the Tohoku ship-based program.

For comparison we also present in Figure 5 calculations for longitude 145°E of the CO_2 concentrations at the surface and 320 mb, from the three-dimensional model with zonally uniform sources. The results differ markedly from the observational data presented by Nakazawa et al. (1991a). We performed a number of additional modelling calculations to explore the possibility that the observed peak in CO_2 is confined to the western Pacific and represents strong vertical mixing of CO_2 from local sources, possibly including deforestatation in south-east Asia. It was found that even with a source strongly concentrated in South-East Asia, the effect on the high altitude profile was only a small fraction of a ppmv. It appears that the discrepancy between model and observations is a reflection of the limited resolution of the model tropopause, allowing surface signals to penetrate higher into the model atmosphere before being attenuated.

7. Interannual Variations

As well as a long-term increase, the annual mean CO_2 concentrations show a significant amount of inter-annual variation. Bacastow (1976) found the anomalous rate of increase for Mauna Loa to be correlated with the Southern Oscillation Index (SOI) which characterises a large-scale interannual variation in the meteorological conditions in the Pacific. This correlation was further explored by Thompson et al. (1986). Elliott et al. (1991) have presented extensive comparisons between CO_2 changes and Pacific sea-surface temperatures.

While this interannual signal may be able to give valuable insights into aspects of the global carbon cycle, for the present we concentrate on the implications of this variability for the problems of deducing surface sources. There are three main possibilities:

i. If the interannual variability has to be treated as uninterpretable noise, it imposes a very severe limitation on the conclusions that can be drawn from the data. To make further progress it is necessary to be able to decompose the interannual variations into contributions from interannual variations in sources and sinks and contributions due to interannual variations in transport.

ii. If the variations in concentration are due to variations in sources and sinks it is possible to deduce these by using an atmospheric transport model forced to follow the concentrations as was done by Tans et al. (1989a).

iii. If the variations are due to variations in transport, it becomes neccessary to model such varying transport in order to improve the effective signal-to-noise ratio. At present this is a somewhat distant possibility — indeed there are few quantitative estimates of the possible size of influences of interannual variation in transport. One exception is the study by Gaudry et al. (1991) where they concluded that the global-scale of the coherent ENSO perturbation of CO_2 concentrations indicated a change in the net source rather than a redistribution of CO_2 by changed atmospheric transport.

As well as variations in the annual mean gradient, there are also interannual variations in the seasonal cycle (see for example Thompson et al., 1986) but these are outside the scope of this chapter.

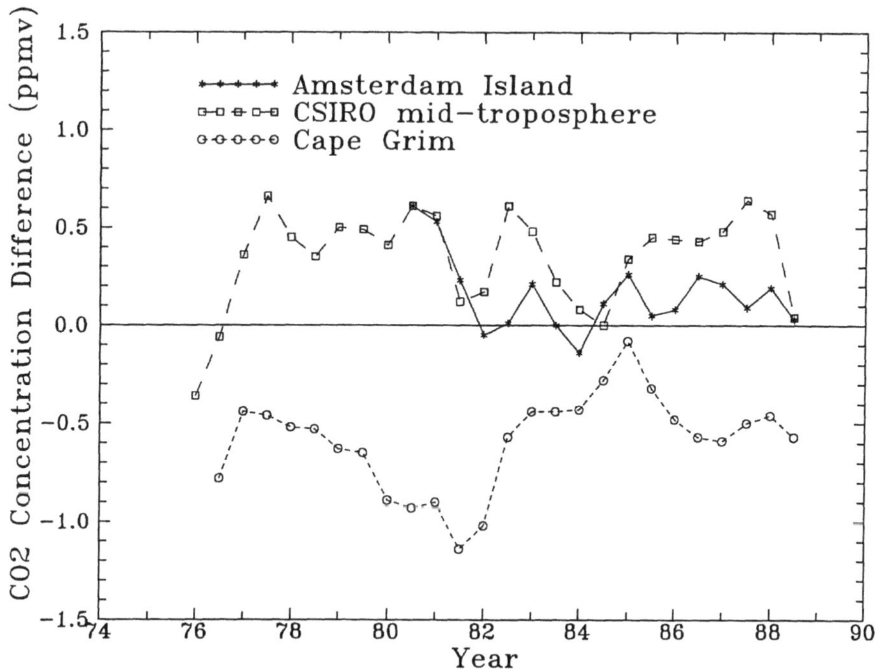

Figure 6. Differences between CO_2 concentrations from the DSIR Baring Head program and other southern sites. Data are 12-month means plotted each 6 months; Years are labelled at their mid-points. The Baring Head data are subtracted from each of the other records. 'o' denotes Cape Grim, □ denotes CSIRO mid-troposphere and '*' denotes Amsterdam Island.

As a final example of interannual variability, we present data for the southern ocean region. Figure 6 shows the difference between the CO_2 records for Bariang Head, NZ and three other sites: Cape Grim, CSIRO mid-troposphere program and Amsterdam Island. (In each case, the baring Head record is subtracted from the other data). For data after 1987, a provisional linear correction has been applied to the CSIRO data because of an apparent drift in the local calibration standards. Clearly, there is a relatively small-scale effect that influences the Cape Grim data on an interannual time-scale but which has much less influence on the other sites. A number of explanations have been considered:

(i). changes in ocean p_{CO_2} due to interannual variations in ocean currents; *or*

(ii). changes in wind strengths leading to changes in the amount of local influence on the selected baseline data; *or*

(iii). changes in wind strength leading to changes in the air-sea gas exchange rate.

More detailed studies, aimed at distinguishing between the various possibilities are currently in progress. For any of the three expanations it would require quite sophisticated modelling (based on meteorological data for the actual years involved) to represent the proposed effects, rather than having to regard the variations of Figure 6 as part of the 'noise'.

8. Concluding Remarks

Observations of the global distribution of atmospheric CO_2 provide the basis for estimating regional carbon budgets. Such regional budgets provide the best prospect for refining the global atmospheric carbon budget beyond the level of uncertainty provided by direct estimates of contributions and/or global-scale modelling. At present, the lack of error estimates in such source-deduction calculations limits the utilisation of regional budgets in analyses of the global budget. The preliminary formalism described in Section 2, above shows that an analysis of errors in sources derived from concentration distributions will require a detailed specification of the causes of the variability in the observational records.

The error analysis formalism serves to emphasise the need for sophistication in the transport modelling. Any variability that can not be modelled, whether synoptic variation, interannual variation or longitudinal variation (if using a two-dimensional model), must be treated as part of the 'noise', acting to reduce the signal-to-noise ratio. The ill-conditioned nature of the source-deduction problem suggests that increases in precision (especially in improved intercalibration between sites) would be more important than additional sites. A possible exception would be additional aircraft data. One important question concerns the optimum altitude for such data in that increasing altitude gives rapid attenuation of source components that are highly variable in space and/or time. The trade-off is between loss of information and reduction of 'aliasing' errors. However none of these questions can be answered quantitatively at present because detailed analyses of optimal observational networks remain to be performed.

Acknowledgements

The authors wish to thank Paul Steele, Ian Galbally, John Garratt and Cathy Trudinger for valuable comments on the manuscript. The three-dimensional model calculations reported here were performed by Cathy Trudinger using the GISS tracer-transport model kindly supplied by Inez Fung. The CSIRO three-dimensional modelling project is largely funded by the State Electricity Commission of Victoria. Henry Granek and Cathy Trudinger prepared the figures. We thank David beardsmore and the staff at Cape Grim for the on-going support of the observational program at that station.

References

Bacastow, R.B. (1976) Modulation of atmospheric carbon dioxide by the Southern Oscilation. Nature 261:116–118

Beardsmore, D. and Pearman, G.I. (1987) Atmospheric carbon dioxide measurements in the Australian region: data from surface observations. Tellus 39B:42–66

Beardsmore, D.J., Pearman, G.I. and O'Brien, R.C. (1984) The CSIRO (Australia) Atmospheric Carbon Dioxide Program: Surface Data. Division of Atmospheric Research, Technical Paper No. 6.

Bischof, W. (1965) Carbon dioxide in the upper troposphere and lower stratosphere. I. Tellus 17:398–402

Bischof, W. (1970) Carbon dioxide measurement from aircraft. Tellus 22:546–550

Bischof, W. (1971) Carbon dioxide in the upper troposphere and lower stratosphere. II. Tellus 23:558–561

Bischof, W. (1973) Carbon dioxide in the upper troposphere and lower stratosphere. III. Tellus 25:305–308

Bischof, W. (1981) The CO_2 content of the upper polar troposphere between 1963–1979. pp 113–116 of Carbon Cycle Modelling: SCOPE 16. Ed. B. Bolin. John Wiley and Sons, Chichester.

Bischof, W., Borchers, R., Fabian, P. and Kruger, B.C (1985) Increased concentration and vertical distribution of carbon dioxide in the stratosphere. Nature 316:708–710

Bolin, B. and Bischof, W. (1970) Variations of the carbon dioxide content of the atmosphere in the northern hemisphere. Tellus 22:431–442

Bolin, B. and Keeling, C.D. (1963) Large-scale atmospheric mixing as deduced from the seasonal and meridional variations of carbon dioxide. J. Geophys. Res. 68:3899–3920

CDIAC (1990). Trends '90: A Compendium of Data on Global Change. Ed. T.A. Boden, P. Kanciruk and M.P. Farrell. Carbon Dioxide Information Analysis Center, Oak Ridge, Tennessee.

Ciattaglia, L. (1983) Interpretation of atmospheric CO_2 measurements at Mt. Cimone (Italy) related to wind data. J. Geophys. Res. 88:1331–1338

Ciattaglia, L., Cundari, V. and Colombo, T. (1987) Further measurements of atmospheric carbon dioxide at Mt. Cimone, Italy: 1979–1985. Tellus 39B:13–20

CMDL (1990) Climate Monitoring and Diagnostics Laboratory. No. 18. Summary Report 1989. Ed. W.D. Kohmyr and R.M. Rosson. (US Dept. Commerce).

Conway, T.J., Tans, P., Waterman, L.S., Thoning, K.W., Masarie, K.A. and Gammon, R.H. (1988) Atmospheric carbon dioxide measurements in the remote global troposphere 1981–1984. Tellus 40B:81–115

Elliott, W.P., Angell, J.K. and Thoning, K.W. (1991) Relation of atmospheric CO_2 to tropical sea and air temperatures and precipitation. Tellus 43B:144–155

Enting, I.G. (1989a) CO_2 measurements by CSIRO (Australia). pp 43–50 of The Statistical Treatment of CO_2 Data Records. Ed. W.P. Elliott. NOAA Technical Memorandum ERL ARL-173.

Enting, I.G. (1989b) Studies of baseline selection criteria for Cape Grim, Tasmania. pp 51–60 of The Statistical Treatment of CO_2 Data Records. Ed. W.P. Elliott. NOAA Technical Memorandum ERL ARL-173.

Enting, I.G. and Mansbridge, J.V. (1989) Seasonal sources and sinks of atmospheric CO_2: Direct inversion of filtered data. Tellus 41B:111–126

Enting, I.G. and Mansbridge, J.V. (1991) Latitudinal distribution of sources and sinks of CO_2: Results of an inversion study. Tellus 43B:156–170

Enting, I.G. and Newsam, G.N. (1990a) Inverse problems in atmospheric constituent studies: II. Sources in the free atmosphere. Inverse Problems 6:349–362

Enting, I.G. and Newsam, G.N. (1990b) Atmospheric inversion problems: Implications for baseline monitoring. J. Atmos. Chem. 11:69–87

Enting, I.G. and Newsam, G.N. (1991) An improved approach to one-dimensional modelling of meridional transport in the atmosphere. Tellus 43B:76–79

Fraser, P.J., Pearman, G.I. and Hyson, P. (1983) The global distribution of atmospheric carbon dioxide: 2. A review of provisional background observations, 1978–1980. J. Geophys. Res. 88C:3591–3598

Fung, I., Prentice, K., Matthews, E., Lerner, J. and Russell, G. (1983) Three-dimensional tracer model study of atmospheric CO_2: response to seasonal exchanges with the terrestrial biosphere. J.Geophys. Res. 88C:1281–1294

Garratt, J.R. and Pearman, G.I. (1973) Large-scale CO_2 fluxes in the Southern Hemisphere troposphere. Nature 242:54–56

Gaudry, A., Monfray, P., Polian, G., Bonsang, G., Ardouin, B., Jegou, A. and Lambert, G. (1991) Non-seasonal variations of atmospheric CO_2 concentrations at Amsterdam Island. Tellus 43B:136–143

GMCC (1983) Geophysical Monitoring for Climatic Change. No. 11. Summary Report 1982. Ed. J.M. Harris and B.A. Bodhaine. (U.S. Dept Commerce.)

GMCC (1986a) Geophysical Monitoring for Climatic Change. No. 13. Summary Report 1984. Ed. E.C. Nickerson. (U.S. Dept Commerce.)

GMCC (1986b) Geophysical Monitoring for Climatic Change. No. 14. Summary Report 1985. Ed. R.C. Schnell and R.M. Rosson. (U.S. Dept Commerce.)

GMCC (1989) Geophysical Monitoring for Climatic Change. No. 17. Summary Report 1988. Ed. J.W. Elkins and R.M. Rosson. (U.S. Dept Commerce.)

Heimann, M., Keeling, C.D. and Tucker, C.J. (1989) A three-dimensional model of atmospheric CO_2 transport based on observed winds: 3. Seasonal cycle and synoptic time scale variations. pp 277–303 of Aspects of Climate Variability in the Pacific and Western Americas. Ed. D.H. Peterson. Geophysical Monograph 55. AGU (Washington).

Keeling, C.D., Bacastow, R.B., Bainbridge, A.E., Ekdahl, C.A., Guenther, P.R. Waterman, L.S., and Chin, J. (1976a) Atmospheric carbon dioxide variations at Mauna Loa observatory, Hawaii. Tellus 28:538–551

Keeling, C.D., Adams, J.A., Ekdahl, C.A. and Guenther, P.R. (1976b) Atmospheric carbon dioxide variations at the South Pole. Tellus 28:552–564

Keeling, C.D., Carter, A.F. and Mook, W.G. (1984) Seasonal, latitudinal and secular variations in the abundance and isotopic ratios of atmospheric CO_2: 2. Results from oceanographic cruises in the tropical Pacific Ocean. J. Geophys. Res. 89D:4615–4628

Keeling, C.D., Guenther, P.R. and Moss, D.J. (1986) Scripps Reference Gas Calibration System For Carbon Dioxide-In-Air Standards: Revision 1985. Environmental Pollution Monitoring and Research Programme, Publication No. 42. (WMO/TD-125). WMO, Geneva.

Keeling, C.D., Bacastow, R.B., Carter, A.F., Piper, S.C., Whorf, T.P., Heimann, M., Mook, W.G. and Roeloffzen, H. (1989a) A three-dimensional model of atmospheric CO_2 transport based on observed winds: 1. Analysis of observational data. pp 165–235 of Aspects of Climate Variability in the Pacific and Western Americas. Ed. D.H. Peterson. Geophysical Monograph 55. AGU (Washington).

Keeling, C.D., Piper, S.C. and Heimann, M. (1989b) A three-dimensional model of atmospheric CO_2 transport based on observed winds: 4. Mean annual gradients and interannual variations. pp 305–363 of Aspects of Climate Variability in the Pacific and Western Americas. Ed. D.H. Peterson. Geophysical Monograph 55. AGU (Washington).

Komhyr, W.D., Gammon, R.H., Harris, T.B., Waterman, L.S., Conway, T.J., Taylor, W.R. and Thoning, K.W. (1985) Global atmospheric CO_2 distribution and variations from 1968–1982 NOAA/GMCC CO_2 flask sample data. J. Geophys. Res. 90:5567–5596

Levin, I. (1987) Atmospheric CO_2 in continental Europe — an alternative approach to clean air CO_2 data. Tellus 39B:21–28

Levin, I., Graul, R and Trivett, N.B.A. (1992) The CO_2 observation network in the Federal Republic of Germany. In: Proceedings of the WMO CO_2 experts meeting, Lake Arrowhead, California, October 1990. (in press). WMO, Geneva.

Lowe, D.C., Guenther, P.R. and Keeling, C.D. (1979) The concentration of atmospheric carbon dioxide at Baring Head, New Zealand. Tellus 31:58–67

Monfray, P., Gaudry, A., Polian, G. and Lambert, G. (1987) Seasonal variations of atmospheric CO_2 in the southern Indian Ocean. Tellus 39B:67–71

Nakazawa, T., Miyashita, K., Aoki, S. and Tanaka, M. (1991a) Temporal and spatial variations of upper tropospheric and lower stratospheric carbon dioxide. Tellus 43B:106–117

Nakazawa, T., Aoki, S., Murayama, S., Fukabori, M., Yamanouchi, T., Murayam, H., Shiobara, M., Hashida, G., Kawaguchi, S. and Tanaka, M. (1991b) The concentration of atmospheric carbon dioxide at the Japanese Antarctic station, Syowa. Tellus 43B:126–135

Navascués, B. and Rus, C. (1991) Carbon dioxide observations at Izaña baseline station, Tenerife (Canary Islands): 1984–1988. Tellus 43B:118–125

Newsam, G.N. and Enting, I.G. (1988) Inverse problems in atmospheric constituent studies: I. Determination of surface sources under a diffusive transport approximation. Inverse Problems 4:1037–1054

Pearman, G.I. (1980) Atmospheric CO_2 Concentration Measurements. A Review of Methodologies, Existing Programmes and Available Data. WMO Project on Research and Monitoring of Atmospheric CO_2. Report No. 3. WMO, Geneva.

Pearman, G.I. and Beardsmore, D.J. (1984) Atmospheric carbon dioxide measurements in the Australian region: Ten years of aircraft data. Tellus 36B:1–24

Pearman, G.I. and Hyson, P. (1980) Activities of the global biosphere reflected in atmospheric CO_2 records. J. Geophys. Res. 85:4457–4467

Pearman, G.I. and Hyson, P. (1986) Global transport and inter-reservoir exchange of carbon dioxide with particular reference to stable isotope distribution. J. Atmos. Chem. 4:81–124

Pearman, G.I., Beardsmore, D.J. and O'Brien, R.C. (1983a) The CSIRO (Australia) Atmospheric Carbon Dioxide Monitoring Program: Ten Years of Aircraft Data. CSIRO Division of Atmospheric Physics Technical Paper No. 45. 113pp.

Pearman, G.I., Hyson, P. and Fraser, P.J. (1983b) The global distribution of atmospheric carbon dioxide: 1. Aspects of observations and modelling. J. Geophys. Res. 88C:3581–3590

Plumb, R.A. and Mahlman, J.D. (1987) The zonally averaged transport characteristics of the GFDL general circulation/transport model. J. Atmos. Sci. 44:298–327

Reiter, R., Sladkovic, R. and H.-J. Kanter (1986) Concentration of trace gases in the lower troposphere, simultaneously recorded at neighbouring mountain stations. Meteorol. Atmos. Phys. 35:187–200

Schmidt, U. and Khedim, A. (1991) In situ measurements of carbon dioxide in the winter arctic vortex and at midlatitudes: An indicator of the age of stratospheric air. Geophys. Res. Lett. 18:763–766

Schmitt, R., Schreiber, B. and Levin, I. (1988) Efects of long-range transport on atmospheric trace constituents at the baseline station Tenerife (Canary Islands). J. Atmos. Chem. 7:335–351

Tanaka, M., Nakazawa, T. and Aoki, S. (1987a) Time and space variations of tropospheric carbon dioxide over Japan. Tellus 39B:3–12

Tanaka, M., Nakazawa, T. and Aoki, S. (1987b) Seasonal and meridional variations of atmospheric carbon dioxide in the lower troposphere of the northern and southern hemispheres. Tellus 39B:29–41

Tanaka, M., Nakazawa, T., Shiobara, M., Ohshima, H., Aoki, S., Kawaguchi, S., Yamanouchi, T., Makini, Y. and Murayama, H. (1987c) Variations of atmospheric carbon dioxide at Syowa Station (69°00'S, 39°35'E), Antarctica. Tellus 39B:72–79

Tans, P.P., Conway, T.J. and Nakazawa, T. (1989a) Latitudinal distribution of the sources and sinks of atmospheric carbon dioxide derived from surface observations and an atmospheric transport model. J. Geophys. Res. 94D:5151–5172

Tans, P.P., Thoning, K.W., Elliott, W.P. and Conway, T.J. (1989b) Background atmospheric CO_2 patterns from weekly flask samples at Barrow, Alaska. Optimal signal recovery and error estimates. pp 112–123 of The Statistical Treatment of CO_2 Data Records. Ed. W.P. Elliott. NOAA Technical Memorandum ERL ARL-173.

Tans, P.P., Fung, I.Y. and Takahashi, T. (1990a) Observational constraints on the global atmospheric CO_2 budget. Science 247:1431–1438

Tans, P.P., Thoning, K.W., Elliott, W.P. and Conway, T.J. (1990b) Error estimates of background atmospheric CO_2 patterns from weekly flask samples. J. Geophys. Res. 95D:14063–14070

Thompson, M.L., Enting, I.G., Pearman, G.I. and Hyson, P. (1986) Interannual variation of atmospheric CO_2 concentration. J. Atmos. Chem. 4:125–155

Thoning, K.W. (1989) Selection of NOAA/GMCC CO_2 data from Mauna Loa Observatory. pp 1–26 of The Statistical Treatment of CO_2 Data Records. Ed. W.P. Elliott. NOAA Technical Memorandum ERL ARL-173.

Trivett, N.B.A., Higuchi, K. and Symington, S. (1989) Trends and seasonal cycles of atmospheric CO_2 over Alert, Sable Island, and Cape St. James, as analysed by a forward stepwise multiple regression technique. pp 27–42 of The Statistical Treatment of CO_2 Data Records. Ed. W.P. Elliott. NOAA Technical Memorandum ERL ARL-173.

Wilson, S.R. and Ayers, G.P. (Eds.) (1990) Baseline Atmospheric Program (Australia) 1988. Dept. of Administrative Services and CSIRO, Melbourne.

WMO (1987) Report of the NBS/WMO Expert Meeting On Atmospheric Carbon Dioxide Measurement Techniques: Gaithersburg, Maryland, 15–17 June 1987. Environmental Pollution Monitoring and Research Programme, Publication No. 51. WMO, Geneva.

WMO (1990) Provisional Daily Atmospheric Carbon Dioxide Concentrations as Measured at Global Atmosphere Watch (GAW) –BAPMoN sites for the year 1989. Environmental Pollution Monitoring and Research Programme: Publication No. 69. WMO, Geneva.

Wong, C.S., Chan, Y.-H., Page, J.S. and Bellegay, R.D. (1984) Trends of atmospheric CO_2 over Canadian WMO background stations at ocean weather station P, Sable Island and Alert. J. Geophys. Res. 89:9527–9539

Appendix: Statistical Aspects of Inversion Calculations

The trade-offs involved in various inversion procedures can be illustrated by analogy with the process of digital filtering of time series. Consider a time series of equi-spaced data, z_k, which is the sum of a signal, x_k, and a noise term, n_k. If these are stationary times series then they can be characterised by spectral densities $f_x(\omega)$ and $f_n(\omega)$. Estimates $\hat{x}_k$ of the signal can be constructed by applying a digital filter of the form $\hat{x}_k = \sum_j \Phi_j z_{k-j}$. If the frequency response of the filter is $\phi(\omega)$ then the mean-square-error in the estimates is

$$\mathrm{E}[(\hat{x}_k - x_k)^2] = \int_{-\pi}^{\pi} |1 - \phi(\omega)|^2 f_x(\omega)\, d\omega + \int_{-\pi}^{\pi} |\phi(\omega)|^2 f_n(\omega)\, d\omega$$

where E[.] denotes a statistical expectation.

The first term in the expression above represents a bias in that the filter distorts the signal. The second term represents an error due to the filter failing to supress all of the noise. The mean-square-error is minimised for the case

$$\phi_{\mathrm{opt}}(\omega) = \frac{f_x(\omega)}{f_x(\omega) + f_n(\omega)} = \frac{1}{1 + f_n(\omega)/f_x(\omega)}$$

The form of the denominator in the second form for ϕ_{opt} emphasises that it is the ratio of signal to noise that is the key quantity that limits our ability to separate a signal from noise.

The same principle applies to the transformations involved in inverting atmospheric concentration data. Any inversion scheme will involve a trade-off between errors arising from distorting the 'signal' that is sought and errors arising from interpreting 'noise' as if it were signal. Each type of error will have the greatest effect in those components of the signal that vary most rapidly in space and/or time. As in the filtering of time series, the choice of an optimal inversion strategy requires a knowledge of the statistical characteristics of both signal and noise.

The general problem considered here is, given a set of concentration data c_i, related to a set of sources and sinks, s_p, by a specified atmospheric transport

$$c_i = \sum_p T_{ip}\, s_p$$

determine estimates, $\hat{s}_p$, of the components of the source/sink distribution. The analysis is presented in terms of an arbitrary discretisation of the source/sink distribution. Some specific possibilities are discussed below.

The general solution to this problem can be expressed in terms of an inversion operator, H, as

$$\hat{s}_p = \sum_i H_{pi}\, c_i$$

Since the number of observations is less than the number of degrees of freedom of the sources, H can never be an exact inverse of T; one must construct approximate inverses, preferably by choosing the approximation to optimise the statistical properties of the estimates, $\hat{s}_p$.

The simplest statistical model is to put

$$c_i = \sum_p T_{ip}\, s_p + \epsilon_i$$

where ϵ_i represents the error in observation i. A more relevant statistical model is based on the recognition that much of the so-called 'noise' in the observations represents contributions from sources that are not properly treated, either due to model error or due to the use of a truncated representation of the source distribution. A refined statistical model uses

$$\epsilon_j = \sum_p (\tau_{jp} - T_{jp})\, s_p + \sum_p{}' \tau_{jp}\, s_p + \eta_j$$

where T represents the 'model' transport, τ represents the 'true' transport, $\sum$ is a sum over those source components used in the model and $\sum'$ is a sum over the remaining source components. The term η_j then becomes an error term associated purely with the measurements of data item j.

The first term, due to discrepancies between actual and model transport can arise in several ways:

(i). inadequacy in modelling the processes, including inadequate resolution;

(ii). failure to represent small-scale variability in transport;

(iii). failure to represent interannual variation in transport;

(iv). use of zonally averaged models.

To some extent, the first point can be checked by validating transport models against other tracers with known sources and sinks and concentrations. In practice the possibilities are severely limited by gaps in our knowledge of sources and sinks. For transport fields derived from general circulation models (GCMs) the transport of heat in the GCM provides a partial validation of the transport.

The second term in the 'error model' represents the effects of small-scale source variations omitted from the model. In these terms the selection of 'baseline' data, designed to be representative of large-scale air masses, represents an attempt to keep this term as small as possible.

Since the number of observations is less that the number of degrees of freedom in the sources (the latter being conceptually infinite), it is clear that the estimates $\hat{s}_\rho$ must represent a 'projection' of the source distribution onto a restricted form of source distribution. From the equations above one has

$$\hat{s}_\rho = \sum_j \sum_\mu H_{\rho j} T_{i\mu}\, s_\mu + \sum_j H_{\rho j} \epsilon_j$$

so that the 'projection' is defined by the operator X where

$$X_{\rho\mu} = \sum_j H_{\rho j} T_{j\mu}$$

The statistical properties of $\hat{s}$ are determined by

$$\text{var}(\hat{s}_\rho) = \text{E}[(\hat{s}_\rho - \text{E}[\hat{s}_\rho])^2] = \sum_i \sum_j H_{\rho i} H_{\rho j}\, \text{E}[\epsilon_i \epsilon_j]$$

where E[.] denotes statistical expectations and assuming $\text{E}[\epsilon_j] = 0$. More generally:

$$\text{cov}(\hat{s}_\rho, \hat{s}_\mu) = \text{E}[(\hat{s}_\rho - \text{E}[\hat{s}_\rho])(\hat{s}_\mu - \text{E}[\hat{s}_\mu])] = \sum_i \sum_j H_{\rho i} H_{\mu j}\, \text{E}[\epsilon_i \epsilon_j]$$

More important is the mean-square-error of the estimates. This is

$$\text{E}[(s_\rho - \hat{s}_\rho)^2] = \sum_\rho r_\rho^2 + \sum_i \sum_j H_{\rho i} H_{\rho j}\, \text{E}[\epsilon_i \epsilon_j]$$

where

$$r_\rho = s_\rho - \sum_\mu X_{\rho\mu} s_\mu$$

represents the 'unresolved' (or incorrectly resolved) source components.

The analysis above is presented in abstract form, without specifying the way in which the source components s_ρ are discretised. For the variation in time, an obvious technique is to use a Fourier series for the cycle, plus some smooth representation of the longer term variations. For the spatial variations, there are two main approaches. The first is to use some form of spectral expansion so that most of the information will relate to a (small) finite set of long-wavelength components. The second is to adopt a process-oriented division, analysing the sources in terms of contributions from regions that are assumed homogeneous, or at least having a fixed spatial structure, (see for example, Keeling et al., 1989b). Use of either approach will involve making some assumptions about the nature of the source components that are neglected.

Abbreviations

The following abbreviations are used in this chapter. See also the 3-letter sites codes used by NOAA (Table 2) and note that DC8, F27, F28, B707 and B727 refer to aircraft types.

AES	Atmospheric Environment Service (Canada).
BAPMoN	Background Air Pollution Monitoring Network, (WMO).
CDIAC	Carbon Dioxide Information Analysis Center (Oak Ridge National Laboratory, Tennessee, USA).
CFR	Centre des Faibles Radioactivités (France).
CMDL	Climate Monitoring and Diagnostics Laboratory (formerly GMCC). (NOAA).
CSIRO	Commonwealth Scientific and Industrial Research Organisation (formerly Commonwealth Scientific and Industrial Research Organization) (Australia). CO_2 observations within CSIRO are conducted by the Division of Atmospheric Research (formerly by the Division of Atmospheric Physics).
DSIR	Department of Scientific and Industrial Research (New Zealand). CO_2 studies within DSIR were conducted by the Institute of Nuclear Sciences, prior to an on-going re-organisation.
ENSO	El Nino/Southern Oscillation (phenomenon).
GFDL	Geophysical Fluid Dynamics Laboratory.
GISS	NASA Goddard Space Flight Center, Institute for Space Studies.
GMCC	Geophysical Monitoring for Climatic Change (NOAA).
INM	Instituto Nacional de Meteorologia (Spain).
NASA	National Aeronautic and Space Administration (USA).
NOAA	National Oceanic and Atmospheric Administration (US Dept Commerce).
SIO	Scripps Institution of Oceanography.
SOI	Southern Oscillation Index.
WMO	World Meteorological Organization.

SEASONAL CYCLES IN ATMOSPHERIC CO_2 CONCENTRATIONS

Martin R Manning[1]
Cooperative Institute for Research in Environmental Sciences
University of Colorado
Boulder, Colorado
USA

1. Introduction

This chapter addresses the seasonal cycles in the concentration of atmospheric CO_2, and the information they contain about the corresponding seasonal cycles of sources and sinks of CO_2. Although the important factors will be covered, this is not intended as an exhaustive review of the subject, and in particular no attempt has been made to review the extensive literature in this area.

Two separate issues are discussed here. The first is the observed cycles themselves, some of their statistical properties, and their geographic and temporal variation. The second is the relationship between seasonal variation in sources and sinks and the seasonal variation of concentration. The determination of sources and sinks which produce observed concentrations is known as the inverse problem and there are limits to how well seasonal sources and sinks can be defined from concentration data alone.

[1]Permanent address: Nuclear Sciences Group, DSIR Physical Sciences, PO Box 31312, Lower Hutt, New Zealand

NATO ASI Series, Vol. I 15
The Global Carbon Cycle
Edited by M. Heimann

2. Background

Since the first year of high precision measurements of atmospheric CO_2 at Mauna Loa (Keeling, 1960) it has been known that the concentration of CO_2 in marine air remote from the continents shows a seasonal cycle. The timing of this cycle is consistent with a decrease in atmospheric CO_2 during the spring and summer seasons when most mid-latitude and high latitude plant growth occurs. This depletion of atmospheric CO_2 is followed by an increase of essentially the same amount during the autumn and winter seasons when plant decay and soil respiration dominate. The regular cyclic variation in atmospheric CO_2 concentrations in the Northern Hemisphere is larger than the secular increase from year to year observed in the annual mean CO_2 concentration.

Although there has been little doubt that the seasonal cycle of the terrestrial biosphere is the dominant cause for seasonal cycles in atmospheric CO_2, the observation of corresponding seasonal cycles in $^{13}C/^{12}C$ ratios confirms the hypothesis. The photosynthetic pathway of carbon into plants favours the lighter ^{12}C isotope. So, during nett uptake of CO_2 by the biosphere, the CO_2 remaining in the atmosphere becomes progressively enriched in the heavier ^{13}C isotope. Isotope measurements are now available at several sites (e.g. Mook *et al*, 1983, Keeling *et al*, 1989, Goodman *et al*, 1990) and show seasonal cycles in the $^{13}C/^{12}C$ ratio which correlate closely with the seasonal cycles in CO_2 concentration. The amplitude of these cycles in the isotope ratio are consistent with the biosphere being the dominant source of the seasonality in CO_2 concentration.

The extensive data now available on atmospheric CO_2 concentrations provide us with a unique oversight of globally averaged nett ecosystem uptake and emissions. Estimates of this flux between the biosphere and atmosphere, and to some extent it geographic distribution, must match the observed seasonal and latitudinal variation in CO_2 at a network of over 50 sites. A major component of the biosphere - atmosphere exchange is nett primary productivity (NPP), and atmospheric CO_2 data proves to be an important constraint on NPP estimates (Fung *et al*, 1983, Heimann *et al*, 1989). There is also some evidence, from long term trends in the seasonal cycles of atmospheric CO_2, that biosphere - atmosphere fluxes of CO_2 are changing (Bacastow *et al*, 1985, Cleveland *et al*, 1983).

The connection between CO_2 source and sink fluxes (henceforth simply called sources) and the observed concentrations in the atmosphere is through atmospheric transport with its accompanying dispersion. In principle given a source distribution and atmospheric transport fields, we should be able to calculate the corresponding CO_2 concentration distribution. In practice we must work with estimates of both sources and transport, and we can use concentration data to improve our understanding of each. For example, as the east-west mixing in the mid to high latitudes is fast compared to changes in sources we can estimate the zonal average source strength from the concentration data at corresponding latitudes. In addition the rate at which the amplitude of the seasonal cycle decreases with distance from the continental source regions, both horizontally and vertically, provides constraints on dispersion rates in the atmosphere.

The seasonal cycle of CO_2 is not due solely to the seasonal cycle of uptake and release by the terrestrial biosphere. The exchange of CO_2 with the oceans has a seasonal component. This because of seasonal variation in the productivity of the marine biosphere, as well as the seasonal cycle in sea surface temperatures and the lower solubility of CO_2 in warmer water. Measurements of the ^{14}C isotope, which is absent in fossil fuels, in atmospheric CO_2 suggest, at least in Europe, a significant seasonal variation in fossil fuel emissions (Levin *et al*, 1989). The other significant anthropogenic emission, biomass burning (Crutzen and Andreae, 1990) is also seasonal in character being confined to the local "dry season". The photolytic oxidation of CO to CO_2 in the atmosphere, although a minor total source of CO_2, is strongly seasonal in mid to high latitudes due to the seasonal variation of UV radiation in the troposphere, and hence of the OH radical which is the primary oxidant. In addition to such seasonal factors in the nett sources of CO_2, the concentrations observed at specific sites are influenced by the seasonal variation in atmospheric transport.

Small variations in the amplitude and shape of seasonal cycles of atmospheric CO_2 concentrations are observed from year to year at all sites. These are generally larger than the precision of the measurements themselves, and so indicate a natural variability. As a result mean amplitudes and shapes for seasonal cycles must be obtained from several years of data. Inter-annual variations in the seasonal cycle will be due in part to small variations in the seasonal sources and sinks of CO_2 from year to year. However, inter-annual variation in atmospheric transport also plays a role. Synoptic scale weather patterns influence the observations in a way that is not repeated uniformly each year. In particular at sites near the equator there can be significant variations in the shape of the observed seasonal cycle from year to year depending on the movement of the inter-tropical convergence zone which acts as a meteorological boundary between the hemispheres.

The inter-annual variations of seasonal cycles are of considerable scientific interest, as by comparing natural variations in forcing and response terms we learn about the processes governing the global carbon cycle. However, we should not lose sight of the fact that the general constancy of these seasonal cycles, observed over many years, is clear evidence of the remarkable balance that is a general feature of the natural global carbon cycle. At the same time we are challenged to understand the small inter-annual variations and trends in these seasonal cycles in terms of corresponding changes in the dynamics of the carbon cycle.

3. Observed Seasonal Cycles in Atmospheric CO_2 Concentrations

3.1 Available data

Since the original use of the non-dispersive infra-red analysis technique to obtain high precision concentrations of CO_2 in the atmosphere in 1957 (Keeling, 1960) the number of sites at which such measurements are made has grown steadily. The development of a CO_2 measurement network coordinated by the World Meteorological Organization, formerly under the Background Atmospheric Pollution Monitoring Network (Manning, 1982) and now under the Global Atmosphere Watch programme, has provided a basis for comparable measurements by a large number of agencies.

A recent review (Tans, 1991) shows that continuous atmospheric CO_2 measurements are being made at 25 sites, and flask sample measurements are being made on a regular basis at 29 additional sites. In addition to these surface measurements, extensive aircraft based measurements have been made in some regions (e.g. Bischof , 1973, Pearman and Beardsmore, 1984, Nakazawa *et al*, 1991).

Most atmospheric CO_2 measurement programmes have focussed on marine boundary layer air. Selection of measurement sites together with data collection and screening procedures have been designed to minimize or remove the influence of local sources or sinks. This emphasis has made data analysis and interpretation easier but means that we have only limited information on regional scale averages of CO_2 over the continental land masses. Modelling studies suggest that the seasonal cycles of CO_2 are up to twice as large over the major land masses as they are over the oceans (Fung *et al*, 1983, Heimann *et al*, 1989), and there is supporting evidence for this in the data available at present. Future extension of the global CO_2 network will focus more on measurements over land, through the use of desert locations, high tower and aircraft sampling to obtain data that reflects CO_2 sources over terrestrial regions (Tans, 1991).

The absolute inter-comparability of CO_2 measurements by different organizations does not yet match existing technical capability (Monfray *et al*, 1991). However, the present calibration methodology, based on calibration gases maintained over several years, means that calibration differences in the data from different measurement programmes are constrained to vary slowly, although perhaps with small discrete steps where calibration gases are changed. Thus the amplitudes and shapes of seasonal cycles from different programmes can be compared with more confidence than annual mean concentrations or the trend in the annual mean over several years.

For the purposes of this study seasonal cycles have been extracted from data collected by: the NOAA Climate Monitoring and Diagnostic Laboratory global flask network (Conway *et al*, 1988, Tans *et al*, 1990); the German Umweltbundesamt in collaboration with the University of Heidelberg (Levin *et al*, 1991); and the New Zealand Department of Scientific and Industrial Research (Manning and Pohl,

1986). This particular selection of data is by no means exhaustive and was dictated largely by ease of access for the author. However, it is globally representative, within the constraint of the bias towards marine boundary layer air mentioned above, and is consistent with other data (e.g. Keeling *et al*, 1989).

Data were used from CMDL flask sites which had produced at least 5 consecutive years of data in the 1981 - 1990 period. Where data were available prior to 1981 for sites that met this criteria, they were also used. The flask data were first subject to standard screening procedures in order to remove contaminated samples and periods not representative of regional scale average CO_2 concentrations (Conway *et al*, 1988). The data were then treated as time series by pooling all values occurring within 14-day periods and treating a year as 26 such periods (the extra day being taken in the last period). Sites where data was missing for more than 10% of the 14-day periods were not considered further.

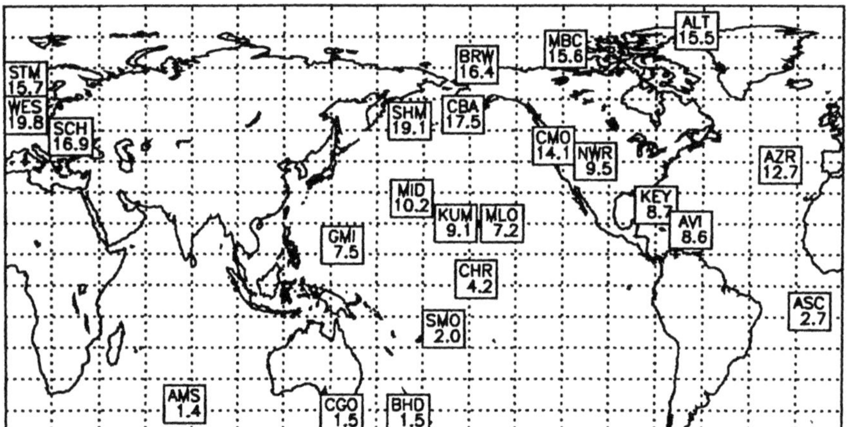

Figure 1. Sites from the NOAA/CMDL flask network and stations in Germany and New Zealand used in this study. Boxes located approximately at each site show the station code used in the text and the (peak to peak) amplitude of the mean seasonal cycle.

Table 1. Sites and CO2 seasonal cycle statistics

Site Code	Name	no yrs	lat (deg)	lon (deg)	height (m)	Cycle Minima (ppm) mean	min	max	Cycle Maxima (ppm) mean	min	max	Cycle Range (ppm) mean	min	max	+/- week	-/+ week	weeks -ve
ALT	Alert, NWT, Canada	5.6	82	-63	6	-9.9	-11.3	-8.7	5.7	4.6	7.0	15.6	14.2	16.5	27	47	20
MBC	Mould Bay, Canada	10.8	76	-119	58	-10.2	-11.0	-9.5	5.8	4.4	7.8	16.0	14.5	18.6	27	48	21
BRW	Barrow, Alaska	14.7	71	-157	11	-10.9	-13.1	-8.7	5.8	4.5	6.9	16.7	13.9	20.0	26	46	20
STM	Ocean Station M	9.8	66	2	7	-10.0	-10.7	-9.0	5.7	4.5	6.7	15.7	14.6	17.0	26	45	19
CBA	Cold Bay, Alaska	12.3	55	-163	25	-11.3	-13.2	-9.2	6.0	4.4	7.8	17.3	14.8	21.1	25	45	20
WES	Westerland, Germany	9.5	55	8	8	-11.4	-14.2	-9.0	8.6	5.5	11.0	20.0	15.8	23.2	23	45	22
SHM	Shemya Is	5.3	53	174	40	-12.6	-14.3	-10.6	6.7	6.1	7.4	19.2	17.0	21.8	25	44	19
SCH	Schauinsland, Germany	18.0	48	8	1200	-9.7	-12.7	-6.3	7.1	4.5	9.8	16.7	11.2	20.9	21	44	23
CMO	C. Meares, Oregon	8.9	45	-124	30	-8.2	-10.0	-7.0	5.7	4.8	6.8	13.9	12.3	16.3	25	46	21
NWR	Niwot Ridge, Colorado	15.0	40	-106	3749	-5.3	-8.5	-3.4	4.2	3.3	5.4	9.4	7.3	12.9	24	46	22
AZR	Azores	9.2	39	-27	30	-6.9	-10.1	-3.3	5.0	3.1	6.6	11.9	6.4	16.7	25	47	22
MID	Sand Is, Midway	5.7	28	-177	4	-6.1	-7.4	-3.2	4.3	3.8	5.2	10.4	7.2	12.6	28	50	22
KEY	Key Biscayne, Florida	9.5	26	-80	3	-5.1	-6.8	-4.0	3.7	3.2	4.4	8.8	7.2	10.9	27	50	23
MLO	Mauna Loa, Hawaii	14.6	20	-156	3397	-3.8	-5.1	-2.7	3.4	3.0	4.5	7.2	6.3	8.6	29	3	26
KUM	Kumukahi, Hawaii	10.6	20	-155	3	-5.0	-5.8	-4.0	4.0	3.2	5.1	9.0	8.0	10.4	28	52	24
AVI	St Croix, Virgin Is	10.0	18	-65	3	-4.8	-6.5	-3.8	3.8	3.2	4.7	8.6	7.7	11.2	28	1	25
GMI	Guam, Mariana Is	10.7	13	145	2	-4.2	-6.1	-2.8	3.3	2.5	4.7	7.5	5.9	9.2	30	3	25
CHR	Christmas Is	6.9	2	-157	3	-2.1	-3.0	-1.3	2.0	1.6	3.2	4.0	3.0	4.6	29	3	26
ASC	Ascension Is	11.4	-8	-14	54	-1.5	-2.7	-1.0	1.3	0.8	2.0	2.8	2.1	4.0	48	22	26
SMO	Matatula Pt, Samoa	16.0	-14	-171	42	-0.9	-1.5	-0.2	1.0	0.4	1.7	2.0	1.3	2.9			
AMS	Amsterdam Is	8.2	-38	77	150	-0.6	-1.1	-0.1	0.7	0.3	1.6	1.3	1.0	2.1			
CGO	C Grim, Tasmania	6.7	-41	145	94	-0.7	-1.0	-0.3	0.8	0.3	1.3	1.4	1.2	1.6	49	26	29
BHD	Baring Head, N Zealand	13.0	-41	174	80	-0.8	-1.2	-0.4	0.8	0.4	1.5	1.6	0.9	2.5	51	26	27
PSA	Palmer Stn, Antarctica	10.0	-65	-64	10	-1.2	-1.7	-0.7	0.9	-0.1	1.8	2.1	0.7	2.9	51	28	29
SPO	South Pole	16.0	-90	0	2810	-0.8	-1.2	-0.2	0.8	0.3	1.5	1.5	1.1	2.5	52	28	28

The German and New Zealand data from continuous analysers were treated similarly. That is all data had first passed the standard procedures used by the originating organizations to remove contaminated periods, and produce a series of CO_2 concentrations for "baseline" events. These were then pooled to provide a time series of values for 14-day periods. In this way CO_2 concentration time series were obtained for 25 sites. Figure 1 and Table 1 give the location of these and Figure 2 shows a representative selection of the data from three sites.

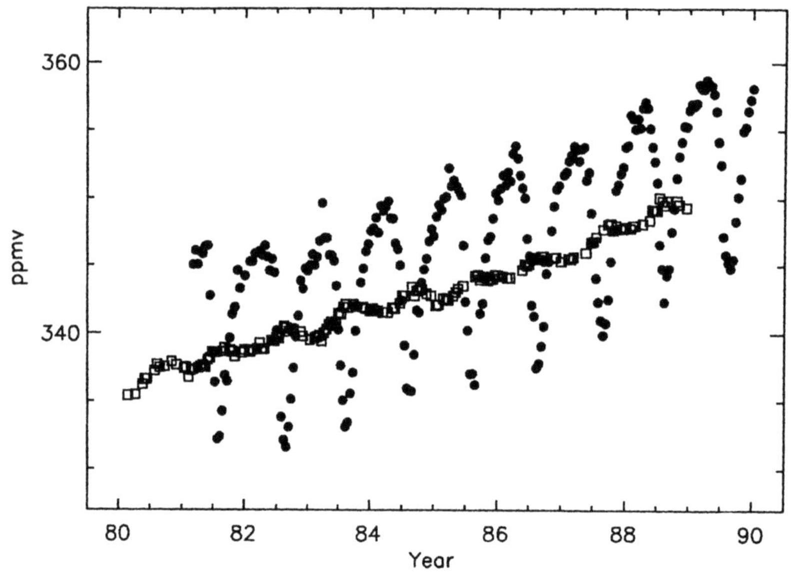

Figure 2. CO_2 concentration at two sites representative of high northern and mid southern latitudes. Data from STM (66°N) are shown as solid circles and data from BHD (41°S) as open boxes.

3.2 Extraction of seasonal cycles

In order to estimate seasonal cycles in CO_2 concentration from the observed data it is necessary to remove the secular trend that is evident in the data. However this trend is apparently not simply linear and does not follow any simple *a priori* formula. The data could be detrended by subtracting a trend proportional to fossil fuel emissions, however, this inherently assumes that the nett effect of other sources and sinks has not changed. It is preferable to use the data themselves to estimate a trend, but this leads to an inherent ambiguity in the separation of the seasonal cycle and trend components. Resolution of this ambiguity has been the subject of many studies (e.g. Cleveland *et al*, 1983).

A specific instance of the ambiguity between seasonal cycles and trend in atmospheric CO_2 occurs during the El Nino Southern Oscillation anomaly of 1982-83. During this period most sites observed a reduction in the normal rate of increase of CO_2 concentration followed by a "bounce back" period when the rate of increase was greater than normal. As the time scale of this anomaly is about one year it is possible to treat the anomaly as being entirely related to annual cycles in which case it would modify the normal cycle for two consecutive years. Alternatively one can absorb most of this anomaly in a trend which is allowed to decrease and then increase its gradient over a timescale of one or two years. There are similar small inter-annual variations throughout the CO_2 concentration records, and to some extent the choice as to whether these are treated as variations in the seasonal cycle or in the trend is a subjective one.

A quantitative analysis of the amount of variation in the CO_2 time series at different frequencies or time scales is provided by the power spectrum of the series (Priestley, 1981). Figure 3 shows such power spectra for 5 representative sites. These are maximal entropy estimates of the power spectra after the data were linearly detrended to reduce the low frequency part of the spectrum. Sites at similar latitudes have very similar power spectra. The Northern Hemisphere sites show a dominant annual component with a harmonic structure that becomes more pronounced at higher latitudes. This harmonic structure is determined by the shape of the mean annual cycle. The sharper draw down in high latitudes involves larger high frequency components. The German continental sites have larger components in the frequency range 5 yr^{-1} to 10 yr^{-1} reflecting the more noisy character of these series. The data from Samoa show only a weak annual cycle with the lower frequency components having the greatest power. The Southern Hemisphere sites again show a clear annual component with relatively weak harmonics.

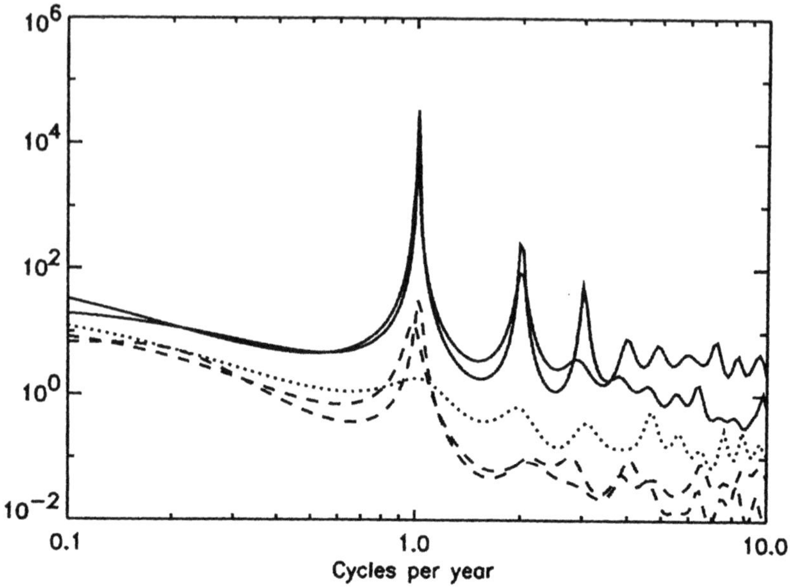

Figure 3. Power spectra of CO2 concentration time series for 5 representative sites determined by the maximal entropy method. The ordinate units are ppm^2 yr^2. The two solid curves are for BRW and SCH, with SCH being higher at high frequencies. The dotted curve is for SMO and the two dashed curves are for BHD and SPO, with SPO showing a slightly more pronounced peak at 1 yr^{-1}.

The differences between power spectra at different locations can be summarized in terms of three features. The differences in the frequency range 1 yr^{-1} and higher are due to differences in the shape of the mean annual cycle. The power spectra at frequencies of 0.2 yr^{-1} and lower are very similar because this part of the frequency spectrum is produced by the curvature in the long term trend, and over these time scales CO_2 is well mixed throughout the troposphere. There is an intermediate frequency range from 0.2 yr^{-1} to 1.0 yr^{-1} over which there is significant variation between sites with the northern sites generally showing more structure. This is the range in which inter-annual variability occurs.

Based on the above analysis of power spectra, I believe that decomposition of CO_2 time series should separate three components, covering: the long term trend (emphasizing components with frequencies of 0.2 yr^{-1} and less); the mean seasonal cycle (emphasizing the frequency range 1.0 yr^{-1} and higher); and an interannual component (emphasizing the intermediate frequency range 0.2 yr^{-1} to 1.0 yr^{-1}). The approach taken here is consistent with this but does not use Fourier components explicitly. The time series are decomposed into 3 separate components as:

$$C(t) = T(t) + S(t) + I(t)$$

where:

C(t) is the original time series,

T(t) is a smooth trend showing average behaviour over timescales of 5 years or longer,

S(t) is a seasonal component which gives the behaviour over timescales of 1 year and less, but is allowed to vary slightly from year to year,

I(t) is the residual, which contains both high frequency noise and components with periods between 1 and 5 years.

The trend curve T(t) used here has been determined by the STL procedure introduced by Cleveland *et al* (1989). This involves a simultaneous estimation of seasonal and trend components each of which is determined by "loess" smoothing of the original data minus the alternate (trend or seasonal) component. The loess smoothing procedure replaces each point with a value determined by a weighted linear least squares fit to data in a window about the point. In this case the smoothing window used to determine the trend has a width of 5 years. In determining S(t) the smoothing widow has been set very wide so as to effectively fit linear trends independently for each 14 day period. In determining both the trend and seasonal components the STL procedure iteratively reduces the weighting of outliers, and so the resulting components are relatively robust to the occurrence of a small fraction of noisy data.

It should be noted that the seasonal component derived by the STL procedure has no imposed smoothness from one 14-day period to the next. Thus the rapid draw down in the high northern latitudes can be followed closely. When Fourier components are used to represent the seasonal components, high frequencies are needed to follow the sharp gradient changes observed at sites such as BRW and these can produce spurious, or overemphasized, wiggles in other parts of the fitte seasonal cycle. In addition, by fitting trend and seasonal components simult neously the STL procedure is insensitive to small patches of missing data.

The properties of seasonal and trend components produced by the STL procedu as used here can be estimated by its effect on signals of different periodicity. Fi ure 4 shows the amplitude of the seasonal and trend components derived by ST from generated time series containing single Fourier components of different fr quencies superimposed on a linear trend. The relative amplitude of the S and components has been determined by their standard deviation relative to that of th Fourier component, and is shown in Figure 4. It can be seen that signals wit periods less than 1 year are resolved entirely into the seasonal cycle S(t), wherea signals with periods greater than 1 yr are strongly damped in S(t). Signals wit period longer than about 4 years become part of the trend T(t), and signals wit intermediate frequencies will appear only in I(t).

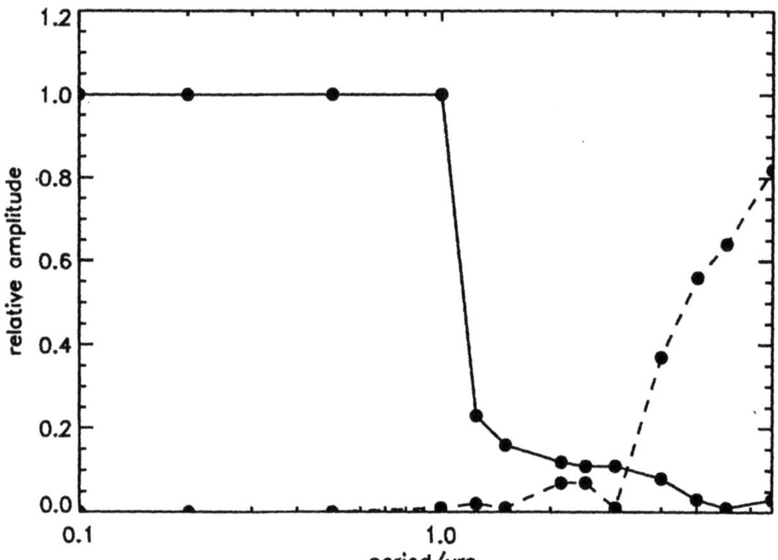

Figure 4. Relative amplitude of seasonal (solid line) and trend (dashed line) components determined by the STL procedure for artificial time series with single Fourier periodic component plotted against the period of that component.

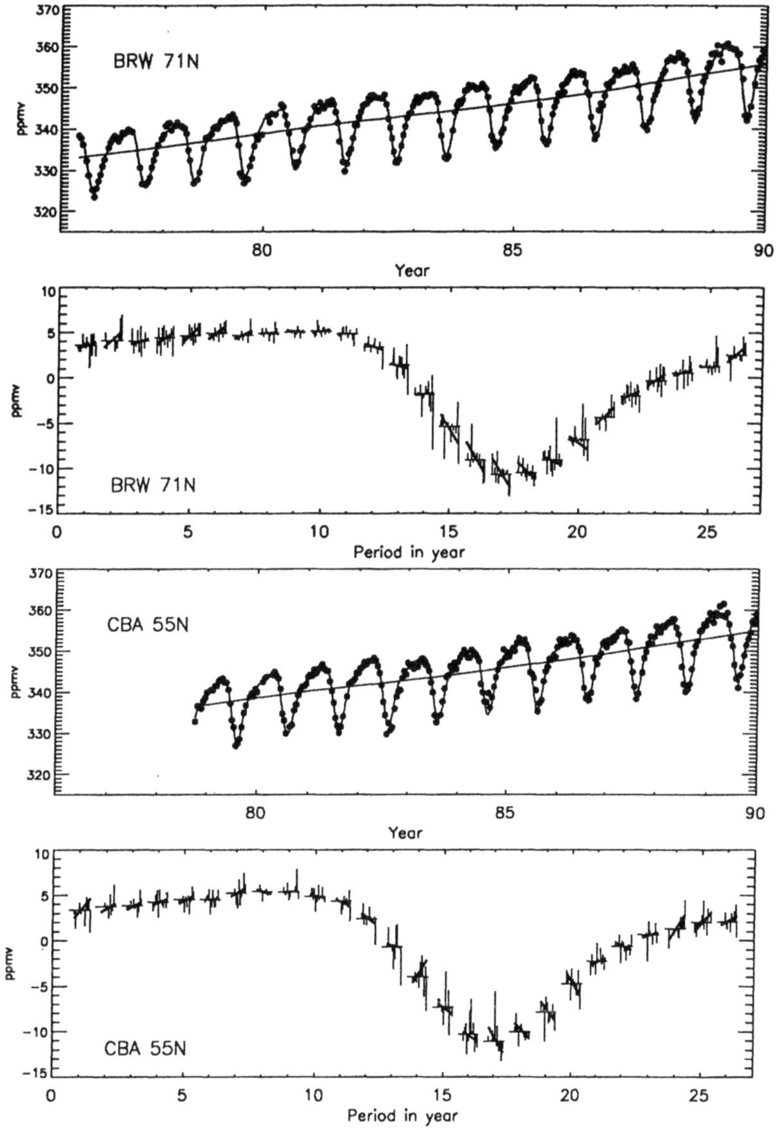

Figure 5. Seasonal and trend analysis as described in the text for two sites. The uppermost panel shows the original data for the BRW site (solid circles) and T(t) and T(t) + S(t) as solid curves. The second panel shows the detrended BRW data as described in the text. The lower two panels show data and fit for the CBA site in the same manner.

Figure 5 shows the results of the above analysis for data from two sites. The seasonal cycle is shown with horizontal bars at the value of the 25% trimmed mean (i.e. mean of values between the upper and lower quartiles) of (S(t) + I(t)) values for each 14-day period and indicate the overall cycle. Vertical bars connect individual detrended values to the 25% trimmed mean level, and in some cases show apparent systematic behaviour from year to year. A thick line is drawn corresponding to the linear trend in each 14-day period determined by STL.

Note that although the STL procedure produces an estimate of slowly varying seasonal cycles, the seasonal cycle statistics in Table 1 are based simply on the detrended data, i.e. on:

$$C(t) - T(t) = S(t) + I(t)$$

and so include all the inter-annual variability of the seasonal cycles.

3.3 Variation of seasonal cycle with location

The seasonal cycles derived as discussed in the previous section for the 25 sites, have peak to peak amplitudes ranging from 1.4 to 19.5 ppmv as shown in Figure 1 and Table 1. The phasing of the seasonal cycle is characterized more consistently by the times of the year at which positive to negative transitions occur, than by the times of the maximum and minimum which are harder to define and more variable at high northern latitudes. The times when the cycle changes sign correspond closely to the times of maximum gradients, and hence to the times of maximum uptake and release of CO_2. Phases defined this way are shown in Table 1. Note the correlation between phase and amplitude.

A dominant factor in the spatial variation of the seasonal cycles is latitude as shown in Figure 6. Thus seasonal cycles have amplitudes of around 1 ppmv at the South Pole, these increase slightly over the Antarctic ocean and then decrease further north to change phase near the equator. The cycle amplitudes increase markedly in the Northern Hemisphere, reaching a maximum of nearly 20 ppmv around 55°N and then decreasing slightly towards the pole. A corresponding zonally averaged view of seasonal cycles as a function of latitude is shown in Figure 7 which was obtained by averaging the cycles from sites in 10 degree latitude bands and smoothing these slightly for a better visual appearance of the surface depicted.

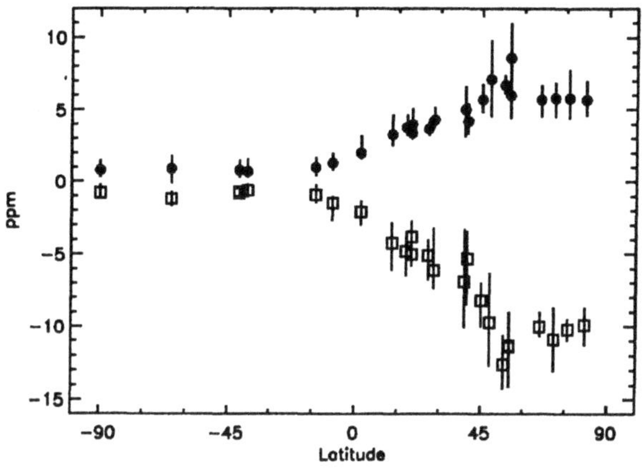

Figure 6. Maxima and minima of CO_2 seasonal variation plotted against latitude. Solid circles show the mean maximum value of the detrended data for each site and open squares show the mean minimum values. Vertical bars show the complete range of values for annual maxima and minima at each site.

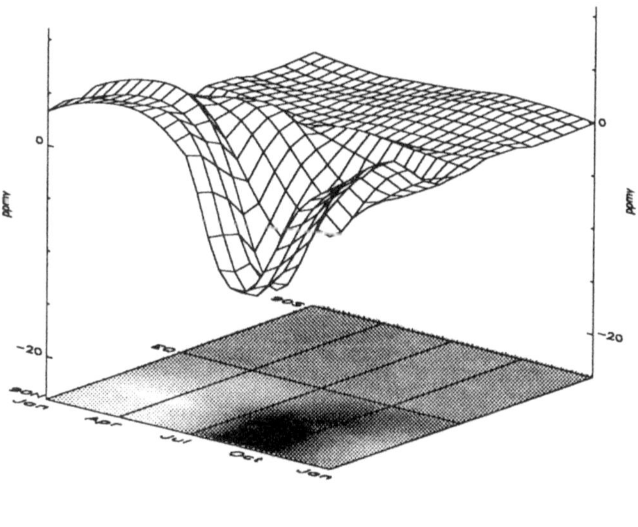

Figure 7. The average seasonal CO2 cycle shown as a surface (z axis in ppmv), against latitude and time of year. The base plane shows a gray-scale view of the same data.

As observed above the dominant cause of these seasonal cycles is the seasonal variation in terrestrial biosphere sources. Models of atmospheric transport coupled with estimates of the biospheric sources (Fung *et al*, 1987, Heimann and Keeling, 1989) are able to describe most of the seasonality observed in CO_2 at marine sites. It should be noted that although the tropical latitudes have a greater NPP than do temperate latitudes, the seasonal variation of biospheric fluxes is much less pronounced there. The maximum in amplitude of concentration cycles at latitudes around 50°N is consistent with estimates of NPP seasonality as a function of latitude obtained either from estimates of photosynthetic uptake based on satellite radiometer measurements or from regional scale ecology and climate considerations.

Despite the dominant overall variation of seasonal cycle with latitude, there is clear evidence in the present network for variation with longitude. A more detailed examination of the geographic variation of seasonal cycles is consistent with the dominant seasonal sources being on the continents. Thus the largest amplitudes and earliest Northern Hemisphere phases are observed between 50° and 55°N. Note that Shemya Is, downwind of Northern Asia in the prevailing westerly airflow at these latitudes, has a larger seasonal cycle than either Cold Bay, Alaska, or Cape Meares, Oregon, where the cycles are smaller and delayed in phase consistent with dispersive damping of the cycle transported from Asia and with there being no large seasonal forcing by the intervening Northern Pacific Ocean.

The seasonal cycle observed in the Atlantic at the Azores is larger than that observed at Cape Meares. This is consistent with less dispersion of the seasonal cycle forced by the North American land mass in the shorter transit across the Atlantic than that forced by the Asian land mass across the Pacific. It may also indicate that there is a greater seasonal effect of the Northern Atlantic ocean itself compared to the Northern Pacific.

There is considerable variation in the seasonal cycles observed at sites near the equator. Thus at Ascension Is, in the Atlantic, the phase of the cycle indicates a forcing by the Southern Hemisphere biosphere. Whereas at Samoa, further south but in the Pacific, the seasonal cycle is essentially undefined. The variation of CO_2

at Samoa over a 3 year period was analysed in detail by Halter *et al* (1988) in terms of wind trajectories. They show that 3 separate components exist in the data record, originating from the Northern Hemisphere, the Southern Hemisphere, and an intermediate zone of easterly winds between the inter-tropical and South Pacific convergences.

3.4 Trends in seasonal cycles

Because of the year to year variability of seasonal cycles, detecting evidence of any trend in amplitude or shape is a non-trivial statistical problem. Analyses of the longer CO_2 records collected at Mauna Loa and South Pole have shown statistically significant increases in the amplitude of the seasonal cycle. Cleveland *et al* (1983) and Bacastow *et al* (1985) determined growth rates of 0.5%/yr and 0.7%/yr in the cycle amplitude using quite different types of analysis. Kohlmaier *et al* (1989) have related this change in seasonal cycle observed at Mauna Loa to changes in net photosynthetic uptake and respiratory emission of CO_2.

The change in the peak to peak amplitude of CO_2 at Mauna Loa over 10 years, indicated by these trend analyses, is ca 0.35 ppmv. This can be compared to the year to year variability of the Mauna Loa seasonal cycle in the CMDL flask data, measured as the standard deviation of the detrended data about the average seasonal cycle which is 0.41 ppmv. Thus it is clear that to determine trends in the seasonal cycle at a specific site, at the level currently indicated, we require records of more than 10 years.

An alternative to using data from a single site is to pool data from geographically related sites in order to improve the confidence level of seasonal cycle trend estimates. One way of doing this is presented here. The linear trend estimates determined separately for each 14-day period by the STL procedure described above give estimates of how the cycle is changing at different times of the year for each time series analysed. The slopes of these trends for each 14-day period have been averaged for the 4 Arctic sites which have time series of more than 9 years (i.e. CBA, BRW, MBC and ALT), and the resulting average trend in the Arctic seasonal cycle is shown in Figure 8. The period covered by this analysis is 1980 to 1990.

The degree of spatial coherence (between different sites) and temporal coherence (between adjacent 14-day periods) indicates a consistent trend in the seasonal cycle at these Arctic sites. The indicated rates of change in the seasonal component are ca +0.1 ppmv/yr at the maximum and ca -0.15 ppmv/yr at the minimum. These rates are about 1.5%/yr in terms of the seasonal cycle amplitude, or about twice the rate of relative increase reported for Mauna Loa. Note that the trend is for the CO_2 draw down to occur earlier in the year, and not simply to change the amplitude of the cycle.

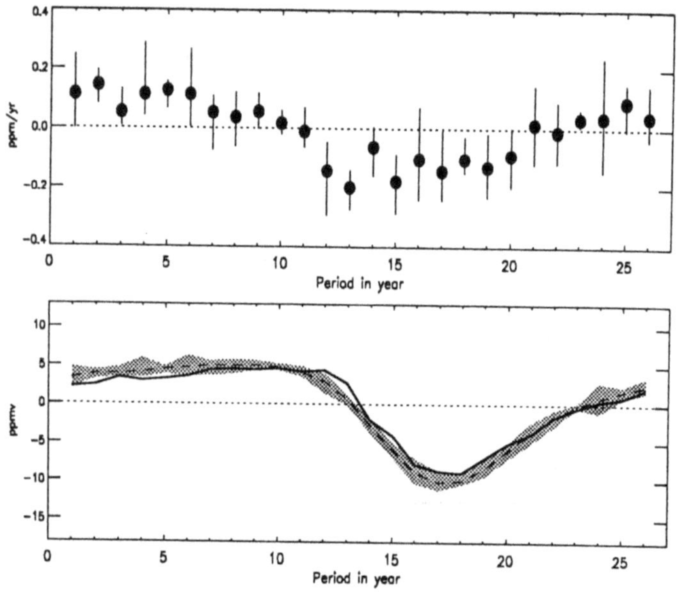

Figure 8. Trend in Arctic seasonal CO_2 cycles. The upper panel shows the average (solid circles) and range (vertical bars) of linear trends at each 14-day period through the year of the detrended data from CBA, BRW, MBC and ALT. The lower panel shows the average cycle at these 4 sites (thick line) and the cycle extrapolated 10 years forward on the basis of the average trend (dashed line) and on the basis of minimum and maximum trends from the different sites (shaded range).

A similar analysis for the continuous analyser data from BRW shows a similar trend pattern. However the data used here for MLO, or other groupings of sites, does not show any significant trend in the seasonal cycle over the 1980 to 1990 period. This result is not inconsistent with the analyses of Bacastow *et al* and

Cleveland *et al*, but does suggest that the rates of change in the seasonal cycle are greatest in the high Northern latitudes.

4. The inverse problem

4.1 Definition of the problem

In order to determine the seasonal variation in sources that will produce the observed variation in atmospheric concentration we must use a model of atmospheric transport and dispersion between the source regions and measurement sites. Such models work with source fluxes as boundary conditions and predict concentrations. The dispersion inherent in atmospheric transport, together with the sparsity of available measurements, place limits on our ability to use such models in an "inverse" mode, determining sources from observed concentrations. In this section a 2-dimensional model (Plumb & Mahlman, 1987) will be used in an investigation of the nature of these limitations.

Atmospheric transport models are discussed in more detail in other chapters, but for the present purposes can be summarized as follows. For a given distribution of source fluxes S, an atmospheric transport model will predict a concentration distribution C. The relationship between sources and concentrations, for any particular model, is linear so that we can write

$$C = T \, . \, S \qquad (1)$$

where T is a matrix summarizing the transport properties of the model.

We need to invert equation (1) in order to solve for S, given C. However, if the problem is set up so that the matrix T is square, i.e. we use the same number of dimensions for the S and C vectors, then T is usually a badly conditioned matrix. This means that small changes in C lead to large changes in the predicted S. If the dimension of the problem is large, the observational errors in C, and representational errors due to a mismatch between the modelled world and real world, usually result in the sources determined by direct inversion being highly variable in space and time.

The generation of spurious or noisy sources can be limited by working with concentration data that is first smoothed in space and time (Enting and Mansbridge, 1989). An alternative technique, that can be used in conjunction with such smoothing, is to treat an overdetermined problem, i.e. one in which the dimension of C is larger than the dimension of S. In this case the source being determined has fewer degrees of freedom than the number of available concentration values and a least squares solution to equation (1) can be used. The approach taken here is a generalization of the least squares method using a singular value decomposition (SVD) of the T matrix, which allows more detailed control over the generation of spurious sources.

Equation (1) expresses the relationship between sources and concentrations in a general way and can be used in different contexts. The remainder of this section deals with a specific model inversion but the treatment can be readily generalized. S is taken to be a vector in a space where the basis functions are specific source flux distributions in space and time. Thus S represents some combination of these. C is a vector in a space where the basis functions represent a unit concentration at a specific time of the year and specific geographic location. Thus the components of C are a set of concentrations at different times and locations.

The model used here (Plumb and Mahlman, 1987) treats zonally averaged transport and predicts zonal average concentrations from zonal average sources. Thus the basis functions of S have been taken as source fluxes equally distributed over 10° latitude bands, modulated with a frequency of 1, 2, 3 or 4 cycles per year, and with two different phasings (either maximum or zero flux on January 1st). The components of C are zonal average concentrations at each of 12 months for the 10° latitude bands.

Sources in the 80° to 90° latitude bands are not considered, so there are 128 source components (4 frequencies by 2 phases by 16 latitude bands), and 216 concentration components (12 months by 18 latitude bands). The sources are therefore overdetermined by a factor of nearly 2. The matrix T is generated by running the model for each of the basic source distributions to determine repeating annual concentration cycles in each 10° latitude band and using the resulting C values as the corresponding column of the 216 x 128 matrix.

4.2 Singular value decomposition

The singular value decomposition (SVD) technique (Lawson and Hanson, 1974, Press *et al*, 1988) applied to equation (1) resolves a set of principal source components and a corresponding set of concentration responses which are mutually independent. Mathematically this corresponds to choosing alternative basis functions in the S and C spaces so that the T matrix becomes diagonal (i.e. the upper 128 x 128 matrix is diagonal and the remainder is zero). Algebraically we write

$$T = U \cdot D \cdot V^T$$

where U is an orthogonal 216 x 128 matrix acting on the C space, D is a 128 x 128 diagonal matrix, and V is an orthogonal 128 x 128 matrix acting on the S space. The columns of U and V will be denoted by $\mathbf{u}_i$ and $\mathbf{v}_j$, and the diagonal elements of D by d_i.

Each of the orthonormal $\mathbf{u}_i$ vectors is a concentration distribution resulting from the corresponding v_i source distribution. There are 128 orthonormal $\mathbf{v}_i$ principal source components, and the corresponding $\mathbf{u}_i$ span the space of all concentration distributions that can be produced from the space of sources being considered. In general the observed concentration C will contain components that cannot be reproduced by any source combination, however, the solution to (1) which uses all the components of C that can be represented this way is the least squares solution:

$$S = V \cdot D^{-1} \cdot U^T \cdot C = \Sigma_{i=1}^{128} \mathbf{v}_i \; (\mathbf{u}_i^T \cdot C) / d_i \qquad (2)$$

Note that the coefficient of the principal source distribution $\mathbf{v}_i$ in (2) is just the component of C along $\mathbf{u}_i$ divided by d_i.

Where the d_i are small, any errors in $(\mathbf{u}_i^T . C)$ resulting from measurement or representational errors in C are amplified in (2). Thus we can consider a series of restricted solutions to (1) in which progressively smaller d_i are considered. From now on we assume the d_i are in order of decreasing magnitude and the restricted solutions are

$$S_k = V . D_k^{-1} . U^T . C$$

where D_k^{-1} is the diagonal matrix whose first k elements are $1/d_i$, i = 1..k, and whose remaining elements are zero. Thus

$$S_k = \Sigma_{i=1}^{k} \ \mathbf{v}_i . (\mathbf{u}_i^T C) / d_i \tag{3}$$

The concentration distribution corresponding to the restricted solution S_k is

$$C_k = \Sigma_{i=1}^{k} \ \mathbf{u}_i . (\mathbf{u}_i^T C) \tag{4}$$

and the sum of squares of differences between this and the observed concentration C is

$$|C - C_k|^2 = (C^T.C) - \Sigma_{i=1}^{k} (\mathbf{u}_i^T C)^2 \tag{5}$$

It can be seen from (5) that reducing k produces a poorer fit to the data, however, the benefit is that the uncertainty in the sources is reduced. This can be seen by considering the variance of S_k, which is related to the variance-covariance matrix G of the observations C, by

$$Var(S_k) = V . D_k^{-1} . U^T . G . U . D_k^{-1} . V^T$$

The total variance in the components of S_k is then

$$\mathrm{Trace}(\mathrm{Var}(S_k)) = \Sigma_{i=1}^{k} h_i / d_i^2 \qquad (6)$$

where h_i is the i'th diagonal element of (U^TGU). Thus, reducing k and removing terms with small d_i in (6), reduces the total variance of the source distribution implied by the variance in the original observations. The advantage of the SVD technique is that it allows the rate at which C_k approaches C, given by (5), and the growth of the variance in S_k, given by (6), to be evaluated as a function of k.

4.3 Application to a 2-dimensional model

The application of SVD in the case of the 2-dimensional model introduced above, allows us to consider several typical features of the inverse modelling problem. For example, the range of source distributions considered is not wide enough to allow a good fit to the observed concentrations. Specifically the sharp draw down of CO_2 in the high northern latitudes, and the implied sharp onset of CO_2 sinks, cannot be closely simulated by the restricted range of Fourier components in the source terms used here. Thus the inverse sources are inherently restricted and cannot produce an ideal fit to the data. On the other hand, the higher frequency source components in the Southern Hemisphere are not likely to be important source terms. Thus the space of allowed sources contains degrees of freedom that should be ignored by the inversion procedure.

Estimates of inverse sources by Tans et al (1989) and Enting and Mansbridge (1991) also use transport models based on Plumb and Mahlman (1978) although in both cases the details of the inversion process are rather different to the procedure used here, and Enting and Mansbridge make significant adjustments to the two-dimensional transport fields used. The present study is intended to demonstrate the limits inherent in the inversion process, rather than to produce alternative estimates of CO_2 surface fluxes.

The observed concentration C has been taken from the smoothed seasonal cycles, shown in figure 7. Errors in C, allowing for both zonal representation discrepancies and measurement errors, are taken by assuming 10% relative errors, but that this error is not less than 0.5 ppmv. Figure 9 shows, the corresponding total variance in the restricted solutions S_k for the source distribution, and the approach of the corresponding C_k to the observed concentrations C. Note that after taking the first 30 principal source components into the solution, there is little extra improvement in the fit to the observed concentrations, while the total variance in the inferred sources starts to rise significantly.

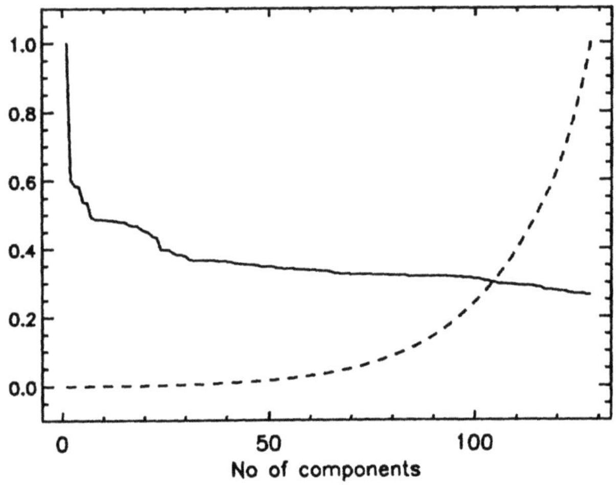

Figure 9. The distance of the restricted solution C_k from observed C (sum of squares of component differences), shown as a solid line and normalized so the maximum distance is 1, and the total variance of the corresponding S_k , shown as a dashed line and also normalized so the maximum variance is 1.

Figure 10 shows the source and corresponding concentration components for a typical dominant component, a marginal component, and a component that only adds noise to the the inverse sources. Note that the dominant components are determined solely from the model, not from observations. Thus it is a feature of the modelled atmospheric transport, that source cycles with low spatial variability and located in the high latitudes are inherently well determined in the inversion process. In contrast source distributions with high spatial variability are not. It is impractical to show all 128 principal source components here, but they reveal that distributions with high temporal variability (e.g. 4 cycles per year) are better determined than distributions with a similar level of spatial variability.

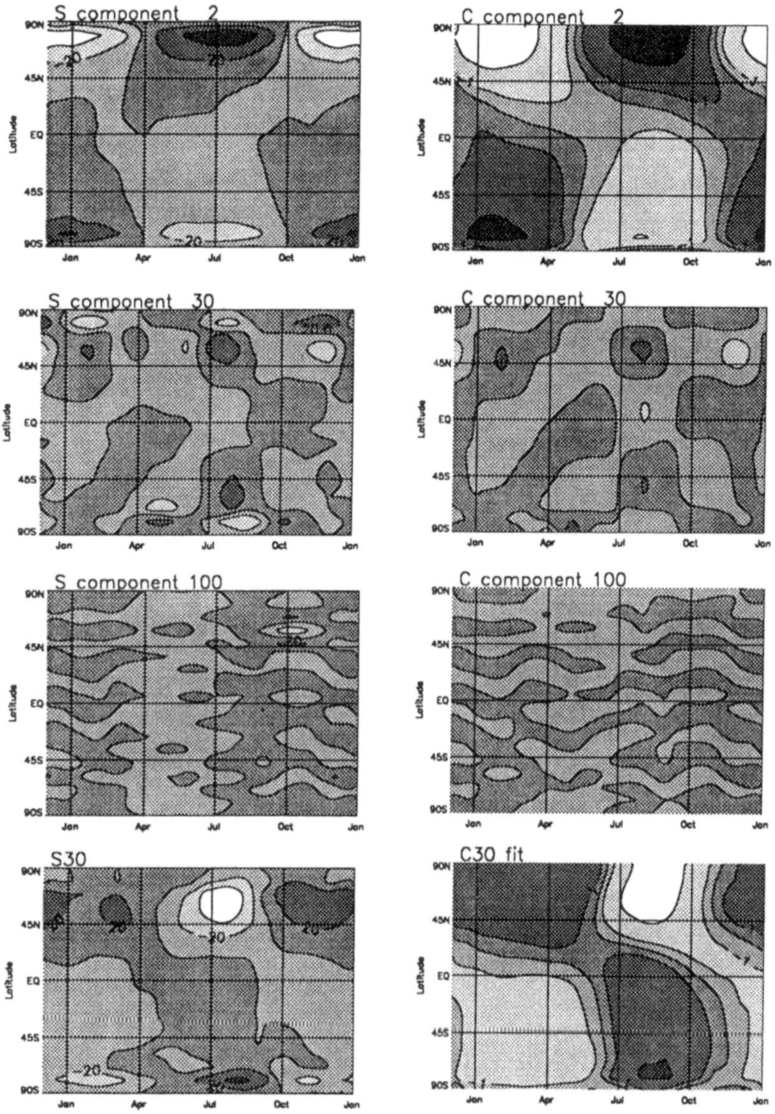

Figure 10. Contour plots of source distributions (left hand panels) and corresponding concentration distributions (right hand panels). The uppermost row shows a dominant (k=2) source component from the SVD analysis; the second row shows a marginal component (k=30); the third row shows a poorly determined component (k=100); and the lowest row shows the solution combining k=1 to k=30. Contour levels for source distributions are at -50, -20, 0, 20, and 50 GtC/ yr/global-area, and for concentrations are at -4, -1, 0, 1 and 4 ppmv.

Figure 10 also shows the inverse sources and the concentrations corresponding to a cutoff level of k=30. Note that the annual CO_2 cycle in the high latitudes is not particularly well reproduced. In particular, the amplitude of the fitted cycle is less than observed. Another feature apparent in the last 2 panels of figure 10, is a pronounced seasonality at the high southern latitudes in the inverse sources and derived concentration. Investigation of this artifact, suggests that it arises from very low horizontal diffusion in the Plumb & Mahlman model in this region, coupled with the need to cancel a seasonal cycle propagating from the Northern Hemisphere. This highlights the susceptibility of source inversion to defects in the transport model.

5. Conclusions

In this chapter I have attempted to summarize the data available on seasonal cycles of CO_2. Such cycles have to be extracted from the raw data which contains a secular trend and interannual variations as well as measurement noise. There are numerous different procedures for extracting the cycles, and these give comparable results for the longer data records. However, careful extraction of an average or slowly varying seasonal cycle should have a clear treatment of the CO_2 data, over three different time scales, viz 0 to 12 months, 1 to 5 years, and 5 or more years.

Space and time preclude a review here of the growing body of $^{13}C/^{12}C$ measurements in atmospheric CO_2, but these clearly indicate that seasonal forcing is mainly due to the biosphere. The geographic variation of seasonal cycles observed at numerous sites is consistent with the dominant seasonal forcing being from the continental regions. Three-dimensional models of atmospheric transport with estimates of biospheric and other CO_2 fluxes (Heimann and Keeling, 1989, Fung *et al*, 1987) give reasonable agreement with the observed seasonal cycles and confirm that most of the seasonality in observed concentrations is due to the terrestrial biosphere. Such studies also show that present CO_2 measurements set constraints on net photosynthetic uptake and respiration averaged over large geographic regions.

Extension of the CO_2 measurement network to desert, high tower, and more aircraft measurements should allow a better characterization of CO_2 sources and sinks. Data from the relatively recent NOAA/CMDL site at Shemya Island, shows a strong seasonal cycle, which is somewhat larger than that predicted by current 3-dimensional models. This suggests that data arising from further extension of the CO_2 network may lead to adjustment of present atmospheric transport models and/or estimates of fluxes.

Determination of CO_2 source and sink fluxes from observed concentrations is an inversion problem common in geophysical and other applications. A detailed analysis of the inverse problem for a 2-dimensional atmospheric transport model has been presented here, and suggests that the present CO_2 network, in conjunction with 2-dimensional atmospheric models, can resolve about 30 independent components of the seasonal variation of CO_2 fluxes over the globe. High frequency temporal variation in the sources can be determined better than high frequency spatial variation.

This result should not be taken as indicating a similar limitation of source determination using 3-dimensional atmospheric models. An important future development will be the detailed investigation of the source inversion problem for 3-dimensional models. This will allow a better assessment of how well CO_2 fluxes can be determined, and will also allow more effective planning of extensions to CO_2 measurement programmes.

6. Acknowledgments

I would like to gratefully acknowledge the assistance of Ingeborg Levin, University of Heidelberg, and Tom Conway, NOAA Climate Monitoring and Diagnostics Laboratory, for access to their data and advice on its interpretation. I would also like to thank Pieter Tans, NOAA Climate Monitoring and Diagnostics Laboratory, for many valuable discussions which have helped to clarify this work.

7. References

Bacastow RB, Keeling CD, Whorf TP (1985) Seasonal amplitude increase in atmospheric CO_2 concentration at Mauna Loa, Hawaii, 1959-1982. J Geophys Res, 90:10,529-10,540

Bischof W (1973) Carbon dioxide concentration in the upper troposphere and lower stratosphere - III. Tellus, 25:305-308

Cleveland WS, Freeny AE, Graedel E (1983) The seasonal component of atmospheric CO_2: Information from new approaches to the decomposition of seasonal time series. J. Geophys Res, 88:10,934-10,946

Cleveland WS, McRae JE (1989) The use of loess and STL in the analysis of atmospheric CO_2 and related data (in) The statistical treatment of CO_2 data records, NOAA Tech memo ERL-ARL-173, (Ed, Elliot WP), Silver Spring, Maryland

Conway TJ, Tans P, Waterman LS, Thoning KW, Masarie KA (1988) Atmospheric carbon dioxide measurements in the remote troposphere. Tellus, 40B:81-115

Crutzen PJ, Andreae MO (1990) Biomass burning in the tropics: impact on atmospheric chemistry and biogeochemical cycles. Science, 250:1669-1678

Enting IG, Mansbridge JV (1989) Seasonal sources and sinks of atmospheric CO_2 direct inversion of filtered data. Tellus, 41B:111-126 (2)

Enting IG, Mansbridge JV (1991) Latitudinal distribution of sources and sinks of CO_2: results of an inversion study. Tellus, 43B:156-170

Fung I, Prentice K, Matthews E, Lerner J, Russell G (1983) Three-dimensional tracer model study of atmospheric CO_2: Response to seasonal exchanges with the terrestrial biosphere. J Geophys Res, 88:1281-1294

Fung IY, Tucker CJ, Prentice KC (1987) Application of advanced very high resolution radiometer vegetation index to study atmosphere-biosphere exchange of CO_2. J Geophys Res, 92:299-3015

Goodman HS, Robbins FJ, Francey RJ (1990) $^{13}C/^{12}C$ and $^{18}O/^{16}O$ in baseline CO_2, 1987-88. (in) Baseline 88, CSIRO Division of Atmospheric Research, CSIRO Melbourne

Halter BC, Harris JM, Conway TJ (1988) Component signals in the record of atmospheric carbon dioxide concentration at American Samoa. J Geophys Res, 93:15,914-15,918

Heimann M, Keeling CD (1989) A three-dimensional model of atmospheric CO_2 transport based on observed winds: 2. Model description and simulated tracer experiments. Geophysical Monograph 55, pp 237-275, American Geophysical Union

Heimann M, Keeling CD, Tucker CJ (1989) A three-dimensional model of atmospheric CO_2 transport based on observed winds: 3. Seasonal cycle and synoptic time scale variations. Geophysical Monograph 55, pp 277-303, American Geophysical Union

Keeling CD, Bacastow RB, Carter AF, Piper SC, Whorf TP, Heimann M, Mook WG, Roeloffzen H (1989) A three dimensional model of atmospheric CO_2 transport based on observed winds: 1. Analysis of observational data. Geophysical Monograph 55, pp 165-236, American Geophysical Union

Keeling CD (1960) The concentration and isotopic abundance of carbon dioxide in the atmosphere. Tellus, 12:200-203

Kohlmaier GH, Sire E-O, Janacek A, Keeling CD, Piper SC, Revelle R (1989) Modelling the seasonal contribution of a CO_2 fertilization effect of the terrestrial vegetation to the amplitude increase in atmospheric CO_2 at Mauna Loa observatory. Tellus, 41

Lawson CL, Hanson RJ (1974) Solving least squares problems, Prentice Hall

Levin I, Graul R, Trivett NBA (1991) The CO_2 observation network in the Federal Republic of Germany. (in) Tans, 1991 (see below)

Levin I, Schuchard J, Kromer B, Muennich KO (1989) The continental European Suess effect. Radiocarbon, 31:431-440 (3)

Manning MR (1982) An assessment of BAPMoN data currently available on the concentration of CO_2 in the atmosphere. Environmental Pollution Monitoring Programme report, World Meteorological Organization, Geneva, February 1982

Manning MR, Pohl KP (1986) Atmospheric CO_2 measurements in New Zealand, 1971-1985. INS-R--350, DSIR Institute of Nuclear Sciences

Monfray P, Ramonet M, Gaudry A, Pearman G, Beardsmore D, Manning M, Pohl P (1991) An intercalibration of CO_2 measurements between France, Australia and New Zealand. (in) Tans, 1991, (see below)

Mook WG, Koopmans M, Carter AF, Keeling CD (1983) Seasonal latitudinal and secular variations in the abundance and isotope ratios of atmospheric carbon dioxide, 1, Results from land stations. J. Geophys Res, 88:915-933

Nakazawa T, Miyashita K, Aoki S, Tanaka M (1991) Temporal and spatial variation of upper tropospheric and lower stratospheric carbon dioxide. Tellus, 43B:106-117

Pearman G, Beardsmore DJ (1984) Atmospheric carbon dioxide measurements in the Australian region: Ten years of aircraft data. Tellus, 36B:1-24

Plumb RA, Mahlman JD (1987) The zonally averaged transport characteristics of the GFDL general circulation/ transport model. J Atmos Sci, 4:298-326

Priestley MB (1981) Spectral analysis and time series, Vol 1, Academic Press

Press WH, Flannery BP, Teukolsky SA, Vetterling WT (1988) Numerical Recipes in C, Cambridge University Press

Tans PP (Ed) (1991) Report of the WMO meeting of experts on carbon dioxide concentration and isotopic measurements techniques, Lake Arrowhead, California, 14-19 Oct 1990. World Meteorological Organization, Geneva

Tans PP, Conway TJ, Nakazawa T (1989) Latitudinal distribution of the sources and sinks of atmospheric carbon dioxide derived from surface observations and an atmospheric transport model. J Geophys Res, 94:5151-5172

Tans PP, Thoning KW, Elliott WP, Conway TJ (1990) Error estimates of background atmospheric CO_2 patterns from weekly flask samples. J Geophys Res, 95:14,063-14,070

WHAT CAUSED THE GLACIAL TO INTERGLACIAL CO_2 CHANGE?

W. S. Broecker and T.-H. Peng[1]
Lamont-Doherty Geological Observatory
of Columbia University
Palisades, NY 10964

ABSTRACT

Scenarios put forward to explain the 80 μatm glacial to interglacial change in atmospheric CO_2 content are evaluated. The conclusion is that no single mechanism is adequate. Rather, contributions from temperature, sea ice, biologic pumping, nutrient deepening, and $CaCO_3$ cycling must be called upon. The observation that the $^{13}C/^{12}C$ ratio for Antarctic foraminifera was 0.9±0.1‰ lower during glacial than during interglacial time constitutes a huge fly in the ointment for all scenarios proposed to date.

INTRODUCTION

Over the last decade numerous attempts have been made to explain the increase in atmospheric CO_2 content from 200 to 280 ppm between glacial and interglacial time, determined from measurements of gas extracted from ice cores (Barnola et al., 1987; Neftel et al., 1988). However, no one of these explanations by itself appears to be capable of simultaneously accounting for both the 80 μatm magnitude of this change and for the carbon isotope record kept by foraminifera shells in marine sediments. So the answer to the question posed in the title to this paper must be "we don't yet know." As all of the possible mechanisms appear to have been explored, the answer probably lies in combining a number of smaller contributions.

---

[1]Environmental Sciences Divison, Oak Ridge National Laboratory, Oak Ridge, TN 37830

NATO ASI Series, Vol. I 15
The Global Carbon Cycle
Edited by M. Heimann

THE ROLE OF ORGANIC RESERVOIRS

During the Holocene, the amount of carbon residing in living wood, in soil humus and in ocean humus totaled about seven times that in the atmosphere as CO_2 (see Table 1). Hence glacial to interglacial changes in the size of this organic matter reservoir are potentially important. Fortunately, a means exists to assess their magnitude. As the carbon isotope composition of organic material ($\delta^{13}C \cong$ - 25‰) is quite different than that for average ocean-atmosphere carbon ($\delta^{13}C \cong$ 0‰), any transfer of carbon between the organic and the inorganic reservoir will lead to a shift in the carbon isotope composition of average ocean-atmosphere carbon. Measurements on benthic and planktonic foraminifera from deep sea sediments suggest that the average ^{13}C content of carbon in this reservoir was 0.3 to 0.4‰ lower during peak glacial than during peak interglacial time (Curry et al., 1988; Duplessy et al., 1988). This requires that the ocean-atmosphere system had during glacial time 1.4 ± 0.2 percent (i.e. 100 x 0.35/25) higher carbon content than interglacials. When equilibrated with the carbon dissolved in the sea, this would create a 47 ppm (i.e 0.014 x 12 x 280) higher glacial atmospheric CO_2 content. Even after $CaCO_3$ compensation (see below) is taken into account, the increase would be about 25 ppm. Thus, if the foraminifera-based estimate of the $\delta^{13}C$ change in ocean-atmosphere carbon is correct, then the glacial atmosphere should have had a CO_2 partial pressure of 305 μatm (compared to the ice core value of 200 μatm). Hence we need to account for a 105 μatm rather than an 80 μatm CO_2 rise at the close of glacial time.

As pointed out below, the decrease in atmospheric CO_2 content attributable to forest growth and soil development at the onset of interglacials may have been counterbalanced by massive coral growth during the rise in sea level which accompanied deglaciation (Berger, 1987).

THE ROLE OF THE SEA'S INORGANIC CARBON RESERVOIR

Prior to the industrial revolution, the ocean contained about sixty times more carbon (in dissolved inorganic form) than the atmosphere (see Table 1). Even when account is taken of the buffer capacity* of sea water, the effective amount of carbon in the ocean is about five times the amount in the atmosphere. As the ocean's mixing time is about 1000 years, on the 10,000 year time scale over which full glacial and full interglacial climates prevailed, the atmosphere and ocean must have come to chemical equilibrium with one another.

Table 1. Estimate of the glacial (G) to interglacial (I) transfer of carbon from the inorganic reservoirs to the organic reservoirs (based on a rise of 0.35‰ in the $^{13}C/^{12}C$ for ocean ΣCO_2 between glacial and interglacial time).

	INTERGL. 10^{15}gC	$\Delta(G \rightarrow I)$ 10^{15}gC	$\Delta(G \rightarrow I)$ %
INORGANIC RESERVOIRS			
Atmosphere (CO_2)	500		
Ocean (HCO_3^-, $CO_3^=$, CO_2)	30,000		
	30,500	-425	-1.4
ORGANIC RESERVOIRS			
Forests (wood + roots)	500		
Soils (liter + humus)	1,500		
Sea (dissolved org.)	1,500		
	3,500	+425	+12.1

As elegantly shown by the Harvardton Bears (Knox and McElroy, 1984; Sarmiento and Toggweiler, 1984; Siegenthaler and Wenk, 1984), the cold parts of the surface ocean dictate the atmosphere's CO_2 content. The reason is that only these parts of the ocean surface communicate with the large deep sea reservoir. Because the rate of air-sea CO_2 exchange is rapid with respect to the rate of mixing within the sea, CO_2 is efficiently transferred through

* The buffer capacity of sea water (often referred to as the Revelle factor) is the ratio of the fractional change in CO_2 partial pressure to the fractional change in ΣCO_2 content. For warm surface sea water (>24°C), the buffer factor averages about 8, and for cold sea water (<3°C), it averages about 13. The average value for the entire ocean is about 12. Hence the effective amount of carbon in the sea is about one twelfth the actual amount.

the atmosphere from the warm to the cold part of the ocean. Hence the assumption can be made that the warm parts of the ocean are largely passive; their CO_2 partial pressure will be dictated by that for the high latitude outcrops of the deep sea. The CO_2 partial pressure of these outcropping areas is dictated by 1) their temperature, 2) their salinity, 3) their nutrient content, 4) the Redfield ratio for particulate and dissolved organic material exported from the sea surface and 5) their $CO_3^=$ ion concentration.

Temperature:

The average temperature of water in the deep sea appears from faunal (CLIMAP, 1981) oxygen isotope (Shackleton, 1987; Labeyrie et al., 1987) evidence to have been about 2°C lower during glacial time than during interglacial time. A similar cooling must then have also occurred in the polar surface waters which ventilated the deep sea at that time. Such a cooling would have drawn down the atmospheric CO_2 content by about 18 μatm (Broecker and Peng, 1986 and 1987a).

Salinity:

The aggregate salinity of the glacial ocean was about 3% higher during glacial time due to storage of fresh water in the ice caps. This would have raised the salt content from its average of 35 gm/liter today to 36 gm/liter then. Such a change would have raised the atmosphere's CO_2 pressure by 11 μatm (Broecker and Peng, 1986).

As pointed out by Broecker and Peng (1987a), changes in sea ice must also be considered. The salinity of deep water generated in today's Antarctic is 1 gm per liter higher than in ambient surface water. This enrichment is driven by brine rejection from seasonally grown sea ice. By contrast, densification by sea ice formation plays only a small role in today's northern Atlantic (see Figure 1). During glacial time, the greater extent of sea ice in the northern Atlantic and in the Antarctic may well have increased the role of brine formation. If so, then the salinity of polar surface waters would have been correspondingly lower relative to the ocean average and could easily have offset the impact of the oceanwide salinity increase.

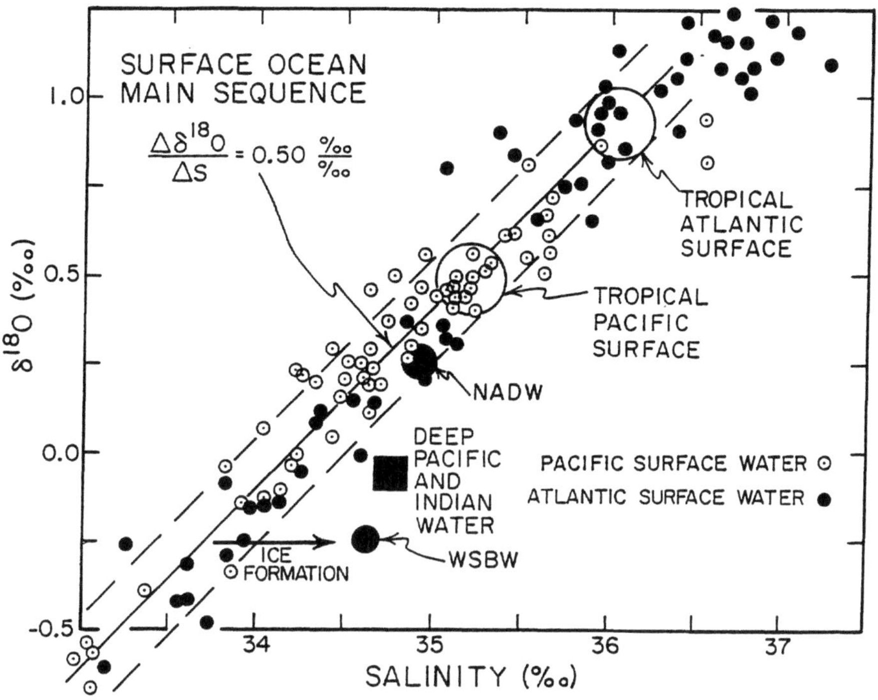

Figure 1. Distribution of $^{18}O/^{16}O$ ratios in the surface ocean as determined from measurements made in the laboratory of Harmon Craig on samples collected as part of the Geochemical Ocean Sections Study Program. The small circles are for individual surface water samples. The two large solid circles are for the two major sources of deep water; North Atlantic deep water (NADW) and Weddell Sea bottom water (WSBW). As can be seen the point for WSBW falls off the main sequence for surface waters reflecting the importance of ice formation in the densification of water in the Antarctic region. By contrast the NADW point lies close to the main sequence. The solid square is for deep water in the Indian and Pacific Oceans (Craig and Gordon, 1965).

Biologic Pumping:

Organisms living in surface water fix CO_2 into organic matter. Part of this organic matter is transported to the sea's interior where it is oxidized and returned to inorganic form. Because of the action of this so-called biologic pump (Volk and Hoffert, 1985), the CO_2 partial pressure in surface waters is less than that in the sea's interior. The strength of the pumping action is limited by the availability of the limiting nutrients, phosphate and nitrate. Microorganisms require about one phosphorus atom per 128 carbon atoms and 16 nitrogen atoms per 128 carbon atoms (Broecker et al., 1985; Takahashi et al., 1985; Broecker and Peng, 1987b). As the molar ratio of nitrate to phosphate in the sea is 16, these two nutrients colimit plant productivity. For today's ocean, were all the nutrients efficiently extracted from surface water by plants, then the atmosphere's CO_2 content would be about 150 ppm. By contrast, were the ocean sterile, then the atmosphere's CO_2 content would be about 470 ppm. Prior to the industrial revolution, the atmosphere's CO_2 content was 280 ppm. The reason for this is that the nutrients NO_3 and PO_4 are not efficiently utilized in high latitude surface waters. As shown in Figure 2, compared to the ocean average PO_4 concentration of 2.2 μmol/kg, Antarctic surface waters have 1.6 μmol/kg, and northern Atlantic surface waters, 0.9 μmol/kg.

So one explanation for the lower atmospheric CO_2 values during glacial time is that the nutrient content of high latitude surface waters was lower than today. In other words, during the glacial time, the biologic pump was more efficient than it is today. Unfortunately, a key prediction made by this hypothesis is not fulfilled. It has to do with the carbon isotope record for polar surface water kept by the shells of planktonic foraminifera. During photosynthesis by marine plants, the heavy isotope of carbon, ^{13}C, is discriminated against by about 20‰. Because of this, the $^{13}C/^{12}C$ ratio in surface ocean water is greater than that in deep water. For example, surface waters in the tropical Pacific have a $^{13}C/^{12}C$ ratio about 2‰ higher than that for the underlying deep water. Hence were the biologic pumping action at high latitudes stronger during glacial time, other factors being equal, the $^{13}C/^{12}C$ ratio in polar waters should have been more positive than now. As reproduced in Figure 3, Charles and Fairbanks (1990) have shown that the opposite is true. The $^{13}C/^{12}C$ ratio in planktonic foraminifera was nearly

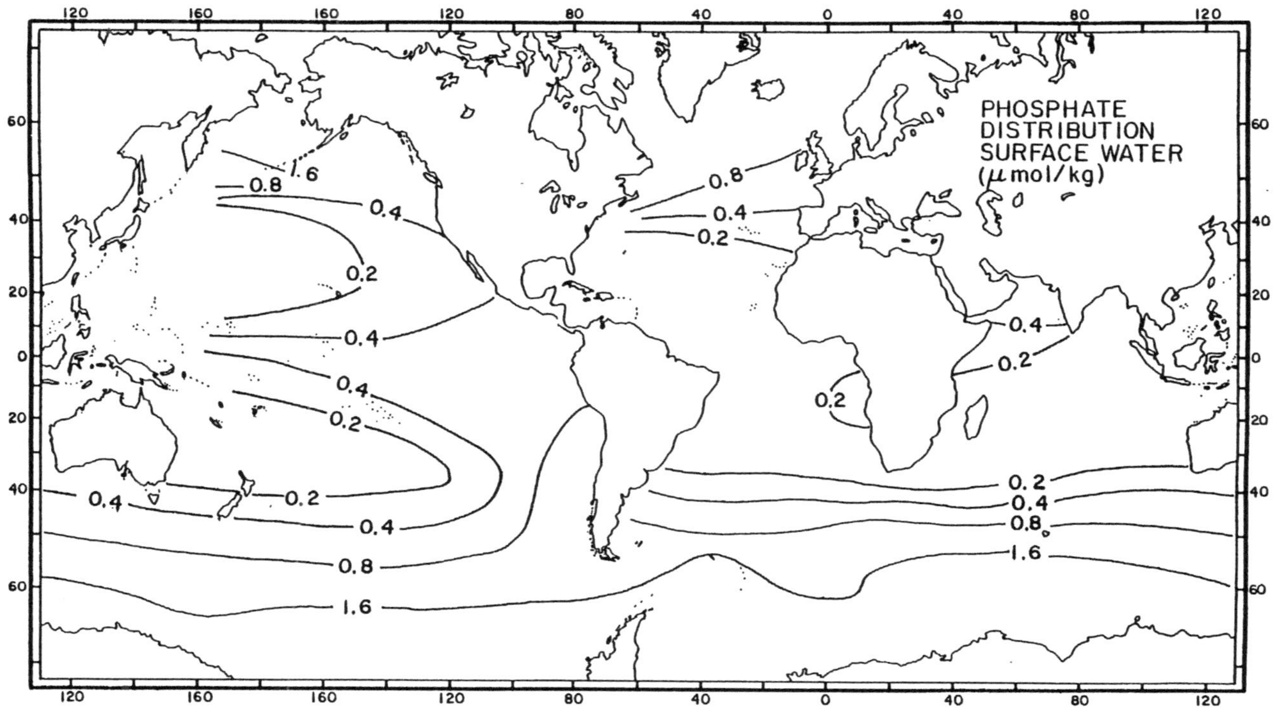

Figure 2. Contours of the phosphate content of surface water. For the northern Atlantic the values are for the winter period. In other high latitude regions they represent summer values (and hence minimum values for the winter period).

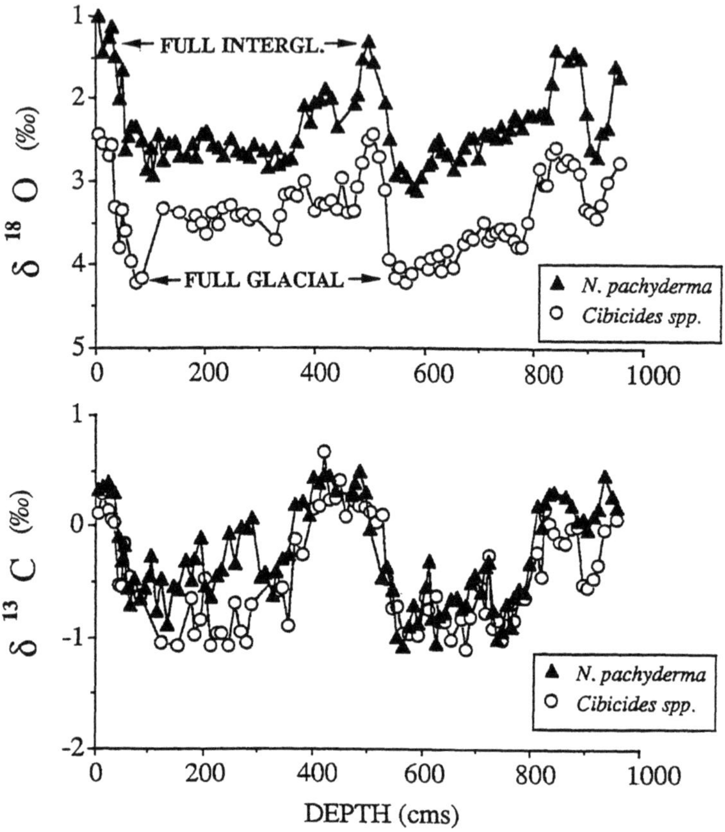

Figure 3. Record of the isotopic composition of *N. pachyderma* (planktonic foraminifera) and *Cibicides* spp. (benthic foraminifera) from Core V22-108 from the Antarctic Ocean (43°S, 3°W; 4171m) for the last 250,000 years (Charles and Fairbanks, 1990).

1.0‰ lower during peak glacial time than during peak interglacial time! Furthermore, the difference between the $^{13}C/^{12}C$ ratio in Antarctic planktonic and benthic foraminifera shells does not increase significantly from glacial to interglacial time. Thus no support exists in the carbon isotope record for stronger pumping action in the Antarctic region during glacial time. Rather, this record creates an enigma in that the glacial to interglacial changes in carbon isotope ratio for the Antarctic Ocean foraminifera are the largest in the world ocean. Only when this enigma has been resolved will a universally acceptable explanation for the glacial to interglacial atmospheric CO_2 change emerge.

Changes in the efficiency of nutrient use in Antarctic surface waters is not the only means by which the biologic pump might be strengthened. Two other possibilities come to mind; the first involves a change in the ratios of carbon to phosphorus and carbon to nitrogen in marine organic material (Broecker, 1982). Were these ratios larger during glacial time, then the biologic pump would have been stronger. While differences in ecology could conceivably produce a change in the chemical composition of marine phytoplankton, the near uniformity of these ratios throughout today's ocean (Broecker et al., 1985) makes this scenario an unlikely one. The second way by which the biological pump might have been strengthened during glacial time is if more phosphate (and nitrate) were present in the sea then than now. As originally proposed by Broecker (1981), phosphate burial in sediments deposited on continental shelves and in estuaries during the marine transgression which accompanied deglaciation might have led to such a removal.

Regardless of the mechanism by which the biologic pump was strengthened, an increase in the carbon isotope difference between surface and deep waters would be generated. As reproduced in Figure 4, evidence for a glacial to interglacial change in the surface to deep carbon isotope is present in records from tropical ocean cores from the Pacific Ocean (Shackleton et al., 1983). However as the change is only a few tenths per mil, biases associated with depth habitat and vital effects of planktonic foraminifera may confuse the situation. However, if the change is taken at face value, then after $CaCO_3$ compensation, about 40 μatm lowering of the CO_2 content of the atmosphere could be attributed to a

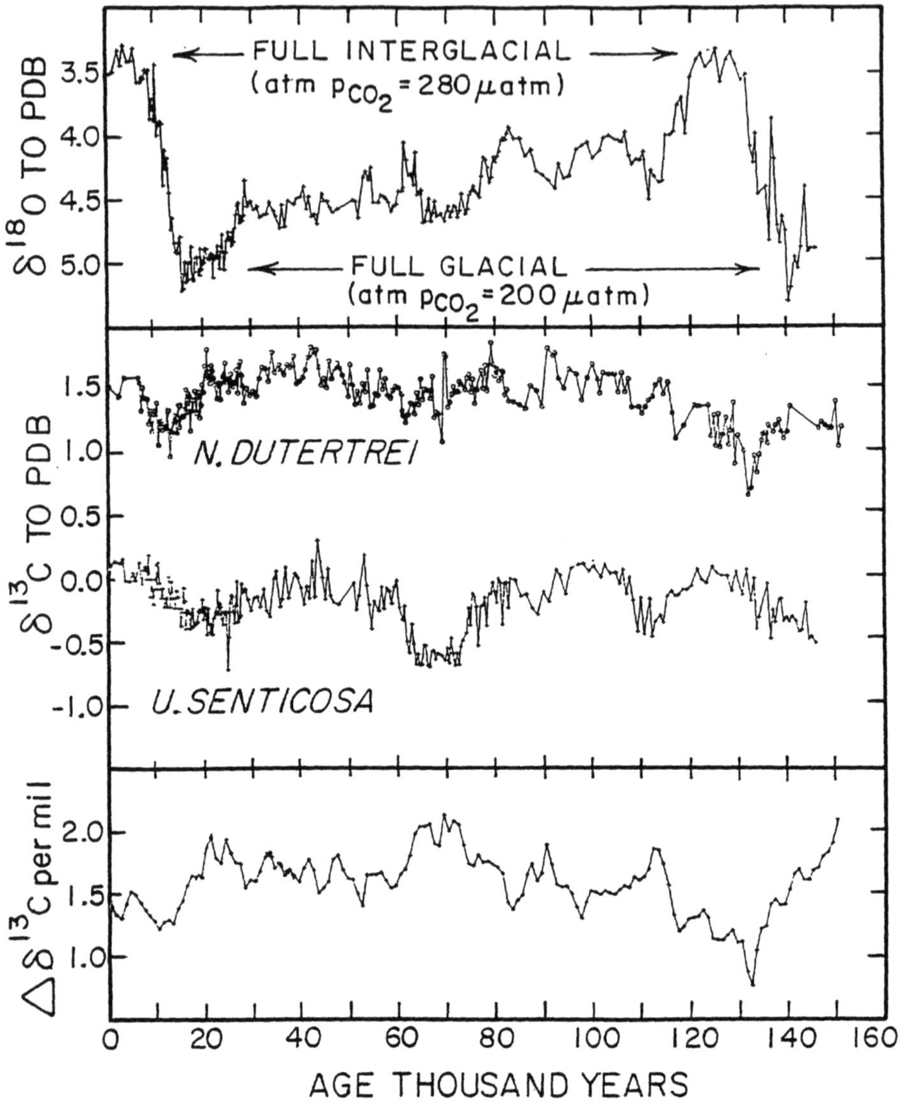

Figure 4. Top: oxygen isotope data for benthic foraminifera, *Uvigerina senticosa*, in V19-30 from the East Equatorial Pacific Ocean (Shackleton et al., 1983). Middle: carbon isotope data for planktonic foraminifera, *Neogloboquadrina dutertrei* (O) and for *U. senticosa* (+) from the same core. Bottom: carbon isotope difference between ocean surface (*N. dutertrei*) and sea floor (*U. senticosa*).

strengthening of the ocean's biological pump during glacial time. It must be emphasized that in using the carbon isotope record for the tropical Pacific instead of the carbon isotope record for the Antarctic, we are violating the principle set out by the Harvardton Bears: the polar ocean must rule.

Carbonate ion concentration in the deep sea:

The CO_2 partial pressure exerted by a given parcel of sea water depends on its alkalinity as well as on its ΣCO_2 concentration. The major cause for changes in alkalinity within the sea is the production and dissolution of $CaCO_3$. This cycle is akin to that for organic material in that $CaCO_3$ is produced in the surface ocean and is destroyed at depth in the sea. However, two important differences exist. First, the generation of $CaCO_3$ raises rathers than lowers the CO_2 content of surface waters. The alkalinity change is twice as great as the ΣCO_2 change. Second, rather than being almost completely destroyed in the sea's interior as is organic matter, roughly one quarter of the $CaCO_3$ produced by marine organisms survives dissolution and accumulates on the sea floor. The role of the $CaCO_3$ cycle on the atmosphere's CO_2 content is multifaceted and complex.

The first point to be made is that the cycle of $CaCO_3$ sets the $CO_3^=$ ion content in the deep sea and in so doing influences the CO_2 content of the atmosphere. The basis of the control is economic; namely, in a steady state the rate at which $CaCO_3$ accumulates in deep sea sediments must match the rate at which the ingredients for $CaCO_3$ (i.e. CaO and CO_2) are supplied to the ocean-atmosphere reservoir. In today's ocean marine organisms manufacture $CaCO_3$ at a rate about four times faster than that at which ingredients are supplied (see Figure 5). In order to compensate for this overproduction, the chemical composition of sea water has adjusted so that three quarters of the $CaCO_3$ redissolves. This compensation is accomplished by changes in the ratio of the area of the sea's floor bathed in water supersaturated with respect to the mineral calcite to that bathed in water undersaturated with respect to the mineral calcite. As the Ca^{++} ion content of sea water is nearly uniform, the tendency toward $CaCO_3$ dissolution is dictated by the $CO_3^=$ ion concentration. In particular, it is related to the ratio of the in situ $CO_3^=$ ion concentration to the calcite saturation $CO_3^=$ ion concentration. When this

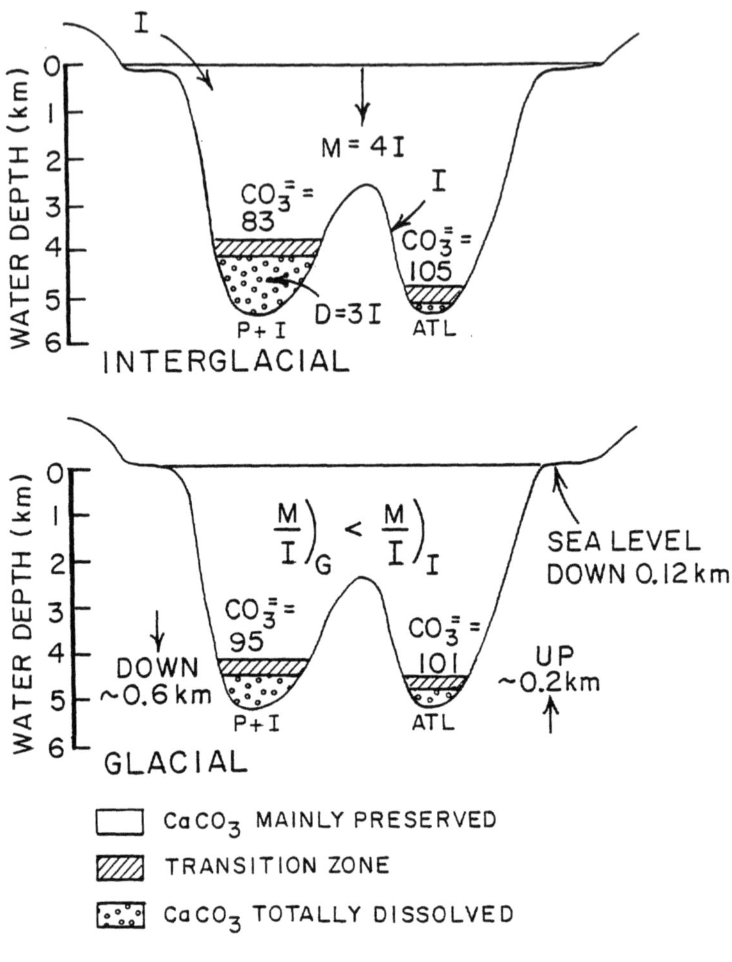

Figure 5. Somewhat idealized depiction of the contrast between the $CaCO_3$ budget in interglacial and glacial oceans. In today's ocean, the transition zone separating sediments which have experienced minor loss of calcite to dissolution from those which have lost virtually all their $CaCO_3$ lies about one kilometer deeper in the Atlantic Ocean than in the Pacific and Indian Oceans. This depth difference is consistent with the 23 µmol/kg higher $CO_3^=$ concentration in deep Atlantic waters. During glacial time, the lysocline in the Pacific Ocean was about 0.6 kilometers deeper and that in the Atlantic about 0.2 kilometers shallower. The average deepening by about 500 meters for the global ocean during glacial time suggests a 10 µmol/kg higher mean deep sea $CO_3^=$ concentration and suggests that the ratio of the rate of manufacture of $CaCO_3$ by marine organisms, M, to the rate of input of the ingredients for $CaCO_3$ to the sea, I, was lower during glacial time. The units of $CO_3^=$ are µmol/kg.

ratio is less than unity, calcite tends to dissolve; the lower the ratio, the more rapid the dissolution rate. The in situ $CO_3^{=}$ concentration is nearly constant with depth at any given locality. However, because of the pressure dependence of the solubility of calcite, the solubility concentration increases by about 20 μmol/kg per kilometer depth (see Figure 6). This leads to a crossover from supersaturation at shallower depths to undersaturation at greater depths. Above the crossover, the in situ $CO_3^{=}$ concentration exceeds the solubility value. Below this crossover, the reverse is true (see Figure 6).

This situation is clearly imprinted in the sediments. Sediments above the crossover are rich in calcite; sediments below the crossover have lost part or all of their calcite to dissolution (see Figure 6). If for some reason the match between ingredient supply and calcite accumulation is disrupted, the $CO_3^{=}$ ion concentration in the deep sea will be driven to a new value at which loss once again matches supply. This will change the crossover depth and hence the ratio of the area over which calcite accumulates to the area over which it dissolves.

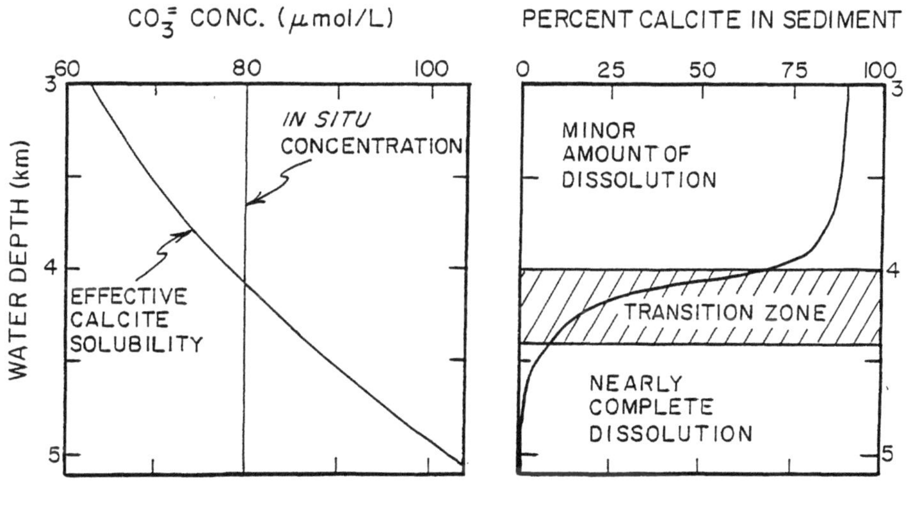

Figure 6. Somewhat idealized depiction of how the insitu and solubility $CO_3^{=}$ ion concentrations change with water depth (left) and of how the calcite content of sediments changes with water depth (right) in the tropical Pacific Ocean.

As the half time for this readjustment is only several thousand years, changes in the rate of delivery of the ingredients for $CaCO_3$ or in the accumulation rate of $CaCO_3$ associated with glacial cycles drive changes in the $CO_3^{=}$ ion content of the deep sea.

Changes in the $CO_3^{=}$ ion concentration in deep water create corresponding changes in the $CO_3^{=}$ ion concentration in surface water. The CO_2 partial pressure for any given batch of surface water is inversely proportional to its $CO_3^{=}$ ion concentration. Hence changes in the $CO_3^{=}$ concentration produce changes in atmospheric CO_2 content.

Firm evidence for a change in crossover depth exists in deep sea sediments. As reproduced in Figure 7, cores from the tropical Pacific and Indian Ocean taken within the transition zone which separates the realm of calcite preservation from the realm of complete dissolution reveal that during glacial time the transition zone was shifted to greater depth (Farrell and Prell, 1989). In other words, dissolution was less intense at any given depth in these oceans. For the Atlantic, the change was in the opposite direction (see Figure 6). As the Atlantic change was smaller and as together the Pacific and Indian Oceans cover a much larger area than the Atlantic, this evidence suggests that during glacial time the ratio of ingredient supply to $CaCO_3$ manufacture was greater than it was during interglacial time. One possible explanation for this change is that coral reef growth was more rapid during interglacials (Berger, 1982; Opdyke and Walker, in press). Also, some evidence exists in support of a higher accumulation rate of $CaCO_3$ in the deep sea during interglacials. Unfortunately, we have as yet no means to reconstruct glacial-interglacial changes in the rate of chemical erosion of the continents.

If the depth of the transition zone separating sediments rich in $CaCO_3$ from sediments nearly free of $CaCO_3$ were on the average 500 meters deeper during peak glacial time, then the $CO_3^{=}$ ion content of deep water must have averaged about 10 μmol/kg greater. The corresponding decrease in atmospheric CO_2 content would have been about 20 μatm (Broecker et al., 1991).

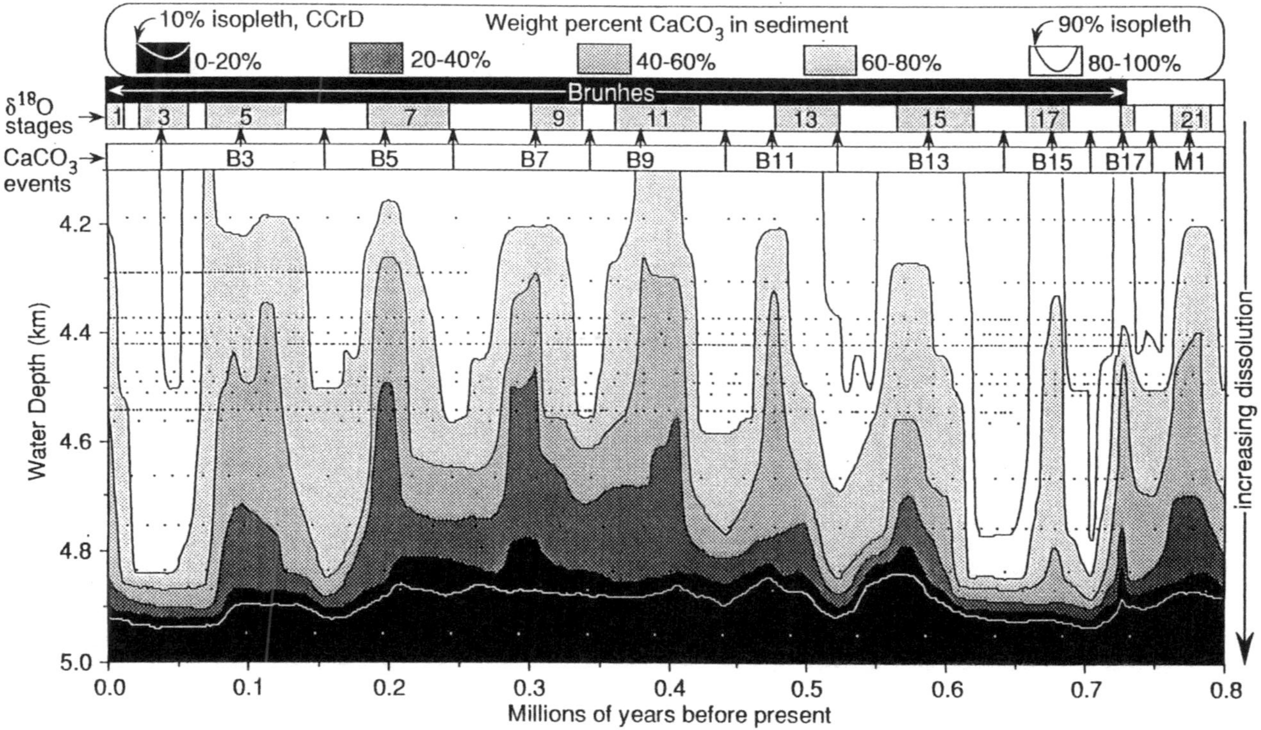

Figure 7. $CaCO_3$ content variations of sediments in the equatorial Pacific as a function of water depth over the last 800,000 years (Farrell and Prell, 1989). As can be seen, while the isopleths for 80 and 90 percent $CaCO_3$ content undergo glacial to interglacial excursions of up to 600 meters, isopleths for the 10 and 20 percent $CaCO_3$ content undergo much smaller excursions. Broecker et al. (1991) attribute this difference to chemical erosion during interglacials. If so, then the excursions of the high $CaCO_3$ isopleths are more indicative of the actual excursion of the deep sea $CO_3^{=}$ concentration.

Carbonate ion concentration in the surface ocean:

While the $CO_3^=$ concentration in the deep sea is set by the requirement that the rate of storage of $CaCO_3$ in marine sediments match the rate of supply of the ingredients required for the manufacture of $CaCO_3$, it is, of course, the concentration of $CO_3^=$ in surface water which directly influences the CO_2 partial pressure in the atmosphere. In today's ocean, the $CO_3^=$ ion concentration in surface water is considerably higher than that in deep water. While the $CO_3^=$ concentration in deep water ranges from a low of about 80 μmol/kg in the deep Pacific to a high of about 110 μmol/kg in the deep Atlantic, the $CO_3^=$ concentration in surface water ranges from a low of about 130 in polar surface waters to a high of about 250 μmol/kg in warm surface waters. This difference is a direct consequence of the biologic pump. However, the situation is more complicated. Changes in the strength of the biologic pump change distribution of $CO_3^=$ between surface and deep waters. Hence they disrupt the balance between $CaCO_3$ accumulation and the supply of ingredients for $CaCO_3$. Compensation will drive the deep water values back to those leading to a balance between supply and loss. So, in addition to knowing how the $CO_3^=$ ion concentration changed between glacial and interglacial time, it is important to know how the difference in $CO_3^=$ concentration between surface and deep water changed. If, as discussed above, it is the polar surface ocean which sets the CO_2 partial pressure in both the atmosphere and the warm surface ocean, then the important $CO_3^=$ difference is that between Antarctic surface water and ambient deep water.

The major source of difference between the $CO_3^=$ ion concentration in surface waters and deep waters is the biological pump. In the simple example shown in Figure 8, the strength of the biologic pump is increased driving up the pCO_2 difference between surface and deep water and also the difference in $CO_3^=$ ion concentration. This is where $CaCO_3$ compensation enters the picture. The resulting $CO_3^=$ ion concentration in the deep sea will be too low; more $CaCO_3$ will be dissolved than is being supplied. As a result of this excess dissolution, the $CO_3^=$ ion concentration will be driven back to its required value. This will raise the $CO_3^=$ ion concentration in surface water and thereby lower the CO_2 partial pressure. Any change which leads to an increase in biological pumping strength will generate such a two step drop in

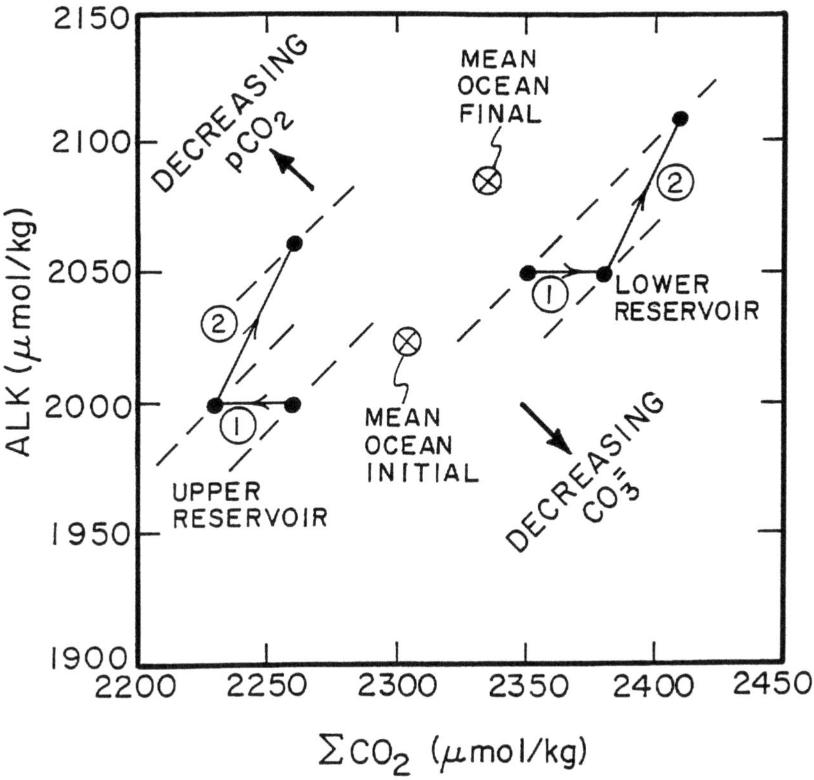

Figure 8. Schematic example of how $CaCO_3$ compensation operates. Consider a hypothetical ocean with two reservoirs of equal volume. An increase in biological pumping power occurs which reduces the ΣCO_2 in the upper reservoir and correspondingly increases it in the lower reservoir (step 1). Then $CaCO_3$ dissolves from sediments lining the deep reservoir bringing the $CO_3^=$ ion concentration back to its initial value (step 2). The addition of this Ca^{++} and $CO_3^=$ ion raises the alkalinity in both reservoirs by two units for every one unit by which it raises the ΣCO_2 concentration. While $CaCO_3$ compensation returns the $CO_3^=$ and pCO_2 of the deep reservoir to their initial value, it decreases the pCO_2 and increases the $CO_3^=$ ion content of the upper reservoir. In this example, the pCO_2 change in the upper reservoir before $CaCO_3$ compensation is only half the change after $CaCO_3$ compensation. Hence in this example, $CaCO_3$ compensation doubles the impact on the atmospheric CO_2 content created by the increase in biologic pumping power. Note that the dashed diagonals are contours of constant $CO_3^=$ ion concentration and also of constant pCO_2.

atmospheric CO_2 content. The first will be the result of the biologic pumping action. The second will be the result of $CaCO_3$ compensation. While the first could be accomplished in centuries, the second would require several thousand years.

Boyle (1988) pointed out a way in which the $CO_3^=$ concentration between surface and deep water could be perturbed which does not involve the biologic pump. Measurements of carbon isotope ratios and trace metal concentrations in benthic foraminifera shells from deep sea sediments suggest that the vertical distribution of nutrient constituents in the sea during glacial time differed from today's. While today's intermediate waters now have slightly higher concentrations than deep waters, during glacial time, the situation was reversed. Deep waters appear to have had higher nutrient concentrations than intermediate waters. The important point is that waters at the depth of the sedimentary $CaCO_3$ content transition had higher nutrient concentration during glacial time. This would have reduced the $CO_3^=$ concentration in deep sea water. However, $CaCO_3$ compensation would have returned the $CO_3^=$ concentration to its control value. This increase in $CO_3^=$ ion concentration would apply to the entire ocean, including surface waters, causing the CO_2 partial pressure in the atmosphere to be drawn down.

$CaCO_3$ deposition in coral reefs during the transgression of the sea at the time of the last deglaciation may have counterbalanced the increase in the size of the organic reservoirs which occurred at the beginning of each interglacial. If roughly one mole of $CaCO_3$ was deposited for each mole of carbon stored in trees and humus, then the atmosphere's CO_2 content would have remained unchanged.

THE INVERSE PROBLEM

Broecker and Peng, 1989, turn the problem around. They assume that the surface waters of the Antarctic set the atmosphere's CO_2 concentration and seek a means to lower the CO_2 partial pressure in these waters without decreasing their nutrient concentration. They suggest that this might have been accomplished by a major increase in the alkalinity of the Antarctic water column during glacial. Unfortunately, we couldn't then, nor can we now, give a hard answer regarding how the required alkalinity change might have been accomplished. As in all such analyses, the large drop in the $\delta^{13}C$ value for the planktonic foraminifera from glacial age Antarctic sediments is impossible to explain.

SUMMARY

Estimates of the contributions by each of the mechanisms discussed above are given in Table 2. In this analysis we adopt the Shackleton et al. equatorial ^{13}C record as a measure of the strengthening of the biologic pump and deepening of nutrients during glacial time. Without increased brine formation from sea ice to compensate for the ocean-wide salinity increase during glacials and without coral growth to compensate for the increase in terrestrial biomass during episodes of deglaciation, only about half of the 80 μatm glacial to interglacial atmospheric CO_2 changes can be explained. Only with the sea ice and coral reef contributions is the 80 μatm magnitude achieved.

Table 2. Two scenarios of possible contributions to the glacial to interglacial CO_2 increase (i.e. from ~200 μatm during glacial time to about 280 μatm during interglacial time).

	ΔpCO_2 μatm	ΔpCO_2 μatm
Increased biomass (1.4%)	-25	0+
Temperature rise (~2°C)	+18	+18
Salinity drop (1 gm/l)	-11	0++
Reduced deep sea $CO_3^=$ (10 μmol/kg)	+20	+20
Increased $\Delta CO_3^=$ surface-deep*	+40	+40
	+42	+78

*As the result of a reduction in the power of the biological pump and the redistribution of nutrients between intermediate and deep water.

+Assumes that the deposition of $CaCO_3$ in coral reefs during the marine transgression which accompanied deglaciation compensated for the growth of the forest and soil organic reservoirs during early interglacial time.

++Assumes that the polar sea ice effect compensated for the ocean-wide salinity change.

ACKNOWLEDGEMENTS

This research is supported by ongoing grants to Broecker from Exxon Corporation, from the NSF Climate Dynamics Program, and from the DOE CO_2 Program LLNL B130547. Lamont-Doherty Geological Observatory Contribution No. 4906.

Peng's research is sponsored by the Carbon Dioxide Research Program, Environmental Sciences Division, Office of Health and Environmental Research, U.S. Department of Energy under contract DE-AC05-84OR21400 with Martin Marietta Energy Systems, Inc. Publication No. 3865, Environmental Sciences Division, Oak Ridge National Laboratory.

REFERENCES

Barnola JM, Raynaud D, Korotkevich YS, Lorius C (1987) Vostok ice core provides 160,000-year record of atmospheric CO_2. Nature 329:408-414

Berger WH (1982) Increase of carbon dioxide in the atmosphere during deglaciation: the coral reef hypothesis. Naturwissenschaften 69:87-88

Boyle EA (1988) Vertical oceanic nutrient fractionation and glacial/interglacial CO_2 cycles. Nature 331:55-56

Broecker WS (1982) Glacial to interglacial changes in ocean chemistry. Prog. Oceanog. 11:151-197

Broecker WS, Takahashi T, Takahashi T (1985) Sources and flow patterns of deep-ocean waters as deduced from potential temperature salinity and initial phosphate concentrations. J. Geophys. Res. 90:6925-6939

Broecker WS, Peng T-H (1986) Carbon cycle: 1985. Glacial to interglacial changes in the operation of the global carbon cycle. Radiocarbon 28:309-327

Broecker WS, Peng T-H (1987a) The oceanic salt pump: does it contribute to the glacial-interglacial difference in atmospheric CO_2 content? Global Biogeochemical Cycles 1:251-259

Broecker WS, Peng T-H (1987b) C/P ratios in marine detritus. Global Biogeochemical Cycles 1:155-161

Broecker WS, Peng T-H (1989) The cause of the glaical to interglacial atmospheric CO_2 change: A polar alkalinity hypothesis. Global Biogeochemical Cycles 3:215-239

Broecker WS, Klas M, Clark E, Bonani G, Ivy S, and Wolfli W (1991) The influence of $CaCO_2$ dissolution on core top radiocarbon ages for deep-sea sediments. Paleoceanography 6:593-608

Charles CD, Fairbanks RG (1990) Glacial to interglacial changes in the isotopic gradients of southern ocean surface water. In: Bleil U, Thiede J (eds) Geological History of the Polar Oceans: Arctic Versus Antarctic. Kluwer Academic Publishers, Netherlands, pp 519-538

CLIMAP PROJECT MEMBERS (1981) Seasonal reconstruction of the Earth's surface at the last glacial maximum. Geol. Soc. Amer., Map and Chart Series 36

Craig H, Gordon LI (1965) Deuterium and oxygen-18 variations in the ocean and the marine atmosphere. In: Tongiorgi T (ed),Stable Isotope in Oceanographic Studies and Paleotemperatures. Consiglio Nazional delle Richerche Laboratorio de Geologia Nucleare, Pisa, Italy, pp 9-130

Curry WB, Duplessy JC, Labeyrie LD, Shackleton NJ (1988) Changes in the distribution of $\delta^{13}C$ of deep water ΣCO_2 between the last glaciation and the Holocene, Paleoceanography 3:317-341

Duplessy JC, Shackleton NJ, Fairbanks RG, Labeyrie LD, Oppo D, Kallel N (1988) Deepwater source variations during the last climatic cycle and their impact on the global deepwater circulation. Paleoceanography 3:343-360

Farrell JW, Prell WL (1989) Climatic change and $CaCO_3$ preservation: An 800,000 year bathymetric reconstruction from the central equatorial Pacific Ocean. Paleoceanography 4:447-466

Knox F, McElroy M (1984) Change in atmospheric CO_2: Influence of the marine biota at high latitude. J. Geophys. Res. 89:4629-4637

Labeyrie LD, Duplessy JC, Blanc PL (1987) Variations in mode of formation and temperature of oceanic deep waters over the past 125,000 years. Nature 327:477-482

Neftel A, Oeschger H, Schwander J, Stauffer B, Zumbrunn R (1982) Ice core sample measurements give atmospheric CO_2 content during the past 40,000 yr. Nature 295:220-223

Opdyke BN, Walker JCG (1991, in press) The return of the coral reef hypothesis: Glacial to interglacial partitioning of basin to shelf carbonate and its effect on Holocene atmospheric pCO_2. Geology

Sarmiento JL, Toggweiler R (1984) A new model for the role of the oceans in determining atmospheric pCO_2. Nature 308:621-624

Shackleton NJ, Hall MA, Line J, Shuxi C (1983) Carbon isotope data in core V19-30 confirm reduced carbon dioxide concentration in the ice age atmosphere. Nature 306:319-322

Shackleton NJ (1987) Oxygen isotopes, ice volume and sea level. Quaternary Sci. Rev. 6:183-190

Siegenthaler U, Wenk T (1984) Rapid atmospheric CO_2 variations and ocean circulation. Nature 308:624-626

Takahashi T, Broecker WS, Langer S (1985) Redfield ratio based on chemical data from isopycnal surfaces. J. Geophys. Res. 90:6907-6924

Volk T, Hoffert MI (1985) Ocean carbon pumps: Analysis of relative strengths and efficiencies in ocean-driven atmospheric CO_2 changes. In: Sundquist ET, Broecker WS (eds) The Carbon Cycle and Atmospheric CO_2: Natural Variations Archean to Present. Geophysical Monograph 32, American Geophysical Union, Washington, D.C., pp 99-110

THE MAGNITUDE AND DISTRIBUTION OF FOSSIL-FUEL-RELATED CARBON RELEASES

Gregg Marland
Tom Boden
Environmental Sciences Division
Oak Ridge National Laboratory
OAK RIDGE, TN. 37831-6335
USA

Like Keeling's plot of atmospheric CO_2 concentrations at Mauna Loa Observatory, the basic plot of CO_2 emissions from fossil fuel burning (Figure 1) is familiar to most people involved with the global carbon cycle. Like the time series of atmospheric concentrations, the time series of anthropogenic emissions is one of the cornerstones of our understanding of the global carbon cycle. As we grapple with the global balance of carbon fluxes, it is thus appropriate to review the data on anthropogenic emissions and to inquire how well the basic emissions numbers are in fact known. Emissions data are often reported to 4 "significant" figures (see Table 1) and convey a sense that they are well established, but it behooves us to have a realistic sense of their accuracy. Marland and Rotty (1984) have suggested that the global values are accurate to ± 10% but the nature of the underlying data precludes a rigorous error analysis. The intent of this paper is to summarize the derivation of estimates for CO_2 emissions from fossil fuel burning and to provide some perspective into the underlying data and assumptions.

Modern CO_2 emissions estimates were first published by Keeling (1973) and have been subsequently re-examined and updated by Marland, Rotty, and Boden in a series of papers beginning in 1984. The emissions time series has evolved over time as the demands placed on it have evolved. Initially only the global total of emissions was estimated, but emissions were estimated by country when it became apparent that country or regional data would improve our ability to make projections into the future. This evolved into a data set on a 5 x 5 degree grid of latitude and longitude when there was a need for data on emissions as a

NATO ASI Series, Vol. I 15
The Global Carbon Cycle
Edited by M. Heimann

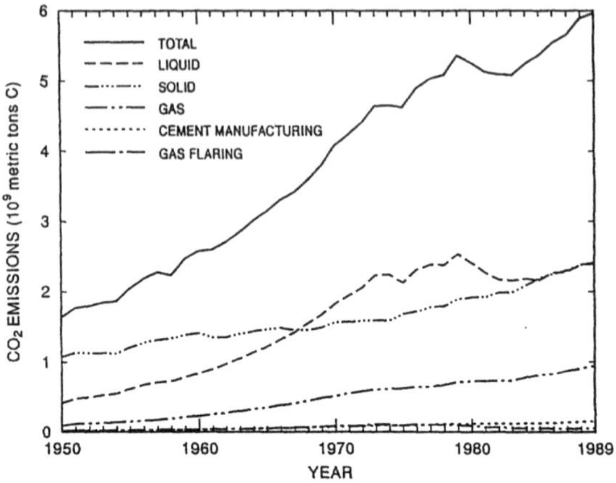

Figure 1: Global emissions of CO_2 from fossil-fuel combustion and cement manufacturing.

function of latitude. At present there are two parallel lines of discussion that will influence the emissions data set. There is increasing political interest in trying to limit the rate of increase of greenhouse gases in the atmosphere and this leads to increasing interest in the sources of, and responsibility for, CO_2 emissions. The focus for this interest seems now to be in the Intergovernmental Panel on Climate Change (IPCC). There is also increasing interest in the geographic and temporal distribution of emissions. This interest is evident at this NATO Advanced Study Institute and an effort within the International Geosphere-Biosphere Program (IGBP) has as its goal an emissions data set on a 1 x 1 degree grid of latitude and longitude. These two lines of interest are leading toward a geopolitical data set and a geochemical data set, respectively, and it will be nice if these are generally concordant and we understand the differences. We will return at the end of this paper to comment briefly on the potential differences between these two CO_2 emissions data sets. The data set described below is for emissions from 1950 to 1989 and mates with the data set described by Keeling (1973) to provide a continuous data set for emissions since 1860.

The starting point is that when fossil fuels (coal, petroleum, and natural gas) are oxidized to provide thermal energy useful to

Year	Total CO_2 Emissions (10^6 tonsC)	CO_2 From Solid Fuels (10^6 tonsC)	CO_2 From Liquid Fuels (10^6 tonsC)	CO_2 From Gas Fuels (10^6 tonsC)	CO_2 From Cement Manufacture (10^6 tonsC)	CO_2 From Gas Flaring (10^6 tonsC)	Per Capita CO_2 Emissions (tonsC)
1950	1638	1077	423	97	18	23	0.65
1951	1775	1137	479	115	20	24	0.69
1952	1803	1127	504	124	22	26	0.69
1953	1848	1132	533	131	24	27	0.70
1954	1871	1123	557	138	27	27	0.69
1955	2050	1215	625	150	30	31	0.74
1956	2185	1281	679	161	32	32	0.78
1957	2278	1317	714	178	34	35	0.80
1958	2338	1344	732	192	36	35	0.80
1959	2471	1390	790	214	40	36	0.83
1960	2586	1419	850	235	43	39	0.86
1961	2602	1356	905	254	45	42	0.85
1962	2708	1358	981	277	49	44	0.86
1963	2855	1404	1053	300	51	47	0.89
1964	3016	1442	1138	328	57	51	0.92
1965	3154	1468	1221	351	59	55	0.95
1966	3314	1485	1325	380	63	60	0.97
1967	3420	1455	1424	410	65	66	0.98
1968	3596	1456	1552	445	70	73	1.01
1969	3809	1494	1674	487	74	80	1.05
1970	4091	1571	1838	516	78	87	1.11
1971	4242	1571	1946	554	84	88	1.12
1972	4409	1587	2056	583	89	94	1.15
1973	4648	1594	2240	608	95	110	1.18
1974	4656	1591	2244	618	96	107	1.16
1975	4629	1686	2131	623	95	93	1.13
1976	4895	1723	2313	647	103	109	1.18
1977	5034	1786	2390	646	108	104	1.19
1978	5082	1802	2383	674	116	107	1.18
1979	5366	1899	2535	714	119	100	1.23
1980	5264	1921	2409	725	120	89	1.18
1981	5129	1930	2272	735	121	72	1.13
1982	5094	1993	2178	734	121	69	1.11
1983	5085	1998	2163	736	125	63	1.09
1984	5243	2088	2186	785	128	57	1.10
1985	5369	2196	2169	819	131	55	1.11
1986	5551	2253	2274	836	136	52	1.12
1987	5661	2310	2285	876	142	48	1.13
1988	5897	2388	2388	916	150	55	1.15
1989	5967	2392	2419	946	152	56	1.15

Table 1: Global total emissions of CO_2 from fossil-fuel combustion and cement manufacture (1 ton = 10^3 kg).

mankind, CO_2 is an essential and necessary product. To estimate the global total of CO_2 emissions from fossil fuel burning we need to know only the amount of fuel produced, its carbon content, and the amount of fuel that was produced but not oxidized. (In any time interval the amount of fuel oxidized will fall short of the amount produced because of losses in handling, increases in stockpiles, inefficiencies at the point of combustion and because some fossil fuels are used to produce non-fuel products like plastics, lubricants, and road surfacing.) If we illustrate the combustion process with methane, the primary component of natural gas, we can see that the relationship between carbon dioxide emission and energy release depends on the carbon to hydrogen ratio of the fuel.

$$CH_4 + 2O_2 \rightarrow CO_2 + 2H_2O \qquad (1)$$

For broad categories of fuels, however, there is a strong relationship between carbon content and energy content. Within these categories we can get a good estimate of the carbon content of produced fuel if we tabulate fuel production in units of energy content and use coefficients for carbon per unit of energy content. Thus, Winschel (1990) has shown that a wide variety of coals can be represented (within ± 3%) with a single coefficient for kg $C/10^9j$. Although a slightly larger coefficient appears to be appropriate for low-grade, brown coals, our data set uses a single factor for converting coal production in "tons coal equivalent" to tons of carbon, where "tons of coal equivalent" is defined in terms of the energy content of the coal (1 tce ≡ 29.3076×10^9j). Marland and Rotty (1984) have shown that there is a similar relationship that is suitable for the bulk of world natural gases. For liquid fuels we have taken a global average for the mass fraction of crude oil which is carbon and find that there is little error in assuming that all liquid fuels are 85% carbon by mass (see Table 2). One caution is that the energy content of fuels, i.e. the "heating value", is sometimes represented in published data by the higher heating value, where

Fuel	Higher Heating Value (Mj/kg)	Carbon Content (kg C/kg Fuel)	Carbon Coefficient (kg C/10^9 joules)
Coal	29.31 (1)	.707 (1)	24.12
Natural gas			13.78 (1,2)
Crude oil	42.62 (2)	.850 (2)	19.94
Kerosine (also jet fuel)	46.5 (3)	.858 (3)	18.45
Gasoline	47.2 (4)	.869 (3)	18.41
Diesel Fuel (also distillate fuel oil)	45.7 (3)	.865 (3)	18.93
Liquid Petroleum Gas (LPG)	50.0 (6)	.818 (6)	16.36
Residual Fuel Oil (#6 fuel oil)	42.5 (4)	.873 (5)	20.53
Light Fuel Oil (#4 fuel oil)	43.9 (4)	.865 (5)	19.70

Data Sources: (1) Marland and Rotty (1984), (2) Marland (1983), (3) Rose and Cooper (1977), (4) Heisler (1984), (5) Perry and Chilton (1973), (6) Ho and Renner (1989).

Table 2: Coefficients for conversion of fuel consumption data in energy or mass units to CO_2 emissions as mass of carbon.

the product water in equation 1 above is as a liquid, and sometimes by the lower heating value, where the product water is in the vapor phase. It is important to assure that the conversion coefficients are in the same units as the production values.

One of the most difficult tasks in refining estimates of CO_2 emissions is establishing the amount and fate of carbon in fossil fuels which is not oxidized by combustion. As mentioned earlier, a small amount of carbon exits the combustion process as unoxidized soot or ash and petroleum has a large number of non-fuel applications. The magnitude of this unoxidized fraction can be closely bounded and will not have a large impact on the global total of emissions, but can be responsible for differences of a few percent in CO_2 emissions estimates. The problem is three-fold: estimating how much of the fossil fuel is not promptly burned, estimating the lifetime of unoxidized products in the environment, and estimating how these factors differ among countries and have changed over the 1950 to 1989 time interval. Our estimates assume that these factors have not changed with time and then incorporate some simplifying assumptions based largely on experience in the U.S. and Western Europe. The rationale and results as they affect estimates of global total emissions are contained in Marland and Rotty (1984). With an awareness that use of fossil fuels for non-fuel purposes can be very different in different countries, we have used global average estimates of unoxidized carbon to calculate global total emissions of CO_2 but have taken a somewhat different approach when calculating emissions by country. In estimating CO_2 emissions by country we assume simply that if the U.N. energy data set lists an application as a non-fuel use (white spirits, lubricants, asphalt, petroleum waxes, etc.), then the carbon is not oxidized. We know that some of these products will be oxidized promptly, that more will be oxidized over time, and that the country estimates of CO_2 emissions are thus short by some small but variable amount. Also, since our global total of emissions tries to make allowance for this oxidation whereas the country estimates do not, the sum of all country emissions will be less than the global total. This is a topic to which we will return and provide some quantitative perspective under the discussion of bunker fuels below.

Making CO_2 emissions estimates for nations, or any other geographically bounded regions, is more complex and thus more uncertain than is coming up with global totals of emissions. Whereas the global total requires that we know only the amount of fuel produced and the fraction not oxidized, to get country totals we need data on international trade. This is trade in not only the primary fuels, but in all of the secondary fuel products as well. Also, while net annual changes in fuel storage are believed not to be a significant factor on the global scale, they could be important at the national level. National estimates of CO_2 emissions are thus based on what the United Nations statistical office calls "apparent consumption", production plus imports minus the sum of exports, stock changes, fuels loaded for use in international commerce (bunker fuels), and production of secondary products that are to be used for non-fuel purposes. The demand on international data is much greater. Exclusion of bunker fuels from national totals is rationalized on geochemical grounds, i.e. that the actual discharge of CO_2 occurs not within the geographic confines of the country where the final fuel transaction actually occurs, but along the corridors of international commerce. On political criteria it is not immediately obvious how one would choose to assign responsibility for fuels used to transport goods and passengers on international routes.

There are several potential sources of data on the production and trade of fuels. We haven chosen to work with the energy data set from the United Nations statistical office (U.N., 1991) because it provides a reasonably complete and homogeneous data set which extends, for most categories of interest, from 1950 to the present. The data series is revised whenever new data become available and the full data set is updated annually on magnetic tape. We have worked with the U.N. statistical office for several years and each year perform our own quality assurance tests on the completeness, reasonableness, and accuracy of the full data set. Of course, errors in basic numeric values often cannot be discovered by this kind of quality assurance analysis. Not surprisingly for such a large data set, we have periodically identified erroneous, missing, or suspect data; but our experience has been that the problems are seldom of sufficient magnitude to significantly affect estimates of global total CO_2 emissions and that the U.N. is responsive in trying to rectify problems that we discover. We are undoubtedly the most intense user of the U.N. energy data set.

According to their formal documentation, the U.N. energy data set is, "compiled primarily from annual questionnaires...and supplemented by official national statistical publications. Where official data are not available or are inconsistent, estimates are made...based on governmental, professional or commercial materials." The U.N. statistical office is not obligated to simply compile data submitted by member nations but calls on a variety of data sources and its own judgement. There is no formalism to obligate nations to respond to the questionnaires and comprehensive reporting, especially for developing countries, is often not achieved. The data set is more complete for years since 1970 than for prior years, but production and trade statistics form the core of the data set and are reasonably complete to 1950. Data on patterns of energy end use are often not complete prior to 1970 (and frequently since 1970 as well) and clearly have lower priority than the production and trade data. Because of the quantity of data and the way in which it is compiled, it is typically 15 to 18 months from the end of a calendar year before the full data set is released. In the basic data file, the U.N. reports quantities of solid and liquid fuels in mass units and quantities of gaseous fuels in energy units. A second file provides detailed conversion factors, by country and year, to convert data on solid fuels from mass to energy units.

There are other energy data sets which might be chosen, but none are as complete or provide as long a consistent time series as does the U.N. data set. Also, they are never truly independent sources of information. We are in the process now of doing detailed comparisons between the U.N. and International Energy Agency (OECD/IEA) data sets with the recognition that the two agencies often use identical inputs but have different internal mechanisms of data interpretation and management.

With this brief summary of the data and philosophy behind the CO_2 emissions estimates, we can look at some of the features of the data which provide insight into data quality. In this context, it is our feeling that the greatest uncertainty in the emissions estimates is inherent in the global data set on production and trade. Considering the uncertainties in these data, one could quickly reach a point of diminishing returns by investing excessively in refining factors for fuel chemistry and combustion efficiency. On the other hand, it is likely that the trends of consumption as a function of time are more accurate than the individual values and we may loose real information by failing to recognize differences in fuel chemistry or conversion efficiency. Also, the nature of the data reporting and collection process suggests to us that fuel production is more likely to be under-reported than over-reported. The question is whether small producers, producers for local markets, and producers in non-market-based economies are fully represented. The consequence of this is that the uncertainty bounds are not symmetric and we suggest that our emissions estimates are more likely on the low side than on the high side of the true values.

Considering the differences in the way global and national emissions are calculated, it is interesting to compare the sum of CO_2 emissions for all nations with the global total. Because there is no coal used for bunker fuels now (during the early part of our time series there was some small usage of coal for bunker fuels), there are three reasons that the sum of emissions from solid fuels for all countries might differ from the global total: 1.) there is a net change in the amount of coal in stockpiles, 2.) the national totals do not attempt to account for the ultimate oxidation of non-fuel products, and 3.) because of accounting difficulties the sum of imports for all countries is not the same as the sum of exports. The latter factor can give us an interesting measure because our inclination is that the fuel data should be best where there is a buyer and a seller and hence two parties interested in accounting for the transaction. For liquid fuels there is an additional factor that can cause the sum of emissions from all countries to differ from the global total: bunker fuel use is included in the global total but is not a part of the national totals. Table 3 shows that for solid fuels there is a good match between the sum of national values and the global total. About 11% of solid fuel production is traded internationally and the sum of all imports has exceeded the sum of all exports by an average of 7 % over the last 4 years. Figure 2 shows for liquid fuels the global total of CO_2 emissions and the sum of emissions for all countries. The global total always exceeds the sum of all countries and Table 3 shows that in 1989 62% of the difference is accounted for by failure to assign bunker fuel consumption to any nation. Some 44% of world crude oil production is now traded internationally and additional trade occurs as secondary petroleum products. For crude oil, over the last 4 years, the total of imports in the U.N. data set has exceeded the total of exports by about 5%.

CO_2 Emissions (10^6 metric tons C)

	Global Total					Sum of All Countries			
Year	Total	Gas Fuels	Liquid Fuels	Solid Fuels	Bunker Fuels	Total	Gas Fuels	Liquid Fuels	Solid Fuels
1950	1638	97	423	1077	35	1590	96	385	1068
1951	1775	115	479	1137	42	1710	113	428	1124
1952	1803	124	504	1127	43	1733	121	453	1111
1953	1848	131	533	1132	44	1783	129	482	1120
1954	1871	138	557	1123	45	1823	136	514	1119
1955	2050	150	625	1215	51	1993	148	573	1212
1956	2185	161	679	1281	55	2120	159	622	1274
1957	2278	178	714	1317	61	2189	175	647	1298
1958	2338	192	732	1344	57	2260	191	681	1318
1959	2471	214	790	1390	58	2389	213	728	1372
1960	2586	235	850	1419	65	2520	234	781	1423
1961	2602	254	905	1356	72	2527	251	825	1363
1962	2708	277	981	1358	75	2619	275	892	1359
1963	2855	300	1053	1404	78	2775	297	966	1413
1964	3016	328	1138	1442	86	2910	324	1040	1439
1965	3154	351	1221	1468	89	3042	347	1119	1463
1966	3314	380	1325	1485	96	3185	379	1207	1476
1967	3420	410	1424	1455	103	3280	406	1291	1452
1968	3596	445	1552	1456	109	3465	443	1411	1467
1969	3809	487	1674	1494	114	3691	483	1532	1521
1970	4091	516	1838	1571	120	3879	508	1628	1578
1971	4242	554	1946	1571	127	4022	548	1728	1575
1972	4409	583	2056	1587	136	4202	580	1865	1572
1973	4648	608	2240	1594	144	4425	600	2002	1618
1974	4656	618	2244	1591	137	4408	612	1974	1620
1975	4629	623	2131	1686	122	4415	614	1958	1655
1976	4895	647	2313	1723	122	4687	647	2101	1727
1977	5034	646	2390	1786	124	4806	637	2177	1779
1978	5082	674	2383	1802	127	4976	670	2259	1824
1979	5366	714	2535	1899	129	5117	707	2297	1895
1980	5264	725	2409	1921	128	5060	722	2208	1922
1981	5129	735	2272	1930	120	4957	725	2108	1932
1982	5094	734	2178	1993	111	4936	725	2155	1966
1983	5085	736	2163	1998	105	4983	739	2035	2022
1984	5243	785	2186	2088	102	5124	787	2044	2107
1985	5369	819	2169	2196	103	5261	820	2030	2226
1986	5551	836	2274	2253	113	5353	825	2089	2251
1987	5661	876	2285	2310	121	5491	868	2113	2320
1988	5897	916	2388	2388	124	5694	910	2170	2409
1989	5967	946	2419	2392	126	5792	942	2216	2425

Table 3: Global total emissions of CO_2 are compared, by fuel, with the sum of emissions from all countries. Emissions from bunker fuels, shown separately, are included in the global total of emissions but are not included in the sum of emissions by country. For cement manufacturing and gas flaring the sum of all countries is identical to the global total except as described in Table 6 and accompanying text. The totals shown do include emissions from cement manufacturing and gas flaring from Table 1.

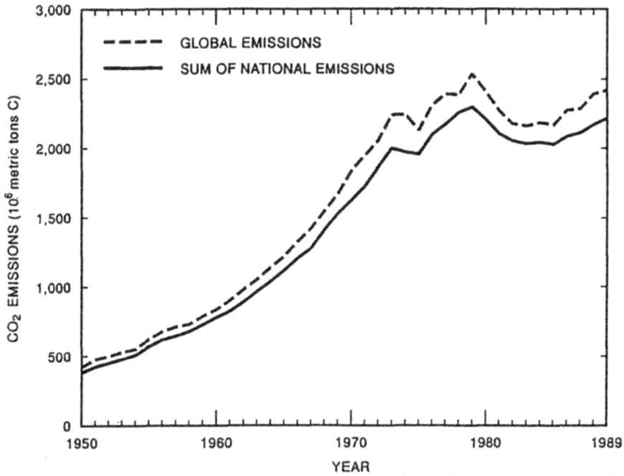

Figure 2: The global total of CO_2 emissions from liquid fuels is compared with the sum of emissions from liquid fuels from all countries. The difference is largely because emissions from fuels used in international commerce are included in the global total but not in the national totals, and because emissions from the oxidation of petroleum products used for non-fuel purposes are treated differently in the national and global accounts. See text for discussion.

Having noted that the U.N. energy data set is continually open to revision and updating, it is interesting to look at the magnitude of revisions that actually occur. In general the release of data for a given year is accompanied by a very large number of updates to data for the last previous year. The number of revisions declines rapidly with time but changes to the early years of the data set are not uncommon. Revisions of older data are usually revisions to an entire time series prompted by some new access to information. In Table 4 we show the global total of CO_2 emissions for a series of years and how those estimates have been altered by subsequent annual revisions of the underlying data. In Table 5 we have taken the U.N. data tape which includes energy data through 1988 (issued in 1990) and compared resulting CO_2 emissions estimates with calculations based on the tape issued in 1991 and including data for year 1989. Only for Africa and Eastern Europe were revisions large enough to alter emissions estimates by as much as 1 million tons of carbon for years prior to 1985. Revisions larger than 5 million tons of carbon occurred in only 4 of the 9 regions and only for the most recent years. Changes in the oldest data arose from revisions in gas production data from Romania and the USSR and from petroleum import data for Senegal. The revisions of the Senegal data series are of particular interest to us because they were in response to our reporting of data problems. Calculation of CO_2 emissions from the 1990 data tape (Figure 3a) produced negative values for a series of years in the late 1960s and subsequent review of the underlying data suggested that data on petroleum imports were missing for those years. The Senegal data series was subsequently revised and Figure 3b shows the CO_2 emissions curve based on the 1991 data tape. It is clear that the U.N. statistics office addressed but did not solve the data problem and that the changes made were just large enough to show up in the regional totals reported in millions of tons carbon. Due to rounding and to compensating changes in the Romanian gas data, the change is not reflected in the estimate of global total CO_2 emissions.

Emissions For Year	Year Of Publication 1973	1984	1985	1987	1988	1989a	1989b	1990
1950	1632	1638	1638	1639	1639	1639	1638	1638
1960	2611	2597	2597	2586	2586	2586	2586	2586
1970		4092	4092	4116	4090	4090	4090	4090
1980		5288	5288	5255	5249	5249	5263	5263
1984				5330	5238	5238	5263	5260
1985					5338	5338	5382	5379
1986						5555	5565	5561
1987							5650	5680
1988								5893

Data Sources: 1973 = Keeling (1973), 1984 = Marland and Rotty (1984), 1985 = Rotty and Masters (1985), 1987 = Rotty (1987), 1988 = Marland et al. (1989), 1989a = Marland (1989), 1989b = Marland and Boden (1989), 1990 = Marland (1990).

Table 4: Effect of annual revisions of the U.N. energy data set on the estimates of global total CO_2 emissions. The rows in the table show estimates of emissions for the year given at the head of the row as revised in subsequent years (columns) as the U.N. released revised versions of the data set. The two columns of values for 1989 reflect a midyear-revision in the way we have treated data for gas flaring.

Data for emissions from Iran show another kind of problem in the data set. The time series of CO_2 emissions yields a small negative value for 1950. Detailed examination of the underlying data suggests that the problem is one of small differences between large numbers. Apparent fuel consumption in Iran in 1950 was essentially the difference between production and exports, both very large numbers. In this case the error in global total CO_2 emissions is very small although the allocation to Iran is clearly not reasonable because of small errors in large numbers. Detailed examination of CO_2 emissions data from Switzerland, for example, reveals an instance where a negative emissions number appears erroneous but is ultimately a correct reflection of our calculation strategy. If coal is used to produce a secondary, gaseous, fuel and the gas is then exported, data for the country of origin will show all of the CO_2 emissions under solid fuels and a small negative number will appear under gas fuels to indicate the export of a secondary product.

	Most Distant Year That CO_2 Emissions Estimate Changed		
Region	by 1 million metric tons C	by more than 1 million metric tons C	by 5 million metric tons C
AFR	1965	1987	-----
AMD	1986	1988	-----
CPA	1985	1988	-----
CPE	1955	1984	1984
FEA	1985	1986	1988
MDE	1987	1987	1987
NAM	1987	1987	-----
OCN	1985	-----	-----
WEU	1986	1986	1988

Table 5: The U.N. energy data tape issued in 1991 has been compared with the energy data tape issued in 1990. The question posed was how far back in the data set does one observe significant changes in estimated CO_2 emissions as a result of revisions in the energy data set. To get more perspective on the magnitude and distribution of changes, the question was posed separately for each of 9 geographic regions (AFR=Africa, AMD=America developing, CPA=Centrally-planned Asia, CPE=Eastern Europe and the USSR, FEA=Far East, MDE=Middle East, NAM=US and Canada, OCN=Oceania, and WEU=Western Europe). A dash indicates that even for the most recent year (1988) there were no changes of this magnitude.

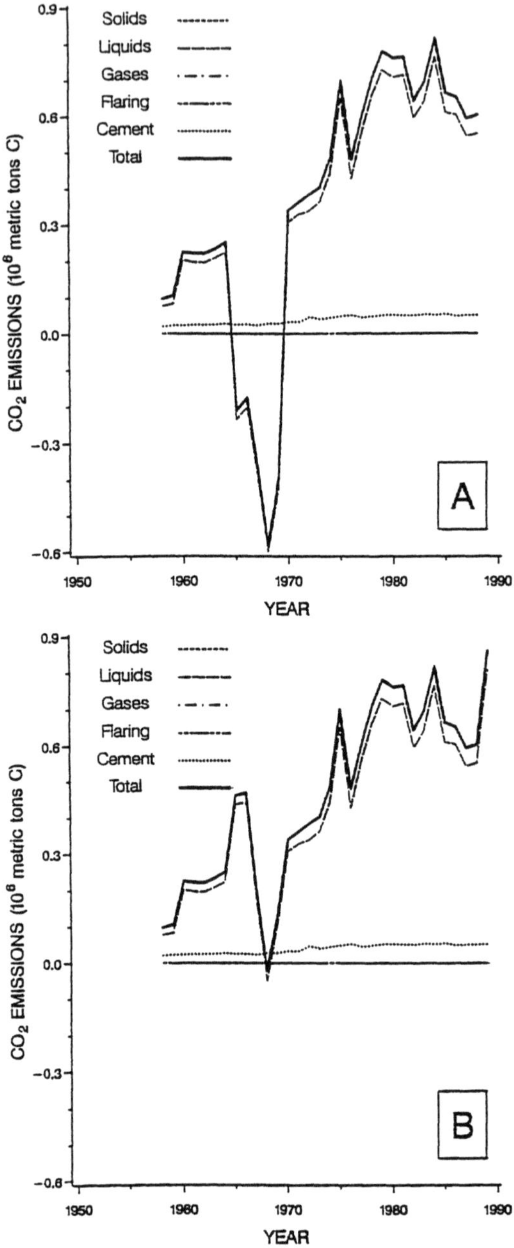

Figure 3: The calculation of CO_2 emissions form Senegal reflects revisions in the U.N. energy data set from the data tape issued in 1990 (Figure 3A) to the data tape issued in 1991 (Figure 3B).

There are two additional components of the CO_2 emissions data set which have not yet been addressed; CO_2 is released during the manufacture of cement and there is some CO_2 produced when gas is flared at oil fields. A comparison of data sets suggests to us that the most complete data set on cement manufacture is that available from the U.S. Bureau of Mines (1990). The CO_2 is released during the calcining of calcium carbonate to produce calcium oxide and amounts to about 2% of global CO_2 emissions from fossil fuels.

Emissions from gas flaring are more difficult to document because there has been no economic incentive to inventory such losses and hence the data are very poor. As recently as the late 1970s, 2/3 of gas production in the Middle East was flared and in Nigeria 90% of production was flared. Flaring has dropped dramatically over the last 15 years and the time series of emissions for some countries is vastly different depending on whether or not emissions from gas flaring are included (Figure 4). Review of the U.N. energy data set provides at least some gas flaring data for 46 countries but no data for the U.S.S.R., for example, which is currently the largest source of gas flaring. Only 4 countries show non-zero values for gas flaring in 1950 although this jumps to 38 countries in 1970 (see Table 6). The gas flaring data set maintained by the U.S. Department of Energy (1991) includes some country coverage not provided in the U.N. data set but is deficient in a great lapse of data from the 1979-1982 interval when the data responsibility was transferred from the Bureau of Mines. In an effort to assemble the most complete data set possible, we have adopted the U.N. data set, supplemented it with some data from the U.S. D.O.E., and interpolated a few missing values, especially for the U.S.S.R. Rotty (1974) created a data set for global-total gas flaring prior to 1970 and we have retained his totals even though all of the emissions are not assigned to specific countries. In Table 6 the first two columns show the number of countries with non-zero values for gas flaring in the U.N. energy data set and the total of gas flared from these countries. The last column in the table shows the global total of gas flaring when our supplemental data is added.

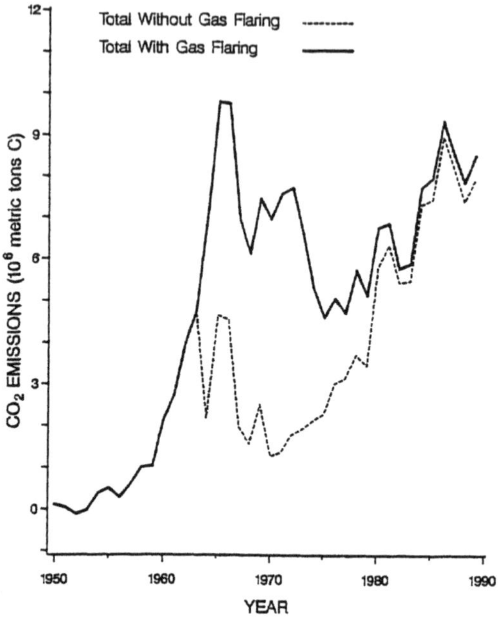

Figure 4: Data from Kuwait illustrate that the time series of CO_2 emissions from some countries can appear quite different in gross form, depending on whether or not emissions from flaring of natural gas are included.

Another interesting qualitative insight on data quality can be gleaned by looking at per capita CO_2 emissions by country. To permit this kind of review our background data file now includes the U.N. population statistics. Table 7 shows the 15 nations with the highest calculated per capita emissions of CO_2 for 1989. The table suggests that for small countries and/or small economies, minor accounting errors, perhaps incomplete

Year	Country Sum of Gas Flaring From U.N. Data Set ($10^9 m^3$)	Number of Countries With Non-zero Gas Flaring in U.N. Data Set	Total Global Gas Flaring This Analysis ($10^9 m^3$)
1950	38	4	44
1955	50	7	58
1960	46	8	75
1965	68	15	104
1970	144	38	167
1975	152	39	177
1980	151	40	169
1985	93	38	104
1988	85	40	105

Table 6: The U.N. energy data tape contains some data on gas flaring for 46 countries but these data have been supplemented by data from other sources in order to estimate the total of CO_2 emissions from gas flaring. Shown here is the total quantity of gas flared for selected years as reported in the U.N. data set, the number of countries which contribute non-zero values to this sum, and the total value used for our CO_2 emissions data set after the insertion of supplemental data.

Country	1989 Per Capita CO_2 Emissions (tons C/person)	Comment on Source of CO_2
Wake Island	10.9	100% liquids
Qatar	10.3	79% natural gas
United Arab Emirates	9.0	7% flaring, 61% natural gas
U.S. Virgin Islands	8.0	100% liquids
Luxembourg	6.8	
Bahrain	6.7	79% liquids
Netherland Antilles	6.2	100% liquids
Falkland Islands	5.6	73% liquids
GDR	5.4	
U.S.	5.4	
Brunei	4.7	91% natural gas
Canada	4.7	
Kuwait	4.3	56% natural gas, 7% gas flaring
Australia	4.2	
St. Pierre & Miquelon	4.1	100% liquids
Global Average	1.15	

Table 7: The 15 nations with highest estimated per-capita emissions of CO_2 from inland fuel use and cement manufacturing in 1989.

identification of bunker fuels as opposed to inland fuel consumption, can lead to significant distortion of the national representation. We can't tell easily whether the extreme values of Table 7 imply data problems or something unusual about the structure of the economy, and this almost surely differs for the different countries. Very high per capita values for the former German Democratic Republic are probably a consequence of great inefficiency in energy use.

We are left, ultimately, with questions on both the global total of CO_2 emissions and on the geographic distribution of those emissions. Both depend on the quality of national statistics on energy production and trade. The data set for 1989 includes emissions data from 194 nations but it is worth noting that when all nations are arranged in descending order of CO_2 emissions, the sum of emissions from the last 120 nations is smaller than the emissions from the 20th largest alone. Figure 5 shows a cumulative frequency plot for the 20 largest CO_2 - emitting nations and reveals that 84% of total emissions come from these 20 nations. The concentration of emissions sources was even more extreme in 1950 when the top 20 nations accounted for 93% of the total. The answer is, thus, that although there are many difficulties with the international data sets, our uncertainty of the total of CO_2 emissions is dominated by the quality of data from the top 20 orso nations. An order of magnitude error for any nation below the top 20 would scarcely affect the global total but a modest percentage error in data from the USA, USSR, or China could have an important effect on the total. With increasing interest in national totals of emissions we are witnessing the appearance of national estimates from government or corporate sources within a number of countries. We are currently in the process of trying to compare these estimates with our own and to understand the differences. Most values published to date do indeed come within 10% of our values, and are usually closer. Again, one should not conclude that these are independent values because many rely on the same sources of data, but they do raise confidence in the broad issues of data handling and synthesis.

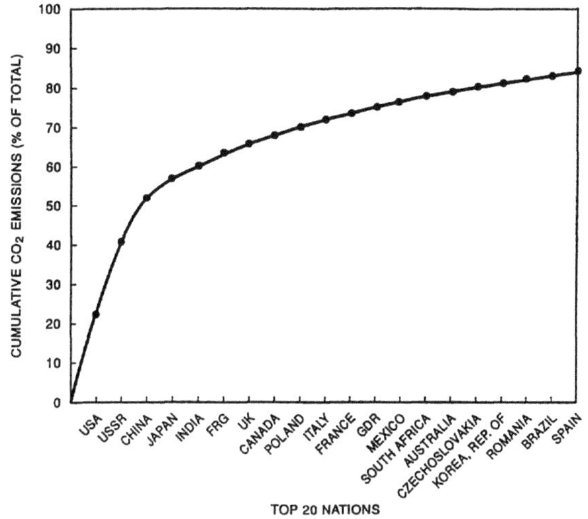

Figure 5: The cumulative amount of CO_2 emissions from the 20 largest CO_2 - emitting nations in 1989 plotted as the fraction of emissions from all countries.

Attempts to allocate CO_2 emissions within countries create additional problems. Many nations are quite compact, especially on a 5 x 5 degree grid spacing, and misallocation within the country can result in only a small displacement of the emissions source. For some of the larger nations, i.e. the U.S. and Canada, data are easily available at the level of states or provinces and we can reduce the problem to that level. For some other large countries it is necessary to use surrogate measures in order to estimate the internal distribution of CO_2 emissions. An obvious choice, and the one used in our earlier paper (Marland et al., 1985), is to use the internal distribution of population as a surrogate for the distribution of energy use. An obvious problem in the current situation is that to extend the distribution of population and energy use to represent CO_2 emissions has the effect of assuming that the fuel mix is the same in all portions of the nation. We know, for example, that in the U.S. hydroelectric power is widely used in the Northwest while natural gas is most extensively used along the Gulf Coast and coal in the Midwest. We are pursuing data sources and collaborators that would permit us to do better, but the population-based distribution provides an initial estimate.

We close with brief comment on an important difference between a geochemical and a geopolitical data set of CO_2 emissions. This concerns the issue of responsibility which was raised above during the discussion of bunker fuels. As an illustration, there are now 2 published inventories of CO_2 emissions from the state of California. An environmental group (Renew America, 1988) has estimated emissions which actually occur from within the state of California. The California Energy Commission (1990), on the other hand, has estimated what they termed emissions "attributable to California economic activity". The latter includes emissions from electric power plants located elsewhere if the electricity is used in California and from petroleum refineries if the refined products are for use in California. When all else is adjusted for, the difference on this account is on the order of 16%. Interestingly, the California Energy Commission did not broach the idea of accounting for energy, and CO_2 emissions, embodied in goods and services outside of the energy sector. The point is that the CO_2 emissions physically discharged from within a nation are of interest to the geochemist but they provide only a first approximation of the emissions for which the nation might be held accountable in a geopolitical context. In this same vein, one might speculate whether there could be politically-motivated misrepresentations in some of the national data reported to the U.N. A 4-year "high" in Chinese coal production data around 1960 comes to mind as a possible candidate, but, in general, we are aware of no evidence for such systematic misrepresentation.

The set of analyses and comparisons summarized here does not provide a rigorous error analysis for the data set of CO_2 emissions from fossil-fuel burning. It does provide some insight into the nature and quality of the available data and emphasizes that the accuracy of the global total emissions data set hinges on the accuracy of fuel production and trade statistics from a small number of countries. Emissions estimates reported to four "significant" figures should be used with a measure of skepticism. These analyses, along with our discussions with people responsible for the basic data sets, lead us to continue to believe that ± 10% is probably an appropriate uncertainty bound for the estimates of global total CO_2 emissions, although the uncertainty is asymmetric and we suggest the results are more likely on the low-side than on the high-side of the true values. Emissions estimates for particular countries have greater uncertainty, especially for countries with weak systems for data collection and processing.

In closing, we offer an initial estimate of global total CO_2 emissions for 1990. The British Petroleum Company issues an annual data set on world energy consumption (British Petroleum, 1991). It has less detail than the U.N. data set but is released months earlier. Using the British Petroleum energy consumption data for 1989 and 1990 to extrapolate our 1989 CO_2 emissions value, it appears that CO_2 emissions for 1990 were marginally lower than in 1989. We estimate total emissions at 5.955 Gt C in 1990. The decline appears largely due to economic disruption in Eastern Europe and the USSR and should be expected again in 1991.

REFERENCES

British Petroleum Company (1991) BP Statistical Review of World Energy, London.
California Energy Commission (1990) 1988 Inventory of California greenhouse gas emissions. California Energy Commission, Sacramento, California
Heisler SI (1984) Wiley Engineer's Desk Reference. J. Wiley, New York, p. 413.
Ho SP and Renner TA (1989) The global warming impact of attainment strategies using alternative fuels. Amoco Oil Company Research and Development Department, Naperville, Illinois.
Keeling CD (1973) Industrial production of carbon dioxide from fossil fuels and limestone. Tellus 35:174-198.
Marland G (1983) Carbon dioxide emission rates for conventional and synthetic fuels. Energy 8:981-992.
Marland G (1989) Fossil Fuels CO_2 Emissions: Three Countries Account for 50% in 1986, p. 1-3, CDIAC Communications, Winter 1989, Carbon Dioxide Information Analysis Center, Oak Ridge National Laboratory, Oak Ridge, Tn.
Marland G (1990) Global CO_2 Emissions. In: Boden TA, Kanciruk P, and Farrell MP, (eds.) Trends '90: A compendium of data on global change, ORNL/CDIAC-36, Carbon Dioxide Information Analysis Center, Oak Ridge National Laboratory, Oak Ridge, Tn., p. 93.

Marland G, Boden TA, Griffin RC, Huang SF, Kanciruk P, and Nelson TR (1989) Estimates of CO_2 emissions from fossil-fuel burning and cement manufacturing, based on the United Nations Energy Statistics and the U.S. Bureau of Mines Cement Manufacturing Data. ORNL/CDIAC-25, NDP-030, Carbon Dioxide Information Analysis Center, Oak Ridge National Laboratory, Oak Ridge, Tn.

Marland G, and Rotty RM (1984) Carbon dioxide emissions from fossil fuels: a procedure for estimation and results for 1950-1982. Tellus 36B:232-261.

Marland G, Rotty RM, and Treat NL (1985) CO_2 from fossil fuel burning: global distribution of emissions. Tellus 37B: 243-258.

Marland G, and Boden TA (1989) Carbon dioxide releases from fossil-fuel burning, testimony before the Senate Committee on energy and natural resources. July 26, 1989, p. 62-84, S. Hrg. 101-235, DOE's National Energy Plan and Global Warming, U.S. Senate, U.S. Government Printing Office, Washington, D.C.

Perry RH and Chilton CH (1973) Chemical Engineer's Handbook. 5th edition, section 9, p. 10, McGraw Hill, New York.

Renew America (1988) Reducing the rate of global warming; the states' role. Prepared by Machado S, and Piltz R, for Renew America, Washington, D.C.

Rose JW and Cooper JR (1977) Technical data on fuel. The British National Committee, World Energy Conference, London.

Rotty RM (1974) First estimation of global flaring of natural Gas. Atmospheric Environment 15:73-86.

Rotty RM (1987) A look at 1983 CO_2 emissions from fossil fuels (with preliminary data for 1984). Tellus 39B:203-208.

Rotty RM and Masters CD (1985) Carbon dioxide from fossil fuel combustion: Trends, Resources, and Technological Implications. In: Trabalka JR (ed.), Atmospheric Carbon Dioxide and the Global Carbon Cycle, DOE/ER-0239, U.S. Department of Energy, Office of Energy Research, Washington, D.C.

United Nations (1991) Energy Statistics Yearbook (1989). United Nations Department of International Economic and Social Affairs, Statistical Office, New York. (And supporting data tape).

U.S. Bureau of Mines (1990) Minerals Yearbook, Vol. 1, Metals, Minerals, and Fuels. Washington, D.C. (and preceding yearbooks).

U.S. Department of Energy (1991) 1989 International Energy Annual. DOE/EIA-0219(89), Energy Information Administration, Washington, D.C. (and preceding volumes.)

Winschel RA (1990) The relationship of carbon dioxide emissions with coal rank and sulfur content. J. Air and Waste Management Association 40:861-865.

CHANGES IN TERRESTRIAL CARBON OVER THE LAST 135 YEARS

R.A. Houghton
The Woods Hole Research Center
P.O. Box 296
Woods Hole, MA 02543 USA

Introduction

About 2100 Pg of organic C are held in the earth's terrestrial ecosystems: 550 Pg in living plants and 1500 Pg in the organic matter of soils (Houghton et al. 1985b). Because this terrestrial reservoir of carbon is about 3 times greater than the atmospheric reservoir of carbon, and because the annual exchanges of carbon between terrestrial ecosystems and the atmosphere are large (100-120 PgC), changes in terrestrial ecosystems are important in affecting the atmospheric concentration of CO_2. Attempts to estimate such changes have been underway for more than a decade (Woodwell and Pecan 1973, Bolin 1977, Woodwell and Houghton 1977, Woodwell et al. 1978).

The amount of carbon held in terrestrial ecosystems may be changed as a result of (1) direct human effects on land use, such as conversion of forests to agricultural lands, and (2) natural variation in, and indirect human effects on, the environment (such as eutrophication, increased concentrations of atmospheric CO_2, or changes in climate) that affect the amount of carbon per unit area of land.

The flux of carbon to or from the atmosphere from changes in land use can be determined with more confidence than the flux resulting from changes in the mass of carbon per unit area where no change in land use has occurred. This is so because of the large difference in carbon stocks between forests and the agricultural ecosystems that replace them (forests hold 20 to 100 times more carbon per unit area), and because changes in agricultural lands and rates of deforestation and reforestation are, to some extent, measured and recorded. Most estimates of the flux of carbon between terrestrial

NATO ASI Series, Vol. I 15
The Global Carbon Cycle
Edited by M. Heimann

ecosystems and the atmosphere have been based on these changes in land use. In contrast, changes in the density of carbon within ecosystems are likely to be small relative to the amounts of carbon held in vegetation and soil, and, hence, difficult to measure. At the same time, small changes per unit area, if spread over a large portion of the earth, may be large enough to affect significantly the atmospheric concentration of CO_2. For example, a 1% change per year in the stocks of carbon in the world's terrestrial ecosystems would yield a flux of about 21 Pg C/yr, almost four times the annual emission of carbon to the atmosphere from worldwide combustion of fossil fuels; yet these stocks of carbon are thought to be known within only $\pm$ 20% (Houghton et al. 1985b).

Two Estimates of a Terrestrial Carbon Flux

The conversion of forests to croplands and pastures releases carbon to the atmosphere either rapidly, as trees are burned, or more slowly, through decomposition of plant material and soil organic matter. Conversely, the regrowth of forests on abandoned agricultural lands withdraws carbon from the atmosphere and stores it again on land. Recent estimates of the global flux of carbon from terrestrial ecosystems in 1980 as a result of deforestation and reforestation range between 0.5 and 2.5 PgC/yr (Detwiler and Hall 1988, Hall and Uhlig 1991, Houghton et al. 1985a, Houghton 1991b). This entire flux was from the tropics; changes in land use outside the tropics are thought to have contributed a net flux near zero in 1980 (Houghton et al. 1987, Melillo et al. 1988). In the 10 years since 1980, increases in the rates of tropical deforestation (Myers 1989; FAO 1990) suggest that the release of carbon from these changes is currently near the higher end of this range (Houghton 1991b).

Early attempts to calculate the cumulative flux of carbon to the atmosphere from changes in land use yielded estimates between 62 PgC and 228 PgC for the 120-year period 1860 to 1980 (Houghton et al. 1983, Richards et al. 1983) (Table 1). Subsequent analyses have included more types of land use, incorporated better estimates of the ecosystems affected and, hence, the stocks of carbon involved, and revised the estimated loss of carbon from soils. The modifications led to a new estimate of the total net release of carbon from changes in land use that is between about 100 and 130 PgC for the period 1850 to 1985 (Fig. 1). The higher estimate is preferred because it includes a net release of carbon from degradation of forests as well as from outright deforestation (Houghton 1991a). The new estimates are intermediate between earlier estimates (Table 1). This flux of carbon from changes in land use is less than, but of the same order of magnitude as, the emission of carbon from fossil fuels (about 190 PgC by 1985).

Table 1. Different estimates of the long-term net release of carbon from terrestrial ecosystems.

Interval	Method	Net Release (PgC)	Reference
1860-1978	Changes in cultivated area	62	Richards et al. 1983
1860-1980	Changes in land use	135-228	Houghton et al. 1983
1850-1985	Changes in land use	98-128	This study
1860-1984	Deconvolution	51-111	Siegenthaler and Oeschger 1987
1770-1984	Deconvolution	90-159	
1740-1980	Deconvolution	93	Keeling et al. 1989

An independent estimate of the total net loss of carbon from terrestrial ecosystems over the last 100 to 200 years has been calculated from models of oceanic uptake (Siegenthaler and Oeschger 1987, Keeling et al. 1989). Normally, ocean models are used to calculate atmospheric concentrations of CO_2 from specified annual emissions of carbon from fossil fuels and deforestation. The models can be used in reverse, however. If the historic concentrations of CO_2 are known, the same models can be used to calculate the annual emissions required to generate those concentrations. The use of the models in this way has been referred to as deconvolution. Subtraction of the annual emissions of fossil fuel carbon from the total emissions calculated from these deconvolutions yields a residual flux, a non-fossil flux of carbon. The flux need not be a terrestrial flux, but it is often

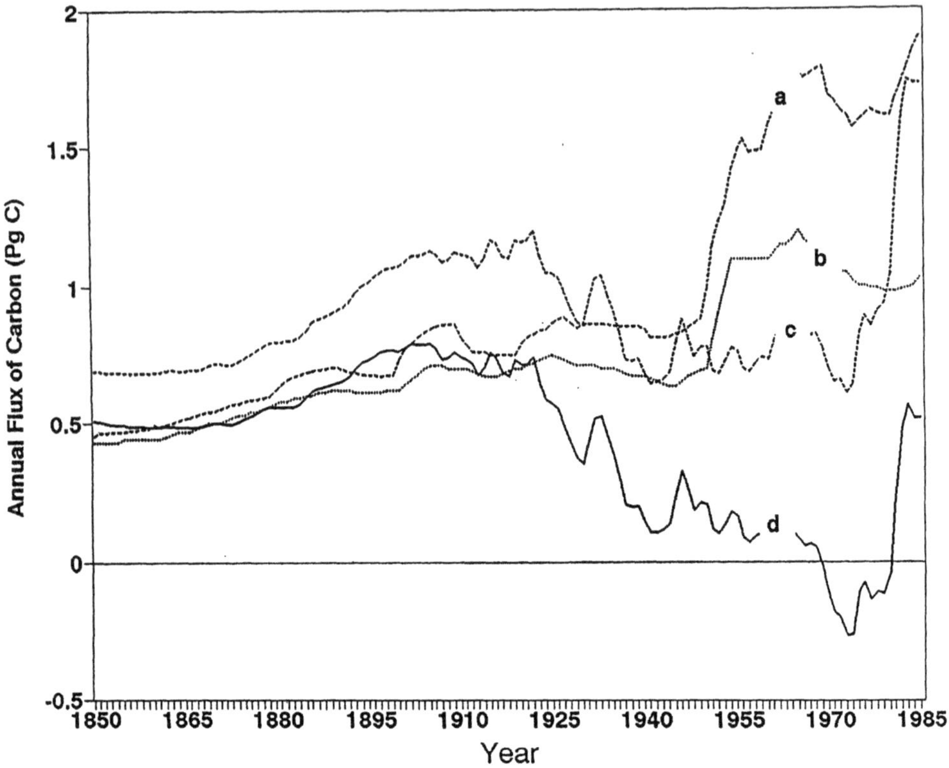

Figure 1. Different estimates of the annual flux of carbon from terrestrial ecosystems to the atmosphere. Lines a and b are estimates of the flux of carbon from changes in land use. Line a is based on high estimates of tropical biomass (preferred estimate); line b is based on low estimates of tropical biomass. Lines c and d are estimates of the net biotic flux, based on deconvolution of CO_2 concentrations with carbon models (the higher curve (c) is based on the outcrop model of the oceans which can absorb more CO_2 than the box-diffusion model (d)) (from Siegenthaler and Oeschger 1987). The box-diffusion model is thought to give a more accurate estimate.

assumed to be the net terrestrial flux of carbon to the atmosphere. At least two analyses have determined this non-fossil flux of carbon over the period since about 1850 (Siegenthaler and Oeschger 1987, Keeling et al. 1989). Two estimates were determined by Siegenthaler and Oeschger (1987) using two different models of oceanic circulation, a box diffusion model and an outcrop model. The outcrop model simulates a direct exchange of carbon between the atmosphere and deeper, colder waters and estimates a greater absorption of carbon than the box diffusion model. Hence, the outcrop model accommodates a greater biotic source. With these two models the non-fossil flux was estimated to have been about 50 and 110 PgC over the period 1860 to 1984 (Table 1). The lower estimate is preferred because it is based on an ocean model (box diffusion) thought to be more representative of the real ocean. In fact, neither of these models simulates oceanic circulation, but their results, including those of Keeling et al. (1989) (Table 1), do not very greatly from results obtained from more recently developed, more sophisticated models that do simulate circulation (Maier-Reimer and Hasselmann 1987, Bacastow and Maier-Reimer 1990, Sarmiento et al. 1992). These new models give somewhat lower estimates of oceanic uptake (1.5 - 1.8 PgC/yr in the 1980's) compared to the simple box models. Despite the limitations of the simpler models, however, their use for deconvolution does not invalidate the analyses described here.

Siegenthaler and Oeschger (1987) referred to the residual, or non-fossil, flux as a net terrestrial flux, and this paper will do the same, using the deconvolutions originally calculated by Siegenthaler and Oeschger (1987). Estimates of the cumulative flux of carbon based on these deconvolutions overlap with those calculated directly from changes in land use (Table 1), but the temporal patterns of annual flux do not agree (Fig. 1). Comparisons of previous estimates of flux have also pointed to this discrepancy (Enting and Pearman 1986, Enting and Mansbridge 1987).

Reconciliation of the Discrepancy

Both estimates of flux (land use and deconvolution) may be correct, and the difference between them may indicate a change in terrestrial ecosystems related to changes in the global environment. Assuming that the entire flux obtained through deconvolution is terrestrial, the deconvolution approach provides an estimate of the <u>net</u> biotic flux, including both the land-use flux and a flux due to other changes in terrestrial ecosystems. Thus, the difference between the net flux and the land-use flux defines a terrestrial flux that has so far been unknown. Here the pattern of this flux over the last century is shown for the first time. It is most likely the result of changes brought about by changes in climate, changes in the concentration of atmospheric CO_2, and changes in the mobilization of nutrients or toxins. The effect of these factors, either singly or together, on the storage of carbon in the world's terrestrial ecosystems is unknown. There has, as yet, been no measurement of even the direction of change.

The errors involved in calculating the flux of carbon from terrestrial ecosystems with both the deconvolution technique and the land-use approach make an analysis of their differences tenuous, but the gross trends may indicate which environmental factors were important in the past and, perhaps, the transient effects of climatic change over the next century. As mentioned above, the trends may not represent changes in terrestrial ecosystems at all, but, instead, changes in the ocean's circulation and its absorptive capacity. Short-term evidence from $^{13}C/^{12}C$ ratios suggests, however, that the changes are of terrestrial rather than oceanic origin (Siegenthaler and Oeschger 1987, Keeling et al. 1989).

The differences between the net biotic flux (from deconvolution) and the land-use flux are shown in Fig. 2. Negative differences indicate that the release of carbon from land-use change was larger than the net flux, and are interpreted here to mean that some terrestrial ecosystems were accumulating carbon independent of land-use change. Positive differences indicate that the net release of carbon was greater than the land-use flux, and that terrestrial ecosystems were releasing carbon in addition to the releases caused by changes in land use.

The four curves indicating this biotic flux (Fig. 2) are the differences between two estimates of the net flux and two estimates of the land-use flux (Fig. 1). The curve showing the largest sink (curve d: deconvolution with a box diffusion ocean model minus land-use flux with high estimates of tropical biomass) is the preferred curve. It accounts for degradation as well as outright clearing of tropical forests (Houghton 1991a), and it is thought to provide a more accurate oceanic flux (Siegenthaler and Oeschger 1987). The

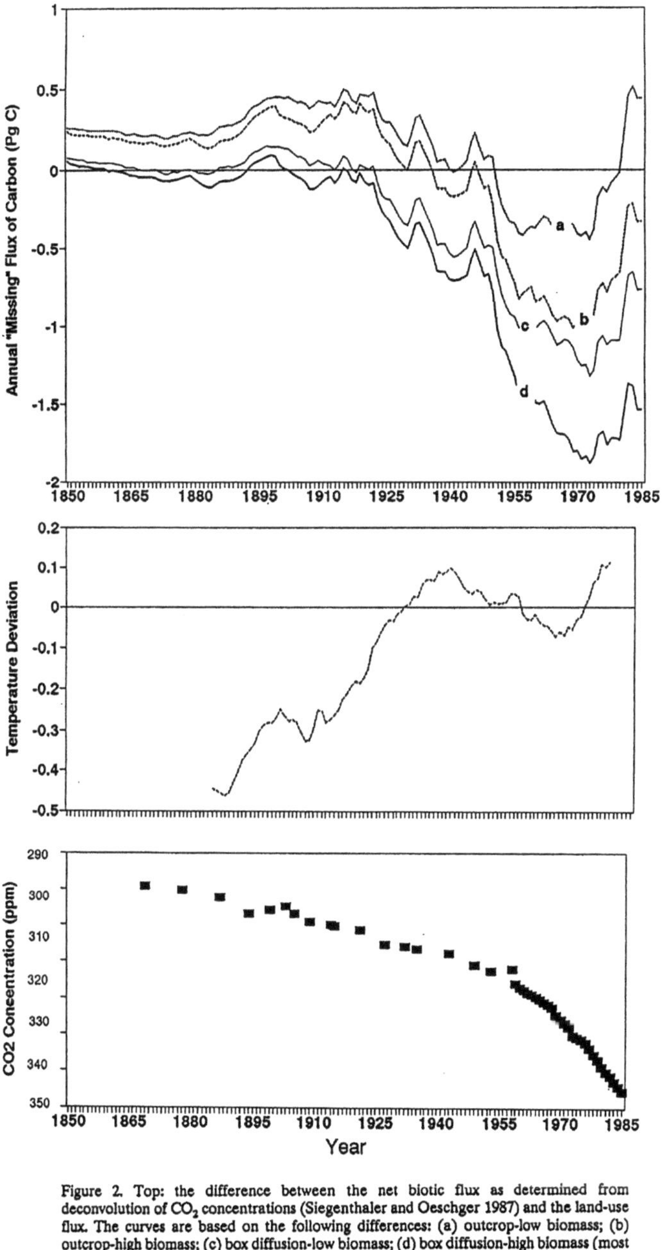

Figure 2. Top: the difference between the net biotic flux as determined from deconvolution of CO_2 concentrations (Siegenthaler and Oeschger 1987) and the land-use flux. The curves are based on the following differences: (a) outcrop-low biomass; (b) outcrop-high biomass; (c) box diffusion-low biomass; (d) box diffusion-high biomass (most likely or preferred case). Negative values indicate a biotic sink not accounted for by land-use change; positive values indicate a biotic source in addition to that from land-use change. Middle: Mean global land surface temperature deviations from the 1951-1980 baseline (Hansen and Lebedeff 1987). Bottom: Concentration of atmospheric CO_2 (from Neftel et al. 1985, Friedli et al. 1986, Keeling et al. 1989), plotted with the ordinate reversed so that the overall trend in concentration parallels the trend in the "missing" biotic flux of carbon (the flux independent of land-use change).

curve showing the least departure from zero (curve a) is the least preferred or least likely curve. All four curves exhibit the same three features, however: first, a period before 1920 of little variation. With the box diffusion model, the residual terrestrial flux is essentially zero before 1920; with the outcrop model, the flux is positive, but small. Second, there is a period between about 1920 and 1975 when some of the world's terrestrial ecosystems were apparently accumulating carbon at an increasing rate. Finally, there is an indication that the rate of accumulation began decreasing after the mid-1970's. The reader is reminded not to interpret Fig. 2 as showing a net terrestrial sink for carbon. The net terrestrial flux has almost always been a source of carbon to the atmosphere (Fig. 1, curve d). Figure 2 shows only the residual flux, a flux, proposed here, to be due to environmental effects on terrestrial ecosystems.

The accumulation of carbon in ecosystems not directly disturbed by changes in land use was about 70 PgC between 1850 and 1984 in the preferred analysis. Over this entire time period, changes in land use released about 126 PgC, so the net flux of carbon between terrestrial ecosystems and the atmosphere was a release of about 56 PgC (Siegenthaler and Oeschger 1987).

Possible Mechanisms

The residual flux discussed here is the "missing" carbon (Broecker et al. 1979). Initially the "missing" carbon was defined as that amount of carbon released from combustion of fossil fuels that neither accumulated in the atmosphere nor was absorbed by the (modeled) oceans. Estimates of the release of carbon from deforestation and other changes in land use increased the size of the "missing" carbon. If terrestrial ecosystems were required as a sink for carbon, an even larger sink was needed to compensate for the net release from the combined effects of deforestation and reforestation. This missing carbon has, of course, never been directly measured; it has only been defined as the difference between two "knowns". Resolution of the missing carbon is important for two reasons. First, the global carbon cycle cannot be balanced at present. If the cycle were well enough understood that emissions of carbon from fossil fuels and terrestrial ecosystems balanced accumulations of carbon in the atmosphere, oceans, and land, scientists would be better able to predict future concentrations of CO_2 in the atmosphere from scenarios of projected energy and land use. The inability to balance the global carbon budget makes predictions of future CO_2 concentration uncertain and, hence, makes predictions of rates of global warming uncertain. The second reason to resolve the "missing" carbon issue is to gain insight into potential feedbacks between climate and the carbon cycle.

The amount of carbon held in terrestrial ecosystems is determined by the balance between photosynthesis and respiration, the latter including decay of soil organic matter. Both of these metabolic processes are affected by many environmental factors, but surface air temperature and concentrations of CO_2 have received the most attention recently. Arguments are made, on the one hand, that terrestrial ecosystems will amplify a global warming (positive feedback) by releasing additional carbon to the atmosphere as a result of a warming; and, on the other hand, that they will reduce the warming (negative feedback) by accumulating carbon in response to increasing concentrations of CO_2 in the atmosphere. The historical variation in missing carbon, obtained here, allows us to ask whether the variation is related to temperature, atmospheric CO_2, or some other environmental factor. A correlation does not demonstrate a cause-effect relationship, of course, but it may reveal which environmental factor has been most important in the past, and it might suggest the direction of response of terrestrial ecosystems to global change in the future. Will they act as a net source or sink for carbon?

Linear regressions were used here to explore correlations between the annual flux of "missing" carbon and two environmental factors available globally over the period 1880 to 1985: concentrations of CO_2 in the atmosphere and surface air temperatures.

Concentrations of CO_2 in the atmosphere

The long continuous increase in "missing" carbon after 1920 parallels the trend in industrial activity. Among the many environmental factors thought to increase the accumulation of carbon on land, the two most often advanced are concentrations of CO_2 in the atmosphere and availability of fixed nitrogen, from both fertilizer use and internal combustion engines. Elevated concentrations of CO_2 have been shown to enhance photosynthesis and growth in many plants (Strain and Cure 1985, Allen et al. 1987), and may increase the storage of carbon in terrestrial ecosystems if the CO_2-enhanced growth is not respired (Grulke et al. 1990, Mooney et al. 1991).

With a linear regression and 11-year running averages for both variables, annual concentrations of CO_2 explained 46 to 85% of the variation in the annual flux of "missing" carbon over the entire 135-year period (Table 2a and Fig. 2). CO_2 concentrations over this period were obtained from Neftel et al. (1985), Friedli et al. (1986), and Keeling et al. (1989). The highest correlation coefficient (r^2 = 0.85) was associated with the preferred estimate (box diffusion model - high estimates of biomass). The coefficients of the regression analysis were used to calculate the effect of CO_2 fertilization on carbon storage. The biotic growth factor (or beta factor) is one index of the response of net primary productivity (NPP) (the annual net uptake of carbon by green plants) to elevated concentrations of CO_2 (C). It was defined by Bacastow and Keeling (1973) as:

$$\text{Beta} = \left(\frac{NPP}{NPP_0} - 1 \right) / \ln (C/C_0) \qquad (1)$$

where subscripts $_0$ refer to the initial (preindustrial) rates or concentrations. Using the regression coefficients relating the residual flux to CO_2 concentration, the annual storage of carbon increased by 1.0 to 3.2 PgC/yr as concentrations of CO_2 increased from 285 ppmv in 1850 to 350 ppmv in 1985. Global NPP is estimated to be between 45 and 60 PgC/yr currently (Houghton et al. 1985b), and would have been 1.0 to 3.2 PgC/yr lower in 1850. Assuming a current NPP of 50, equation (1) gives a beta factor of between 0.10 and 0.33, the higher estimate being the one based on the preferred oceanic uptake model (box diffusion) and the preferred estimate of land-use flux (higher estimates of tropical biomass) (Table 2a). Because the increased flux between 1850 and 1985 (an increase of 3.2 PgC/yr) represents a net storage of carbon and not an increase in NPP, and because only a portion of NPP may be stored, the beta factor calculated for NPP would be higher than 0.33. If half of an increase in NPP were respired, for example, and only half accumulated in either wood or soil, the increase in NPP would be twice the increase in

Table 2a. "Missing carbon" (PgC) and results of regression analyses over the period 1885-1980. Negative values of missing carbon indicate an accumulation of carbon on land.

	a	b	c	d
	Outcrop-Low biom.	Outcrop-High biom.	Box diffusion-Low biomass	Box diffusion-High biomass (Preferred analysis)
Cumulative flux of missing carbon 1850-1985	+21	-8	-43	-71
Missing flux in 1985	0.44	-0.32	-0.76	-1.53

REGRESSION COEFFICIENTS Missing Flux (PgC/yr) = A + B(Temp. Dev.)* (°C) + C[CO_2]** (ppmv)

1885-1980

	Outcrop-Low biom.	Outcrop-High biom.	Box diffusion-Low biomass	Box diffusion-High biomass (Preferred analysis)
A (constant)	4.546	9.917	9.105	14.476
B	-0.228	-0.064	0.302	-0.010
C	-0.015***	-0.033***	-0.031***	-0.049***
r^2	0.458	0.683	0.804	0.847
Additional flux in 1985 for a 65 ppmv increase in CO_2 above 1850 levels	-0.97	-2.14	-2.02	-3.18
CO_2 fertilization factor (beta)	0.10	0.22	0.20	0.33

* Mean global land surface temperature deviations from the 1951-1980 baseline (Hansen and Lebedeff 1987).

** [CO_2] refers to the absolute concentration of CO_2, not its deviation from preindustrial levels.

*** Coefficients significant at the 99% level.

storage, or 6.4 PgC/yr, and the beta factor would be 0.71. When these same calculations are based on a global terrestrial NPP of 60 PgC/yr, rather than 50 PgC/yr, the beta factors are somewhat lower, 0.27 and 0.58 assuming all and half the NPP is accumulated, respectively.

The use of linear regression coefficients in a non-linear equation is not strictly correct. It may be justified here, however, because the error it introduces is small relative to other errors and unknowns in the analysis, and because the increase in atmospheric CO_2 has been relatively small, about 25% since 1850.

Mobilization of nutrients

The apparent increase in carbon storage between 1850 and 1985 may not have been driven by increased concentrations of CO_2. Increased availability of nitrogen (Mayewski and Legrand 1990) or other nutrients from human activities may have increased the storage of carbon in vegetation and soils. The historical pattern of newly available nitrogen, like atmospheric concentrations of CO_2, parallels industrial activity, and correlations cannot determine which of these two factors, or, indeed, which of any number of factors related to industrial activity, might have increased the annual storage of carbon on land. On the other hand, a previous analysis of the increased loading of nitrogen and phosphorus on land and in coastal waters (Peterson and Melillo 1985) estimated the enhanced storage of carbon to be no more than 0.2 PgC annually. Thus, eutrophication is unlikely to explain the sink strength determined here (1.0 to 3.2 PgC/yr). The apparent reduction in the annual accumulation rate since 1975 despite continued increases in atmospheric CO_2 and in sources of nitrogen may be the result of nitrogen saturation of ecosystems (Aber et al. 1990) or of accumulation of $[H^+]$ or other toxins in ecosystems. Such a saturation seems unlikely to have happened globally as abruptly as suggested by this analysis, however.

Surface temperature

Another possible explanation for the pattern of "missing" carbon is variation in the mean surface temperature of the earth's land surface (Hansen and Lebedeff 1987). Respiration (including decomposition) is generally more temperature sensitive than photosynthesis, so the gradual warming since 1880 should have caused an increasing release of carbon rather than the increasing accumulation shown in Fig 2 (Schleser 1982, Woodwell 1989). The expected relationship is apparent after 1940, however, when the increasing sink for carbon between 1940 and 1970 matches the decrease in global temperature over this period, and when the recent reduction in storage corresponds to the warming since 1970 (Table 2b and Fig. 2).

The regression coefficients for the temperature-flux relationship (Table 2b) suggest that the 0.5°C warming between 1880 and 1985 may have been responsible for an additional release that is currently 1.2 to 1.8 PgC/yr (1.8 PgC/yr preferred). These values allow computation of a temperature coefficient (Q_{10}), defined as the factor by which the respiration rate increases when temperature is increased by 10°C. By assuming an exponential relationship between respiration and temperature, one can derive the equation:

$$Q_{10} = \left(\frac{k_2}{k_1}\right)^{\frac{10}{T_2 - T_1}} \qquad (2)$$

where k_1 and k_2 are rates of carbon flux at temperatures T_1 and T_2, respectively. Total respiration of terrestrial ecosystems, globally, is about 100 PgC/yr. Heterotrophic respiration (respiration of animals and microbes) is about 50 PgC/yr. If temperature is assumed to have affected only the rate of heterotrophic (largely soil) respiration (Rs_h) [k_2 = 50, k_1 = 48.2 to 48.8], then the Q_{10} is between 1.6 and 2.0, consistent with values commonly measured in the laboratory and field. If the change in temperature is assumed to affect also the rate of plant respiration (autotrophic + heterotrophic = total respiration of the ecosystem, Rs_e), the values of Q_{10} range between 1.3 to 1.4 (Table 2b). In all cases the preferred estimate of "missing" carbon yields Q_{10}'s at the high ends of these ranges. These values of Q_{10} assume that only respiration is affected by temperature. If rates of photosynthesis have also been increased by warmer temperatures, the Q_{10} for respiration alone is higher.

Table 2b. Results of regression analyses over the period 1943-1980. Negative values of missing carbon indicate an accumulation of carbon on land.

REGRESSION COEFFICIENTS Missing Flux = A + B(Temp. Dev.)* + $C[CO_2]$**
(PgC/yr) (°C) (ppmv)

1943-1980

	a	b	c	d
	Outcrop- Low biom.	Outcrop- High biom.	Box diffusion- Low biomass	Box diffusion- High biomass (Preferred analysis)
A (constant)	-0.165	4.129	4.677	8.971
B	2.568***	3.528***	2.323***	3.284***
C	-0.000	-0.015***	-0.018***	-0.033***
r^2	0.820	0.822	0.834	0.889
Additional flux in 1985 for a 0.5°C increase above 1880 temperature	1.28	1.76	1.16	1.64
Q_{10} (Rs_h=50 PgC)	1.68	2.05	1.60	1.95
Q_{10} (Rs_e=100 PgC)	1.29	1.43	1.26	1.39

* Mean global land surface temperature deviations from the 1951-1980 baseline (Hansen and Lebedeff 1987).

** $[CO_2]$ refers to the absolute concentration of CO_2, not its deviation from preindustrial levels.

*** Coefficients significant at the 99% level.

Over the period 1940 to 1985, temperature and CO_2 concentrations, together, explain 82 to 89% of the variance in the flux of "missing" carbon (Table 2b). Over this period, the preferred estimate of a beta factor is 0.20 (assuming all the increase in NPP is stored) and the preferred estimate of the Q_{10} is 2.0 (assuming only heterotrophic respiration is affected by temperature).

If the environmental factors addressed here are indeed the ones responsible for variations in the annual flux of carbon from terrestrial ecosystems, the effects of temperature and CO_2 may have largely canceled each other thus far and may continue to do so in the future. On the other hand, one of the most important parameters, soil moisture, has been ignored. Furthermore, interactions of environmental factors are likely to be more complicated than presented here. An increase in temperature, for example, will not only increase respiration and release additional carbon but will also increase mineralization of nutrients in soil, thereby possibly enhancing the growth of trees and the storage of carbon.

Implications

Other analyses have shown a positive correlation between mean global temperature and atmospheric concentrations of CO_2 (Keeling et al. 1989, Kuo et al. 1990, Marston et al. 1991). These previous correlations were based on data since 1958, however, and the results of the work reported here suggest that before 1940 the relationship between temperature and carbon flux was reversed (Fig. 2). If the increase in global temperature since about 1970 has been responsible for the reduced sink of carbon shown here, other factors must have obscured the warming effect between 1880 and 1940. Perhaps the geographic distribution of the warming was different. Perhaps the effect of warming on the ratio of respiration to photosynthesis depends on the rate of warming or on the initial temperature.

The apparent decline in accumulation since the early 1970's may be an artifact of the spline used in the deconvolution analysis (Enting 1987), but the increased emission (or decreased accumulation) of carbon starting in the mid-1970's has been observed by others (Keeling et al. 1989). The cause remains uncertain: clearly, there has been an increased release of carbon from tropical deforestation over the last decade (Houghton 1991b); from the analysis reported here, there appears also to have been an increased release (or reduced accumulation) elsewhere. The analysis cannot distinguish geographic variations in the flux.

The recent analysis by Tans et al. (1990) suggests a net annual accumulation of 2 to 3.6

PgC/yr in the lands of the northern temperate zone for the years 1981 to 1987. The global sink determined here is less, 1.5 PgC/yr at most (Table 2a), but could be larger in the temperate zone if compensated by a net source, independent of land-use change, elsewhere. It is important to recall that the land-use flux includes the accumulation of carbon on land from regrowth of logged forests and from reforestation of abandoned agricultural lands. Thus, the near-zero flux of carbon from land-use change in the northern temperate zones (Houghton et al. 1987, Melillo et al. 1988) represents a balance between the oxidation of wood products and the accumulation of carbon in regrowing forests. According to those analyses, the annual accumulation in regrowing forests in northern mid-latitudes was about 1 PgC in 1980. It is unlikely that the net land-use flux has exceeded ± 0.5 PgC/yr in the years since then (unpublished).

A comparison of these results with those given by Tans et al. (1990) is also influenced by the estimate of oceanic uptake determined by Tans et al. Their estimate of oceanic uptake was considerably less than the uptake obtained with the oceanic models used for the deconvolutions shown in Fig. 1. If the models are correct, the terrestrial sink required by Tans et al. will be less.

These analyses suggest that the terrestrial sink for carbon in the early 1980's was larger than at any time in the last 135 years (Fig. 2). If the same trend applies to the temperate zone, the sink reported by Tans et al. (1990) for the period 1981 to 1987 was also near a maximum. Before the middle of this century, the oceans were apparently absorbing all of the carbon released from both fossil fuels and changes in land use.

The analysis reported here also suggests that the net biotic flux around 1985 (0.5 to 1.7 PgC/yr according to the deconvolution, 0.5 preferred) (Siegenthaler and Oeschger 1987) was the result of three offsetting fluxes: a flux from changes in land use, releasing 0.5 to 2.5 PgC/yr to the atmosphere; a CO_2-, nutrient-, or cooling-enhanced storage, annually storing 1.0 to 3.2 PgC/yr on land (Table 2a); and a warming-enhanced respiration, releasing on the order of 1.2 to 1.8 PgC/yr (Table 2b). Again, the preferred analysis yields high estimates of each of these fluxes. These calculated sources and sinks of carbon and the factors responsible are, of course, suggestive only. This analysis did not consider influences due to changes in precipitation. Neither did it account for changes in oceanic temperature. In particular, the analysis could not distinguish between temperature-induced releases from land or the ocean.

Nevertheless, terrestrial ecosystems, overall, seem to have been a net source of carbon to the atmosphere in 1985 (about 0.5 PgC/yr). Despite suggestions that undisturbed terrestrial ecosystems may have been a sink (for as much as 3.2 PgC/yr), this sink was not enough to offset the release from deforestation (0.5 to 2.5 PgC/yr) and the possible release of carbon (another 1.8 PgC/yr) from a warming-enhanced respiration.

Acknowledgements

I thank Inez Fung, G.M. Woodwell, and several reviewers for critical reviews of earlier drafts of the manuscript; J. Hackler for his help with analysis and graphics; and U. Siegenthaler for providing the results of his deconvolution. The research was supported by the National Science Foundation (BSR-8704749) and the U.S. Department of Energy (subcontract 19X-SB830C for Oak Ridge National Laboratory, Oak Ridge, Tennessee 37831-6335 operated by Martin Marietta Energy Systems, Inc. for the U.S. Department of Energy under subcontract DE-AC05-84OR21400).

References

Aber JD, Nadelhoffer KJ, Steudler P, Melillo JM (1990) Nitrogen saturation in northern forest ecosystems. BioScience 39:378-386

Allen LH, Boote KJ, Jones JW, Jones PH, Valle RR, Acock B, Rogers HH, Dahlman RD (1987) The response of vegetation to rising carbon dioxide: Photosynthesis, biomass, and seed yield of soybean. Global Biogeochemical Cycles 1:1-14.

Bacastow R, Keeling CD (1973) Atmospheric carbon dioxide and radio-carbon in the natural carbon cycle. II. Changes from A. D. 1700 to 2070 as deduced from a geochemical model. In: Woodwell GM, Pecan EV (eds) Carbon and the Biosphere. U.S. Atomic Energy Commission, Symposium Series 30, National Technical Information Service, Springfield, Virginia, pp 86-135.

Bacastow R, Maier-Reimer E (1990) Circulation model of the oceanic carbon cycle. Climate Dynamics 4:95-125.

Bolin B (1977) Changes of land biota and their importance for the carbon cycle. Science 196:613-615.

Broecker WS, Takahashi T, Simpson HH, Peng T-H (1979) Fate of fossil fuel carbon dioxide and the global carbon budget. Science 206:409-418.

Detwiler RP, Hall CAS (1988) Tropical forests and the global carbon cycle. Science 239:42-47.

Enting IG (1987) On the use of smoothing splines to filter CO_2 data. J Geophys Res 92:10977-10984.

Enting IG, Mansbridge JV (1987) The incompatibility of ice-core CO_2 data with reconstructions of biotic CO_2 sources. Tellus 39B:318-325.

Enting IG, Pearman GI (1986) The use of observations in calibrating and validating carbon cycle models. In: Trabalka JR, Reichle DE (eds) The Changing Carbon Cycle. A Global Analysis. Springer-Verlag, New York, pp 425-458.

FAO (1990) Interim Report on Forest Resources Assessment 1990 Project, FAO COFO-90/8(a). FAO, Rome, Italy.

Friedli H, Lotscher H, Oeschger H, Siegenthaler U, Stauffer B (1986) Ice core record of the $^{13}C/^{12}C$ ratio of atmospheric CO_2 in the past two centuries. Nature 324:237-238.

Grulke NE, Riechers GH, Oechel WC, Hjelm U, Jaeger C (1990) Carbon balance in tussock tundra under ambient and elevated atmospheric CO_2. Oecologia 83:485-494.

Hall CAS, Uhlig J (1991) Refining estimates of carbon released from tropical land-use change. Can J For Res 21:118-131.

Hansen J, Lebedeff S (1987) Global trends of measured surface air temperature. J Geophys Res 92:13345-13372.

Houghton RA (1991a) Releases of carbon to the atmosphere from degradation of forests in tropical Asia. Can J For Res 21:132-142.

Houghton RA (1991b) Tropical deforestation and atmospheric carbon dioxide. Climatic Change 19:99-118.

Houghton RA, Hobbie JE, Melillo JM, Moore B, Peterson BJ, Shaver GR, Woodwell GW (1983) Changes in the carbon content of terrestrial biota and soils between 1860 and 1980: A net release of CO_2 to the atmosphere. Ecol Monogr 53:235-262.

Houghton RA, Boone RD, Melillo JM, Palm CA, Woodwell GM, Myers N, Moore B, Skole DL (1985a) Net flux of CO_2 from tropical forests in 1980. Nature 316:617-620.

Houghton RA, Schlesinger WH, Brown S, Richards JF (1985b) Carbon dioxide exchange between the atmosphere and terrestrial ecosystems. In Trabalka JR (ed) Atmospheric Carbon Dioxide and the Global Carbon Cycle. DOE/ER-0239, US Dept Energy, Washington, DC, pp 113-140.

Houghton RA, Boone RD, Fruci JR, Hobbie JE, Melillo JM, Palm CA, Peterson BJ, Shaver GR, Woodwell GM, Moore B, Skole DL, Myers N (1987) The flux of carbon from terrestrial ecosystems to the atmosphere in 1980 due to changes in land use: Geographic distribution of the global flux. Tellus 39B:122-139.

Keeling CD, Bacastow RB, Carter AF, Piper SC, Whorf TP, Heimann M, Mook WG, Roeloffzen H (1989) A three-dimensional model of atmospheric CO_2 transport based on observed winds: 1. Analysis of observational data. In: Peterson DH (ed) Aspects of Climate Variability in the Pacific and the Western Americas. Geophysical Monograph 55, American Geophysical Union, Washington, DC, pp 165-236.

Kuo C, Lindberg C, Thomson DJ (1990) Coherence established between atmospheric carbon dioxide and global temperature. Nature 343:709-714.

Maier-Reimer E, Hasselmann K (1987) Transport and storage in the ocean - An

inorganic ocean-circulation carbon cycle model. Climate Dynamics 2:63-90.
Marston JB, Oppenheimer M, Fujita RM, Gaffin SR (1991) Carbon dioxide and temperature. Nature 349:573-574.
Mayewski PA, Legrand MR (1990) Recent increase in nitrite concentration of Antarctic snow. Nature 346:258-260.
Melillo JM, Fruci JR, Houghton RA, Moore B, Skole DL (1988) Land use change in the Soviet Union between 1850 and 1980: causes of a net release of CO_2 to the atmosphere. Tellus 40B:116-128.
Mooney HA, Drake BG, Luxmoore RJ, Oechel WC, Pitelka LF (1991) Predicting ecosystem responses to elevated CO_2 concentrations. BioScience 41:96-104.
Myers, N. (1989) Deforestation Rates in Tropical Forests and Their Climatic Implications. Friends of the Earth, London, UK.
Neftel A, Moor E, Oeschger H, Stauffer B. (1985) Evidence from polar ice cores for the increase in atmospheric CO_2 in the past two centuries. Nature 315:45-47.
Peterson BJ, Melillo JM (1985) The potential storage of carbon caused by eutrophication of the biosphere. Tellus 37B:117-127.
Richards JF, Olson JS, Rotty RM (1983) Development of a Data Base for Carbon Dioxide Releases Resulting from Conversion of Land to Agricultural Uses. ORNL/TM-8801. Oak Ridge National Laboratory, Oak Ridge, TN.
Sarmiento JL, Orr JC, Siegenthaler U (1992) A perturbation simulation of CO_2 uptake in an ocean general circulation model. J Geophys Res, in press.
Schleser GH (1982) The response of CO_2 evolution from soils to global temperature changes. Z. Naturforsch. A. 37:2037/1-2037/5.
Siegenthaler U, Oeschger H (1987) Biospheric CO_2 emissions during the past 200 years reconstructed by deconvolution of ice core data. Tellus 39B:140-154.
Strain BR, Cure JD (eds)(1985) Direct Effects on Increasing Carbon Dioxide on Vegetation. DOE/ER-0238, US Dept Energy, Washington, D.C.
Tans PP, Fung IY, Takahashi T (1990) Observational constraints on the global atmospheric CO_2 budget. Science 247:1431-1438.
Woodwell GM, Houghton RA (1977) Biotic influences on the world carbon budget. In: Stumm W (ed) Global Chemical Cycles and their Alterations by Man. Dahlem Konferenzen, Berlin, Germany, pp 61-72.
Woodwell GM, Pecan EV (eds)(1973) Carbon and the Biosphere. U.S. Atomic Energy Commission, Symposium Series 30, National Technical Information Service, Springfield, Virginia.
Woodwell GM, Whittaker RH, Reiners WA, Likens GE, Delwiche CC, Botkin DB (1978) The biota and the world carbon budget. Science 199:141-146.
Woodwell GM (1989) The warming of the industrialized middle latitudes 1985-2050: Causes and consequences. Climatic Change 15:31-50.

IMPLICATIONS OF CO_2 EFFECTS ON VEGETATION FOR THE GLOBAL CARBON BUDGET

Roger M. Gifford
CSIRO, Division of Plant Industry
GPO Box 1600, Canberra, ACT 2601
Australia

INTRODUCTION

On a geological timescale, the carbon dioxide concentration in the global atmosphere was, over the few millennia before industrialisation, about as low as it has ever been. It has probably been within that low range of 200 to 300 ppmv throughout the 2 million years of human evolution. To understand the role of vegetation in modulating recent anthropogenic global atmospheric change, examination of the reasons for the low pre-industrial CO_2 concentration is instructive.

While many details of the evolution of atmospheric CO_2 concentration remain to be deduced, there now seems little doubt about the main features of change over the last 500×10^6 years. Knowledge of key features of the geological history of the carbon cycle is pertinent to understanding the role of vegetation in the carbon cycle over time periods of decades to centuries. These features are encapsulated in the recent model of Berner (1990, 1991). In that model the small amount of carbon (C) in the atmosphere, biosphere and oceans, relative to the 99.9% of the global C resident in the rocks and magma, is represented as a single well-mixed pool. In such a model, for which the time-step is ten million years, those pools are essentially well-mixed. The simulated history of atmospheric CO_2 concentration is shown in Figure 1.

The two-humped pattern of atmospheric CO_2 concentration shown in Figure 1 reflects variations in the operation of the carbonate-silicate cycle. Variation in atmospheric CO_2 concentration reflects the tiny residual difference between two large opposing processes: 1) the removal of CO_2 from the atmosphere/ocean system by limestone formation and its incorporation into the magma by tectonic subduction, and 2) the emission of CO_2 from the lithosphere to the atmosphere/ocean system by volcanism. The subducted carbonates react with silica in the magma to produce molten calcium silicate, which later becomes new

NATO ASI Series, Vol. I 15
The Global Carbon Cycle
Edited by M. Heimann

igneous rock and gaseous CO_2 which leaks out. Over that long timescale, the two processes controlling atmospheric CO_2 concentration are the rate of CO_2 leakage, which is related to the intensity of plate tectonic movements, and the rate of solid calcium carbonate formation which, over such timescales, is determined not by ocean processes but by the activity of land plants.

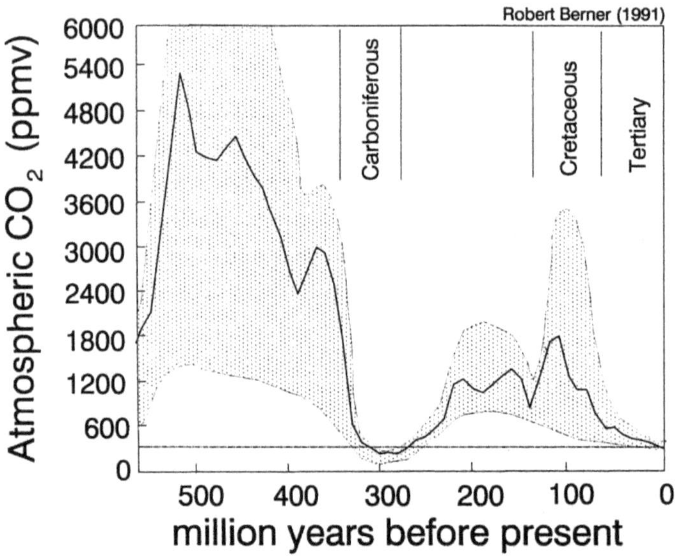

Figure 1. History of global atmospheric CO_2 concentration over the past 500 million years according to a model of the carbonate-silicate cycle by Berner (1991). The shaded area indicates the zone within which the true time-course is expected to fall. The solid line is the result of the standard run of the model.

The reason that terrestrial vegetation is so important in the carbonate-silicate cycle is that in the long term, calcium and magnesium carbonates are precipitated on the ocean floor at the rate that calcium and magnesium ions enter the ocean from the land. The Ca^{++} and Mg^{++} ions are released from land via weathering of the silicate rocks by CO_2 in solution (carbonic acid). The rate of rock weathering is enhanced several orders of magnitude by the presence of the vegetative cover.

Land plants enhance rock weathering by several mechanisms. At first colonisation, their organs of attachment penetrate cracks and help break up the rock thereby increasing the surface area available to the action of carbonic acid in releasing Ca^{++} and Mg^{++} ions and silica from calcium/magnesium silicate. Secondly, the roots foster soil formation by holding the particles together. This reduces particle loss by erosion as does the over-arching canopy of leaves by reducing raindrop impact. Thirdly, in that process of soil formation, a water-holding organic matter "sponge" is layered over the land which keeps the broken-down rock surface wet and in contact with carbonic acid solution much longer than occurs with bare rock. Fourthly, roots and organic matter decomposition produce organic acids which may hasten the release of ions for plant uptake. Fifthly, root and microbial respiration increases the CO_2 concentration in the soil atmosphere by at least an order of magnitude thereby greatly hastening ion release by weathering (Schwartzman and Volk 1989).

It is not vegetation productivity alone that is responsible for the enhanced rate of weathering. Decomposition of dead organic matter is important too. For, when roots and microbes release ions, plants and microbes absorb most of them, later to be released into the soil solution when the organisms die. As a result of this ion turnover, associated with organic matter turnover, there is a gradual ion loss to water courses by leaching.

Berner (1990) and others (eg Volk 1989) interpret some key features of Figure 1 as being attributable to land plant acceleration of rock weathering. The precipitous disappearance of atmospheric CO_2 just before and during the carboniferous period is believed to have been substantially due to the evolution of vascular land plants at that time. Although accumulation of carbon in coal contributed somewhat to that removal of carbon from the atmosphere, it was mostly accelerated rock weathering and consequential limestone formation that was responsible. It is unclear why CO_2 concentration rebounded back up to several thousand ppmv over the ensuing two hundred million years. It may have been partly to do with an increased rate of plate tectonic activity causing continental drift and leading to faster formation and emission of CO_2 from the lithosphere (Berner 1991). But the next cycle of decline starting about 100 million years ago is believed to have been attributable, at least in part, to the widespread displacement of gymnosperms by angiosperms during that period. Decline in the vigor of tectonic activity might also have been involved. Volk (1989) suggested that the common deciduous habit of angiosperms leads to much more rapid loss of ions from the plant-soil system and hence more rapid movement of Ca^{++} and Mg^{++} ions to the ocean. The result of this was that atmospheric CO_2

concentration was gradually drawn down to the low levels that we now see and the conditions were set for the geologically current series of glaciations, just as there were Ice Ages in the previous period of low CO_2 concentration during the carboniferous period.

On the timescale of decades to centuries the rate of turnover of the carbonate-silicate cycle (approximately 0.1 Gt C yr^{-1}) is too low to be of direct significance to explaining the present changes in atmospheric composition that are of current socio-economic concern. However, appreciation of the role of plants in the carbonate-silicate cycle, and of the history of atmospheric CO_2 is important to our understanding of shorter term sub-cycles for several reasons.

Throughout virtually the entire period of the evolution of land plant photosynthesis, atmospheric CO_2 concentration has been many times greater than it is at present. At present, the CO_2 concentration is less than a third of that needed to saturate photosynthesis. As global atmospheric CO_2 concentration and temperature declined over the last 50-100 million years, vegetation productivity must have declined too because C is a principal nutrient in plant dry matter, and higher plant photosynthesis is CO_2 concentration dependent up to about 1500ppmv. Thus the rate of rock weathering must have declined in parallel since vegetation productivity is so closely instrumental in weathering. With reduced weathering there is a reduced rate of CO_2 removal from the atmosphere by carbonate formation. That is, there is a built-in negative feedback in which atmospheric CO_2 decline, caused by plant productivity-induced weathering, limits that productivity to a low value at which CO_2 removal matches tectonic emissions. If atmospheric CO_2 were to increase for some extraneous reason, then in the long term of the carbonate-silicate cycle, stimulated plant productivity via weathering would again pull it down. The result of this dynamic equilibrium is that the atmospheric CO_2 concentration prior to human industrialisation was probably about as low as it could possibly be. For if anything tended to reduce the CO_2 level, say further reduction of tectonic activity, vegetation productivity would be so severely curtailed (being on the steep part of the CO_2 response function) that the rate of weathering and hence limestone formation would decline and equilibrium would thereby eventually be re-established.

Although that equilibrium may hold in the long term of millions of years, it is not applicable in the short term of decades to centuries that are appropriate to the present human concerns. To understand the role of vegetation on that timescale, the *legacy* of the interactive evolution of photosynthesis, vegetation functioning, soil, atmospheric composition and the operation of the carbonate-silicate cycle is nevertheless pertinent. I hypothesise that key features of that legacy are:

1) The response of terrestrial phyto-production to atmospheric CO_2 concentration, having evolved for hundreds of millions of years in much higher CO_2 concentrations than occur at present, is now well below saturation. That is, there has been a negative feedback in which vegetation growth has depleted its key reactant - CO_2 - to rate limiting levels.

2) Atmospheric oxygen, generated over billions of years by the water splitting reaction of marine and terrestrial photosynthesis, has climbed to such a high level (21%) that it is significantly inhibitory to plant production owing to specific well-characterised biochemistry of photosynthesis involving competitive inhibition of CO_2 fixation by O_2 and the associated photorespiration. The build up of atmospheric O_2 has been another, interrelated, negative feedback on primary production.

3) Global average temperature has been reduced, owing mainly to the declining CO_2 concentration, to sub-optimal values for terrestrial productivity. Therefore temperature too has mediated a negative feedback between plant production and the global environment.

4) The mineral nutrients that are turning over between live plants and dead soil organic matter were made available to the biosphere from the rocks largely by the functioning of plants. There is a long term positive feedback in which plant growth fosters nutrient solubilization by weathering and the essential minerals thereby made available foster plant production. Although, in the long term, nutrient availability to the biosphere is largely driven by the suitability of the environment for photosynthesis, there is no doubt, that within any one growing season, vegetation productivity is often restricted by nutrient, especially nitrogen, availability because the nutrients are tied up in non-soluble forms in soil organic matter and micro-organisms.

Furthermore, there is now a huge atmospheric pool of gaseous nitrogen, probably created by the functioning of the biosphere, in dynamic equilibrium with biosphere and soil N-compounds. The conversion of gaseous nitrogen to plant-available combined nitrogen is mainly dependent on the oxidation of photosynthetically fixed carbon compounds. Hence on a time-frame of decades to centuries, variation in the amount of nitrogen in terrestrial ecosystems is probably paced by variation in their carbon economy (Gifford 1992). In the modern world anthropogenic sources of nitrogen, as fertiliser and atmospheric pollution, also contribute to biosphere N. This source is also dependent on the oxidation of organic C in the form of fossil fuels.

Thus fixation of C by the terrestrial biosphere, while being driven by incident solar energy is now ultimately restricted by the atmospheric CO_2 concentration, by the low global average temperature (about 15 °C) and by the high level of atmospheric oxygen, all of which are vegetation generated restrictions. The secondary restriction by mineral nutrient supply is mostly a reflection of competition within the terrestrial biosphere. If the opportunity for increased productivity is increased by a lifting of some of the primary restrictions, then mineral nutrient supply will gradually be dragged up with it except where there is an absolute deficiency of some mineral in the parent rock such as a trace element.

On a global scale the above hypothesis is suggested by the fact that ecosystem primary productivity around the world can be satisfactorily predicted by models that are based solely on annual precipitation and temperature and/or net radiation (Rosenzweig 1968, Lieth 1975, Uchijima and Seino 1985). On a time scale of centuries, the hypothesis is also suggested by the changes in the quantity of carbon stored in global terrestrial vegetation since the last major glacial retreat. Diverse palaeo-evidence has been used to construct palaeobiomaps from which palaeophytomass has been deduced. 18,000 years ago global temperature was 5-7 °C cooler than now and global CO_2 concentration was less than 200ppmv (Watson *et al.* 1990). According to Adams *et al.* (1990), the global phytomass 18,000 years ago was only half of the phytomass estimated for 1975AD and almost a third of what present biomass would have been now in the absence of man's modifications to the land. While the cooler, drier conditions were partly responsible for this, I suggest that the lower CO_2 concentration also played a major part in reducing global phytomaass during glacial periods.

With this background I have constructed a simple model of the global terrestrial carbon cycle, responsive to CO_2 concentration and temperature, which is appropriate to the time scale of modern human affairs.

CQUEST – A SIMPLE MODEL OF THE GLOBAL TERRESTRIAL C-CYCLE OVER DECADES TO CENTURIES.

New vegetation growth (net primary production) may by year's-end contribute to new live phytomass, be eaten by herbivores, or be deposited as litter that contributes to the soil C pool. Soil organic matter is continuously being oxidised to CO_2 by soil micro-organisms that use it as a source of elements to build their own structure and as a source of energy. Turnover of soil micro-organisms is very rapid and dead microbes contribute to soil organic matter. This process of C turnover, involving large numbers of different groups of micro-organisms with different substrate preferences, leads to a spectrum of soil organic matter fractions having a wide range of resistances to further microbial processing. After each decomposition cycle, the remaining humic fraction is more resistant to further decomposition. Such a spectrum of organic matter components can be modelled either as a series of discrete fractions with different turnover times, such as is done in the Rothamsted soil organic matter turnover model and others (Jenkinson 1990), or as comprising a continuum of resistance to decay as suggested by Bosatta and Ågren (1985). In ***CQUEST***, which runs on annual time-steps, soil plus litter C is modelled as comprising two pools, a fast decay one that is oxidised within the year of deposition, and a single exponentially decaying "slow" pool.

Similarly ***CQUEST*** aggregates all the diverse ecosystems and climates of the world. The world may be divided into an arbitrary number of biomes or ecosystems and climatic zones. Vegetation productivity and organic matter decay functions can be applied to each. However, no matter how many regions are defined, in practice each region is still highly heterogeneous, perhaps having almost as large a coefficient of variation as the world has overall when the unit grid-cell within each system is a similar proportion of the respective wholes. Thus no matter how finely one compartmentalises the C cycle in practice, one can always make the criticism that aggregate coefficients are being applied to highly heterogeneous entities. If, then, a simplified highly aggregated model of one aspect of the C-cycle can be parameterised to reproduce the output of a more complex but less aggregated model of that aspect, then perhaps the simpler model can be usefully used in other models dealing with other functional features of the C-cyle while the operation of the model remains transparent and understandable through its structural simplicity. That is my objective with ***CQUEST***. A list of symbol definitions and units is presented as an Appendix.

The soil plus litter dead organic C pool in year (y+1) is calculated from the value in year y plus the change during one year:

$$S_{y+1} = S_y + (L_y - k_2 S_y) \quad (1)$$

where S_y is the soil plus litter carbon pool and L_y is the shoot and root litter carbon deposited on and in the ground in the yth year. Soil organic C decay rate is modelled as being proportional to the amount of organic C present with a first order decay coefficient, k_2.

The amount of tissue death is modelled as being proportional to the standing biomass, B, with proportionality coefficient k_1. The bigger the ecosystem, the greater the mass of organs and whole plants that die each year. The change in standing biomass in a year (B_{y+1}-B_y) equals the gain from annual primary production, P_y, less the loss from death, k_1B. Thus

$$B_{y+1} = B_y + (P_y - k_1 B_y) \quad (2)$$

Not all dead vegetation becomes next year's litter. Some plant matter dies by herbivory (vertebrate as well as invertebrate). Some oxidises to CO_2 by fire. And some of the litter decomposes rapidly in the year of death. A fraction, f_1, of the total death therefore does not contribute to the litter input to the following year's soil pool. Thus

$$L_y = (1-f_1)k_1 B_y \quad (3)$$

This formulation is equivalent to treating the dead organic matter as a two compartment system, one compartment being fast oxidising, the other being slow.

To use the model to simulate the effect of global temperature increase on C emission from soils, it is necessary to include a temperature dependence of decomposition of the slow decomposing pool. For the small changes involved in greenhouse warming (2-3°C) a linear response of k_2 to temperature is adequate and can be formulated as

$$k_2 = k_{2,0}(1+k_3(T_y-T_0)) \tag{4}$$

where T_y is the temperature in year y, and $k_{2,0}$ is k_2 at the temperature, T_0, of a reference year used to start the model during a period of stable temperature. k_3 is the relative sensitivity of the decomposition coefficient to temperature.

Combining (1),(3), and (4) gives

$$S_{y+1} = S_y + [(1-f_1)k_1B_y] - [k_{2,0}(1+k_3(T_y-T_0))S_y] \tag{5}$$

Not only the decay rate of the soil organic pool, but also the productivity of the vegetation that sustains it, changes with temperature. It is necessary to include the vegetation response to temperature in the model in order to predict the time-course of soil organic matter content as greenhouse effect warming occurs, otherwise net CO_2 loss from soil will be over-estimated if warming increases productivity. Furthermore, instrumental in any greenhouse effect temperature increase, there has been and will continue to be prior increase in the atmospheric CO_2 concentration which stimulates vegetation growth.

If $P_{0,0}$ is the global net primary production for reference-year values of CO_2 concentration (C_0) and temperature (T_0), then for small deviations (T_y-T_0) of temperature from the reference temperature, one can represent the net primary production P_0 at C_0 as

$$P_0 = P_{0,0}(1+k_4(T_y-T_0)) \tag{6}$$

where k_4 is the relative temperature sensitivity of P_0 for the restricted temperature range involved.

The CO_2 fertilizing effect on net photosynthesis (A) is known to be a saturating response which can be empirically fitted by a rectangular hyperbola (Hunt et. al 1991). In global C-cycle modelling, a logarithmic function {viz: $A = A_0 (1 + \beta \ln(C/C_0))$ } is commonly adopted following its initial tentative introduction by

Bacastow and Keeling (1973). The parameter, ß, is the ratio of the relative change of A to the relative change of CO_2 concentration that caused that change. However, that function does not effectively simulate the photosynthetic response to CO_2 (Gifford 1980): inevitably ß varies with CO_2 concentration. That situation is unsatisfactory, especially when the independent variable varies by over 50%, as CO_2 concentration does in the model runs used in this paper. It is preferable to use a function in which the parameters are essentially independent of the variables in the equation in the range over which they will be used. For the response of vegetation productivity to CO_2, it is more realistic mechanistically to use a function in which one can have some prospect of establishing an intercept and an asymptote. It is here represented as a rectangular hyperbola that does not pass through the origin:

$$A = k_5(C-C_b) / (1+k_6(C-C_b)) \quad (7)$$

where C_b is the CO_2 compensation concentration at which net daytime photosynthetic CO_2 fixation is zero. While there is no direct experimental field evidence on the exact shape of the function in the long term, it is inevitable that the saturation of leaf photosynthesis observed in the short term must be expressed at the ecosystem level in the long term. C_b arises from the presence of oxygen in the atmosphere (Ogren and Hunt 1978). The oxygen reacts with the same molecule (ribulose bisphosphate, RuBP) that CO_2 reacts with in the process of photosynthesis. The reaction with oxygen leads to a series of biochemical steps known as photorespiration that involve the re-emission of some recently fixed CO_2 to the atmosphere. CO_2 and oxygen interfere with each other's reaction competitively such that if CO_2 concentration is progressively reduced there comes a point at which CO_2 emission from the oxygenase reaction matches CO_2 fixation by the carboxylase reaction. Then there is no net CO_2 exchange. C_b does not include the effect of true ("dark") respiration because it is introduced into the model in another way.

True respiration, R, is an essential plant process, distinct from photorespiration, which in so far as the C-balance is concerned, offsets CO_2 fixation.

$$P = A - R \quad (8)$$

Setting the ratio $R/A = f_2$, equations (7) and (8) rearrange to

$$P = \frac{(1-f_2)k_5(C - C_b)}{(1+k_6(C-C_b))} \quad (9)$$

Assuming that f_2 does not change with CO_2 concentration and temperature (see below), equation (9) becomes

$$P = \frac{k_5'(C_y - C_b)}{(1+k_6(C_y-C_b))} \quad (10)$$

where $k_5' = k_5/(1-f_2)$

The ***CQUEST*** model, then, reduces to five equations (see the box below) which are applied annually starting from a reference year for which the model is parameterised.

Summary of the equations of *CQUEST*

$$B_{y+1} = B_y + (P_y - k_1B_y) \quad (2)$$

$$S_{y+1} = S_y + [(1-f1)k_1B_y]-[k_{2,0}(1+k_3(T_y-T_0))S_y] \quad (5)$$

$$P_0 = P_{0,0}(1+k_4(T_y-T_0)) \quad (6)$$

$$P_y = \frac{k_5'(C_y - C_b)}{(1+k_6(C_y-C_b))} \quad (10)$$

$$W_{y+1} = W_y + [(B_{y+1}-B_y) + (S_{y+1}-S_y)]$$

INPUT DATA FOR *CQUEST*

The model was applied in annual time-steps from 1750 to 2050AD. For each year there was a different combination of temperature and CO_2 concentration as follows.

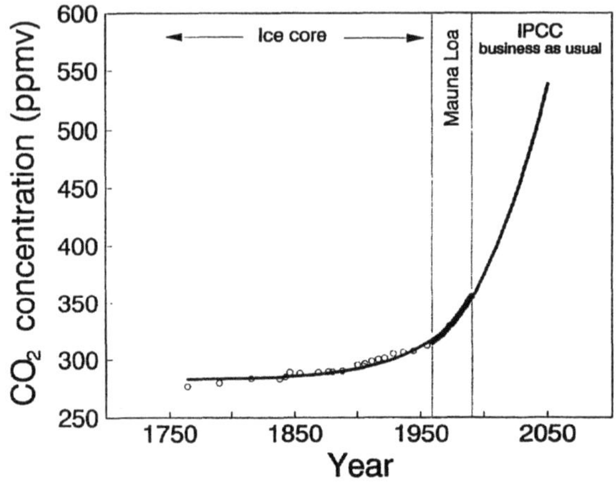

Figure 2. History of global atmospheric CO_2 concentration from 1750 AD to the present and projection to the future. Open circles are ice-core data from Siple Station, Antarctica (Friedli *et al.* 1986, Boden *et al.* 1990). Closed circles are the Mauna Loa record (Boden *et al.* 1990). The curve from 1750 to 1990 is a non-linear fit to the data points. The curve after 1990 is a non-linear fit to the "business-as-usual" scenario of the IPCC (Houghton *et al.* 1990).

CO_2 concentration: Atmospheric CO_2 concentration was essentially stable from about 1700 to 1860AD. It has increased faster than exponentially ever since (Figure 2). The smooth curve in Figure 2 is in fact a plot of two equations. The first was fitted to the combined data for ice core results from Siple Station, Antarctica from 1744 to 1958 and Keeling's annual average CO_2 record from Mauna Loa from 1959 to 1989 (Boden, Kanciruk and Farrell 1990). The second curve was fitted to the "business-as-usual" scenario presented in the Policymaker's Summary in the first report of the Intergovernmental Panel on Climate Change (Houghton, Jenkins and Ephrams 1990). The equations are:

$$C=279.25+2.860e^{[0.0041009+((0.00003716(y-1744))][y-1744]} \quad \text{for } 1744<y<1990, \text{ and}$$

$$C=259.03+96.1715e^{(0.02046+((-0.0000400(y-1991)))(y-1991)} \quad \text{for } 1990<y<2100.$$

For the initial year of the model (1750AD) $C = C_0 = 282.2$ ppmv

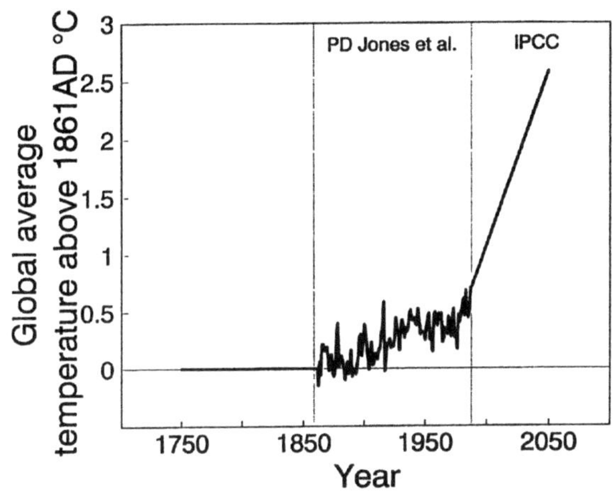

Figure 3. Assumptions used in ***CQUEST*** about the time course of global average temperature variations 1750-2050AD. For 1861 to 1989 the global average instrumental record of temperature according to the Climate Research Unit, Norwich (Boden *et al.* 1990) is plotted. It is assumed that before 1861 there were no temperature anomalies. After 1989 the IPCC "business-as-usual" scenario (Houghton *et al.* 1990) of 0.03 °C yr^{-1} is plotted.

Temperature: Palaeoclimatic reconstructions indicate that temperature was fairly stable from about 1700 to 1850AD (Folland *et al.* 1990). A value of T_0 = 17 °C was adopted for the annual average temperature of the world's vegetated land surface in 1750. The exact value is not important to the present exercise because we are assuming that the range of temperature change between 1750 and 2050AD is sufficiently small that soil organic matter decay and plant productivity change are linearly related to temperature change. For the period 1861 to 1988AD the global average temperature deviations from the 1861 value (uncorrected for El Nino effects), from Jones *et al.* at the Climatic Research Unit, University of East Anglia Norwich (Boden *et al.* 1990), were added to 17 °C. From 1989 to 2050AD the IPCC "business-as-usual" scenario of temperature increasing by 0.03 °C yr^{-1}

(Houghton *et al.* 1990) was assumed (Figure 3). These assumptions led to the 2050AD global average temperature being 19.58 °C, a value 1.8 °C above that of 1990AD.

Carbon in live vegetation (B_0): Estimates of the amount of C in live vegetation have been progressively reduced. Early estimates of around 1000 Gt C by Rodin *et al.* (1975) are believed to have involved extrapolation of data from unrepresentatively perfect examples of ecosystems occupying only limited areas, to larger areas. Furthermore, land use has progressively reduced standing biomass in many areas such that it is difficult to interpret data gathered over many decades in terms of present biomass. Nevertheless, Olson et al (1983) have compiled a global phytomass database and conclude that the carbon in live biomass, taking account of human disturbance up to the 1980s, was 565±100 Gt. In 1750AD, C in live biomass was probably higher than this because it pre-dated the deforestation of the Industrial Revolution and subsequent land exploitation. However ***CQUEST*** is not concerned with biomass loss from deforestation. Deforestation is considered separately (Houghton, this volume; and see below). Thus, since the outcome of the model runs (below) is to conclude that increasig CO_2 and temperature is increasing standing biomass, relative to that which would otherwise pertain, all else equal, a figure of 500 Gt C has been adopted here as a reasonable round figure for 1750 for purposes of seeding the model runs. Initiating the simulations with higher values of B_0 leads to higher estimates of current C sequestration into the landscape owing to the CO_2 fertilizing effect and temperature increase.

Carbon in soil organic detritus (S_0): The wide range of early approximate estimates of the C in global surface litter and soil organic matter, ranging from 700 Gt (Bolin 1970) to 9000 Gt (Reiners 1973), has now been reduced by systematic evaluations to a narrow range from 1456 Gt C (Schlesinger 1977) through 1395±200 Gt C (Post *et al.* 1982) to 1272 Gt C. For the present model a value for S_0 of 1200 Gt C in 1750AD was adopted.

Net primary production (P_0): Net primary production of an ecosystem can be defined as its annual gross photosynthetic carbon fixation less the annual plant respiration above- and below-ground. In practice this attribute cannot be measured directly. Usually it is estimated indirectly by trying to measure annual growth. However, that approach ignores losses of C via volatilization of hydrocarbons, leaching of soluble compounds like sugars and fructans, and exudation and sloughing of materials from roots. Furthermore, root growth is impossible to determine directly in the field. Spatial variability, even within apparently uniform simple ecosystems, is large. Thus the uncertainty is substantial and itself uncertain.

However, in association with his work compiling evidence on the global terrestrial C pool, Olson *et al.* (1983) has arrived at a figure for global net primary production around the 1980s of 60 Gt C. This figure allows for the fact that many ecosystems are now degraded through land use practices. For the present model, a conservative rounded figure of 50 Gt C has been adopted for 1750AD as applicable to the present vegetation back-dated to 1750AD CO_2 and temperature conditions.

PARAMETERISATION OF THE MODEL

Equation 2, k_1. It was assumed that in 1750AD temperature and CO_2 had been stable long enough for primary production, standing biomass and soil organic matter to be at equilibrium values (P_0, B_0 and S_0) with outputs matching inputs. From equation 2, in 1750AD

$$k_1 = P_0/B_0 = 50/500 = 0.10$$

Thus the average turnover time for standing biomass was $1/k_1 = 10$ years. This turnover time is in fact a composite figure for individual components ranging from days for root hairs to centuries for some live tree stems.

Equation 5, f_1, k_3, $k_{2,0}$. The temperature sensitivity coefficient for soil organic matter decay (k_3) and the proportion of prompt oxidation (f_1) was parameterised against a global soil organic matter decomposition study based on the Rothamsted decay model (Jenkinson, Adams and Wild 1991). The latter model has two input pools, decomposable plant material and resistant plant material each having a different decay coefficient. Microbial decay of these pools converts them into three components: CO_2, humified organic matter and microbial biomass. Humified organic matter and microbial biomass are also transformed into those three C forms when they in turn undergo microbial decay. In addition the

Rothamsted model includes a small pool of inert organic matter that does not decay on the timescale of decades. Each of the decomposable pools has a different decay constant which is sensitive to soil temperature and moisture. The model was developed for agricultural soils but was modified and parameterised for each of 16 global biomes using biological, soil and climatic data for each such life zone. The model runs with monthly increments. It was assumed by Jenkinson *et al.* that the global soil C pool was 1394 Gt. Their model required that the global litter input must be 75.8 Gt C yr^{-1} to sustain the assumed global C pool constant at constant temperature. Assuming those figures applied to 1990, the temperature regime was incremented at the rate of 0.03°C yr^{-1} -- equivalent to the IPCC business-as-usual scenario. The litter input was held constant at 75.8 Gt C yr^{-1} throughout their 60 year simulation. Consequently global soil organic matter levels declined in their model because warmer conditions accelerate decomposition. By 2050AD about 60 Gt soil C had been lost in the Rothamsted simulation. In reality, rising temperature and CO_2 concentration also increases the litter input, as is modelled in the present paper. Consequently, the Rothamsted simulation oversestimated the *net* emission from the soil under the climate change scenario.

Equation 5 was parameterised against the global decomposition simulation using the Rothamsted model. To achieve that parameterisation, $L = (1-f_1)k_1B$ was first set at 75.8 Gt C yr^{-1} (as used in the Rothamsted model) and $f_1=0$ was assumed, thereby eliminating the fast decay pool and treating the soil as a single component decay system. The parameter $k_{2,0}$ = 75.8/1394 = 0.0544 was used. Then k_3 was varied until equation 5, applied at annual time-steps over the 60 year period of warming from 1990 to 2050AD, gave results that most closely matched the time course of soil C loss predicted by the Rothamsted model. When f_1 = 0, the value of k_3 giving the best overall fit over the range of temperature appropriate to 1990 to 2050AD was k_3 = 0.031. The result is shown, in terms of its deviation from the Rothamsted model, as curve A in Figure 4. The maximum deviation from the Rothamsted model output was 13% by 2050AD. As expected; the fit of equation 5 to the result of the Rothamsted model was improved with $f_1 > 0$. The most exact fit found was for the combination f_1=0.33 and k_3=0.040 (Figure 4, curve C). For that combination the fit was near perfect over the range of temperature examined. With f_1 set to 0.67 it was necessary to increase k_3 to 0.07 to obtain a reasonable fit to the Rothamsted model (Figure 4, curve B), but it was not as good a fit as for f_1 = 0.33. Thus the Rothamsted model responds, for relatively small temperature changes at least, as a two compartment global soil C pool with a temperature coefficient for decay for the slow pool, 0.04 $^\circ C^{-1}$. As standard values in ***CQUEST*** it was assumed that k_3 = 0.040 and f_1=0.33. Values for k_3 of 0.050 and 0.030 were adopted as upper and lower plausible limits, and values of 0.53 and 0.13 were assumed as upper and lower plausible values of f_1. However, in the light of a view expressed by some colleagues that the proportion of prompt decomposition might be as high as 0.67 and that the Q10 of soil organic matter decomposition could be 2 or higher, I have also included runs of the model with f_1 = 0.67 and k_3 = 0.070 (corresponding to Q_{10}=2).

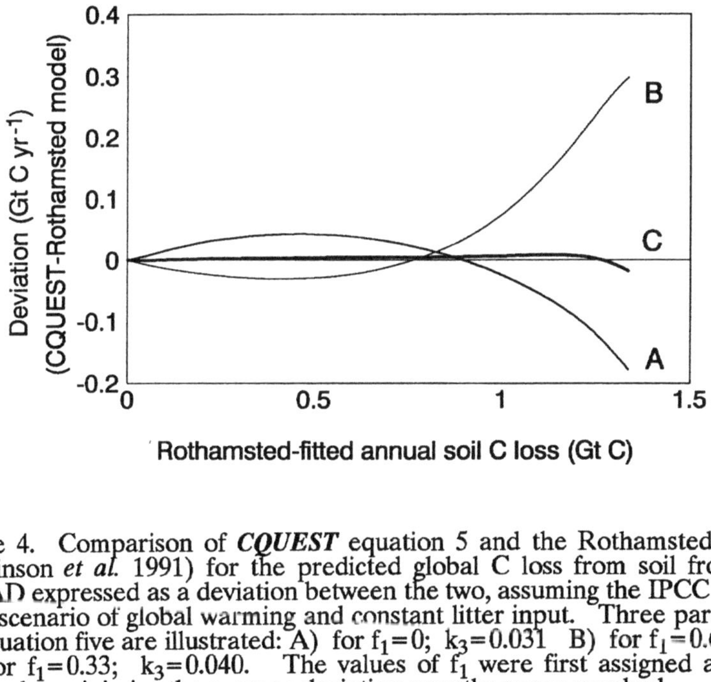

Figure 4. Comparison of ***CQUEST*** equation 5 and the Rothamsted simulation (Jenkinson *et al.* 1991) for the predicted global C loss from soil from 1990 to 2050AD expressed as a deviation between the two, assuming the IPCC business as usual scenario of global warming and constant litter input. Three parameter sets for equation five are illustrated: A) for f_1=0; k_3=0.031 B) for f_1=0.67; k_3=0.07 C) for f_1=0.33; k_3=0.040. The values of f_1 were first assigned and then k_3 adjusted to minimize the average deviation over the range graphed.

As for k_1, the decay coefficient for soil organic matter ($k_{2,0}$) was evaluated assuming that in 1750 the C pool sizes were in equilibrium. Thus, from equation 5

$[k_{2,0} = L_0/S_0 = ((1-f_1)k_1B_0)/S_0]$

for $f_1 = 0.13$, $k_{2,0} = 0.0363$

for $f_1 = 0.33$, $k_{2,0} = 0.0279$

for $f_1 = 0.53$, $k_{2,0} = 0.0196$

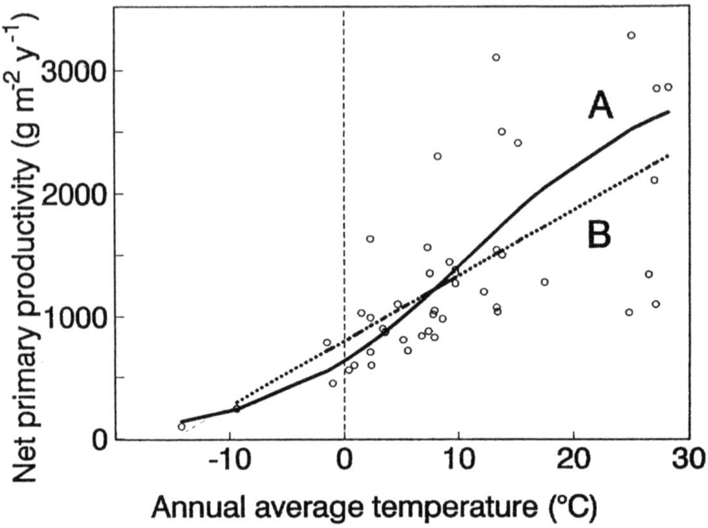

Figure 5. Net primary productivity of diverse ecosystems of the world plotted against their annual average air temperature (from Lieth 1975). The dashed line is a linear regression to the points. The solid curve is the "Miami" model fitted by Lieth to the data (P=3000/(1+exp(1.315-0.119T))) where T is in °C.

Equation 6, k_4. To allow for the impact of global temperature increase on net primary production, P_0, we can draw upon data compiled by Lieth (1975) from ecosystem productivity research papers (Figure 5). Lieth fitted the sigmoidal function

$$P = 3000/(1+e^{(1.315-0.119T)})$$

to the scatter-plot (curve A in Figure 5), where T is annual average temperature (°C). This function has a slope of 0.04 $°C^{-1}$ at 17°C. A linear regression to the same data-set (Figure 5) has a slope of 0.031 $°C^{-1}$ and an zero-productivity intercept close to -15°C. The positive effect of annual average temperature on productivity in that data-set is probably partly attributable to precipitation because there is a positive correlation of annual average temperature with annual rainfall within the data set. Linear regression analysis shows that at 17°C the increase of annual rainfall with annual average temperature in the Lieth data set is at the rate of about 0.03 $^{\circ}C^{-1}$. Fortuitously and conveniently, this rate is similar to that which General Circulation Models predict for the increase in global average precipitation with greenhouse effect warming (Schlesinger and Mitchell 1985). Thus by using the correlation between temperature and productivity for diverse ecosystems around the world as a measure of the greenhouse warming effect on productivity at equilibrium, it is likely that the impact of associated increase in global average rainfall is approximately taken into account. A multiple linear regression of productivity against both precipitation and temperature yields a relationship that ascribes only 0.01 $^{\circ}C^{-1}$ to temperature at 17 °C rather than 0.03 $^{\circ}C^{-1}$ obtained with simple linear regression. For ***CQUEST*** a temperature coefficient of k_4=0.03 $^{\circ}C^{-1}$ was adopted as a standard value, with the assumption that that figure implicitly incorporates the effect of an associated rainfall increase with global warming. For the maximum and minimum C sequestration estimates k_4 values of 0.05 and 0.01 were used.

Equation 10, the CO_2 response of P_0.

The CO_2 compensation point, C_b. C_b is determined by the balance between the oxygenase activity and the carboxylase activity of photosynthetic carbon fixation. It is determined by the kinetic properties of the central CO_2 fixing enzyme, ribulose bisphosphate carboxylase/oxygenase. The carboxylase/oxygenase ratio is conservative among species of the predominant C_3 type and is largely independent of environmental factors except for CO_2 and O_2 concentration and temperature. Oxygenation increases relative to carboxylation with increasing temperature and oxygen concentration, and decreases with CO_2 concentration (Jordan and Ogren 1984). As an approximation, M. Kirschbaum (pers. comm.) has found that C_b for C_3 species in normal atmospheric oxygen can be described by the relation

$$C_b = (0.5116 \mathrm{x} 10^6)\, e^{(-2815.5/(T+273.2))} \qquad (11)$$

where T is in °C. This gives C_b = 31 at 17 °C. The model, however, is not very sensitive to variation in C_b in the range of temperature and CO_2 concentration considered here.

The ratio of plant respiration to gross photosynthesis, f_2. As a broad generalization respiratory CO_2 loss from vegetation consumes about 50% of the gross photosynthetic C fixed. For young plants that are the usual subject of controlled environment research, the value may often be smaller than 0.5. For trees that must maintain a large mass of tissue relative to the annual CO_2 fixed, the value may be higher perhaps up to 0.7. There is a strong short term (i.e. within season) temperature sensitivity of respiration (Paembonan 1991). However that effect is is not necessarily expressed in terms of the R/A ratio (f_2) especially when

TABLE 1. The ratio of daily total respiration to gross photosynthesis ($R/A=f_2$) for cotton (*Gossypium hirsutum*; average age 45 days) pine (*Pinus radiata*; average age 273d) and river red gum (*Eucalyptus camaldulensis*; average age 114d) grown and measured at four temperatures. For each species there were 4 replicate plants per temperature; standard deviations are shown in parenthesese. RGR is the relative growth rate on the day of measurement of f_2.

	Cotton		Pine		Eucalyptus	
Growth temp. °C	RGR (d^{-1})	f_2	RGR (d^{-1})	f_2	RGR (d^{-1})	f_2
15	-	-	0.0059	0.42 (0.03)	0.007	0.47 (0.01)
20	0.018	0.43 (0.02)	0.0070	0.41 (0.03)	0.011	0.40 (0.01)
25	0.024	0.43 (0.00)	0.0061	0.52 (0.01)	0.014	0.40 (0.06)
30	0.027	0.40 (0.01)	0.0062	0.52 (0.03)	0.012	0.39 (0.01)

plants have grown at different temperatures (R.M Gifford, unpubl). For example, data for f_2 evaluated over 24 hour periods for cotton (*Gossypium hirsutum*), river red gum (*Eucalyptus camaldulensis*) and pine (*Pinus radiata*), that were both grown from seed and measured, at different controlled temperatures do not show a consistent or substantial temperature effect on f_2 despite large differences in age, size and relative growth rate between species (Table 1). Photosynthesis and respiration vary approximately in parallel with variation in temperature. Another series of experiments (Gifford 1991) has shown a small but consistent decrease in f_2 when plants are grown in 2x normal CO_2 concentration for diverse species. In the light of that evidence, f_2 was assumed in present model to be independent of CO_2 and temperature. A later refinement to the model, when more experimental evidence is available, may be to make f_2 slightly temperature and CO_2 dependent, but it is not expected that such a refinement would have much impact on the conlusions from the model.

Coefficient k_6. k_6 is the coefficient that determines the CO_2 sensitivity of net primary production. It is the most important parameter in ***CQUEST*** for determining the rate of carbon sequestration. It cannot be deduced solely from the biochemistry of photosynthesis because several feedbacks - both negative and positive - can interpose between the primary biochemical responses and the final expression as acclimated annual production. Most experiments on the long term CO_2 response of growth and productivity are carried out at normal and twice normal atmospheric CO_2 concentration. For experiments conducted in the 1970s to 1980s the normal CO_2 concentration was about 340ppmv. A wide range of seasonal growth responses to CO_2 doubling under controlled environment conditions have been recorded, ranging from slightly negative to a six-fold increase, with a most common and average value being a 30% increase (Kimball 1983, Cure 1985). Idso (1991) argues that this figure is far too small for natural vegetation in the field. In contrast, there has been a traditional reluctance among terrestrial ecologists to accept the existence of a CO_2 fertilizing effect for natural vegetation (King *et al.* 1992) arguing that, because other environmental factors like temperature, water, light and nutrient supply limit field productivity, CO_2 concentration cannot also impose a limitation. This latter view is a simplistic application of the "law of limiting factors" that is rarely applicable (Gifford 1992).

Over a season of growth several environmental factors usually co-limit growth and organic C accumulation; and alleviation of one such restriction can lead to more effective acquisition of other limiting resources. For CO_2 co-limiting with other factors, that principle has been demonstrated with respect to limiting light and water supplies (Gifford 1979, Gifford 1980, Gifford 1992). With regard to the CO_2 x temperature interaction, while there are plausible physiological reasons for expecting relative CO_2 responsiveness of productivity to increase from small at annual average temperature close to 0°C to high at >25°C (Gifford 1992), the experimental evidence is equivocal (reviews by Kimball 1985, Cure 1985, Rawson

1992). Sometimes the CO_2 effect on growth at low temperatures seems too high, other times too low to be readily explicable. The explanation may lie in the species-specific "source-sink" interactions with respect to photo-assimilate supply.

The situation is also not yet fully clear for terrestrial productivity that is restricted by nitrogen and phosphorus supply because of the paucity of experimental evidence (Cure 1985). There is an important distinction between terrestrial and marine productivity. Whereas the species composition of marine biota is dependent on temperature and perhaps CO_2 concentration (Fogg 1991), marine primary productivity is essentially uninfluenced by atmospheric CO_2 concentration and temperature. Rather it is dependent on the upwelling or inflow of mineral nutrients (Platt and Subba Rao 1975), especially phosphate and available nitrogen, that are constantly sedimenting out, as organic particulates, from the illuminated zone. Ocean biological productivity is tightly coupled to the nutrient supply.

Carbon accumulation in terrestrial plants is not so mineral limited as in phytoplankton because dead material stays within the root zone where minerals are recycled. Furthermore, unlike for marine organisms, a large and variable fraction of the organic carbon content is as mineral-free cell wall material. The most abundant organic molecule on earth is probably cellulose. Cellulose, containing no mineral elements, probably comprises about 30% of the terrestrial productivity (Albersheim 1965). Also lignin, another mineral-free molecule which is in the plant cell wall especially in wood, is thought to be about 60% as abundant as cellulose (Freudenberg 1964) and comprises 17-33% of wood (Higuchi 1980). Thus, where there is a strictly limited amount of mineral available for uptake in any one season, CO_2-stimulated growth can be invested in these molecules, as well as soluble carbohydrates and starch, thereby widening the C:mineral ratio of phytomass. The limited available evidence suggests that elevated CO_2 concentration can increase the efficiency of use of available nitrogen (Wong 1980, Hocking and Meyer 1991) and phosphate (Conroy *et al.* in press; Norby and O'Neill 1991) in plant growth. I hypothesise that as the relative carbon content of terrestrial ecosystems gradually increases with the gradual CO_2 enrichment of the atmosphere, net nitrogen capture by ecosystems will follow because N acquisition, be it by symbiotic or free-living N fixing micro-organisms, is largely at the expense of organic carbon oxidation by plants and microbes (Gifford, 1992). It is also possible that the increased availability of organic carbon may increase the efficiency of recovery of available nutrients, increase the conversion of unavailable nutrients to available ones, and reduce gaseous nitrogen losses from ecosystems.

While it is still premature to be categorical about the CO_2 doubling effect on long term global productivity over the time scale of years to centuries, I currently judge, based on the above cited literature and personal research experience, that a realistically conservative best guess is a 25% increase from a 340ppmv CO_2 base (r=1.25). That figure has been used in the model as the standard, with 40% (r=1.40) and 10% (r=1.10) as maximum and minimum credible values for the CO_2 doubling response of global terrestrial primary production. I see the lower value as being appropriate for the situation in which the higher CO_2 concentration solely widens the C:mineral ratio of plant tissues without any tendency for the increased chemical energy supply in the environment (as carbohydrates) and root activity leading to increased mobilization of mineral resources such as by increased nitrogen fixation and myccorrhizal associations. If the generally water-limited condition of the terrestrial biosphere leads to strong expression of the CO_2 fertilizing effect, then the higher figure (r=1.4) may be the more appropriate.

From equation (10)

$$k_6 = \frac{(680 - C_b) - r(340 - C_b)}{(340 - C_b)(680 - C_b)(r - 1)}$$

In the present study r was not varied with temperature. To the extent that r actually does increase with temperature, the present version of ***CQUEST*** slightly understates the rate of increase of net C sequestration as greenhouse effect warming occurs.

Coefficient, k_5'. After establishing P_0 (from Equation 6) and k_6 for the temperature of a year, k_5' was determined as

$$k_5' = \frac{P_0(1 + k_6(282.2 - C_b))}{(282.2 - C_b)}$$

Figure 6 shows the form of the curves for P_0, for the maximum, standard and minimum assumptions, with its value fixed at 50 Gt C at 282.2 ppmv CO_2

appropriate to 1750AD conditions. Also shown is the curve for conditions 2.58°C warmer and otherwise standard conditions.

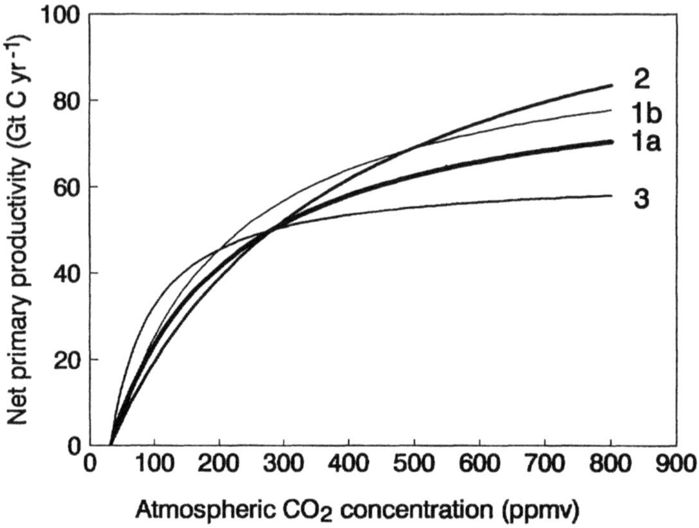

Figure 6. Simulated global net primary production as a function of CO_2 concentration according to equation 10 with parameter values as used in the model for 1a) standard 2) maximum and 3) minimum carbon sequestration into the biosphere. A fourth curve (1b) is the same as the standard curve except temperature is 2.58°C higher. 2.58°C is the temperature assumed for 2050AD above the 1860AD value.

THE RESULTS OF THE SIMULATIONS

The model was run with the various values of the parameters f_1, k_3, k_4 and r indicated above, as summarised in Table 2. The first three simulation runs use what the author considers, in the light of experimental evidence reviewed previously, to be reasonable bounds on the parameter values for estimating 1) standard 2) maximum and 3) minimum global rates of carbon sequestration into organic C. There are also extra simulation runs using some parameter values outside those bounds. These are to cover the range of possibilities advocated by commentators on this work.

Figures 7 and 8 show the repercussions of using the standard values of the parameters in ***CQUEST*** when it is initialised with 1750AD conditions and run to 2050AD. The results in Figure 7 are for CO_2 concentration alone increasing according to Figure 2 with temperature fixed at 17°C and using the standard assumptions (Table 2, Run 1). The annually increasing net primary production owing to the CO_2 fertilizing effect is partitioned into phytomass (Figure 7a) and soil (plus litter) C (Fig. 7b). In 1991 the modelled global phytomass C increased by 0.9 Gt C yr^{-1} while soil C increased by about 1.3 Gt yr^{-1} on account of the CO_2 effect alone. The total annual rate of C sequestration in the standard run at constant temperature is 2.2 Gt C yr^{-1} in 1991AD and that annual rate is itself increasing (Figure 7c).

TABLE 2. Values of parameters for the simulation runs of ***CQUEST***. All runs of the model assume B_0 = 500 Gt C, S_0 = 1200 Gt C, P_0 = 50 Gt C yr^{-1}, and f_2 = 0.5.

Model run	f_1	k_3	k_4	r
1. Standard sequestration	0.33	0.04	0.03	1.25
2. Maximum sequestration	0.13	0.03	0.05	1.40
3. Minimum sequestration	0.53	0.05	0.01	1.10
4. Standard with k_3 = 0.07	0.33	0.07	0.03	1.25
5. Absolute minimum	0.67	0.07	0.01	1.10
6. Standard with k_4 = 0.07	0.33	0.04	0.07	1.25
7. Standard with f_1 = 0.67	0.67	0.04	0.03	1.25

Figure 8 shows the results of keeping CO_2 concentration constant at 282ppmv while allowing temperature to increase according to Figure 3 with the standard parameter values. Between 1860 and 1989 annual increase in biomass C oscillated between positive and negative with no trend. After 1990 when a continuing annual greenhouse effect warming is assumed, without interannual variability, there is a steady net annual sequestration of C into the vegetation of about 0.4 Gt C yr^{-1}.

Unlike output from the Rothamsted model (Jenkinson *et al.* 1991), temperature increase alone does not cause a decrease in global soil C in ***CQUEST*** because the warming conditions increase primary production and hence litterfall by an amount that compensates for the increased soil organic matter oxidation. Although not explicitly modelled in ***CQUEST***, the increased rate of nitrogen mineralisation of organic matter at warmer temperatures would provide available N that would support the temperature-stimulated productivity and litter input.

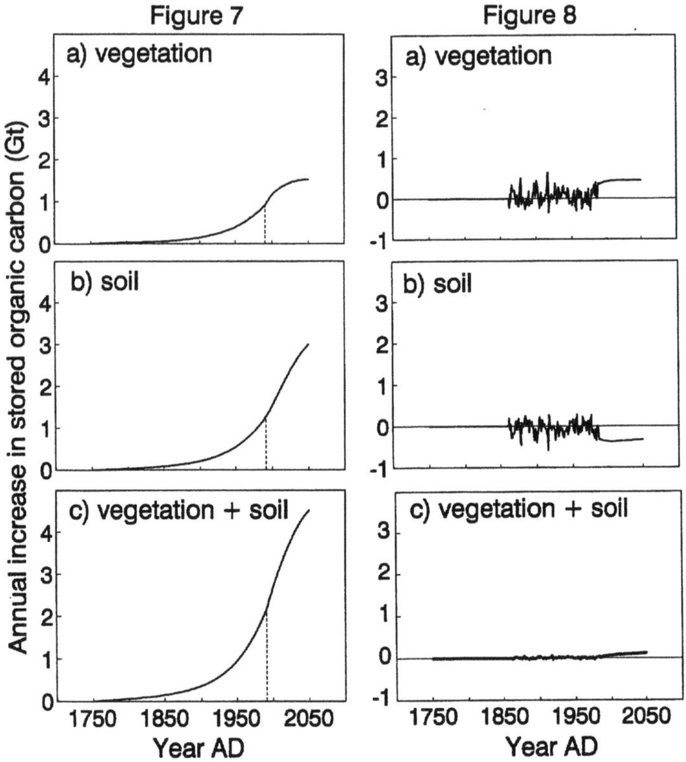

Figure 7. ***CQUEST*** simulation of terrestrial C sequestration as a function of atmospheric CO_2 increase from 1750 to 2050AD, temperature held constant, assuming the standard values of parameters listed in Table 2. a) C accumulation into live phytomass b) carbon sequestration into litter and soil organic matter c) the sum of a) and b). The vertical dashed line marks 1991AD.

Figure 8. ***CQUEST*** simulation of terrestrial C sequestration as a function of annual average temperature from 1750 to 2050AD, CO_2 concentration held constant, assuming the standard values of parameters listed in Table 2. a) C accumulation into live phytomass b) carbon sequestration into litter and soil organic matter c) the sum of a) and b). The vertical dashed line marks 1991AD.

Under the standard assumptions, the two opposing processes of faster soil C oxidation and higher litter input are in approximate balance. Because the noise in the temperature record from 1860 to 1989 acts in opposite senses in vegetation compared with soil C, the oscillation in the sum of the two (Fig. 8c) is much smaller than is apparent for each individually (Figures 8a, 8b). The total C net sequestration into the plant and soil pools combined as a result of the warming scenario alone is +0.03 Gt C yr^{-1} in 1991.

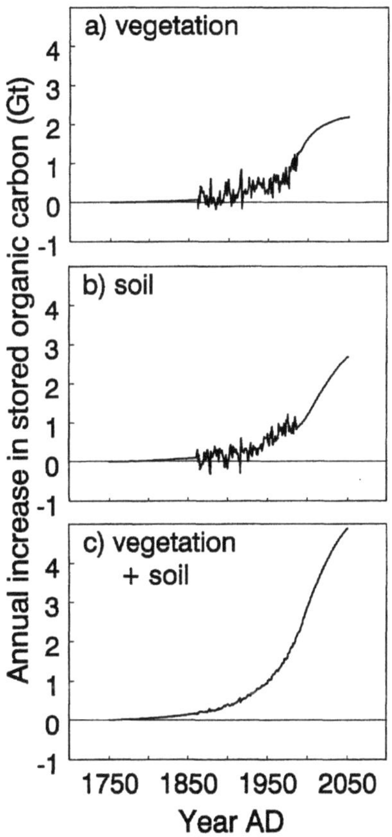

Figure 9. ***CQUEST*** simulation of terrestrial C sequestration as a function of atmospheric CO_2 and temperature change combined from 1750 to 2050AD assuming the standard values of parameters listed in Table 2. a) C accumulation into live phytomass b) carbon sequestration into litter and soil organic matter c) the sum of a) and b).

Comparison of the results in Tables 7 and 8 indicates that the direct impact of the CO_2 concentration increase on carbon storage is probably much greater than the indirect effect via global warming. The combined interactive effects of both CO_2 concentration and temperature increase are shown in Figure 9. The total annual rate of C sequestation into soil and vegetation is, for the standard parameterisation of ***CQUEST***, 2.3 Gt C yr^{-1}.

The maximum and minimum assumptions for the various parameters in Table 2 (Runs 2 and 3) represent ranges which are plausible. The results of applying those values in ***CQUEST*** are shown in Figure 10. The assumptions for minimum C sequestration assumptions yield a net sequestration into soil plus plant of 0.5 Gt C yr^{-1} in 1991. The assumptions for maximum C sequestration lead to a simulated net sequestration of 4.1 Gt C yr^{-1} in 1991.

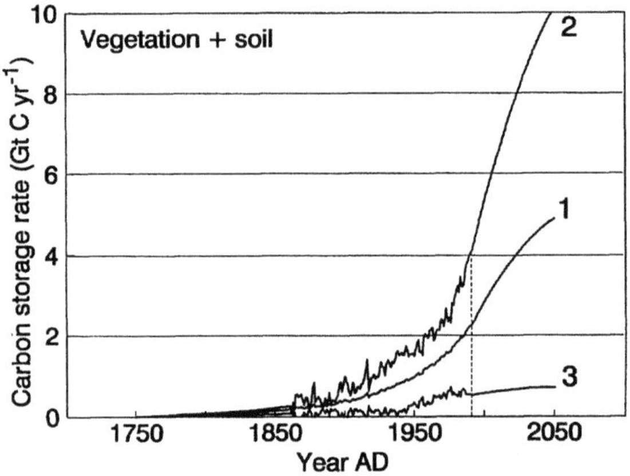

Figure 10. ***CQUEST*** simulations of annual C sequestration into global soil plus vegetation from 1750 to 2050AD for parameterisations that give rise to standard (Run 1), maximum (Run 2), and minimum (Run 3) global sequestration. The vertical dashed line marks 1991AD.

The results of the standard simulation (Run 1) are plotted in terms of the time course of the absolute size of the global soil and vegetation C pools in Figure 11. The modelled increase in the size of these pools since 1750AD is rather small relative to their absolute size. The change would be undetectable by direct measurement of pools in any particular ecosystem given the paucity of data on

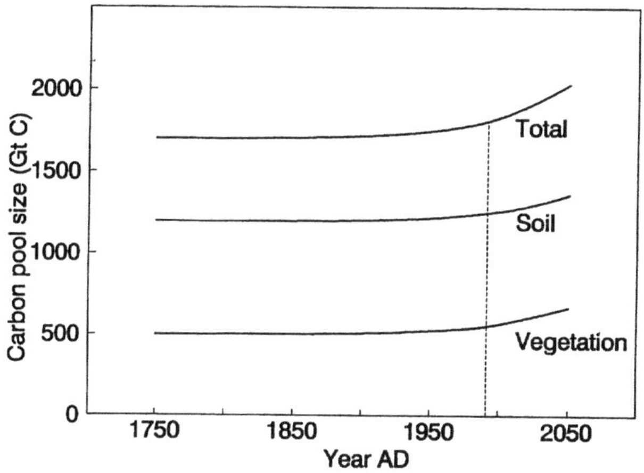

Figure 11. ***CQUEST*** simulation of the global C pool sizes in vegetation, soil and vegetation + soil from 1750 to 2050AD. The vertical lines marks 1991AD.

even current pool sizes, quite apart from the absence of detailed pre-industrial data with which to compare the present.

Although the parameter values for simulation runs 1, 2 and 3 are my currently favored range of values on the basis of global-level biosphere information as well as mechanistic and small scale experimental findings reviewed above, there is nevertheless scope for differing views given the paucity of field evidence and the impossibility of long term controlled CO_2 enrichment experiments on mature tall forests. Simulation runs 4, 5, 6 and 7 cater for some parameter combinations outside that favored range (Table 2). The results of those runs in terms of annual C sequestration rates in 1991 are summarised in Table 3. The time-courses of annual sequestration are shown in Figure 12.

Run 4 is the same as the standard run except that a much higher coefficient for the temperature sensitivity of soil organic matter decomposition is used. The value assumed (k_3 = 0.07) is equivalent to a Q_{10} of 2. This value is one traditionally assumed for thermochemical reactions in general and for soil organic matter decomposition in particular (Tinker and Ineson 1990). However, as for plant respiration (see Table 1) the respiratory sensitivity to temperature of soil is not necessarily as high when the soil microbe populations have developed at different temperatures as when they are just measured at different temperatures. It was not possible to fit equation 5 of ***CQUEST*** to the Rothamsted model output using $k_3 = 0.07$ (see above). A distinction needs also to be made between the decomposition characteristics of the bulk of the soil organic matter pool and those of typical experimental fine litter samples, like cereal straw, or litter surrogates like pure cellulose or glucose. Much of the data on kinetics of decomposition relate to the latter. When the higher k_3 value is used (Run 4) the total C sequestration rate is 1.8 Gt C yr^{-1} (Table 3).

Table 3. The ***CQUEST***-simulated rates of C sequestration into vegetation and soil in 1991AD for seven parameter sets (see Table 2).

Model run	1991AD C sequestration rate (Gt C yr^{-1})		
	Vegetation	Soil	Total
1. (Standard)	1.31	0.94	2.26
2. (Maximum)	2.07	2.02	4.08
3. (Minimum)	0.52	0.00	0.52
4.	1.31	0.52	1.83
5.	0.52	-0.15	0.37
6.	1.89	1.27	3.16
7.	1.31	0.71	2.02

The fifth parameterisation could be seen as an absolute minimum one for C sequestration. It assumes what the author regards as the minimal plausible CO_2 sensitivity of global net primary production ($r = 1.1$), a high temperature sensitivity of soil C decomposition ($k_3 = 0.07$), a low temperature sensitivity of primary production ($k_4 = 0.01$), and also assumes that two thirds of the litter production (including large litter like tree-trunks) is oxidised within the year of its production. Such a low value of k_4 is tantamount to ignoring the annual rainfall increase that correlates with annual temperature increase. With these extreme assumptions, vegetation stores only 0.52 Gt C yr^{-1} while soil loses 0.15 Gt C yr^{-1} in 1991. Overall, the absolute minimum model run sequesters 0.37 Gt C yr^{-1} in 1991.

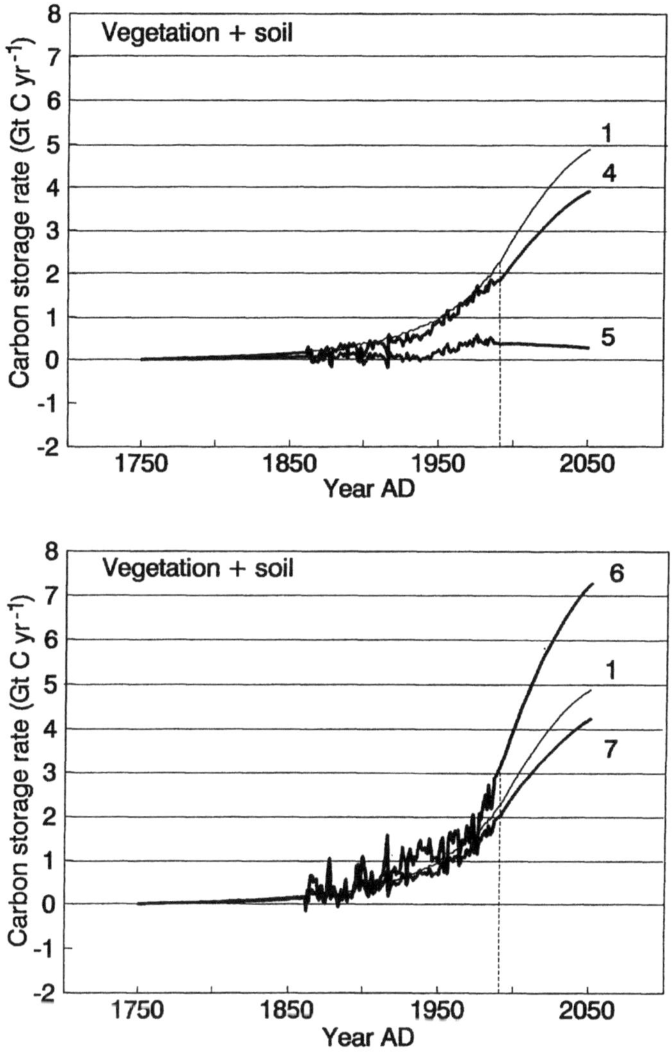

Figure 12. CQUEST simulation of annual C sequestration into the global vegetation plus soil from 1750 to 2050AD for model runs 4, 5, 6, and 7 compared with standard run 1 (see Table 2). The vertical lines mark 1991AD.

The sixth set of parameters is complementary to the fourth in that it is the standard set except the temperature sensitivity of primary production is assumed to be equivalent to a Q_{10} of 2 (k_4 = 0.07). For such a high value of k_4 to be realistic it is probably necessary to regard it as indirectly embracing the impact of a substantial increase in annual rainfall correlated with the annual temperature increase. Then both soil and vegetation become larger sinks than in the standard run and the total 1991AD sequestration rate is 3.2 Gt C yr^{-1}.

Although the best fit of equation 5 to the output of the Rothamsted model occurred with f_1 = 0.33, there is a view that more than 50% of litter production may oxidise in the year of its production. While the currently growing recognition that large litter (branches, trunks, coarse roots) represents a significant part of total litter (Harmon and Hua 1991) suggests that an even lower value of f_1 than 0.33 might be appropriate, Run 7 examines the impact of assuming that two thirds of litter is oxidises within the year of its formation (f_1 = 0.67). Then the total terrestrial sink in 1991AD declines from the standard run value of 2.3 to 2.0 Gt C yr^{-1}.

THE RESULTS EXPRESSED IN TERMS OF THE "BIOTIC GROWTH FACTOR"

Primary production as a function of CO_2 concentration cannot be represented satisfactorily by a logarithmic relationship with a single coefficient (Gifford 1980). The rectangular hyperbolic form, with three parameters, comes closer to reality. Nevertheless, many terrestrial C cycle models include a logarithmic function having a single parameter, the "biotic growth factor", β. The logarithmic function can be applied to net primary production (Goudriaan and Ketner 1984), or to annual CO_2-dependent C storage. The coefficient is a relative response ratio, i.e. the relative change in net primary production (or annual C storage) per unit relative increase in atmospheric CO_2 concentration (or atmospheric CO_2 inventory). For purposes of permitting readers to compare the present model's CO_2 sensitivity with that of others, relative response ratios are here calculated, using results of the standard parameterisation, with respect to:

a) net primary production (as in Figure 6), i.e. (C/P)(dP/dC)
b) annual global C storage in the absence of any temperature changes (from Figure 7c), i.e. (C/W)((dW/dt)/(dC/dt))
c) cumulative carbon storage since 1750AD, i.e. $(C/W)((W-W_0)/(C-C_0))$
d) the method of approximation for annual carbon storage developed by Keeling *et al.*, namely $[dW/dt]/[(C-C_0)x2.12x0.04]$ where 2.12 is the Gt C per ppmv in the

global atmosphere (Piper pers. comm., Keeling *et al.* 1989). This definition relates to accumulation in the "long-lived" biosphere.

The results for the first definition based on net primary production are plotted in Figure 13. The other three are plotted in Figure 14. The discontinuity in the curve A in Figure 14 is an artifact deriving from a slight difference in the absolute value and slope for the two equations comprising the curve in Figure 2 when evaluated for 1991AD. The relative response ratios in 1991AD for definitions a), b), c), are 0.41, 0.25, and 0.25. For (d) β=0.35. Whereas the ratio calculated by method a) is uniquely dependent on CO_2 concentration, its value in the other methods depends also on the time-course of CO_2 concentration. Consequently, in curves A and B (Figure 13) the initial relative response to a CO_2 increment in 1750AD is close to zero, but through time it builds up as the pool sizes adjust.

SIGNIFICANCE

The question of why the global C budget does not seem to balance has challenged global C-cycle modellers for about two decades (e.g. Kerr 1977, Siegenthaler and Oeschger 1978, Broecker *et al.* 1979, Gillis 1991). Much of the discrepancy in the

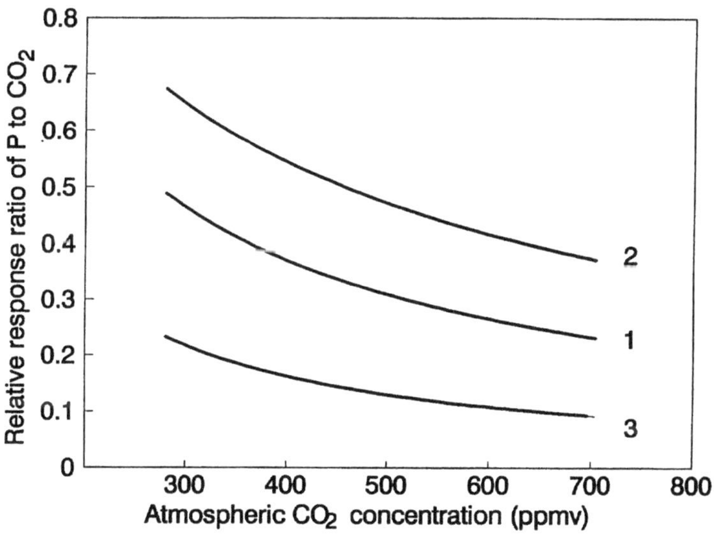

Figure 13. The relative response ratio of net primary production to CO_2 increase [(C/P)/(dP/dC)] for curves 1a, 2, and 3 in Figure 6.

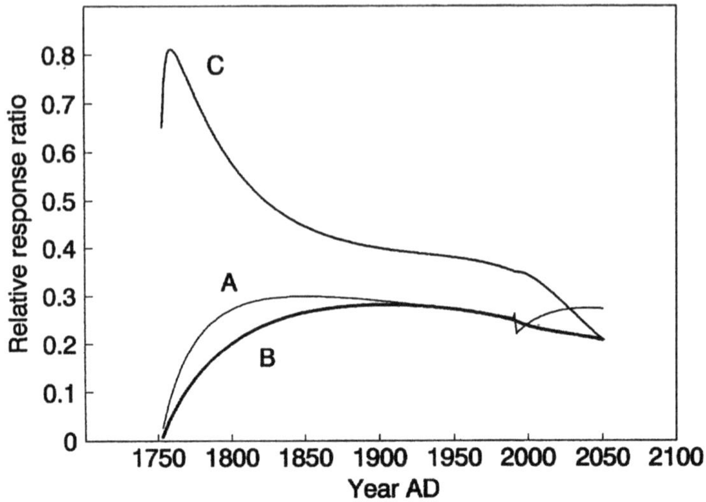

Figure 14. The relative response ratios for carbon storage in the terrestrial biosphere calculated in two ways from Figure 7c, A) as (C/W)((dW/dt)/(dC/dt)); B) as $(C/W)((W-W_0)/(C-C_0))$; and C) ß = $[dW/dt]/[(C-C_0)x2.12x0.04]$ according to the approximation method of Keeling et al. 1989.

late 1970s between estimates of fossil fuel plus deforestation-related sources of CO_2 and known sinks has been resolved by progressive reduction in early high estimates of deforestation rates that at least matched fossil fuel emissions (Woodwell and Houghton 1977), to values in the range 0.9-2.7 Gt C yr^{-1} in 1980 (Houghton *et al.* 1987) and 1.5-3.0 in 1989 (Houghton 1991 and in this volume). Similarly, Myers (1989, 1990) estimated that tropical deforestation was causing release of about 2.4 Gt C yr^{-1} in 1989, an increase from an estimated 1979 release of 1.7 Gt C yr^{-1}.

Thus at present, assuming that fossil fuel C release is about 6 Gt C yr^{-1} (Marland, this volume) and tropical deforestation is as high as 2.4 Gt C yr^{-1}, there is a total of 8.4 Gt C yr^{-1} to account for. The current annual atmospheric CO_2 increase of 1.8 ppmv is equivalent to 3.8 Gt C yr^{-1}. The oceans appear to be removing about 2 Gt C yr^{-1} (Siegenthaler, this volume). Thus there is a further 2.6 Gt C yr^{-1} (approx) to be accounted for. Net C sequestration in the ***CQUEST*** model using the standard set of parameters (Run 1) is 2.3 Gt C yr^{-1} (Table 3), a value similar to the "missing carbon". If the set of parameters for the minimum plausible estimate are assumed, then the model net C sequestration of 0.5 Gt C yr^{-1} falls short of matching the "missing carbon". Assuming the maximum biologically reasonable estimate (run 2) with its 4.1 Gt C yr^{-1} sequestration, would create a problem of a missing source rather than a missing sink. An implication of C removal from the atmosphere by terrestrial C pools is that much of it will be in tropical areas, where net primary production is high; this would tend to cancel out tropical deforestation in terms of its effect on the latitudinal variation in atmospheric CO_2 concentration. Also most of the non-tropical net sequestration into terrestrial sinks must be in the northern hemisphere where most of the temperate vegetation is.

Carbon dioxide is not the only anthropogenic emission that is likely to be stimulating the terrestrial biosphere. There are also large emissions into the atmosphere of nitrogen and sulphur oxides which are in part deposited as sulphates and nitrates. While such pollutants may be at sufficiently high concentrations in some areas near population centers to have toxic effects on plants, perhaps in association with high levels of atmospheric ozone, they are nevertheless plant fertilisers which augment the purposeful additions of fertilizer to the environment. Gifford (1987) concluded that the phytomass C storage generated by the CO_2 and nutrient enrichment, together with several minor C sinks were of sufficient magnitude potentially to balance the global C budget. These sinks are concentrated in the northern hemisphere. However, in the latter evaluation it was assumed that ocean uptake was up 2.3 to 2.8 Gt C yr^{-1} (Bolin 1983, Broecker *et al.* 1979). Ocean uptake is now being set at about 2 Gt C yr^{-1}. This laboratory's original estimate of C sequestration into the terrestrial biosphere on account of the CO_2 fertilizing effect of about 1 - 1.6 Gt C yr^{-1} (Gifford 1980) is here being replaced with the ***CQUEST*** model estimate of 0.5 - 4 Gt C yr^{-1} with 2 Gt C yr^{-1} involving, on present evidence, the mechanistically most plausible assumptions.

The conclusion that the terrestrial biosphere response to the gradually increasing CO_2 concentration, alone or together with anthropogenic mineral nutrient enrichment of Earth, can account for the seemingly missing C in global budgets that ignore it, adds weight to several other publications with a similar conclusion or indication (e.g. Goudriaan and Ketner 1984, Goudriaan 1991, Esser 1987, Thornley *et al.* 1991). It also supports Keeling *et al.* (1989) who found that it was necessary to include a CO_2 fertilization response of the terrestrial biosphere, amounting to 2.1 Gt C yr^{-1} in 1980AD (Keeling et al. 1989), in order to balance the global C budget using a three dimensional box diffusion model of atmospheric CO_2 transport.

There are several important areas requiring further investigation to validate the parameterisation of ***CQUEST*** and to reduce the range of uncertainty concerning the impact of CO_2 and temperature change on C storage by the terrestrial biosphere. First there is a great need to improve the estimates of C in live biomass and net primary production of Earth. Although estimates of these attributes have declined over the years it is possible that they are now underestimated. Recent determination of the productivity of tropical grasslands for example revealed that it might have been previously underestimated 3 or more-fold (Garcia-Moya *et al.* 1988). Second, validation of the magnitude of the long term CO_2 fertilizing effect on plant productivity needs to be carried out on a range of plant communities in conditions as close to field conditions as possible. Third, the long term temperature coefficient of soil organic matter decomposition, plant respiration and the respiration/photosynthesis ratio requires investigation for native species in field or field-like conditions. Also a clearer understanding of the mutual dependence of the C cycle and the N cycle on the timescale of decades to centuries is needed.

Man is now committed to managing the CO_2 content of the atmosphere, be it by default at present. With international moves afoot for nation states to discuss agreed protocols of net CO_2 emissions, that are binding under international law, the need for detailed accurate information and prediction systems of anthropogenic C sources and sinks is pressing.

SUMMARY

CQUEST is a simple model of the global terrestrial carbon cycle, developed to encompass, in as transparent a way as possible, key features of what we know about CO_2 concentration and temperature effects on terrestrial biosphere productivity, decomposition and carbon accumulation on the land. The model is run with various plausible and also extreme parameter combinations using published information on atmospheric CO_2 concentration and globally averaged annual temperatures since 1750AD. It is concluded that the most likely rate at which the global vegetation and soils were accumulating organic carbon in 1991AD, by virtue of actual CO_2 concentration and temperature changes considered alone, is about 2 Gt C yr^{-1}. Rates outside this range (from about 0.5

to 4 Gt C yr^{-1} are not excluded by this analysis but negative values are highly unlikely.

The most likely rate of carbon sequestration calculated, is similar in magnitude, but opposite in sign, to estimates of the flux deriving from net deforestation of the Earth.

REFERENCES

Adams JM, Faure H, Faure-Denard L, McGlade JM, Woodward FI (1990) Increases in terrestrial carbon storage from the last glacial maximum to the present. Nature 348:711-714

Albersheim, P (1965) Biogenesis of the cell wall. In: Bonner J, Varner JE (eds) Plant Biochemistry. Academic Press, New York, pp298-321

Bacastow, R, Keeling CD (1973) Atmospheric carbon dioxide and radiocarbon in the natural carbon cycle. II. Changes from A.D. 1700 to 2070 as deduced from a geochemical model. In: Woodwell GM, Pecan EV (eds) Carbon and the biosphere. United States Atomic Energy Commission, Washington (CONF-720510), pp86-135

Berner, R (1990) Atmospheric carbon dioxide levels over Phanerozoic time. Science 249:1382-1386

Berner, R (1991) A model for atmospheric CO_2 over Phanerozoic time. American J Science 291:339-376

Boden, TA, Kanciruk P, Farrell MP (1990) Trends '90: A Compendium of Data on Global Change. United States Department of Energy, Washington,D.C.

Bolin B (1970) The carbon cycle. Scientific Amer 223:124-132

Bolin B (1983) The carbon cycle. In: Bolin B, Cook RB (eds) The major biogeochemical cycles and their interactions. (SCOPE 21) John Wiley, Chichester, pp41-45

Bosatta E, Agren GI (1985) Theoretical analysis of decomposition of heterogeneous substrates. Soil Biol Biochem 17:601-610

Broecker WS, Takahashi T, Simpson HJ, Peng T-H (1979) Fate of fossil fuel carbon dioxide and the global carbon budget. Science 206:409-418

Conroy JP, Milham PJ, Barlow EWR (1992) Effect of phosphorus availability on the growth response of *Eucalyptus grandis* to high CO_2. Plant Cell Environ (in press)

Cure JD (1985) Carbon dioxide doubling responses: A crop survey. In: Strain BR and Cure JD (eds) Direct effects of increasing carbon dioxide on vegetation. DOE/ER-0238. United States Department of Energy, Washington, D.C., pp99-116

Esser G (1987) Sensitivity of global carbon pools and fluxes to human and potential climatic impacts. Tellus 39B:245-260

Fogg GE (1991) Changing productivity of the oceans in response to a changing climate. Annals of Botany 67 (Suppl 1):57-60

Folland CK, Karl TR, Vinnikov KYA (1990) Observed climate variations and change. In: Houghton JT, Jenkins GJ, Ephraums JJ. (eds) Climate Change: The IPCC Scientific Assessment. Intergovernmental Panel on Climate Change/WMO/UNEP/Cambridge University Press, Cambridge pp196-238

Freidli H, Lotscher H, Oeschger H, Siegenthaler U, Stauffer B (1986) Ice-core record of $^{13}C/^{12}C$ ratio of atmospheric CO_2 in the past two centuries. Nature 324:237-238

Freudenberg K (1964) The formation of lignin in the tissue and *in vitro*. In: Zimmermann MH (ed) The Formation of Wood in Forest Trees. Academic Press, New York, pp 203-218

Garcia-Moya E, Imbamba SK, Kamnalrut A, Evenson JP, Hall DO, Long SP, Scurlock JMO (1988) Primary productivity of natural grass ecosystems of the tropics: a re-assessment. In: Holm M (ed) Ecology of arable lands. Martinus Nijhoff, Dordrecht

Gifford RM (1979) CO_2 and plant growth under water and light stress: implications for balancing the global carbon budget. Search 10:316-318.

Gifford RM (1980) Carbon storage by the biosphere. In: Pearman, GI (ed) Carbon dioxide and climate: Australian Research. Australian Academy of Science, Canberra, pp167-181

Gifford RM (1987) Global photosynthesis, atmospheric carbon dioxide and man's requirements. In : Giovannozzi-Sermanni G, Nannipieri P (eds) Current perspective in environmental biogeochemistry. C.N.R.-I.P.R.A, Rome, Italy, pp413-444

Gifford RM (1991) Impact of increasing atmospheric carbon dioxide concentration on the carbon balance of vegetation. Australia, Energy Research and Development Corporation Project Report No. ERDC 37, 56p

Gifford RM (1992) Interactions of carbon dioxide and growth limiting environmetal factors in vegetation productivity: Implications for the global carbon cycle. Advances in Bioclimatology 1:24-58

Gillis AM (1991) Why can't we balance the globe's carbon budget? BioScience 41:442-447

Goudriaan J (1991) Atmospheric CO_2, global carbon fluxes and the biosphere. In: Rabbinge R, Goudriaan J, Keulen H van, Penning de Vries FWT, Laar HH van (eds) Theoretical production ecology: Reflections and prospects. Pudoc, Wageningen. pp 17-40

Goudriaan J and Ketner P (1984) A simulation study for the global carbon cycle, including man's impact on the biosphere. Climatic Change 6:167-192

Harmon, ME and Hua, C (1991) Coarse woody debris dynamics in two old-growth ecosystems. Bioscience 41:604-610

Higuchi T (1980) Lignin structure and morphological distribution in plant cell walls. In: Kirk TK, Higuchi T and Chang H-M (eds) Lignin biodegradation: Microbiology, chemistry and potential applications, Vol 1. CRC Press, Boca Raton, Florida, pp2-19

Hocking PJ, Meyer CP (1991) Carbon dioxide enrichment decreases critical nitrate and nitrogen concentrations in wheat. J of Plant Nutrition 14:571-584

Houghton JT, Jenkins, Ephraums JJ (eds) (1990) Climate Change: The IPCC Scientific Assessment. Intergovernmental Panel on Climate Change, WMO/UNEP/Cambridge University Press, Cambridge 365p.

Houghton RA (1991) Tropical deforestation and atmospheric carbon dioxide. Climatic Change 19:99-118

Houghton RA, Boone RD, Fruci JR, Hobbie JE, Melillo JM, Palm CA,-Peterson BJ, Shaver GR, Woodwell GM, Moore B, Skole DL, Myers N (1987) The flux of carbon from terrestrial ecosystems to the atmosphere in 1980 due to changes in land use: geographical distribution and global flux. Tellus 398:122-139

Hunt R, Hand DW, Hannah MA, Neal AM (1991) Response to CO_2 enrichment in 27 herbaceous species. Functional Ecology 5:410-421

Idso SB (1991) Comment on "Modelling the seaonal contribution of a CO_2 fertilization effect of the terrestrial vegetation to the amplitude increase in the atmosphere in atmospheric CO_2 at Mauna Loa observatory: by G.H. Kohlmaier *et al.*" Tellus 43B:338-341

Jenkinson DS (1990) The turnover of organic carbon and nitrogen in soil. Phil Trans R Soc Lond 329:361-368

Jenkinson DS, Adams DE, and Wild A (1991) Model estimates of CO_2 emissions from soil in response to global warming. Nature 351: 304-306.
Jordan DB, Ogren WL (1984) The CO_2/O_2 specificity of ribulose 1,5-bisphosphate carboxylase/oxygenase: Dependence on ribulose bisphosphate concentration, pH and temperature. Planta 161:308-313
Keeling , C.D., Bacastow, R.B., Carter, A.F., Piper, S.C., Whorf T.P., Heimann, M., Mook, W.G., Roeloffzen, H. (1989) A three-dimensioanl model of atmospheric CO_2 transport based on observed winds: 1. Analysis of observational data. In: Peterson DH (ed) Aspects of climate variability in the Pacific and the Western Americas. American Geophysical Union, Geophysical Monograph 55, pp165-236
Kerr RA (1977) Carbon dioxide and climate: carbon budget still unbalanced. Science 197:1352-1353
Kimball BA (1983) Carbon dioxide and agricultural yield: An assemblage and analysis of 770 prior observations. United States De[artment of Agriculture, Water Conservation Lab., Phoenix, Arizona, WCL Report 14, 71p
Kimball BA (1985) Adaptation of vegetation and management practices to a higher carbon dioxide world. In:Strain BR, Cure JD (eds) Direct effects of increasing carbon dioxide on vegetation. DOE/ER-0238. United States Department of Energy, Washington D.C., pp185-204
King AW, Emanuel WR, Post WM (1992) Projecting future concentrations of atmospheric CO_2 with global carbon cycle models: The importance of simulating historical changes. Environmental Management 16:91-108
Lieth H. (1975) Modelling the primary productivity of the world. In: Lieth H, Whittaker RB (eds) Primary productivity of the biosphere. Springer-Verlag, New York, pp237-263
Myers N (1990) Tropical forests. In: Leggett J (ed) Global Warming: The Greenpeace Report. Oxford University Press, Oxford, pp372-399
Myers N (ed) (1989) Deforestation rates in tropical forests and their climatic implications. Friends of the Earth, London
Norby RJ, O'Neill EG (1991) Leaf area compensation and nutrient interactions in CO_2-enriched seedlings of yellow-poplar (*Liriodendron tulipifera* L.). New Phytol 17:515-528
Ogren WL, Hunt LD (1978) Comparative biochemistry of ribulose bisphosphate carboxylase in higher plants. In: Siegelman HW, Hind G (eds) Photosynthetic carbon fixation. Plenum Press, New York, London, pp 127-138
Olson JS, Watts JA, Allison LJ (1983) Carbon in live vegetation of major world ecosystems. ORNL-5862. Oak Ridge National Laboratory, Oak Ridge, Tennessee. 180p
Paembonan SA, Hagihara A, Hozumi K (1991) Long-term measurement of CO_2 release from the aboveground parts of a hinoki forest tree in relation to air temperature. Tree Physiology 8:399-405
Platt T and Subba Rao DV (1975) Primary productivity of marine microphytes. In: Cooper JP (ed) Photosynthesis and productivity in different environments. Cambridge University Press, Cambridge, pp 249-280
Post WM, Emanuel WR, Zinke PJ, Stangenberger AG (1982) Soil carbon pools and world life zones. Nature 298:156-159
Rawson HM (1992) Plant responses to elevated CO_2 under different environmental conditions: responses to temperature. Aust J Bot (in review)
Reiners WA (1973) Terrestrial detritus and the carbon cycle. Brookhaven Symp Biol 24:303-327
Rodin LE, Basilevich NI, Rozov NN (1975) Productivity of the World's main ecosystems. In: Reichle DE, Franklin JF, Goodall DW (eds) Productivity of the World ecosystems. National Academy of Science, Washington D.C. pp 15-17,20,22
Rosenzweig, ML (1968) Net primary productivity of terrestrial communities: Prediction from climatological data. Amer. Naturalist 102:67-74

Schlesinger WH (1977) Carbon balance in terrestrial detritus.. Annual Reviews of Ecology and Systematics 8:51-81
Schlesinger ME, Mitchell JFB (1985) Model projections of the equilibrium climatic response to increased carbon dioxide. In: MacCracken MC, Luther FM (eds) Projecting the climatic effects of increasing carbon dioxide. DOE/ER-0237. United States Department of Energy, Washington DC. pp 81-147
Schwartzman and Volk (1989) Biotic enhancement of weathering and the habitability of Earth. Nature 340:457-460
Siegenthaler, U, Oeschger H (1978) Predicting future atmospheric carbon dioxide levels: Review of approaches. Science 199:388-395
Tans PP, Fung IY, Takahashi T (1990) Observational constraints on the global atmospheric CO_2 budget. Science 247:1431-1438
Thornley JHM, Fowler D, Cannell MGR (1991) Terrestrial carbon storage resulting from CO_2 and nitrogen fertilization in temperate grasslands. Plant Cell and Environment 14:1007-1012
Tinker PB and Ineson P (1990) Soil organic matter and biology in relation to cclimate change. In: Scharpenseel HW, Schomaker M, Ayoub A (eds) Soils on a warmer Earth. Elsevier, Amsterdam, pp71-87
Uchijima Z and Seino H (1985) Agroclimatological evaluation of net primary productivity of natural vegetation (1) Chikugo model for evaluating net primary productivity. Journal of Agricultural Meteorology 40:343-352
Volk, T (1989) Rise of angiosperms as a factor in long term climatic cooling. Geology 17:102-110
Watson RT, Rodhe H, Oeschger H, Siegenthaler U (1991) Greenhouse gases and aerosols. In: Climate Change: The IPCC scientific assessment, Intergovernmental Panel on Climate Change/WMO/UNEP/Cambridge University Press, Cambridge, pp 1-40
Whittaker RH (1975) Communities and Ecosystems, 2nd edition. Macmillan, New York, 387p
Wong SC (1980) Elevated atmospheric partial pressure of CO_2 and plant growth I. Interactions of nitrogen nutrition and photosynthetic capacity in C_3 and C_4 species. Oecologia 44:68-74
Woodwell GM, Houghton RA (1977) Biotic influences on the world carbon budget. In: Stumm W (ed) Global chemical cycles and their alterations by man, (Proceedings of the Dahlem Conference, Berlin), pp61-72

APPENDIX 1. Summary of symbols used inthe simple global terrestrial carbon cycle model, ***CQUEST***.

A	Gross CO_2 fixation by land plants	Gt C yr^{-1}
A_0	A at base CO_2 conc.(282ppmv)	Gt C yr^{-1}
$A_{0,0}$	A at 282ppmv and base temp. (17°C)	Gt C yr^{-1}
B	Global terrestrial biomass	Gt C
B_0	B at base CO_2 conc (282ppmv)	Gt C
$B_{0,0}$	B at 282ppmv and base temp. (17°C)	Gt C
C	Global av atmos CO_2 concentration	ppmv
C_0	C at base year (y=0); 282.2	ppmv
C_b	CO_2 compensation concentration	ppmv
L	Global plant litter deposition	Gt C yr^{-1}
L_0	L at y=0	
P	Global net primary production	Gt C yr^{-1}
P_0	P at base CO_2 conc (282ppmv)	Gt C yr^{-1}
$P_{0,0}$	P at 282ppmv and base temp. (17°C)	Gt C yr^{-1}
R	Global plant respiration rate	Gt C yr^{-1}
R_0	R at y=0	
S	Global soil+litter organic C	Gt C
S_0	S at y=0	Gt C
T	Global average temperature	°C
T_0	T at y=0	
W	Soil plus phytomass C	Gt C
y	Year of simulation. y=0 in 1750AD. y is used as a subscript	
f_1	Fractional prompt oxidation of annual plant tissue death	
f_2	R/A ratio	
k_1	Terrestrial biomass turnover rate	yr^{-1}
k_2	Soil organic C turnover rate	yr^{-1}
$k_{2,0}$	k_2 at 1750AD global av. temp.	yr^{-1}
k_3	Relative temp. sensitivity of k_2	$°C^{-1}$
k_4	Relative temp. sensitivity of P	$°C^{-1}$
k_5	Coef. in P v C response function	Gt y^{-1} $ppmv^{-1}$
k_6	Coef. in P v C response function	$ppmv^{-1}$
r	CO_2 doubling enhancement factor for P (for 680ppmv/340ppmv comparison)	

MODELING TERRESTRIAL CARBON CYCLES AT VARYING TEMPORAL AND SPATIAL RESOLUTIONS

Steven W. Running
School of Forestry
University of Montana
Missoula, Montana 59812
U.S.A.

INTRODUCTION

This chapter will illustrate simulation modeling of terrestrial carbon cycles at varying resolutions. I will give three examples from our research showing first, a forest carbon model for a single tree or stand, FOREST-BGC; second a regional ecosystem modeling system, RESSys, for areas of thousands of square kilometers; and finally a simple satellite driven model using the AVHRR Normalized Difference Vegetation Index, NDVI, that can be executed globally. Each modeling logic has advantages for different applications. I will illustrate that the integration of all three models to an organized package provides the power to do carbon cycle modeling at varying spatial and temporal scales with significant versatility and accuracy.

FOREST-BGC, 1-DIMENSIONAL MODELING

The finest resolution is represented by simulations done with the FOREST-BGC model, which originated as a single tree water balance simulation in the early 1970s. FOREST-BGC (BioGeochemical Cycles) is a process-level simulation model that calculates the cycling of carbon, water and nitrogen through forest ecosystems (Figure 1, Running and Coughlan 1988, Running and Gower 1991). The model requires daily data for standard meteorological conditions: maximum-minimum temperatures, dew point, incident shortwave radiation and precipitation, and definition of key site and vegetation variables: forest LAI and soil water holding capacity.

NATO ASI Series, Vol. I 15
The Global Carbon Cycle
Edited by M. Heimann

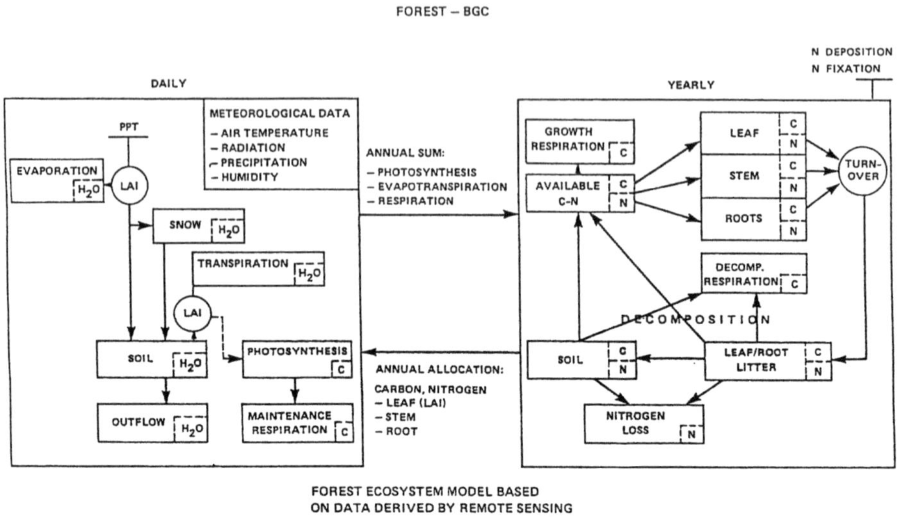

Figure 1. A compartment flow diagram of the FOREST-BGC model of carbon, water and nitrogen cycles in forest ecosystems. This is an example of a 1-dimensional model for modeling small plots. (from Running and Coughlan 1988, Running and Gower 1991).

The following processes are calculated by FOREST-BGC:

<u>Hydrologic</u>
precipitation, snow vs rain partitioning
snowmelt
canopy/litter interception and evaporation
surface runoff vs soil storage
transpiration
physiological water stress and surface resistance
subsurface outflow

<u>Carbon</u>
photosynthesis
maintanence respiration
growth respiration
carbon allocation (leaf/stem/root)
net primary production
litterfall
decomposition
(trace gas emissions, CH4, NMHC)

<u>Nitrogen</u>
N-mineralization
N-allocation (leaf/stem/root/available)

As a brief summary of the model's function, daily precipitation is routed to snowpack or soil dependent on air temperature. A canopy interception fraction based on leaf area index is subtracted and evaporated, and the remaining water goes to a soil compartment where it is available for transpiration. Forest soils generally have a high surface infiltration capacity, so overland flow is rare.

Transpiration is calculated with a Penman-Monteith equation incorporating radiation and vapor pressure deficit drivers for evaporation. The canopy conductance term is a complex function of air temperature, vapor pressure deficit, incident radiation and leaf water potential. Air temperatures below 0° C reduce canopy conductance to cuticular values, and are an important determinant of the growing season length. Canopy conductance is linearly reduced to a default cuticular value when either average daily vapor pressure deficit exceeds 16 mb or pre-dawn leaf water potential, estimated from soil water availability, decreases below -1.65 M Pa. Aerodynamic conductance is normally fixed at 0.2 m/s for coniferous forests in the Penman-Monteith equation.

Canopy photosynthesis is calculated by multiplying a CO_2 diffusion gradient by a mesophyll CO_2 conductance and the canopy water vapor conductance, both of which are controlled by radiation and temperature. The light response surface for mesophyll conductance is asymptotic, with half maximum photosynthesis at 9720 $kJ/m^2/day$. A light attenuation of -0.5/LAI is used for Beers law attenuation of incident radiation to produce canopy average radiation. The inverse parabolic temperature response surface has an optimum of 20° C, with high and low compensation points of 0° and 40° C, respectively. Net photosynthesis (PSN) is computed by subtracting a maintenance respiration term, calculated as an exponential function of air temperature with a $Q_{10} = 2.3$ for all stem and root biomass.

In the yearly half of the model, growth respiration is subtracted from the annual net photosynthesis to define carbon available for net primary production. This carbon is partitioned into leaf, stem and root production based on an optimization logic comparing water, nitrogen and photosynthate availability (Running and Gower 1991). Leaf and root litterfall is deposited into the litter compartment for decomposition. Carbon and nitrogen is then released from litter and soil compartments by a water and temperature controlled decomposition algorithm, and nitrogen is routed to the avialable pool for future growth.

In simplifying earlier ecosystem models, I eliminated a number of seemingly important details. No internal physiology is represented, of cellular water stress, of phloem carbohydrate transport etc. No canopy structure, leaf age class or leaf angular distribution is defined, only simple LAI. No belowground details of rooting processes, root water or nutrient uptake capacity is explicitly defined. Some of these variables are virtually unmeasurable in the field even for small intensive study sites, how would we estimate them at regional or global scales? What we really want is to relate belowground activity to canopy responses that can be more directly measured. The resulting model, neglecting many ecosystem attributes, and relating most key variables to LAI and climate, while oversimplified to many ecologists, provides the only means I see to bring ecosystem modeling to regional scales.

Time Scale Definition

The ecosystem process model FOREST-BGC has a dual time resolution; calculating key canopy hydrologic processes such as interception and ET, and carbon process variables such as photosynthesis and maintenence respiration, daily (Figure 1). The carbon allocation to biomass is executed annually, as are the litterfall, decomposition, and nitrogen budget processes. I found the dual time step essential for adequately and efficiently simulating these ecosystem processes. The hydrologic partitioning of precipitation into instantaneous interception and evaporation versus longer term snow or soil water storage and transpiration or hydrologic outflow proved to be the key process that demands daily timing. Accurate hydrologic partitioning is essential in arid areas for simulating seasonal soil drought and plant water stress, and is essential for producing correct timing on stream discharge hydrographs. In areas of frequent showers, even higher time resolution is useful.

However, I have compromised on a daily timestep for our models after many years of modeling at hourly time steps (see Knight et al. 1985). Because on average, diurnal climatology is very predictable, we found that going from hourly to daily meteorological drivers cut the climate files to 1/24 original size, yet we could simulate the diurnal conditions well enough to predict seasonal canopy process rates from a daily model to within 3% of the hourly simulations. However, given that GCMs require high time resolution, nothing precludes our ecosystem model from matching their time step except computational load.

The carbon growth, litter and decomposition processes, and nitrogen budgets do not require daily time scales. In temperate evergreen ecosystems, yearly calculations matched the timing of field measurements, making initial model development and validation convenient. In more seasonal biomes, such as grasslands, the carbon/nitrogen computations need to be done monthly or even weekly to describe certain processes.

The commonly stated principle relating small space to short time resolution and progressively larger spatial scale to long time resolution makes a nice graph, but has many exceptions. GCMs define a huge spatial scale at a very fine temporal scale, on the order of 12-30 minutes. For atmospheric dynamics, this is an appropriate space-time definition for the processes being represented. Likewise in ecology, events that are large in spatial scale can occur in a very short time. Ecosystem processes that are driven by meteorological conditions, notably the canopy gas exchange processes, are just as dynamic temporally as is climate. A critical problem of ecological representation though, is that while a canopy process, such as photosynthesis occurs over a large region, the absolute control is still exerted within the physiology of the individual organism. However, this does not argue that we then must represent those individual organisms explicitly.

Many important ecosystem processes are triggered by occasional, instantaneous extreme events. Major freezes can alter regional vegetation cover and activity in one night and take years to recover. The meteorological conditions that triggered the Yellowstone fires of 1988 were of a short time scale, yet the results will be felt for a century. Both our ecosystem models, and our climate models tend to be central tendency simulators, representing mean conditions and mean responses and missing important extreme events.

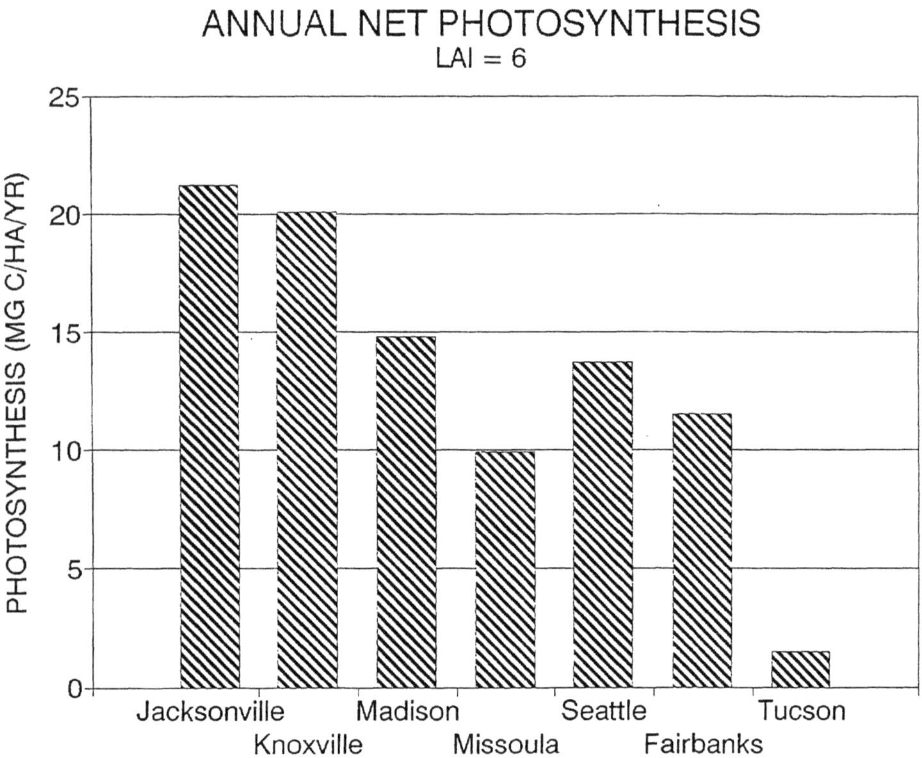

Figure 2. An example of an application of the ecosystem model to make 1-d simulations of a carbon cycle process, annual net photosynthesis for hypothetical forests in different climates around North America (from Running and Coughlan 1988).

FOREST-BGC simulates a wide range of annual photosynthetic carbon fixation for sites across a broad continental climate range (Figure 2). A number of validations of the carbon balance in the FOREST-BGC model have been published. We found a correlation of $R^2 = 0.94$ between annual photosynthesis simulated by FOREST-BGC and measured stem volume growth for six Pinus ponderosa stands in Montana (from Running et al 1989). Recently we simulated the 5-year growth of 176 pine trees in Canada, and found a correlation of $R^2 = 0.95$ between predicted and observed growth. While these studies have found FOREST-BGC simulations to adequately track field measurements of small forest stands, validations at regional scales cannot be based on similar ecological data.

REGIONAL ECOSYSTEM SIMULATION SYSTEM, RESSys

To begin the evolution of "point scale" ecological models to GCM scales, my colleagues and I have developed a Regional Ecosystem Simulation System (RESSys, Figure 3) over the last six years (Band et al 1991). RESSys has three core models, which use four additional core datasets. A topographic model from Band (1986) inputs digital elevation data and outputs effectively a terrain map. This provides a template for the definition of the rest of the system, and can be defined to differing levels of topographic complexity and spatial resolution. Hence, it allows us to zoom in on a small 10 km^2 watershed, or pan out to a 10,000km^2 region. A mountain climatological model uses this topographic file to extrapolate point measured meteorological data across the region two-dimensionally for slope, aspect and elevation. The output file of this model generates the input file of daily meteorological conditions, across the landscape for FOREST-BGC, the ecosystem process model. FOREST-BGC then simulates the ecosystem processes of importance, which are then mapped back to the region on the topographic template. TOPMODEL then allows topographically defined hydrologic routing.

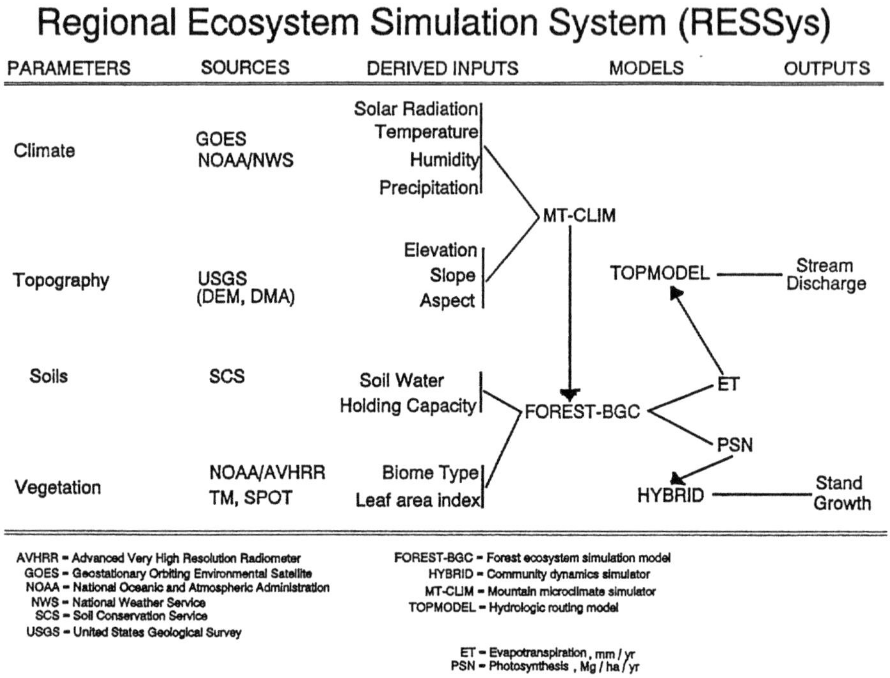

Figure 3. The flow chart for a 2-dimensional ecosystem model, illustrating the additional data needed to describe regional landscape heterogeneity of critical biophysical variables.

Agreement of the key processes required for a global terrestrial ecosystem model will be important when defining the essential classification logic for global partitioning. From the origins as a water balance model, FOREST-BGC emphasizes canopy gas exchange processes, and system water storages in snow and soil. Currently in RESSys we require definition of only leaf area index (LAI) and soil water holding capacity (SWC) of the landscape, with defaults for all other parameters. The FOREST-BGC model was specifically designed to be sensitive to the parameterization of canopy processes by LAI, and soil processes by SWC. These choices of one key vegetation and one edaphic variable then direct our landscape classification, which otherwise could have near endless layers of variables. *I emphasize that an optimal global vegetation classification scheme cannot be devised until a modeling logic has been defined*; then the classification can be based on the variables, such as LAI in our case, that the model has been designed to be particularly sensitive to. We have recently added total canopy nitrogen as a similar, simplifying definition of forest nutrient status.

The other control on landscape definition is the topographically induced microclimate variability. Sensitivity of the ecosystem model to climate then helps to define what resolution of elevational and topographic detail needs be defined. For example, we have found daily temperature resolution of ± 1 deg and precipitation of ± 2mm, to be adequate for driving the ecosystem model.

We have found in RESSys development that the limiting factor is not the accuracy of our "point" models, but the accuracy with which we can define key parameters across the landscape. First, one must decide upon the key variables, such as LAI, then one must devise ways to map them across the region. A breakthrough in our regional logic was the demonstration that LAI could be defined by satellite (Peterson et al. 1987).

By far our greatest difficulty in landscape definition has been in soils data. The soil reservoirs of water, carbon, and nitrogen are significant components of the ecosystem cycles, and have fundamentally different (usually 10-100X slower) time constants than canopy processes. Historically, soil science has emphasized the taxonomic classification of soils into a system that qualitatively defines the temperature and moisture conditions of soil development. However, standard soils maps, such as by the US Soil Conservation Service, do not define soil physical structure, depth, water potential release curves or chemistry quantitatively. At global scales the problem of non-quantitative, spatially coarse soils data is even more acute. We are working on a hydrologic equilibrium theory that integrates climate, LAI and soil water holding capacity, and allows one to infer the soil characteristics from observable climate and maximum LAI (Nemani and Running 1989a).

For global scale work, we suspect some type of similar logic will be necessary.

Standardized Meteorological Data

Ecological modeling has typically been developed around intensive study sites, where a central meteorological station was a first priority. Because one cannot install custom meteorological stations everywhere, the logic of using a routine weather database, with extrapolations was necessary for RESSys. It became clear that daily meteorological data is routinely available, from standard sources like the U.S. National Weather Service, or the World Meteorological Organization, but hourly, weekly or any other time was not. This became a critical design criteria for RESSys, use of standard available meteorological data, even if is not exactly what was wanted. Part of the purpose of the mountain climate (MT-CLIM, Running et al. 1987) simulator was to calculate humidity and insolation from the daily maximum-minimum temperature and precipitation that is the standard "weather" data recorded at the greatest number of stations. Humidity and radiation data is available from only about 5% of the reporting weather stations in the U.S. The International Geosphere-Biosphere Program has recently called for development of "weather generators" as a key component of global ecological research. Advances are also being made in running mesoscale meteorological models nested within GCMs (Segal et al 1988, Dickinson et al 1989). These meteorological models function at a spatial resolution quite compatible for regional ecology.

Regional Simulations and Validations

Simulation of annual photosynthesis of a 800km^2 forested mountain valley in Montana (Figure 4) illustrates the application of RESSys for regional carbon cycle modeling. Sub-watersheds are simulated separately, such as the 10km^2 Soup Creek highlighted, yet the computations can be simplified to a single regional average if desired for larger scale products.

For many years the only "ecological" data that could be directly measured at regional scales was an occasional land cover

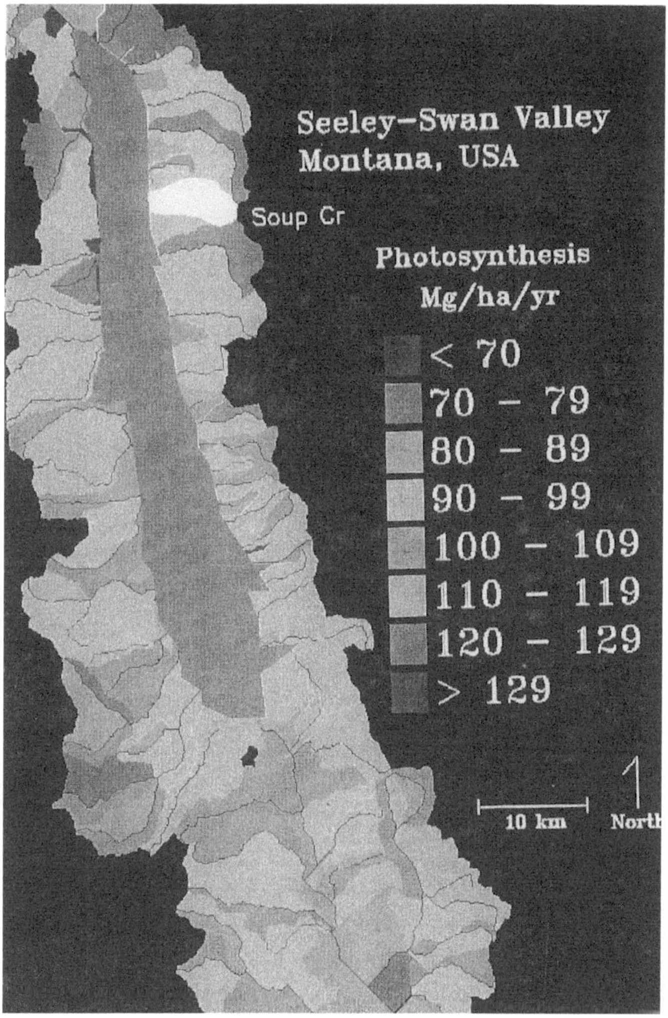

Figure 4. An application of the RESSys model to produce a regional map of annual photosynthesis for a $800km^2$ basin in Montana, USA.

classification from Landsat. Because of the expense and computer requirements needed, these regional maps were produced as one-time static products, much like vegetation maps in atlases. The development of the AVHRR/NDVI products produced weekly, began to add some temporal dynamics to our regional view of ecosystems thru a variable called a Normalized Difference Vegetation Index. The observable continental scale dynamics were dramatic, following the seasonal "green wave" northward in the spring and then back south in the fall (Justice et al. 1985, Goward et al. 1985,1987). However, the NDVI was still one long step removed from the ecosystem process rates of greatest interest, carbon, water, nutrient cycling by the surface. More recently, our first generation RESSys products (Running et al. 1989) have developed the computational ability to calculate these process rates over progressively larger areas. We are now finding that there is no established methodology for validating these regional ecosystem process maps, particularly from the classical field measurements that ecologists consider "hard" validation.

One notable exception is the global scale atmospheric CO_2 concentration analyses (Houghton 1987, Fung et al. 1987, Keeling et al 1989, Tans et al. 1990). These CO_2 data coupled with the seasonal NDVI satellite data are the closest thing to observing a global biospheric "heartbeat" I have yet seen. At a smaller scale, the CO_2 flux measurements taken from aircraft (Wofsy et al. 1988), or micrometeorological tower systems (Baldocchi 1989) over multiple kilometer scales, if done repetitively could provide an integrated regional measure of carbon cycle activity.

The one other regional scale database we have for interpreting ecosystem processes are the hydrologic discharge and balances of gauged watersheds, providing a large spatial measurement of a key ecosystem variable, water cycling. Both these CO_2 and water balance data may provide an integrated spatial measurement for validation, but only at restricted, rather long monthly to yearly time scales. The aircraft and tower flux measurements are the only regional calculations that could provide daily validations.

Another potential source of validation for carbon cycle simulations are the variety of crop yield, forest growth, range forage production etc standard measurements taken by land management agencies. These again are point samples, and not the exact variables needed, but are a huge network of data that could be used for validation of regional primary production simulations.

APPLYING ECOSYSTEM MODELING GLOBALLY WITH NDVI

We currently cannot invision taking RESSys level mechanistic process modeling to the global scale. Alternatively, we see that these ecosystem process models could be used to "calibrate" highly simplified global models in a variety of different biome/climate situations. Many of the following ideas are incorporated in our NASA Earth Observing System MODIS (Moderate Resolution Imaging Spectrometer) research plan. The NASA EOS is the observational cornerstone of the United States Global Change Research Program, and MODIS is the sensor planned for regular global land surface monitoring.

The best current candidate for this global modeling logic is the NDVI or Normalized Difference Vegetation Index, time integrated thru the year. Due to the fact that the NDVI integrates both surface characteristics, (vegetation greenness and LAI) and key meteorological factors such as solar intensity to the surface, the ΣNDVI provides a simple yet surprisingly versatile global measure (Tucker et al. 1985). We tested the correlation between seasonal NDVI and ecosystem process rates for seven locations around North America (Running and Nemani 1988). For continental scale dynamics, we found amazingly good correlations between ΣNDVI and seasonal photosynthesis (Figure 5), transpiration and net primary production (R^2 0.72-0.87). Even weekly ecosystem activity was fairly well described by the seasonal NDVI trace. However, we also found that the NDVI does not change in evergreen forests that are drought or temperature stressed, indicating that the NDVI alone may be able to define structure, but only fortuitously correlates with canopy function.

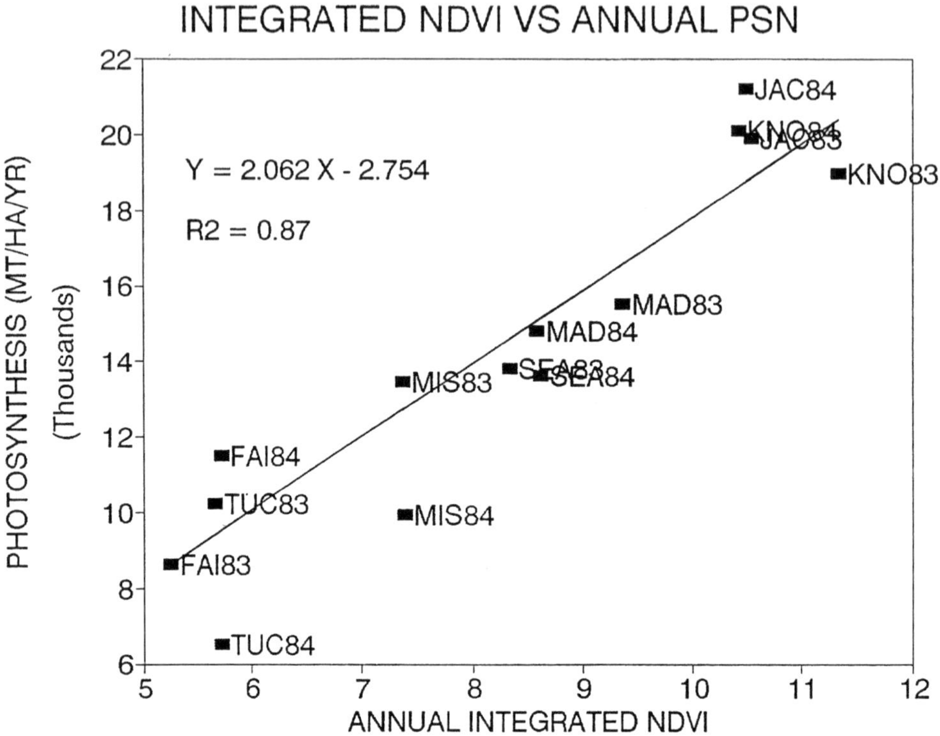

Figure 5. The relationship found between annual integrated Normalized Difference Vegetation Index from the AVHRR satellite and annual photosynthesis simulated for forests in seven contrasting environments of North America (from Running and Nemani 1988). This AVHRR/NDVI based logic can be used to extrapolate ecosystem modeling to global scales.

While details of these ideas are available in the references, the general conclusion we have drawn is that the time integrated NDVI can be a very robust global measure of vegetation activity, if calibrated well to varying conditions and biomes. We are planning to do precisely that for our 10 year MODIS project. We plan to first define a very limited number of biome types based on very simplified structural and functional characteristics. The global vegetation would then be pre-classified by these characteristics, illustrating why we feel the model logic must come before the classification scheme. Next we will build a mechanistic ecosystem process model for each of these biomes, designated collectively BIOME-BGC (Running and Hunt 1991), which then would be used to calibrate the simple global NDVI, biome specific conversion factors, and surface temperature data. After the launch of EOS, this global simulation of biome processes would be done weekly from MODIS data.

GCMs define the land surface in only a spatially coarse (cells of $\approx 20{,}000 km^2$) and static way. Each cell is defined initially as a single vegetation type, and no dynamic interaction occurs between the resulting climate and surface vegetation definition, a limitation that is particularly unfortunate for long time simulations of 10-100 years (Emmanuel et al 1984). Connection of this modeling to a GCM would require at minimum a weekly redefinition of the surface characteristics of the GCM cells, which in principle should not be difficult. A more dynamic coupling that would allow real time system feedbacks might require the NDVI model to be run within the atmospheric model so daily fluxes would enter the atmosphere, and responses would influence the vegetation. Then important responses like continental biome shifts resulting from climate change could be explicitly simulated. However, at some point it is difficult to combine real time global NDVI measurements interfaced to an internally self-sufficient simulation, unless for retrospective testing.

ACKNOWLEDGEMENTS

I thank NATO for sponsoring this Institute, and Dr. Martin Heimann for the invitation. Much of this research has been funded by U.S. National Aeronautics and Space Administration, grant #NAGW-952, and U.S. National Science Foundation #BSR 88-17965. Some of this text was first presented at the UCAR Global Change Institute in Snowmass, CO in 1990.

REFERENCES

Baldocchi, D.D. (1989) Turbulent transfer in a deciduous forest. Tree Physiology. 5:357-377

Band, L.E. (1986) Topographic partitioning of watersheds with digital elevation models. Water Resources Research 22:15-24.

Band, L.E., D.L.Peterson, S.W.Running, J.Coughlan, R.Lammers, J. Dungan, R.Nemani. (1991) Forest ecosystem processes at the watershed scale: Basis for dstributed simulation. Ecological Modeling 56:171-196.

Dickinson, R.E., R.M.Enrico, F.Giorgi, and G.T.Bates. (1989) A regional climate model for the western United States. Climatic Change 15:383-422.

Emanuel W.R., Killough G.G., Post W.M., Shugart, H.H. (1984) Modeling terrestrial ecosystems in the global carbon cycle with shifts in carbon storage capacity by land-use change. Ecology 65:970-983

Fung, I., C.J.Tucker, and K.C.Prentice. (1987) Application of Advanced Very High Resolution Radiometer Vegetation Index to study atmosphere-biopsphere exchange of CO2. J. Geophysical Research. 92:2999-3015.

Goward S.N., Kerber A., Dye D.G., Kalb V. (1987) Comparison of North and South American biomes from AVHRR observations. Geocarto 2:27-40.

Goward S.N., Tucker C.J., Dye D.G. (1985) North American vegetation patterns observed with the NOAA-7 advanced very high resolutionradiometer. Vegetatio 64:3-14.

Houghton, R.A. (1987) Biotic changes consistent with the increased seasonal amplitude of atmospheric CO2 concentrations. J. Geophysical Research 92:4223-4230.

Justice C, Townshend J, Holben B, Tucker C. (1985) Analysis of the phenology of global vegetation using meteorological satellite data. International Journal of Remote Sensing 6:1271-1318

Knight, D.H., Fahey, T.J., Running, S.W. (1985) Factors affecting water and nutrient outflow from lodgepole pine forests in Wyoming. Ecological Monographs 55:29-48.

Keeling, C.D. R.B.Bacastow, A.F.Carter, S.C.Piper, T.P.Whorf, M.Heimann, W.G.Mook, H.Roeloffzen. (1989) A three dimensional model of atmospheric CO2 transport based on observed winds. 1. Analysis of observational data. pp 165-236. IN: Geophysical Monograph #55, American Geophysical Union, Washington D.C.

Nemani R., and Running, S.W. (1989) Testing a theoretical climate-soil-leaf area hydrologic equilibrium of forests using satellite data and ecosystem simulation. Agricultural and Forest Meteorology 44:245-260.

Peterson D.L., Spanner M.A., Running S.W., Teuber K.B. (1987) Relationship of Thematic Mapper Simulator data to leaf area index of temperate coniferous forests. Remote Sensing of Environment 22: 323-341.

Running S.W., Nemani R.R., and Hungerford, R.D.(1987) Extrapolation of synoptic meteorological data in mountainous terrain, and its use for simulating forest evapotranspiration and photosynthesis. Canadian Journal of Forest Research 17:472-483

Running, S.W, and Nemani, R.R. (1988) Relating seasonal patterns of the AVHRR vegetation index to simulated photosynthesis and transpiration of forests in different climates. Remote Sensing of Environment 24:347-367

Running S.W., and Coughlan, J.C. (1988) A general model of forest ecosystem processes for regional applications. I. Hydrologic balance, canopy gas exchange and primary production processes. Ecological Modeling 42:125-154

Running, S.W., Nemani, R.R., Peterson, D.L., Band, L.E., Potts, D.F., Pierce, L.L., Spanner, M.A. (1989) Mapping regional forest evapotranspiration and photosynthesis by coupling satellite data with ecosystem simulation. Ecology 70:1090-1101.

Running, S.W. and E.R.Hunt Jr. (1991) Generalization of a forest ecosystem process model for other biomes, BIOME-BGC, and an application for global-scale models. IN: Scaling Processes Between Leaf and Landscape Levels. J.R.Ehleringer and C.Field eds. Springer-Verlag. (in press).

Running,S.W. and S.T.Gower. (1991) FOREST-BGC, A general model of forest ecosystem processes for regional applications. II. Dynamic carbon allocation and nitrogen budgets. Tree Physiology. 9:147-160.

Segal, M.R. Avissar, M.C. McCumber, and R.A. Pielke. (1988) Evaluation of vegetation effects on the generation and modification of mesoscale circulations. J. Atmospheric Sciences 45:2268-2292.

Tans, P.P., Fung, I.Y., and Takahashi, T. (1990) Observational constraints on the global atmospheric CO2 budget. Science 247:1431-1438.

Tucker C.J., Townshend, J.R.G. Goff, T.E. (1985) African land cover classification using satellite data. Science 227:369-374.

Wofsy, S.C., Harriss, R.C., Kaplan, W.A. (1988) Carbon dioxide in the atmosphere over the Amazon Basin. Journal of Geophysical Research 93:1377-1388.

BIOME MODELLING AND THE CARBON CYCLE

I. Colin Prentice
Department of Plant Ecology
Lund University
Östra Vallgatan 14
S-223 61 Lund
Sweden

1. BIOMES

Structurally and functionally distinct ecosystem complexes, known as biomes, occur in different climates. There are several global biome classifications but most recognize 15-20 biomes, and all agree on the nature and approximate extent of a number of such easily identifiable entities as the tundra, hot deserts, temperate deciduous forests and tropical rain forests. Differences between biomes are important for the carbon cycle because they account for much of the spatial variation in the size of the biomass and soil carbon stores. Biomass carbon storage is typically < 1 kg m^{-2} in tundra and deserts, ca 10 kg m^{-2} in many temperate forests and ca 20 kg m^{-2} in tropical rain forest (Olson et al. 1983). Soil carbon storage is typically < 3 kg m^{-2} in hot deserts, ca 10 kg m^{-2} in tropical rain forest, ca 15 kg m^{-2} in many temperate forests and > 15 kg m^{-2} in the tundra (Post et al. 1982).

Boundaries between adjacent biomes may be gradual, as in the transition between boreal conifer and temperate deciduous forests on the northern continents, or more abrupt, as in the transitions between temperate deciduous forests and grasslands (Walter 1979). The sharpness of the boundary depends on the steepness of the underlying climatic gradient, and in some cases on additional feedback mechanisms tending to keep the boundary more or less sharp. For example, a sharp natural prairie-forest boundary was maintained in pre-settlement time by a positive feedback involving fire (Grimm 1984). Prairie

NATO ASI Series, Vol. I 15
The Global Carbon Cycle
Edited by M. Heimann

burns easily but recovers fast; temperate deciduous forests burn less easily but regrow slowly. Thus, the physical conditions created by each vegetation type favour the persistence of that type. The general location of the boundary was controlled by climate but its detailed location was related to topographic barriers to fire spread; climate changes shifted the location of the boundary with a hysteresis attributable to the feedback (Grimm 1983, 1984). This example illustrates the general concept that spatial and temporal patterns in biome distribution are primarily under climatic control, while feedbacks from vegetation to environment fine-tune the locations of the boundaries between biomes and may cause hysteresis in the response of these boundaries to climate change (Prentice 1986, Prentice and Sarntheim 1992).

Global-scale correlations between the distributions of biomes and climate were already known in pre-Darwinian times. The global patterns of vegetation structure and function cut across the patterns of macroevolutionary affinity or "biogeographic realms" that were discovered later, in the mid-nineteenth century, by Wallace. For example, boreal forest is species-poor and taxonomically relatively homogeneous; much the same plant genera and families are represented through the Holarctic realm, to which the boreal climate is confined. On the other hand, Mediterranean shrubland is species-rich with structurally and functionally similar vegetation formed from taxonomically unrelated floras in disjunct regions with mediterranean climates within the Holarctic, Neotropical, Ethiopian and Australasian realms (Walter 1979). Such observations imply that the distribution of biomes must be understood in terms of biophysical and physiological constraints that have been conservative during the evolution of higher plants. This idea leads naturally to the development of more mechanistically based models that use data from plant physiology, the discipline that has contributed most to present understanding of the causes of the observed correlations between vegetation and climate (Woodward 1987; 1990).

2. EQUILIBRIUM MODELS

The responses of biome distributions and terrestrial carbon storage to changes in climate involve processes with a range of time constants. For example, all such shifts involve local death and extinction of some plant types, immigration and growth of others, but the rates of these processes vary considerably. The time taken for trees to grow imposes a lag of decades to centuries on many climate-induced vegetation changes, such as transitions between forest types (Davis 1989; 1991; Prentice et al. 1991). Changes in soil carbon storage may involve either net oxidation (fast) or accumulation (slow); equilibration between addition and erosion of soil organic matter may take thousands of years (Schlesinger 1990).

The long-term carbon flux from the atmosphere to the terrestrial biosphere is determined by <u>net ecosystem production</u>, the annual balance of net primary production and decomposition. The span of time scales involved make it difficult to build mechanistic models that could predict transient changes in this flux. Many of the component processes are quite well understood but a major effort is required before this understanding can be integrated into operational, global-scale models. Some progress can be made in the meantime by modelling the situation in which global biome distributions and terrestrial carbon storage are assumed to be in equilibrium with climate (Prentice and Solomon 1991). If the dynamics of the processes are ignored then the modelling problem reduces to specifying the biophysical constraints that ultimately limit the global distributions of different functional types of plant, and simple ecological rules that specify the effects of these plant types upon one another and their environment. Models based on these principles are known as <u>biome models</u>.

Biome models predict <u>potential natural vegetation</u>, i.e. vegetation that would occur at equilibrium with climate and in the absence of human intervention. One might perhaps argue that the utility of biome models is restricted: because future climate changes may outstrip the rate at which vegetation can respond to the changes (Prentice 1986; Davis 1989), or simply because a large part of the world's vegetation is not natural (Shugart et al. 1986). However:

(1) Large areas of vegetation -- including vast tracts of the boreal forest, where a substantial fraction of terrestrial carbon resides -- are, although not unaffected by humans, in an essentially unmanaged state (Shugart et al. 1986) and will presumably remain so for a long time.

(2) If equilibrium models are exercised under changed-climate scenarios, they will indicate the directions in which vegetation will change. Such an analysis could be refined by taking into account the causes of change. For example, the impact on the carbon cycle of a widespread change from one forest type to another with higher biomass could depend on whether the original trees will stop growing or die in the new climate, or be destroyed in extensive fires, or be gradually replaced through competition with a new, more productive type. A semi-quantitative assessment of this kind should be possible even on the basis of a equilibrium model, if some reasonable assumptions are made about rates of change associated with different mechanisms of change.

(3) Equilibrium models will indicate possible future constraints on land-use. Forests will not grow in the steppe, nor unirrigated crops in the desert: land-use potential is closely related to potential natural vegetation, and depends on climate.

(4) Equilibrium models are useful in modelling the differences of global vegetation patterns and terrestrial carbon storage between now and other periods of the Quaternary with substantially different global climates (e.g. Guetter and Kutzbach 1990; KC Prentice and Fung 1990). The differences on a Quaternary time scale may be large enough to be considered for all practical purposes as differences between the biospheric equilibria corresponding to the past and present climatic regimes.

(5) Equilibrium models provide the essential framework of constraints within which dynamic models can operate (Prentice

et al. 1989). It is necessary to specify what kinds of plants can thrive in a given environment before a dynamic model can simulate their growth, demography and interactions.

3. HOLDRIDGE LIFE-ZONES

The relationship between plant forms and climate was a major interest of the early ecologists and plant geographers such as Warming, Raunkiaer and Brockmann-Jerosch during the early part of this century. The subject became largely neglected later, as ecosystem dynamics, population biology and evolutionary ecology successively came to dominate ecological thought (see e.g. Crawley 1986). Recent concern with the impacts of global climate change on ecosystems has helped to revitalize the subject (Woodward 1990; Solomon and Shugart 1992). In the meantime, however, scientists concerned with global change had discovered a need for models to predict biome distributions from climate and were obliged to use the only available products: empirical models, such as those of Köppen (1936) and Holdridge (1947), that predate modern plant physiology, ecology and climatology. The popularity of these schemes is due to their simplicity and off-the-shelf availability, not to any particular scientific basis or predictive skill.

The Holdridge life-zone scheme (Holdridge 1947; Emanuel et al. 1985; Leemans 1990) is simultaneously the most popular and the least well-founded of these early models. It is commonly represented by a triangular diagram, representing a linear projection of a two-dimensional climate space defined by annual "biotemperature" (the normalized annual sum of temperatures over 0°C) and annual precipitation after both have been transformed to logarithmic scales. The third axis, called "potential evapotranspiration ratio", is merely the difference of the other two axes.

Hexagonal cells represent the regions nearest to each of a set of evenly spaced points in the projection. An additional criterion, the so-called "frost line", is defined at a biotemperature of 17°C, bisecting one row of cells. Any given climate can be located in this space and assigned the vegetation name corresponding to the cell in which it falls.

Unfortunately the vegetation names assigned to each point (e.g. "boreal dry scrub", "rain tundra") are not always identifiable biomes. Where they are identifiable, they are often wrongly located in climate space. The zig-zagging boundaries between hexagonal cells have bizarre consequences, e.g. an alternation of "tropical" and "subtropical" vegetation units along moisture gradients in the tropics. Nevertheless, by judicious lumping of types and/or adjustments to the boundaries between types, the Holdridge scheme can be forced to make a passable global vegetation map (KC Prentice 1990; Leemans 1990).

A more fundamental problem with the Holdridge scheme is that the climate variables used are not closely related to the variables that actually control the distributions of different types of plants. For example, plant-available moisture is only extremely roughly represented by the ratio of annual precipitation and biotemperature, and the 17°C isotherm of annual biotemperature is only very approximately and indirectly related to the incidence of frost. The lack of any consideration of seasonality generally limits the predictive value of the scheme because of the key importance of seasonal extremes for the biota (Woodward 1987).

Any agreement achieved between the Holdridge scheme (or variants of it) and vegetation thus depends to a large extent on the existence of correlations between different climate variables, e.g. between winter and summer temperatures. But the relationship between these variables was different in the past and is likely to change in the future too, because of climatological mechanisms that cause differential changes in the climate of different seasons (Schlesinger and Mitchell 1985, Kutzbach and Guetter 1986). Such problems cannot be avoided completely, because we do not know exactly what variables control plant distributions. Even if we did, we would probably not be able to measure these variables directly -- for example, plant-available moisture is much easier to simulate than to measure. Nevertheless, it is possible to construct simple biome models that depend on climate variables in a more realistic way and which incorporate data and insights from plant physiology. Woodward (1987) seems to have been the first to attempt this although some physiological and life-history considerations were taken into account in the more elaborate empirical model of Box (1981).

4. BIOME MODELLING

An important consideration in biome modelling is that the basic units of the model should not be biomes, but plants. Biomes represent assemblages of functional plant types and their distribution should "emerge" from the distributions of these types. There are two reasons for this. First, key information on the mechanisms controlling biome distributions comes from physiological investigations on different types of plants. If these experimental data are to be used in the model then the basic units of the model must be plants. Second, palaeoecological analyses have shown that the distributions of plant taxa respond independently to climate change (see e.g. Webb 1988; Davis 1991; Prentice and Solomon 1991), implying that biomes are not fixed in time -- they have come and gone in the past, and no doubt will do so in the future, in response to kaleidoscopic changes in the various climate factors to which plants are sensitive.

Prentice et al. (1992a) developed a biome model based on this principle of emergence. The model was designed to work with global data sets of monthly mean temperature, precipitation and cloudiness (Leemans and Cramer 1991) and soil texture (Zobler 1986) at 0.5° resolution. The model determines which of 13 plant types can persist in a given environment by applying limiting values of five bioclimatic indices -- mean temperature of the coldest month (T_c), growing degree-days on a 5°C base (GDD) and on a 0°C base (GDD_o), mean temperature of the warmest month (T_w) and a moisture index (α)

-- and determines which of those present will dominate by comparing values of a dominance index (D):

	T_c min	T_c max	GDD min	GDD_0 min	T_w min	α min	α max	D
Trees								
tropical evergreen	15.5					0.80		1
tropical raingreen	15.5					0.45	0.95	1
warm-temperate evergreen	5					0.65		2
temperate summergreen	-15	15.5	1200			0.65		3
cool-temperate conifer	-19	5	900			0.65		3
boreal evergreen	-35	-2	350			0.75		3
boreal summergreen		5	350			0.65		3
Non-trees								
sclerophyll/succulent	5					0.28		4
warm grass/shrub					22	0.18		5
cool grass/shrub			500			0.33		6
cold grass/shrub				100		0.33		6
hot desert shrub					22			7
cold desert shrub				100				8

Note that many elements of the matrix are intentionally missing, because different plant types are limited by different variables. The reasoning behind the various bioclimatic indices and limiting values is as follows:

Mean temperature of the coldest month. Low-temperature tolerances of overwintering buds and evergreen leaves fall into classes corresponding to different cold tolerance mechanisms (Woodward 1987). For example, temperate deciduous trees use supercooling, which implies the ability to tolerate temperates below 0°C but not below ca 40°C, the anucleate freezing point of water (Sakai and Weiser 1973). Boreal trees tolerate colder temperatures than this, by progressive dessication of exposed tissues during the onset of winter (Sakai 1979). At the other extreme, most tropical broad-leaved trees cannot tolerate frost and may even be damaged by temperatures below 10°C. Lacking sufficiently dense global coverage of data on absolute minimum temperature, Prentice et al. (1992a) translated these physiological limiting temperatures into minimum values of the mean coldest-month temperature, using an empirical relationship from data in Müller (1982). Some plant types were also assigned maximum values of the mean coldest-month temperature, related to other mechanisms. Winter-deciduous trees require a sufficiently cold winter as a phenological trigger (Box 1981), and boreal evergreen conifers fail to regenerate unless adequately chilled in winter.

Growing degree days (5°C base). The poleward limits of boreal and temperate trees are determined by a total heat requirement for completion of their annual cycle of growth, regeneration, dormancy and budburst (Woodward 1987). This is expressed in the model as a minimum requirement for growing degree days, i.e. the annual integral of temperatures above 5°C. The 5°C base is used because the threshold for tree growth is around 5°C. (Many conifers can show positive net assimilation at subzero temperatures, but higher temperatures are needed for cell division and growth; see e.g. Larcher 1983).

Growing degree days (0°C base). The most cold-tolerant (non-woody) plants were assumed to start growing at 0°C (Woodward 1987).

Mean temperature of the warmest month. Warm-climate grasses and shrubs were assumed to require summers warmer than 22°C.

Moisture index. The Priestley-Taylor coefficient, i.e. the ratio of actual to equilibrium evapotranspiration over the year, provides an index of relative moisture availability. All plant types were assigned a minimum value of this ratio to represent the limit of their drought tolerance, and drought-deciduous trees were assigned a maximum value to represent a phenological trigger. The index was calculated using a simple physically based water-balance model (Prentice et al. 1992b). In the model, equilibrium evapotranspiration is calculated from net radiation and temperature; net radiation is estimated from calculated daily insolation and observed mean monthly cloudiness using semi-empirical formulas, keeping the land-surface albedo constant. Stomatal conductance is assumed to exert no control over evapotranspiration except when stomata close in response to a soil moisture deficit. This occurs when the atmospheric demand (equated with the equilibrium evapotranspiration) exceeds a supply rate which is maximally 1 mm h^{-1} and reduced in proportion to soil wetness (Federer 1982). An analytical solution is used for the daily integral of the lesser of the supply and demand functions. The model does not require data on atmospheric humidity and wind speed, nor does it involve modelling leaf area, surface roughness or stomatal conductance. Because of the boundary-layer humidity feedback (Jarvis and MacNaughton 1986; MacNaughton 1986), the assumption of no stomatal control is a not unreasonable approximation for areas as large as the model grid cells. The balance of precipitation minus evapo-transpiration is added to (taken from) the soil moisture store on a quasi-daily timestep; runoff occurs when the store is full. The size of the store is estimated from soil texture class. The model is iterated until a stable seasonal course of soil moisture is reached.

The temperature and moisture index calculations are made on each grid cell, then the matrix of limiting values (the <u>environmental sieve</u>) is applied to determine which plant types can persist there. The list of types is further reduced by the application of a <u>dominance hierarchy</u> (Box 1981). Only those types in the highest dominance class present qualify for dominance. For example, if trees are present then the presence of grasses is ignored; if tropical trees are present then all temperate trees are ignored; and so on. This is a considerable oversimplification because vegetation is a mosaic in which different plant types may dominate different patches of vegetation (e.g. grasses and trees in savana, or deciduous trees and conifers in boreal forest). It also means throwing away potentially useful information on the presence of non-dominant types (e.g. grasses in xerophytic shrublands, or warm-temperate conifers in some tropical moist forests).

However, some such simplifying procedure is needed to allow direct comparison with global vegetation maps.

Biomes emerge from the model as the various distinct combinations of dominant plant types. These plant types assort into combinations of dominants that can readily be identified with biomes; for example, tropical evergreen trees + tropical raingreen trees = tropical seasonal forest; cold grass/shrub only = tundra.

Prentice et al. (1992a) showed that this biome model can reproduce most major features of the vegetation patterns in Olson et al. (1983). The model also produced some real vegetational distinctions not resolved in the Olson et al. data, such as the areas of dominance by boreal deciduous trees within the east Siberian taiga (Larix) and along the dry southern margins of the taiga in Eurasia and North America. The overall kappa statistic for the comparison was 0.49, when calculated on a grid cell by grid cell basis. (The kappa statistic is the proportion of grid cells in common between two maps, reduced by a correction for the expected agreement due to chance: Monserud and Leemans 1992.) This value compares favourably with the value of 0.40 obtained for an aggregated and recalibrated version of the Holdridge scheme (Monserud and Leemans 1992). The kappa statistic rose to 0.59 when calculated for 10° x 10° blocks, i.e. at a spatial resolution of the order used in most atmospheric general circulation models (AGCMs). Values larger than this were obtained for some biomes: cool mixed forest, taiga (including the Larix forest), tropical dry forest/savana, hot desert, tundra and polar desert. The smallest values were for temperate deciduous forest, where most of the predicted distribution area is now under agriculture.

5. LONG-TERM CHANGES IN TERRESTRIAL CARBON STORAGE

Shackleton (1977) first noted that changes in the $\delta^{13}C$ of deep ocean water, shown in benthic foraminifera, might indicate changes in the total amount of carbon stored in biomass and soils. Terrestrial organic carbon is isotopically light relative to atmospheric CO_2 because of fractionation in photosynthesis, predominantly by C_3 plants (see e.g. Farquhar et al. 1989). Shackleton inferred from measurements on benthic foraminiferal remains that carbon in the ocean at the last glacial maximum (LGM) was depleted in ^{13}C relative to today, and attributed this to a reduction in the terrestrial organic carbon reservoir. The magnitude of the reduction implied by these measurements seemed plausible given the widespread evidence already available at that time for reduced forest biomass, particularly in the tropics. A recent reconstruction of the ocean-wide $\delta^{13}C$ anomaly at the LGM (Duplessy et al. 1988) is consistent with a total terrestrial carbon storage on the order of 500 Pg C less than present (Crowley 1991, Prentice and Sarnthein 1992).

The response of carbon storage to climate is not straightforward. For example, high biomass (carbon storage in living plants) is found not only in tropical forests but also in tall conifer forests, growing in warm-temperate climates with a significant drought season (Olson et al. 1983). High soil carbon storage is found in tundra, where a low rate of net primary production is incompletely balanced by a still lower rate of organic matter decomposition (Post et al. 1982). The response of regional climates to changes in global climatic controls is also not straightforward. The world at the LGM was by no means uniformly cold and dry (COHMAP Members 1988; Crowley and North 1991). Prediction of changes in terrestrial carbon storage, even assuming equilibrium with climate, must involve modelling both the geographic pattern of changes in biome distribution in response to climate -- biome modelling -- and the response of regional climates to changes in global climatic controls.

Palaeoclimatological studies have established in broad outline how and why regional climates change on a glacial-interglacial time scale (Bartlein 1988; COHMAP Members 1988; Crowley and North 1991; Kutzbach et al. 1992) and this understanding provides a basis for modelling past regional climates. Slow variations in the earth's orbital parameters and the configuration of ice sheets are transmitted to the land by the general circulation of the atmosphere. The resulting patterns of seasonal wind fields, temperature, precipitation and radiation can be explored with the help of AGCMs. The atmospheric circulation adjusts rapidly to these changing conditions, so the standard approach is to carry out "snapshot" AGCM simulations in which the earth's orbit, the ice sheets, sea-surface temperatures, sea-level and atmospheric composition are treated as fixed boundary conditions characteristic of particular times in the past.

At the LGM, ca 18 ka (1 ka = 1000 ^{14}C years before present. 18 ka = ca 21 x 10^3 astronomical years ago: Bard et al. 1990) the essential differences from present were in the surface boundary conditions -- large ice sheets, cold sea-surface temperatures and low sea level -- and in atmospheric composition; atmospheric CO_2 stood at ca 190 ppm (Barnola et al. 1987). Snapshot simulations for 18 ka (e.g. Manabe and Hahn 1977; Kutzbach and Guetter 1986; Broccoli and Manabe 1987) have produced major features of land climates that are verified by palaeoclimatic and palaeoecological data (COHMAP Members 1988; Kutzbach et al. 1992). The global mean temperature was colder than today. Land temperatures were generally cool and precipitation low in many regions. The Laurentide ice sheet also had a dynamic effect on the westerly circulation, producing a split jet stream which advected cold air southwards and further cooled the North Atlantic and Eurasia. Superficial consideration of these data and model results suggests that global terrestrial carbon storage might have been substantially reduced. During the glacial-interglacial transition large continental ice sheets gave way to boreal forests and tropical savanas to tropical forests. Such changes would be at least qualitatively consistent with the ca 500 Pg increase inferred from the marine $\delta^{13}C$ record.

KC Prentice and Fung (1990) however came to a very different conclusion. They used the results of an 18 ka simulation with the GISS AGCM (Hansen et al. 1984) to drive a variant of the Holdridge model, and associated alternative sets of soil and vegetation carbon storage estimates with each biome. Depending on which sets of estimates were used, Prentice and Fung's figures suggest that in the 18 ka climate and with a sea-level lowering of 120 m (Fairbanks 1989), the total carbon storage anomaly (18 ka minus present) was in the range of -20 to +15 Pg C. Reduced soil carbon storage due to the ice sheets was balanced by soil carbon storage in extensive tundra areas and by additional forest biomass on the exposed continental shelves in the tropics.

Adams et al. (1990), in response to KC Prentice and Fung (1990), attempted to reconstruct the distributions of biomes at 18 ka by extrapolation from a variety of geological data. They reached a conclusion opposite to that of Prentice and Fung, namely that the increase in terrestrial carbon storage after the LGM was much <u>larger</u> than that inferred from the marine record; as much as 1350 Pg. They pointed out that the simulation of Prentice and Fung was inconsistent with some of the data, and in particular that it overestimated the extent of tropical forests and underestimated the extent of deserts at the LGM. On the other hand, it is not clear what uncertainty should be attached to the figure obtained by Adams et al. (1990). Reliably dated reconstructions of LGM vegetation are sparse to absent in key areas such as Indonesia and the Amazon lowlands (Crowley and North 1991), where errors in the assumed vegetation distribution would have the largest effects on the total carbon budget.

Thus there are now three qualitatively distinct estimates of the carbon transfer from ocean to land during the glacial-interglacial transition: zero from an AGCM-based study, <u>ca</u> 500 Pg from oceanic $\delta^{13}C$ and <u>ca</u> 1350 Pg from an interpretation of the geological data. The reasons for these differences are not properly understood (Crowley 1991). We are currently investigating the problem using the biome model of Prentice et al. (1992a), various AGCM results for 18 ka including that used by KC Prentice and Fung (1990) to drive the model, and palaeoecological data summarized in Kutzbach et al. (1992) to check the predictions of the model. The use of models in this context allows a rational procedure for extrapolation into areas where palaeodata are sparse, while the data provide an essential test of the procedure in areas where they are abundant.

6. BIOME DYNAMICS

The biome model of Prentice et al. (1992a) represents a first step towards a mechanistic approach to modelling terrestrial biosphere dynamics. It is a qualitative model in the sense that it is based entirely on the constraints that ultimately limit the geographic distribution of different types of plants. Feedback effects of vegetation on climate are ignored. Interactions between different plant types are considered only through the crude device of the dominance hierarchy. Ecosystem dynamics are not simulated and in consequence carbon storages can only be treated as fixed biome properties. Although the success of such a simple model in reproducing world vegetation patterns is encouraging, there is considerable scope for further development of this type of model to incorporate quantitative climatic effects on the growth of different plant types, vegetation-climate feedbacks, explicit carbon fluxes among vegetation and ecosystem compartments, and dynamic interactions among plant types.

One possible line of development would be to incorporate one-dimensional hydrological feedbacks by making certain variables that are currently prescribed (including surface albedo and maximum transpiration rate) into interactive variables affected by the current vegetation state. This would allow for negative feedbacks that operate along aridity gradients: where water is in short supply the vegetation is sparse, absorbs less solar radiation and transpires less, thus retaining more soil moisture than is implied by a model without such feedback (Woodward 1987). Subsequent developments along this line might include coupling with climate models in order to simulate large scale interactions between vegetation and climate. For example, it has been proposed that high albedo in deserts may act as a positive feedback (counter to the negative feedback mentioned above) causing increased sinking motion and reduced precipitation, thus maintaining desertified lands as deserts (Rasool 1984).

That some effect of this type exists is suggested by the AGCM study of Street-Perrott et al. (1990) concerning northern Africa at 9 ka, when grasslands covered much of what is now the Sahara. The global climate at 9 ka was by no means identical with today, one key difference being the high northern-hemisphere summer and low northern-hemisphere winter insolation which enhanced the land-sea contrast and consequently produced an increase in the strength and extent of the northern monsoons (Kutzbach and Street-Perrott 1985; Kutzbach and Guetter 1986; COHMAP Members 1988). Street-Perrott et al. (1990) ran an AGCM in two versions, one with the modern distribution of surface albedo and one with lower albedo values estimated from a reconstruction of the former vegetation pattern. In the second of these simulations, the summer precipitation in the central Sahara was > 1 mm d^{-1} greater than in the first and the spatial pattern of the precipitation was more consistent with the palaeoclimatic data. This sensitivity experiment with fixed surface albedo values suggests that much could be learned about large-scale vegetation-climate interactions by coupling a biome model to an AGCM, so that the land surface properties would interact dynamically with the general circulation. The absence of explicit vegetation dynamics would presumably not be important in the context of differences between equilibrium solutions of the coupled model.

Interactions between vegetation and climate through the hydrological cycle are indirectly relevant to carbon cycle studies because they may affect the extent of different biomes. More directly relevant are the dynamics of successional adjustment where biome shifts involve the accumulation and/or destruction of biomass. Some ideas for the development of a global vegetation <u>dynamics</u> model are discussed by Prentice et al. (1989). One key problem in incorporating vegetation dynamics into a static biome modelling framework is in bridging the gap between the characteristic scale of plant-plant interactions (< 1 to ca 30 m) and the finest practical spatial resolution for biome modelling based on global data sets (≥ 50 km). Scaling problems of this kind are familiar enough to meteorologists and oceanographers; they represent important new challenges to terrestrial ecologists. Prentice et al. (1989) outlined a method based on statistical sampling of "patches" within grid cells, with a highly simplified representation of patch-scale dynamics. Such a method can call on well-established principles used to model vegetation dynamics at local to regional scales.

ACKNOWLEDGMENTS

The biome model was developed with collaborators Wolfgang Cramer, Sandy Harrison, Rik Leemans, Bob Monserud and Al Solomon under the auspices of the Biosphere Dynamics Project at IIASA, Laxenburg, while Al Solomon was leader of the project. Bill Emanuel and Sandy Harrison commented helpfully on the manuscript.

REFERENCES

Adams JM, Faure H, Faure-Denard L, McGlade JM, Woodward FI (1990) Increases in terrestrial carbon storage from the Last Glacial Maximum to the present. Nature 348:711-714

Bard E, Hamelin B, Fairbanks RG, Zindler A (1990) Calibration of the ^{14}C timescale over the past 30,000 years using mass spectrometric U-Th ages from Barbados corals. Nature 345:405-410

Barnola JM, Raynaud D, Korotkevich YS, Lorius C (1987) Vostok ice core provides 160,000-year record of atmospheric CO_2. Nature 329:408-414

Bartlein PJ (1988) Late-Tertiary and Quaternary palaeoenvironments. In: Huntley B, Webb III T (eds) Vegetation history. Kluwer, Dordrecht, p 113

Box EO (1981) Macroclimate and plant forms: an introduction to predictive modelling in phytogeography. Junk, The Hague

Broccoli AJ, Manabe S (1987) The influence of continental ice, atmospheric CO_2, and land albedo on the climate of the last glacial maximum. Climate Dynamics 1:87-99

COHMAP Members (1988) Climatic changes of the last 18,000 years: observations and model simulations. Science 241:1043-1052

Crawley MJ (1986) Plant ecology. Blackwell Oxford

Crowley TJ (1991) Ice age carbon. Nature 352:575-576

Crowley TJ, North GR (1991) Paleoclimatology. Oxford University Press, New York

Davis MB (1989) Lags in vegetation response to greenhouse warming. Clim Change 15:75-82

Davis MB (1991) Research questions posed by the paleo-ecological record of global change. In: Bradley RS (ed) Global Changes of the Past. UCAR/OIES, Boulder, p 385

Duplessy JC, Shackleton NJ, Fairbanks RG, Labeyrie L, Oppo D, Kallel N (1988) Deepwater source variations during the last climatic cycle and their impact on the global deepwater circulation. Paleoceanography 3:343-360

Emanuel WR, Shugart HH, Stevenson MP (1985) Climate change and the broad-scale distribution of terrestrial ecosystem complexes. Clim Change 7:29-43

Fairbanks RG (1989) A 17,000-year glacio-eustatic sea level record: influence of glacial melting rates on the Younger Dryas event and deep-ocean circulation. Nature 342:637-642

Farquhar GD, Ehleringer JR, Hubick KT (1989) Carbon isotope discrimination and photosynthesis. Ann Rev Pl Physiol Mol Biol 40:503-537

Federer CA (1982) Transpirational supply and demand: plant, soil and atmospheric effects evaluated by simulation. Water Resources Res 18:355-362

Grimm EC (1983) Chronology and dynamics of vegetation change in the paririe-woodland region of southern Minnesota. New Phytol 93:311-335

Grimm EC (1984) Fire and other factors controlling the Big Woods vegetation. Ecol Monogr 54:291-311

Guetter PJ, Kutzbach JE (1990) A modified Köppen classification applied to model simulations of glacial and interglacial climates. Clim Change 16:193-215

Hansen J, Lacis A, Rind D, Russell G, Stone P, Fung I, Ruedy R, Lerner J (1984) Climate sensitivity: analysis of feedback mechanisms. In: Hansen JE, Takahashi T (eds) Climate processes and climate sensitivity. American Geophysical Union, Washington, p 130

Holdridge LR (1947) Determination of world formations from simple climatic data. Science 105:367-368

Jarvis, PG, MacNaughton KG (1986) Stomatal control of transpiration: scaling up from leaf to region. Adv Ecol Res 15:1-49

Köppen W (1936) Das Geographisches System der Klimate. In Köppen W, Geiger, R (eds) Handbuch der Klimatologie, I(C). Borntraeger, Berlin

Kutzbach JE, Guetter PJ (1986) The influence of changing orbital parameters and surface boundary conditions on climate simulations for the past 18 000 years. J Atm Sci 43:1726-1759

Kutzbach JE, Street-Perrott FA (1985) Milankovitch forcing of fluctuations in the level of tropical lakes from 18 to 0 kyr BP. Nature 317:130-134
Kutzbach JE, Street-Perrott FA, Ruddiman WF, Webb III T, Wright Jr HE (eds) (1992) Global climates since the last glacial maximum. University of Minnesota Press, Minneapolis, in press
Larcher W (1983) Physiological plant ecology, 2nd edn. Springer, Berlin
Leemans R (1990) Possible changes in natural vegetation patterns due to a global warming. WP-90-08, IIASA, Laxenburg
Leemans R, Cramer W (1991) The IIASA database for mean monthly values of temperature, precipitation and cloudiness on a terrestrial grid. RR-91-18, IIASA, Laxenburg
MacNaughton KG (1986) Regional evaporation models. In: Rosenzweig C, Dickinson R (eds) Climate-vegetation interactions. UCAR/OIES, Boulder, p 103
Manabe S, Hahn DG (1977) Simulation of the tropical climate of an ice age. J Geophys Research 82:3889-3911
Monserud RA, Leemans R (1992) The comparison of global vegetation maps. Ecol Modelling, in press
Müller JM (1982) Selected climatic data for a global set of standard stations for vegetation science. Junk, The Hague
Olson JS, Watts JA, Allison LJ (1983) Carbon in live vegetation of major world ecosystems. ORNL-5862, Oak Ridge National Laboratory, Oak Ridge
Post WM, Emanuel WR, Zinke PJ, Stangenberger AG (1982) Soil carbon pools and world life zones. Nature 298:156-159
Prentice IC (1986) Vegetation response to past climatic variation. Vegetatio 67:131-141
Prentice IC, Sarnthein M (1992) Self-regulatory processes in the biosphere in the face of climate change. In Eddy J (ed) Global changes in the perspective of the past. Wiley, Chichester, in press
Prentice IC, Solomon AM (1991) Vegetation models and global change. In: Bradley RS (ed) Global changes of the past. UCAR/OIES, Boulder, p 365
Prentice IC, Sykes MT, Cramer W (1991) The possible dynamic responses of northern forests to greenhouse warming. Global Ecol Biogeog Lett 1:129-135
Prentice IC, Cramer W, Harrison SP, Leemans R, Monserud RA, Solomon AM (1992a) A global biome model based on plant physiology and dominance, soil properties and climate. J Biogeog, in press
Prentice IC, Sykes MT, Cramer W (1992b) A simulation model for the transient effects of climate change on forest landscapes. Ecol Modelling, in press
Prentice IC, Webb RS, Ter-Mikhaelian MT, Solomon AM, Smith TM, Pitovranov SE, Nikolov NE, Minin AA, Leemans R, Lavorel S, Korzukhin MD, Hrabovsky JP, Helmisaari HO, Harrison SP, Emanuel WR, Bonan GB (1989) Developing a global vegetation dynamics model: results of a summer workshop. RR-89-7, IIASA, Laxenburg

Prentice KC (1990) Bioclimatic distribution of vegetation for general circulation model studies. J Geophys Res 95:11 811-11 830
Prentice KC, Fung IY (1990) The sensitivity of terrestrial carbon storage to climate change. Nature 346:48-51
Rasool SI (1984) On dynamics of deserts and climate. In: Houghton JT (ed) The global climate. Cambridge University Press, Cambridge, p 107
Sakai A (1979) Freezing avoidance mechanism of primordial shoots of conifer buds. Plant Cell Physiol 20:1381-1390
Sakai A, Weiser CJ (1973) Freezing resistance of trees in North America with reference to tree regions. Ecology 54:118-126.
Schlesinger ME, Mitchell JFB (1985) Model projections of the equilibrium climatic response to increased carbon dioxide. In: MacCracken MC, Luther FM (eds) Projecting the climatic effects of increasing carbon dioxide. ER-0237, US-DOE, Washington, p 81
Schlesinger WH (1990) Evidence from chronosequence studies for a low carbon-storage potential of soils. Nature 348:232-234
Shackleton NJ (1977) Carbon-13 in Uvigerina: tropical rain forest history and the equatorial Pacific carbonate dissolution cycles. In: Anderson NR, Malahoff A (eds) The fate of fossil fuel in the ocean. Plenum, New York, p 401
Shugart HH, Antonovsky MYa, Jarvis PG, Sandford AP (1986) CO_2, climatic change and forest ecosystems. In: Bolin B, Döös BR, Jäger J, Warrick RA (eds) The greenhouse effect, climatic change, and ecosystems. Wiley, Chichester, p 475
Solomon AM, Shugart HH (eds) (1992) Vegetation dynamics and global change. Cambridge University Press, Cambridge, in press
Street-Perrott FA, Mitchell JFB, Marchand DS, Brunner JS (1990) Milankovitch and albedo forcing of the tropical monsoons: a comparison of geological evidence and numerical simulations for 9000 yBP. Trans Roy Soc Edinburgh 81:407-427
Walter H (1979) Vegetation of the earth and ecological systems of the geo-biosphere, 2nd edn. Springer, New York
Webb III T (1988) Eastern North America. In: Huntley B, Webb III T (eds) Vegetation history. Kluwer, Dordrecht, p 385
Woodward FI (1987) Climate and plant distribution. Cambridge University Press, Cambridge
Woodward FI (1990) Global change: translating plant ecophysiological responses to ecosystems. Trends Ecol Evol 5:308-311
Zobler L (1986) A world soil file for global climate modeling. NASA Tech Memo 87802

A DYNAMIC MODEL OF TERRESTRIAL CARBON CYCLING

William R. Emanuel
Anthony W. King
Wilfred M. Post
Environmental Sciences Division
Oak Ridge National Laboratory
P.O. Box 2008, MS-6335
Oak Ridge, Tennessee 37831-6335
USA

INTRODUCTION

Many human activities tend to reduce the amount of carbon in plants and soils (Woodwell *et al.* 1983, Houghton *et al.* 1983, Houghton and Skole 1990). Houghton *et al.* (1983) estimate that forest clearing and harvest released 180 Pg (1 Pg = 1×10^{15}g) of carbon into the atmosphere between 1860 and 1980 compared to about 163 Pg from fossil fuel emissions over the same period (Marland *et al.* 1989)—terrestrial releases exceeded fossil fuel emissions until about 1959, according to these estimates. The 1990 net flux of carbon into the atmosphere, mostly due to tropical deforestation, may have approached 3 Pg/year (Houghton, this volume). Thus the use and manipulation of terrestrial ecosystems probably caused a significant part of the observed increases in atmospheric CO_2 concentration (Bolin 1986, Post *et al.* 1990).

But oceanic carbon uptake is apparently not large enough to accommodate such large releases from land pools in addition to fossil fuel emissions (Siegenthaler and Oeschger 1987, Enting and Mansbridge 1987). Estimates of the decreases in terrestrial carbon storage derived from land-use data are about two-times larger than ocean modeling studies imply are consistent with changes in the carbon content of the atmosphere indicated by ice-core data (Friedli *et al.* 1986, Neftel *et al.* 1985) and by direct measurements (Keeling 1986). Furthermore, while tropical deforestation seems to be accelerating (Houghton and Skole 1990), estimates of oceanic uptake permit no source in addition to fossil fuels since about 1970 or even require an additional sink for carbon from the atmosphere. These inconsistencies in our understanding of rising atmospheric CO_2 levels force substantial skepticism about projections of future CO_2 increases due to further fossil fuel use.

Several explanations are being pursued. The oceans may remove more carbon from the atmosphere than current models indicate. Fossil fuel emissions estimates may be high. Estimates of carbon releases from vegetation and soils derived from land-use and

NATO ASI Series, Vol. I 15
The Global Carbon Cycle
Edited by M. Heimann

population data may be incorrect, or increases in carbon pools in some ecosystems due to the stimulation of primary productivity by rising atmospheric CO_2 concentration may compensate partially for terrestrial releases. This CO_2 fertilization effect is observed in laboratory and field experiments with small plants growing under controlled environmental conditions but has not been verified in natural ecosystems occurring over significant areas (Eamus and Jarvis 1989, Bazzaz 1990). Many carbon cycle models include a CO_2 fertilization effect (e.g., Keeling *et al.* 1989). Each of these factors probably contributes to some extent to inconsistencies in our understanding of past changes in atmospheric CO_2, but the role of terrestrial ecosystems is most uncertain.

The net flux between the atmosphere and terrestrial ecosystems is small compared with global uptake by primary productivity or loss by decomposition and fire—these naturally occurring fluxes are each about 50 PgC/year (Post *et al.* 1990). But forest clearing or harvest strongly perturb the ecosystems they affect. Both cause large, immediate changes in plant carbon pools, and because the production of dead organic matter is altered, they affect carbon pools in litter and soil as well. Depending on the management practices and types of ecosystems involved, responses to a land-use event can continue for decades to several centuries. Following clearing or similar disturbances, net productivity generally exceeds losses, and disturbed ecosystems act as sinks for carbon from the atmosphere. In many instances, ground vegetation rather than trees is established after clearing, and by management this conversion may continue indefinitely, with lower carbon storage than in the original ecosystems. Carbon in wood products such as building materials or in wood disposed of by burial may not return to the atmosphere for some time. A mathematical model that tracks areas affected by land use and simulates these ecosystem responses is essential to estimating the consequences of human activities for atmospheric CO_2 levels.

Moore *et al.* (1981) and Houghton *et al.* (1983) describe one model for analyzing land-use change. That model tracks each land unit affected by forest harvest or clearing. Synoptic response functions, Fig. 1, describe changes in the density of carbon storage per unit land area following clearing or harvest. A set of compartments return carbon associated with human pools such as building materials to the atmosphere with a delay.

This paper describes an alternative model for analyzing changes in terrestrial carbon pools in response to land use. A compartment model simulates responses to forest harvest, clearing, or other land-use activities. An area distribution function, which summarizes the areal extent of each ecosystem type according to carbon density in vegetation per unit area between minimum and maximum densities, further describes carbon in vegetation. Forest harvest and clearing as well as plant growth alter this area distribution function through time. Simulations using this approach agree reasonably

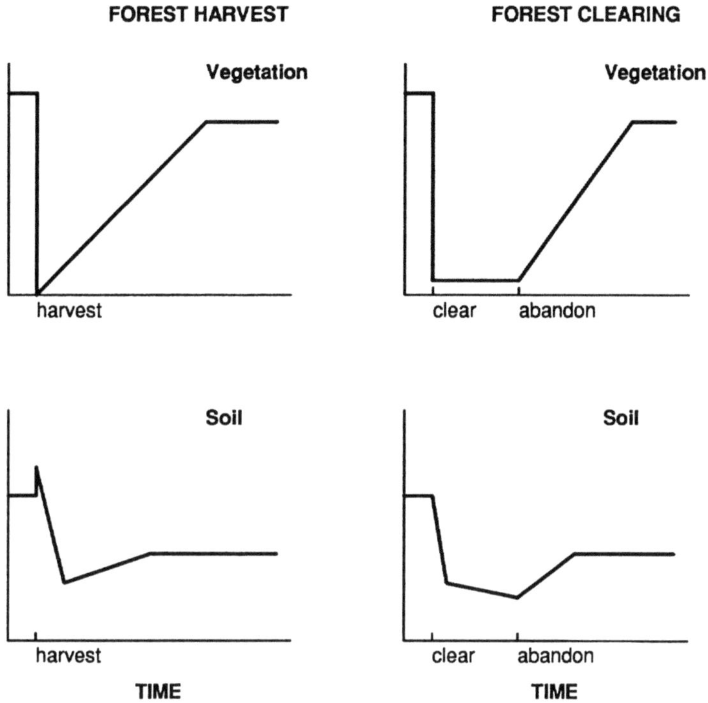

Figure 1. Synoptic response functions used by Houghton *et al.* (1983) to describe changes in carbon density per unit area at a site after forest harvest or clearing.

well with estimates of the net terrestrial flux into the atmosphere reported by Houghton *et al.* (1983).

MODEL DESCRIPTION

A model for analyzing the impacts of land use on terrestrial carbon cycling must track changes in areal extents of ecosystem complexes affected by land-use change and simulate the responses of terrestrial carbon pools to land-use perturbations through any period of human management or recovery that follows. In our model, the land surface is divided into land units occupied by one or more ecosystem types. We combine a function describing the areal distribution of the density of carbon in vegetation per unit area with a compartment representation of carbon cycling to form a model for each ecosystem. Land use transfers carbon between compartments and alters the area distribution functions. Forest harvest, for example, decreases the area of forest at high carbon density on a land unit and increases the area at low density by the same amount.

Forest clearing for agriculture decreases the area in the forest distribution function and adds it to the distribution function for crops.

Carbon in a terrestrial ecosystem is divided into vegetation, litter, and soil compartments, Fig. 2. Vegetation assimilates carbon from the atmosphere by primary production. Litterfall and belowground mortality transfer carbon from the vegetation compartment into the litter and soil compartments respectively. Carbon moves from litter into soil as surface dead organic matter decomposes, while decomposition releases carbon from the litter and soil compartments back into the atmosphere.

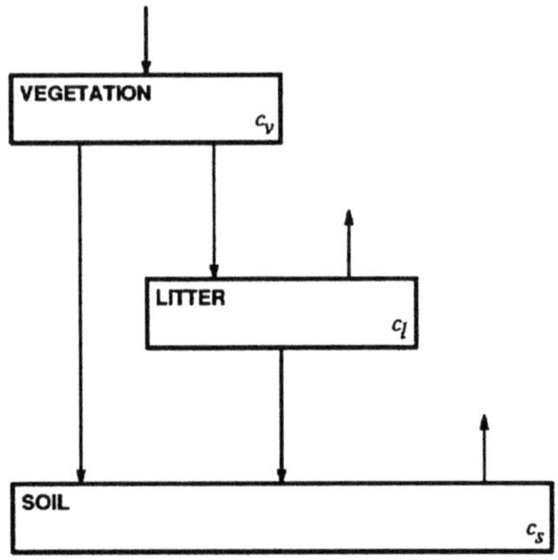

Figure 2. A three-compartment representation of carbon in terrestrial ecosystems. Carbon in dead organic matter and decomposers is divided between litter and soil compartments.

Let c_v, c_l, and c_s represent the densities of carbon per unit area in the vegetation, litter, and soil compartments. The uptake of carbon from the atmosphere f_{av}, corresponding to net primary productivity, is assumed to depend on the size of vegetation measured by its carbon density. A logistic function describes this dependence:

$$f_{av} = \nu c_v - \rho c_v^2, \tag{1}$$

where ν is an intrinsic growth rate, and ρ determines the asymptotic level to which the carbon density in vegetation converges. All other fluxes correspond to transfers or to the decomposition of dead organic matter and are linearly proportional to the content of their donor compartments. Thus, a coupled system of differential equations, each describing material balance for a compartment, describes carbon turnover in the three compartments:

$$\frac{dc_v}{dt} = \nu c_v - \rho c_v^2 - (\alpha_{vl} + \alpha_{vs})c_v \tag{2}$$

$$\frac{dc_l}{dt} = \alpha_{vl}c_v - (\alpha_{ls} + \alpha_{la})c_l \tag{3}$$

$$\frac{dc_s}{dt} = \alpha_{vs}c_v + \alpha_{ls}c_l - \alpha_{sa}c_s. \tag{4}$$

The symbol α_{ij} represents the rate coefficient for the flux from compartment i into compartment j, where the subscripts v, l, and s refer to the vegetation, litter, and soil compartments respectively. Subscript a indicates the atmospheric carbon reservoir.

Equation (2) is independent and can be solved:

$$c_v(t) = \frac{c_v(t_o)e^{(\nu-\alpha_{vl}-\alpha_{vs})(t-t_o)}}{1 - c_v(t_o)\frac{\rho}{\nu-\alpha_{vl}-\alpha_{vs}}\left(1 - e^{(\nu-\alpha_{vl}-\alpha_{vs})(t-t_o)}\right)}, \tag{5}$$

where $c_v(t_o)$ is the density of carbon in vegetation at an initial time t_o. Equation (2) has two equilibrium points:

$$c_v^\circ = 0, \tag{6}$$

$$c_v^\circ = \frac{1}{\rho}(\nu - \alpha_{vl} - \alpha_{vs}). \tag{7}$$

The first-derivative test, applied to Eq. (2), indicates that the maximum rate of carbon accumulation in vegetation occurs at one-half the nonzero equilibrium point, Eq. (7). The time required to reach that point of maximum accumulation rate can be derived from Eq. (5):

$$(t^* - t_o) = \frac{1}{\nu - \alpha_{vl} - \alpha_{vs}} \ln\left(\frac{c_v^\circ - c_v(t_o)}{c_v(t_o)}\right), \tag{8}$$

where $t^* > t_o$ is the time at which maximum accumulation rate occurs.

The intrinsic growth rate ν can be calculated from Eq. (8) for specified initial conditions, equilibrium carbon density, and the point at which maximum accumulation rate occurs:

$$\nu = \frac{1}{t^* - t_o} \ln\left(\frac{c_v^\circ - c_v(t_o)}{c_v(t_o)}\right) + \alpha_{vl} + \alpha_{vs}. \tag{9}$$

The parameter ρ is then given in terms of the nonzero equilibrium point:

$$\rho = \frac{1}{c_v^\circ}(\nu - \alpha_{vl} - \alpha_{vs}). \tag{10}$$

Since $c_v^\circ = 0$ is a trivial, stable equilibrium point of Eq. (2), carbon does not accumulate in vegetation from zero initial conditions. Equation (9) must be evaluated for a small $c_v(t_o)$ that can be viewed as the carbon content of vegetation at establishment.

The remaining rate coefficients can be estimated from an assumed partitioning of dead organic matter fluxes into litter and soil, specified net primary productivity per unit area, and equilibrium carbon densities in vegetation, litter, and soil. These can be derived from global summaries of terrestrial carbon (e.g., Whittaker and Likens 1973, Schlesinger 1977, Post *et al.* 1982, Olson *et al.* 1983, Ajtay *et al.* 1979) if the compartment model represents an ecosystem typical of a major biome, although this approach does not contend with the substantial variability that occurs within any biome.

The equilibrium fluxes into litter and soil can be derived from the net primary productivity flux:

$$f_{vs}^{\circ} = \theta_v f_{av}^{\circ} \tag{11}$$

$$f_{vl}^{\circ} = f_{av}^{\circ} - f_{vs}^{\circ} = (1 - \theta_v) f_{av}^{\circ} \tag{12}$$

$$f_{ls}^{\circ} = \theta_l f_{vl}^{\circ} = \theta_l (1 - \theta_v) f_{av}^{\circ} \tag{13}$$

$$f_{la}^{\circ} = f_{vl}^{\circ} - f_{ls}^{\circ} = (1 - \theta_l(1 - \theta_v)) f_{av}^{\circ} \tag{14}$$

$$f_{sa}^{\circ} = f_{ls}^{\circ} = \theta_l (1 - \theta_v) f_{av}^{\circ}, \tag{15}$$

where f_{ij}° is the equilibrium flux per unit area from compartment i into compartment j. The symbol θ_v represents the fraction of the total loss from vegetation that enters the soil compartment directly due to belowground mortality; θ_l is the fraction of the loss from the litter compartment that contributes to soil carbon. The rate coefficients are then:

$$\alpha_{ij} = \frac{f_{ij}^{\circ}}{c_i^{\circ}}. \tag{16}$$

For sample calculations, assume that equilibrium carbon densities in vegetation, litter, and soil ($c_v^{\circ}, c_l^{\circ}, c_s^{\circ}$) are 20, 0.81, and 10.89 kg/m^2 respectively. Let 50% of the total loss from vegetation enter the soil compartment (θ_v), and 10% of the loss from litter enter soil (θ_l). Assume that equilibrium net primary productivity (f_{av}°) is 0.9 kg/m^2/year, and that maximum productivity occurs 25 years into recovery from setting carbon density in vegetation to its minimum value. These values are typical of those used by Houghton *et al.* (1983) for tropical moist forest. Table 1 lists values of the three-compartment model parameters calculated using these equilibrium values in the procedure outlined above.

Table 1. Three-compartment model parameters.

Parameter	Value
ν	0.257
ρ	0.0106
α_{vl}	0.0225
α_{vs}	0.0225
α_{ls}	0.0556
α_{la}	0.5
α_{sa}	0.0454

Figure 3 displays the response of the three-compartment model to initial conditions representing a forest harvest event as defined by Houghton *et al.* (1983). The initial density of carbon in vegetation is 0.1 kg/m^2. The initial densities in litter and soil are higher than equilibrium values, representing dead plant material that remains after harvest. Two cases are considered. In case A, 50% of this material is assigned to each dead organic matter pool; while in case B, all of the material is assigned to the litter compartment. The corresponding response functions assumed by Houghton *et al.* (1983) are shown for comparison; however, vegetation and dead organic matter are both assumed to return to equilibrium levels.

The vegetation response is as expected for a logistic equation calibrated for an equilibrium value at the recovered carbon density of the linear response function. The important point is that setting the time to maximum rate of carbon accumulation, Eq. (8), to one-half of the recovery time of the linear response function yields consistent overall recovery times between the two approaches. The curvilinear response of the logistic equation is otherwise more realistic than the linear response function.

The response for dead organic matter approximates the linear response function best for case B in which all residual plant material is assigned to the litter compartment. Even in that case, the initial decrease in dead organic matter is more rapid than in the linear response function, while the minimum occurs about 20 years later than in the linear response function, which is near recovery at that point. The decrease in carbon in litter and soil compartments of the three-compartment model is caused only by the decrease in the inputs of dead organic matter from vegetation. Houghton *et al.* (1983), however, considered an additional decrease in dead organic matter due to direct disturbance to litter and soil by land-use activities.

If the geographic distribution of a variable such as carbon density is ignored, its spatial heterogeneity within a land unit can be summarized by a distribution function. Consider a land unit of area A_T. The function $A(c,t)$ describes the area within this land unit at time t that has carbon density c in a reservoir such as vegetation or soil.

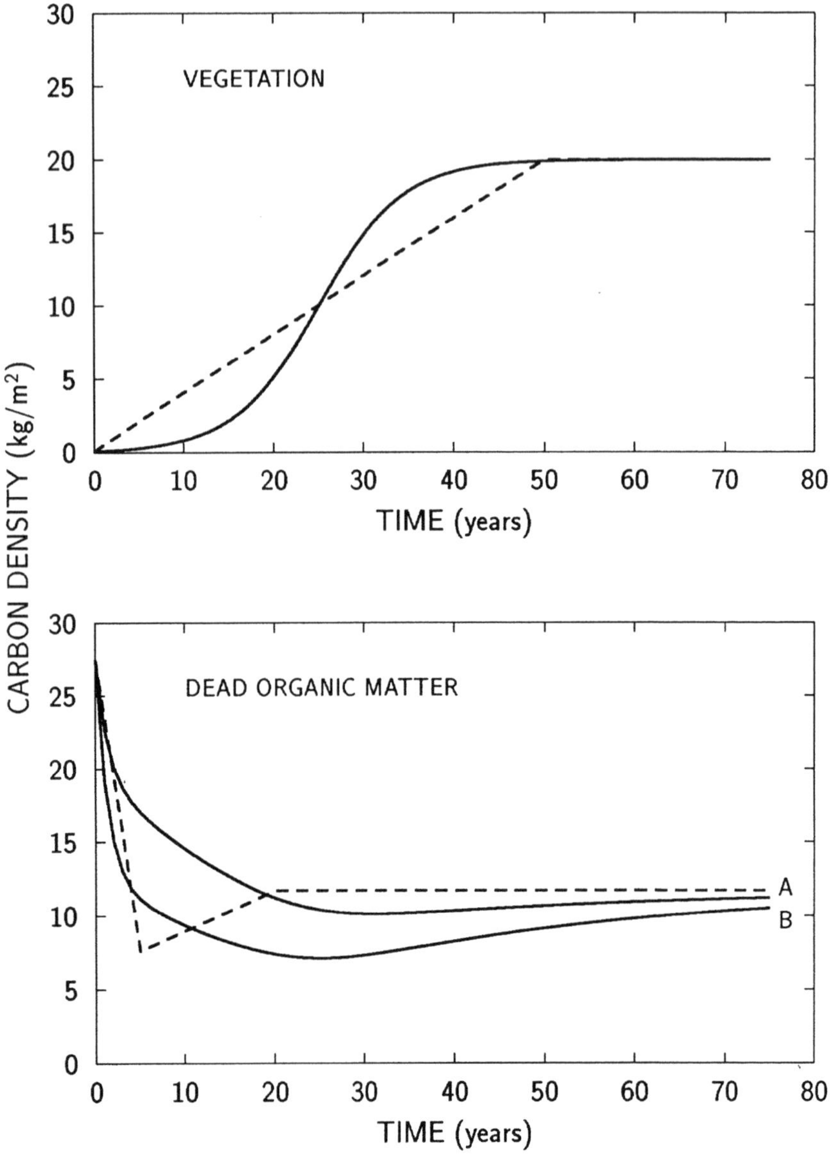

Figure 3. Solution to the three-compartment model of carbon turnover in terrestrial ecosystems with small initial carbon content of vegetation as would occur following forest harvest. In case A, 50% of the harvest residual is assigned to each dead organic matter pool; while in case B, all of the residual is assigned to the litter compartment. Dashed lines are the corresponding response functions from Houghton *et al.* (1983).

Thus the total area of the land unit, which remains constant, is

$$A_T = \int_{c_{min}}^{c_{max}} A(c,t)dc, \tag{17}$$

and the carbon content of the reservoir is

$$c_T = \int_{c_{min}}^{c_{max}} cA(c,t)dc, \tag{18}$$

where carbon density ranges between c_{min} and c_{max}. Hackney *et al.* (1980) use a similar approach to describe size distribution in fish populations.

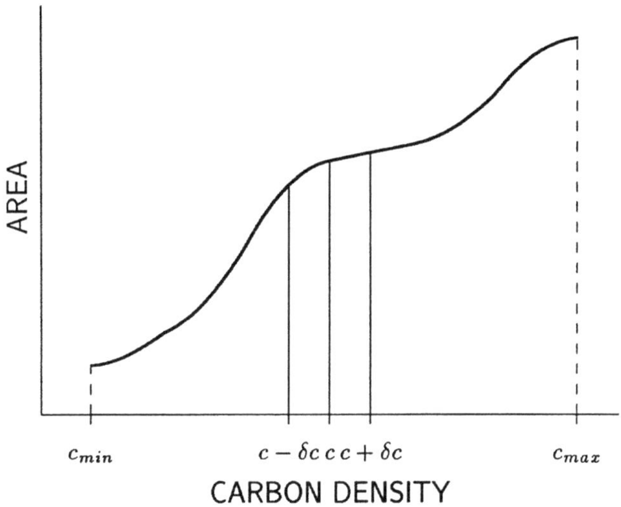

Figure 4. Area distribution function dynamics associated with plant growth and mortality.

If an area distribution function describes carbon density in vegetation, then plant growth and mortality can affect the distribution function. Assume that a nonlinear differential equation describes variations in the density of carbon in vegetation at any point within a land unit:

$$\frac{dc}{dt} = g(c). \tag{19}$$

This generalizes the logistic growth equation, Eq. (2), above. Consider changes in area over an interval in carbon density about c during the time interval t to $t + \delta t$ (Fig. 4). Increases from lower density $c - \delta c$ contribute to the area at density c. Similarly increases to higher density $c + \delta c$ decrease the area at density c. The area added to density c is proportional to the area at $c - \delta c$ and to the differential growth implied by Eq. (19) $g(c - \delta c)\delta t$ compared to the interval δc. Thus,

$$\text{area from } c - \delta c \doteq A(c - \delta c, t)\frac{g(c - \delta c)\delta t}{\delta c}. \tag{20}$$

Similarly the area advanced from carbon density c to $c + \delta c$ is

$$\text{area to } c + \delta c = A(c,t)\frac{g(c)\delta t}{\delta c}. \tag{21}$$

and thus the change in area at density c is

$$A(c,t+\delta t) - A(c,t) = A(c-\delta c,t)\frac{g(c-\delta c)\delta t}{\delta c} - A(c,t)\frac{g(c)\delta t}{\delta c}. \tag{22}$$

Reorganizing,

$$\frac{A(c,t+\delta t) - A(c,t)}{\delta t} = \frac{g(c-\delta c)A(c-\delta c,t) - g(c)A(c,t)}{\delta c}. \tag{23}$$

As $\delta c \to 0$ and $\delta t \to 0$,

$$\frac{\partial A(c,t)}{\partial t} = -\frac{\partial(g(c)A(c,t))}{\partial c}. \tag{24}$$

Equation (24) can be solved from an initial area distribution $A(c,t_o)$ in order to simulate the response through time of the vegetation carbon pool to plant growth.

The carbon content of litter and soil pools is described by Eqs. (3) and (4) only in terms of mean carbon densities per unit area. The fluxes from vegetation into the litter and soil compartments are linearly proportional to the mean carbon density in vegetation:

$$\bar{c}_v = \frac{\int_{c_{min}}^{c_{max}} cA(c,t)dc}{\int_{c_{min}}^{c_{max}} A(c,t)dc}. \tag{25}$$

Land use forces further changes in the area distribution. These perturbations are introduced into Eq. (24) as forcing functions that can depend on carbon density and time:

$$\frac{\partial A(c,t)}{\partial t} = -\frac{\partial(g(c)A(c,t))}{\partial c} + \sum_{i=0}^{n_l-1} f_i(c,t). \tag{26}$$

The character of the land-use forcing functions $f_i(c,t)$ is different for each kind of activity. In the case of forest harvest at maximum carbon density,

$$f(c,t) = \begin{cases} -f_h(t) & c = c_{cmax} \\ f_h(t) & c = c_{min} \\ 0 & \text{elsewhere.} \end{cases} \tag{27}$$

where $f_h(t)$ is the area harvested at time t. Forest clearing for agriculture at maximum carbon density is described by a function of the form:

$$f(c,t) = \begin{cases} -f_c(t) & c = c_{max} \\ 0 & \text{elsewhere.} \end{cases} \tag{28}$$

But this clearing function for the forest area distribution is paired with an establishment function at minimum carbon density in the area distribution function for the ecosystem replacing forest:

$$f(c,t) = \begin{cases} f_c(t) & c = c_{min} \\ 0 & \text{elsewhere.} \end{cases} \tag{29}$$

Land use, including forest harvest and clearing, also causes immediate transfers of carbon between the three compartments at the time when perturbations occur. These perturbation fluxes are usually specified as a fraction of the carbon affected by harvest or clearing that is left as dead organic matter and the portions of this carbon assigned to the litter and soil compartments.

Carbon removed by land use can be transferred immediately into the atmosphere or into decay compartments representing carbon in wood products such as paper or building material. Losses to the atmosphere from these compartments are linearly proportional to their contents:

$$\frac{dc_i}{dt} = -\frac{c_i}{\tau_i} + \sum_{j=0}^{n_l-1} f_d^{ij}(t) \tag{30}$$
$$i = 0, 1, \ldots, n_d - 1,$$

where c_i is the carbon content of decay compartment i. The parameter τ_i is the time constant associated with compartment i, and n_d is the number of decay compartments. The flux f_d^{ij} is the input into decay compartment i associated with the land-use forcing function f_i, Eq. (26).

The general description of carbon cycling for an ecosystem combines the equations for an area distribution of vegetation carbon density driven by land-use forcing functions and differential equations for the density of carbon in litter and soil compartments and for the carbon contents of decay compartments:

$$\frac{\partial A(c,t)}{\partial t} = -\frac{\partial(g(c)A(c,t))}{\partial c} + \sum_{i=0}^{n_l-1} f_i(c,t) \tag{31}$$

$$g(c) = \nu c_v - \rho c_v^2 - \alpha_{vl} c_v - \alpha_{vs} c_v \tag{32}$$

$$\frac{dc_l}{dt} = \alpha_{vl}\bar{c}_v - (\alpha_{ls} + \alpha_{la})c_l + \sum_{i=0}^{n_l-1} f_l^i(t) \tag{33}$$

$$\frac{dc_s}{dt} = \alpha_{vs}\bar{c}_v + \alpha_{ls}c_l - \alpha_{sa}c_s + \sum_{i=0}^{n_l-1} f_s^i(t) \tag{34}$$

$$\bar{c}_v = \frac{\int_{c_{min}}^{c_{max}} cA(c,t)dc}{\int_{c_{min}}^{c_{max}} A(c,t)dc} \tag{35}$$

$$\frac{dc_j}{dt} = -\frac{c_j}{\tau_j} + \sum_{i=0}^{n_l-1} f_d^{ij}(t) \tag{36}$$
$$j = 0, 1, \ldots, n_d - 1.$$

The relationships between the land-use forcing functions $f_i(c,t)$ and the corresponding carbon fluxes into litter $f_l^i(t)$, soil $f_s^i(t)$, and the jth decay compartment $f_d^{ij}(t)$ depend on the kind of land-use perturbation that $f_i(c,t)$ describes, for example forest harvest versus clearing, and these fluxes are not generalized further here.

The system of equations Eqs. (31–36) is solved simultaneously with initial conditions $A(c,t_o)$, $c_l(t_o)$, $c_s(t_o)$, and $c_j(t_o)$, $j = 0,1,\ldots,n_d - 1$. For arbitrary land-use forcing functions $f_i(c,t)$, numerical solution is necessary. The partial differential equation in the area distribution $A(c,t)$, Eq. (31), can be approximated by a system of ordinary differential equations in time that are then solved simultaneously with Eqs. (33), (34), and (36) by a classical method such as Runge-Kutta.

Partition the area distribution function $A(c,t)$ by the carbon densities $\hat{c}_i$, $i = 0,1,\ldots,n_a - 1$. Let $\hat{c}_0 = c_{min}$, the minimum carbon density, and $\hat{c}_{n_a-1} = c_{max}$, the maximum density. Approximate the partial derivative with respect to carbon density in Eq. (31) at the midpoint of each interior partition by a finite difference:

$$\frac{\partial(g(c)A(c,t))}{\partial c} \approx \frac{g(c_i)A_i - g(c_{i-1})A_{i-1}}{c_i - c_{i-1}} \qquad i = 1,2,\ldots,n_a - 2, \tag{37}$$

where $c_i = (\hat{c}_{i+1} + \hat{c}_i)/2$, and $A_i = A(c_i,t)$. At minimum carbon density:

$$\frac{\partial(g(c)A(c,t))}{\partial c} \approx \frac{g(c_0)A_0}{c_1 - c_0}, \tag{38}$$

and at maximum density:

$$\frac{\partial(g(c)A(c,t))}{\partial c} \approx \frac{g(c_{n_a-2})A_{n_a-2}}{c_{n_a-1} - c_{n_a-2}}. \tag{39}$$

Substituting into Eq. (31) reduces that partial differential equation to a system of ordinary differential equations in the areas at the partition midpoints:

$$\frac{dA_i}{dt} = k_i A_i - k_{i-1}A_i + \sum_{j=0}^{n_l-1} f_j(c,t) \qquad i = 1,2,\ldots,n_a - 2 \tag{40}$$

$$\frac{dA_0}{dt} = -k_0 A_0 + \sum_{j=0}^{n_l-1} f_j(c_0,t) \tag{41}$$

$$\frac{dA_{n_a-1}}{dt} = k_{n_a-2}A_{n_a-2} + \sum_{j=0}^{n_l-1} f_j(c_{n_a-1},t), \tag{42}$$

where $k_i = g(c_i)$.

For large-scale simulations, the land surface—this may be the entire earth's surface or a region such as a continent—is divided into land units designated $k = 0,1,\ldots,p-1$. The shape of these land units is arbitrary—ten continents might be considered for example, or a grid on latitude and longitude might divide each continent into land units. Each land unit contains ecosystems indexed by $j = 0, 1, \ldots, m-1$. Carbon cycling in each ecosystem and its response to land-use perturbations are represented by the model just described. Land units are independent of each other. Ecosystems within a particular land unit are coupled in that land use alters their relative areal extents and transfers carbon between their compartments. This coupling between ecosystems is expressed in terms of the land-use forcing functions applied to each. Thus each ecosystem model can be solved independently.

BENCHMARK SIMULATIONS

Initial testing of the model described in the preceding sections focused on the analyses reported by Houghton *et al.* (1983). Given reasonable agreement between Houghton *et al.*'s (1983) estimates of past changes in terrestrial carbon and solutions to this model forced by the same land-use functions, we plan sensitivity tests against responses to environmental conditions including temperature, moisture, and atmospheric CO_2 concentration. These tests will bound the degree to which responses to past climatic variations and rises in atmospheric CO_2 levels may have compensated in part for carbon losses due to land use.

Ideally, analyses of the impacts of land use on terrestrial carbon storage should begin with natural conditions and consider the entire history of human activities. Because ecosystem structure and function are so strongly correlated with environmental conditions, particularly climate, carbon storage and turnover on the natural landscape can be more accurately estimated than can conditions on a disturbed landscape prior to the development of tools such as areal photography and satellite remote sensing. But a complete quantitative history of land-use, such as areas of forest cleared for agriculture, is not available to drive model solutions through the entire human era.

Houghton *et al.* (1983) set initial conditions on terrestrial carbon pools in 1700. As seen in ice-core records, atmospheric CO_2 concentration was reasonably constant for about 500 years prior to that time. In fact human activities prior to 1800 appear to have had an insignificant effect on atmospheric CO_2. A South Pole ice core (Sigenthaler *et al.* 1988) shows only minor variations in CO_2 concentration prior to 1800, after which the Siple ice core (Friedli *et al.* 1986, Neftel *et al.* 1985) indicates a clearly increasing trend.

Following Houghton *et al.* (1983), Table 2 summarizes the extents of major biomes within 10 continental regions. This biome distribution was derived by distributing estimates of global biome extents by Whittaker and Likens (1973) across continental

Table 2. Biome areas ($10^{12}m^2$) within continental-scale regions of the world.†

Biome	North America	Europe	USSR	Pacific	China	Latin America	Middle East	Tropical Africa	South Asia	Southeast Asia
Tropical Moist Forest				1.66	1.40	5.00	0.19	4.26	0.12	2.15
Tropical Seasonal Forest						1.75		4.81	1.34	0.38
Temperate Evergreen Forest	2.36	0.85	1.08	0.32	0.40	0.26	0.19			
Temperate Deciduous Forest	1.57	0.65	1.48	0.49	0.81	1.12				
Boreal Forest	3.94	0.35	7.50							
Tropical Woodland				0.20		2.85		4.51	1.89	
Temperate Woodland	2.29	0.45	1.32			3.47				
Tropical Grassland				4.97		0.74	0.90	4.59	1.82	1.23
Temperate Grassland	5.68	0.88	1.75		7.23	4.97				
Tundra and Alpine Meadow	3.43	0.27	3.36							
Desert Scrub			5.04	1.18	1.45		9.79	4.06		
Cultivated Land	0.03	0.67	0.33	0.05	0.29	0.07	0.20	0.44	0.53	0.04
Pasture Land	0.04	0.75	0.53	0.24	0.83	0.37	0.54	1.87	0.07	0.02
Total	19.34	4.87	22.39	9.11	11.15	20.60	11.81	24.54	5.77	3.82

†Houghton *et al.* (1983)

Table 3. Terrestrial carbon storage.[†]

Biome	Area ($10^{12}m^2$)	Carbon Density in Vegetation (kg/m^2)	Carbon in Vegetation (Pg)	Carbon Density in Soil (kg/m^2)	Carbon in Soil (Pg)
Tropical Moist Forest	13.52	20.0	270.4	11.7	158.2
Tropical Seasonal Forest	8.28	16.0	132.5	11.7	96.9
Temperate Evergreen Forest	5.46	16.0	87.4	13.4	73.2
Temperate Deciduous Forest	6.12	13.5	82.6	13.4	82.0
Boreal Forest	11.79	9.0	106.1	20.6	242.9
Tropical Woodland and Shrubland	9.45	2.7	25.5	6.9	65.2
Temperate Woodland and Shrubland	7.53	2.7	20.3	6.9	52.0
Tropical Grassland	14.25	1.8	25.7	4.2	59.9
Temperate Grassland	20.51	0.7	14.4	18.9	387.6
Tundra and Alpine Meadow	7.06	0.3	2.1	20.4	144.0
Desert Scrub	21.52	0.3	6.5	5.8	124.8
Cultivated Land	2.65	0.5	1.3	6.0	15.9
Pasture Land	5.26	0.7	3.7	18.9	99.4
Total	133.40		778.4		1602.0

[†]Houghton *et al.* (1983)

regions according to Küchler's (1964) natural vegetation map with areas in crops and pasture, which account for only 6% of the land surface in 1700, adjusted on the basis of population estimates by McEvedy and Jones (1978). The total area in Table 2 is significantly less than the land surface assigned by Whittaker and Likens; the areas not included are assumed not to be affected by land use.

Table 3 lists the carbon inventory assumed by Houghton *et al.* (1983) for 1700. Carbon densities per unit area in vegetation for each biome are from Whittaker and Likens (1973). Houghton and co-workers attribute the estimates of soil carbon density to Schlesinger (1977); however, the values in Table 3 are generally larger than Schlesinger's, presumably because of the addition of estimates of carbon in surface litter. Estimates of net primary productivity that are consistent with the carbon inventory in Table 3 can also be taken from Whittaker and Likens (1973) and are summarized in Table 4.

Table 4. Net primary productivity of major biomes.

Biome	Area ($10^{12}m^2$)	Net Primary Production kg/m^2/year	Annual Production Pg/year
Tropical Moist Forest	13.52	0.900	12.17
Tropical Seasonal Forest	8.28	0.675	5.59
Temperate Evergreen Forest	5.46	0.585	3.19
Temperate Deciduous Forest	6.12	0.540	3.30
Boreal Forest	11.79	0.360	4.24
Tropical Woodland and Shrubland	9.45	0.315	2.98
Temperate Woodland and Shrubland	7.53	0.270	2.03
Tropical Grassland	14.25	0.315	4.49
Temperate Grassland	20.51	0.225	4.61
Tundra and Alpine Meadow	7.06	0.065	0.46
Desert Scrub	21.52	0.032	0.69
Cultivated Land	2.65	0.290	0.77
Pasture Land	5.26	0.290	1.53
Total	133.40		46.05

Figures 5, 6, and 7 summarize the global-scale response of our model to nominal estimates of forest harvest and clearing summarized by Houghton *et al.* (1983) in their Appendix 2. Houghton and co-workers' nominal estimates are derived from relationships between land use and population. Their forest harvest and clearing functions cause a decrease of 162.4 Pg of carbon in terrestrial pools between 1860 and 1980 in our model solutions. This agrees reasonably well with 180 Pg reported by Houghton and co-workers. As shown in Fig. 5, the net flux of carbon into the atmosphere from vegetation

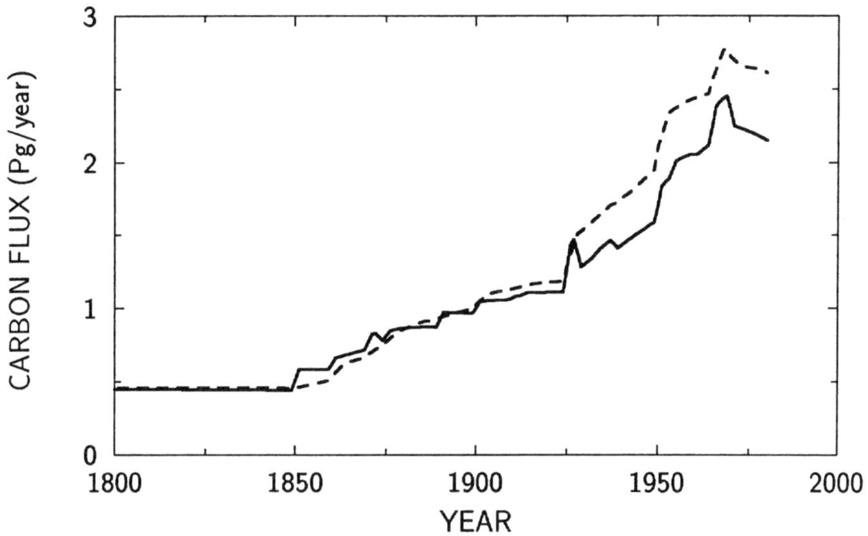

Figure 5. Simulated net flux of carbon into the atmosphere from terrestrial ecosystems. Forest harvest and clearing functions are as in the nominal case from Houghton *et al.* 1983).

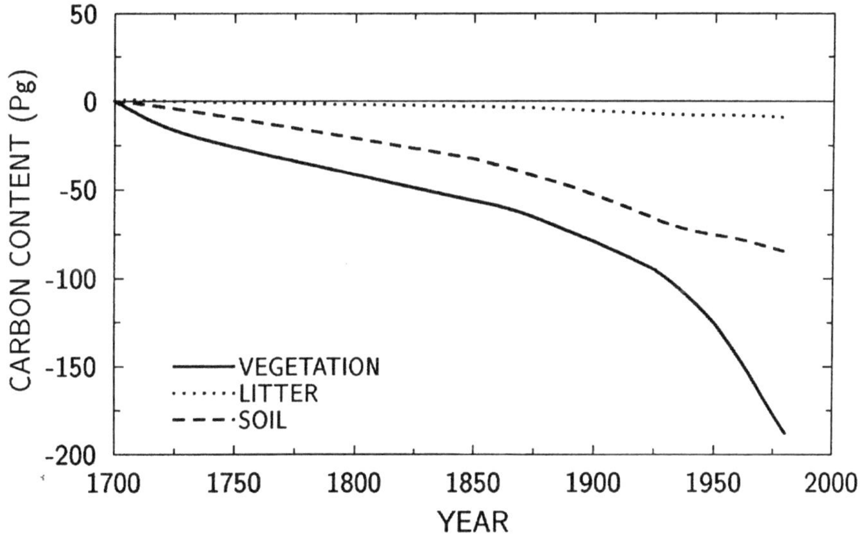

Figure 6. Changes in the carbon content of vegetation, litter, and soil globally over their initial values.

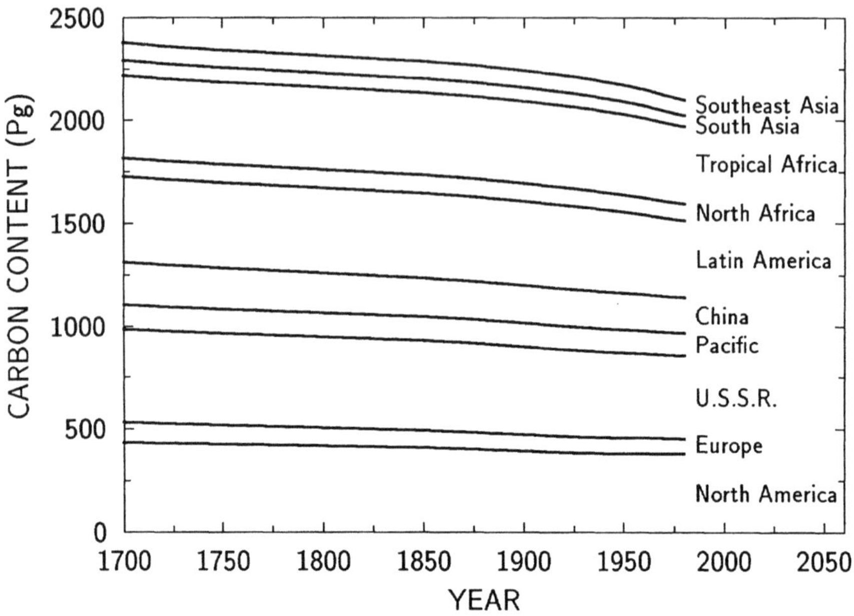

Figure 7. Simulated carbon storage by continental region. At any point in time, the curves are cumulative so that the area between curves corresponds to the carbon storage in that region.

and dead organic matter is less than the Houghton *et al.* (1983) estimates after about 1925.

DISCUSSION AND CONCLUSION

The simulations reported in this paper were conducted in order to test the consistency of our model with the analysis reported by Houghton *et al.* (1983) and refined by them in subsequent studies (e.g., Houghton and Skole 1990). Our results agree reasonably well with those in Houghton *et al.* (1983) with respect to the global net flux of carbon from terrestrial ecosystems into the atmosphere. Forced by the same land-use histories, our alternative model simulates somewhat lower net terrestrial releases, but with the same temporal pattern (Fig. 5). Our simulated net flux is systematically lower than that derived by Houghton and co-workers after about 1925.

Globally, the carbon content of soil and vegetation decrease at somewhat the same rate until about 1925, after which time the decline of carbon in vegetation is substantially faster, while soil carbon loss is slightly slower (Fig. 6). Changes in the fast responding litter pool are negligible compared to those in vegetation and soil. Rates of tropical forest clearing driving these simulations increase sharply in 1925 and are probably responsible for the faster decrease in the standing stock of carbon in vegetation, but it is the slower decline in the soil carbon pool that suggests reasons for the difference in the global net terrestrial flux into the atmosphere that we simulate compared to the results reported by Houghton *et al.* (1983).

Houghton *et al.* (1983) incorporate soil carbon losses due to tilling and erosion into their response functions for dead organic matter following forest clearing for agriculture. We do not yet consider these effects—the carbon contents of the litter and soil compartments in our model decline after clearing only as a result of the decreases in dead organic matter inputs. Preliminary sensitivity tests show that a small, constant loss from the soil compartment of crop ecosystems can bring our simulated net flux into the atmosphere into closer agreement with the results of Houghton and co-workers.

The slow response of the soil carbon compartment in our three-compartment model compared to the Houghton *et al.* (1983) response functions is also important. While we believe that the dead organic matter recovery times assumed by Houghton and co-workers are too rapid, differences in our results suggest an important elaboration of our model. An additional dead organic matter compartment corresponding to the root layer is needed. This compartment would receive all of the belowground input from the vegetation compartment and the flux from litter. Its time constant is only a little longer than that of the litter compartment but substantially less than that of the soil compartment. In such a representation, the large soil carbon pool will be much less responsive to perturbations to the vegetation compartment that decrease the inputs of dead organic matter. The litter and root layer dead organic matter compartments will recover from such perturbations relatively rapidly.

The use of a compartment model instead of specified response functions to represent the response of ecosystems to land use permits the incorporation of descriptions of the processes that are responsible for terrestrial carbon cycling and their dependence on environmental conditions. Mounting evidence that rising atmospheric CO_2 levels are stimulating primary productivity and have done so in the past strongly motivates the incorporation of representations of the processes involved in such an effect into our model. The influences of temperature and moisture must be treated as well.

Our compartment model formulation is flexible in terms of such enhancements. Recall that the area distribution function that describes carbon density in vegetation in each ecosystem is altered through time by plant growth, but only a very general model is assumed—an ordinary nonlinear differential equation, Eq. (19). The parameters in this equation can depend on variables that vary with time including atmospheric CO_2 concentration, temperature, and moisture. Some additional computation is required to produce numerical solutions because the coefficients k_i in Eq. (40–42) then vary with time and must be evaluated as the solution proceeds.

Maintaining a dynamic area distribution function for carbon in the vegetation of each ecosystem provides a powerful way of treating spatial heterogeneity in terrestrial carbon cycling at resolutions finer than the land units that form the basic geographic structure of a model. The use of biome level mean carbon densities in vegetation affected by land use may significantly bias estimates of net releases into the atmosphere—carbon density in vegetation may have a very complicated distribution across such large areas. Such problems can be addressed by initializing the area distribution functions based on data on the factors such as topography and soil characteristics that are responsible for spatial variability in ecosystem carbon cycling.

Finally, we note that the utility of global-scale data summaries available for calibrating terrestrial carbon cycle models is exhausted. The model we propose requires minimum data to set parameter values, but we find that biome-level mean values derived from few samples, and in many cases simply based on the judgement of investigators with extensive field experience in various regions, continue to serve as the only practically available data for this purpose. We believe that progress toward resolving inconsistent views as to the role of terrestrial ecosystems in the global carbon cycle requires not only better data on the areas affected by human activities but reliable and systematically organized data on carbon cycling in terrestrial ecosystems as well.

LITERATURE CITED

Ajtay GL, Ketner P, and Duvigneaud P (1979) Terrestrial primary production and phytomass. In: Bolin B, Döös BR, Jäger J, Warrick RA (eds) *The greenhouse effect, climatic change, and ecosystems*, SCOPE 29. John Wiley, New York, p 129–181.

Bolin B (1986) How much CO_2 will remain in the atmosphere? In: Bolin B, Döös BR, Jäger J, Warrick RA (eds) *The greenhouse effect, climatic change, and ecosystems*, SCOPE 29. John Wiley, New York, p 93–155.

Bazzaz FA (1990) The response of natural ecosystems to the rising global CO_2 levels. *Annu Rev Ecol Syst* 21:167–196.

Eamus D, Jarvis PG (1989) The direct effects of increase in the global atmospheric CO_2 concentration on natural and commercial temperate trees and forests. *Adv Ecol Res* 19:1–55.

Enting IG, Mansbridge JV (1987) The incompatibility of ice-core CO_2 data with reconstructions of biotic CO_2 sources. *Tellus* 39B:318–325.

Friedli, H, Lötscher H, Oeschger H, Siegenthaler U, Stauffer B (1986) Ice core record of $^{13}C/^{12}C$ ratio of atmospheric carbon dioxide in the past two centuries. *Nature* 324:237–238.

Hackney, PA, McDonough TA, DeAngelis DL, Cochran ME (1980) *A partial differential equation model of fish population dynamics and its application in impingement impact analysis*, EPA-600/7-80-068. Industrial Environmental Research Laboratory, U.S. Environmental Protection Agency, Research Triangle Park, North Carolina.

Houghton RA, Hobbie JE, Melillo JM, Moore B, Peterson BJ, Shaver GR, Woodwell GM (1983) Changes in the carbon content of terrestrial biota and soils between 1860 and 1980: A net release of CO_2 to the atmosphere. *Ecol Monogr* 53:235–262.

Houghton RA, Skole DL (1990) Carbon. In: Turner RL, Clark WC, Kates RW, Richards JF, Mathews JT, Meyer WB (eds) *The earth as transformed by human action.* Cambridge University Press, Cambridge, U.K. p 393–408.

Keeling CD (1986) *Atmospheric* CO_2 *concentrations—Mauna Loa Observatory, Hawaii 1958–1986*, NDP-001/R1. Carbon Dioxide Information Center, Oak Ridge National Laboratory, Oak Ridge, Tennessee.

Keeling CD, Bacastow RB, Carter AF, Piper SC, Whorf TP, Heimann M, Mook WG, Roeloffzen H (1989) A three-dimensional model of atmospheric CO_2 transport based on observed winds: 1. Analysis of observational data. In: Peterson DH (ed) *Aspects of climate variability in the pacific and the Western Americas.* American Geophysical Union, Washington, D.C. p 165–236.

Küchler AW (1964) World natural vegetation map. In: Epenshade EB (ed.) *Goode's world atlas.* Rand-McNally, Chicago. p 16–17.

Marland G, Boden TA, Griffin RC, Huang SF, Kanciruk P, Nelson TR (1989) *Estimates of* CO_2 *emissions from fossil fuel burning and cement manufacturing, based on the United Nations energy statistics and the U.S. Bureau of Mines cement manufacturing data*, ORNL/CDIAC-25. Oak Ridge National Laboratory, Oak Ridge, Tennessee.

McEvedy C, Jones R (1978) *Atlas of world population history.* Penguin, Middlesex, U.K.

Moore, B, Boone RD, Hobbie JE, Houghton RA, Melillo JM, Peterson BJ, Shaver GR, Vörösmarty CJ, Woodwell GM (1981) A simple model for analysis of the role of terrestrial ecosystems in the global carbon budget. In: Bolin B (ed) *Carbon cycle modelling*, SCOPE 16. John Wiley, New York. p 365–385.

Neftel A, Moor E, Oeschger H, Stauffer B (1985) Evidence from polar ice cores for the increase in atmospheric CO_2 in the past two centuries. *Nature* 315:45–47.

Olson JS, Watts JA, Allison LJ (1983) *Carbon in live vegetation of major world ecosystems*, ORNL-5862. Oak Ridge National Laboratory, Oak Ridge, Tennessee.

Post WM, Emanuel WR, Zinke PJ, Stangenberger AG (1982) Soil carbon pools and world life zones. *Nature* 298:156–159.

Post WM, Peng TH, Emanuel WR, King AW, Dale VH, DeAngelis DL (1990) The global carbon cycle. *Am Sci* 78:310–326.

Schlesinger WH (1977) Carbon balance in terrestrial detritus. *Annu Rev Ecol Syst* 8:51–81.

Siegenthaler U, Oeschger H (1987) Biospheric CO_2 emissions during the past 200 years reconstructed by deconvolution of ice core data. *Tellus* 39B:140–154.

Siegenthaler U, Friedli H, Loetscher H, Moor E, Neftel A, Oeschger H, Stauffer B (1988) Stable-isotope ratios and concentrations of CO_2 in air from polar ice cores. *Ann Glaciol* 10:1–6.

Whittaker RH, Likens GE (1973) Carbon in the biota. In: Woodwell GM, Pecan EV (eds) *Carbon and the biosphere.* National Technical Information Service, Springfield, Virginia. p 281–302.

Woodwell GM, Hobbie JE, Houghton RA, Melillo JM, Moore B, Peterson BJ, Shaver GR (1983) Global deforestation: Contribution to atmospheric carbon dioxide. *Science* 222:1081–1086.

ACKNOWLEDGEMENT

The authors thank D. L. DeAngelis and J. B. Drake, Oak Ridge National Laboratory, for assistance with model derivation and implementation. Research sponsored by the U.S. Department of Energy, Carbon Dioxide Research Program, Atmospheric and Climate Research Division, Office of Health and Environmental Research, under contract DE-AC05-84OR21400 with Martin Marietta Energy Systems.

CARBON EXCHANGE BETWEEN THE TERRESTRIAL BIOSPHERE AND THE ATMOSPHERE

Gerd Esser
Institute for Plant Ecology
Justus-Liebig-University
Heinrich-Buff-Ring 38
D-W-6300 Giessen
Germany

1 Introduction

Carbon dioxide is presently the most important anthropogenic greenhouse gas, and in contrast to the other greenhouse gases, the human potential to rise its future atmospheric concentration seems to be very high, since the fossil carbon resources (coal, oil, natural gas) probably exceed 6.5×10^{12} tons (6,500 Gt) of carbon. The preindustrial CO_2 concentration in the atmosphere of ca. 280 $\mu l \cdot l^{-1}$ will double within the next 5–6 decades. The predicted global mean warming will then probably be 3.5 ± 1.5 °C, according to climate models (HANSEN et al. 1988).

The atmospheric CO_2 level has already raised considerably since preindustrial times (KEELING 1986, FRIEDLI et al. 1986). In the same period the mean global temperature has increased by about 0.4–0.6 °C. The atmospheric CO_2 level and the temperature rise seem to be correlated. But this correlation does not confirm a causal relationship.

The probable major source for the elevated CO_2 in the atmosphere is the burning of fossil fuels (MARLAND and ROTTY, 1983). On the other side, human activities through deforestation and reforestation, land–use changes, and the secondary effects of these activities on vegetation and soils have contributed to the change in the atmospheric CO_2 concentration. The carbon pools in the terrestrial biosphere contain about 3× the amount of carbon in the atmosphere and the fluxes between those pools and their exchange with the atmosphere are high enough to be able to influence the atmospheric CO_2 considerably (Table 1).

The most important questions concerning the biosphere–atmosphere exchange and its perturbation by humans include:

- How can the direct human impacts to the natural and agricultural systems be quantified. Those impacts are mainly through
  - Land–use changes and deforestation,
  - Burning of semi–natural vegetation (savanna and others),
  - Immissions of fertilizing compounds (i.e., NO_x);
- What is the importance of the CO_2 fertilization complex and how important are the interactions with the water balance of the vegetation;
- How important are small "irreversible" fluxes (charcoal formation by fires, transport of dissolved and particulate organic carbon (DOC, POC) to the ocean).

NATO ASI Series, Vol. I 15
The Global Carbon Cycle
Edited by M. Heimann

Table 1: Major global pools of carbon and the respective annual fluxes which contribute to the global carbon budget. Values are given in Gt (10^9 t) respectively Gt·yr^{-1} of carbon. Sources: MAIER–REIMER & HASSELMANN (1987), ESSER (1990), based on model studies. From ESSER 1991.

pools	ocean	39,000	Gt
	atmosphere	750	
	fossil fuel reserves	6,500	
	soil organic carbon	1,500	
	litter	100	
	live terrestrial phytomass	650	
fluxes	atmosphere → ocean	70–87	Gt·yr^{-1}
	ocean → atmosphere	70–87	
	net primary productivity (atmosphere → green plants)	~45	
	litter production (green plants → litter)	~45	
	soil organic carbon production (litter → SOC)	~10	
	litter depletion (litter → atmosphere)	~35	
	soil organic carbon depletion (SOC → atmosphere)	~10	
	fossil fuel burning (fuel → atmosphere)	~5.5	
	deforestation (green plants → atmosphere)	~1	

Recently some evidence for the probable existence of carbon sinks in the terrestrial biosphere was derived from comparing the results of atmospheric transport models with data for carbon dioxide uptake in the ocean and fossil emissions (KEELING et al. 1989, TANS et al. 1990). It is important to compare these results with predictions from biological models.

In this paper we use a global carbon cycle model of the terrestrial biosphere (Osnabrück Biosphere Model, OBM) to find at least preliminary answers from an ecological point of view to the important questions of the carbon cycle.

2 The carbon cycle model of the terrestrial biosphere

The Osnabrück Biosphere Model (OBM) was developed as an instrument to investigate the carbon balance of the terrestrial biosphere, and the impacts of the rising atmospheric CO_2 level and of climatic changes. It had to meet, therefore, the following demands: (1) include the major carbon fluxes and pools of the terrestrial biosphere; (2) describe the fluxes by means of equations, which consider the deterministic relations to the environment; (3) make the equations valid in the entire span of the environmental variables in the terrestrial biosphere; (4) include any important indirect effects which may influence the global carbon budget.

The biospheric carbon pools in the model are balanced by the carbon fluxes, which are functions of the vector of driving variables. Figure 1 shows the structure of the model. In Table 2 the principal model variables and the procedures of their computation are listed.

The model is regionalized on a 2.5 degrees grid. The mass–balance of the model pools is carried out in the usual way, by integrating the system of differential equations by a 4th–order Runge–Kutta method, or by balancing the integrated fluxes explicitly at each time step, depending on the problem. The pool values necessary as initial conditions to start the model

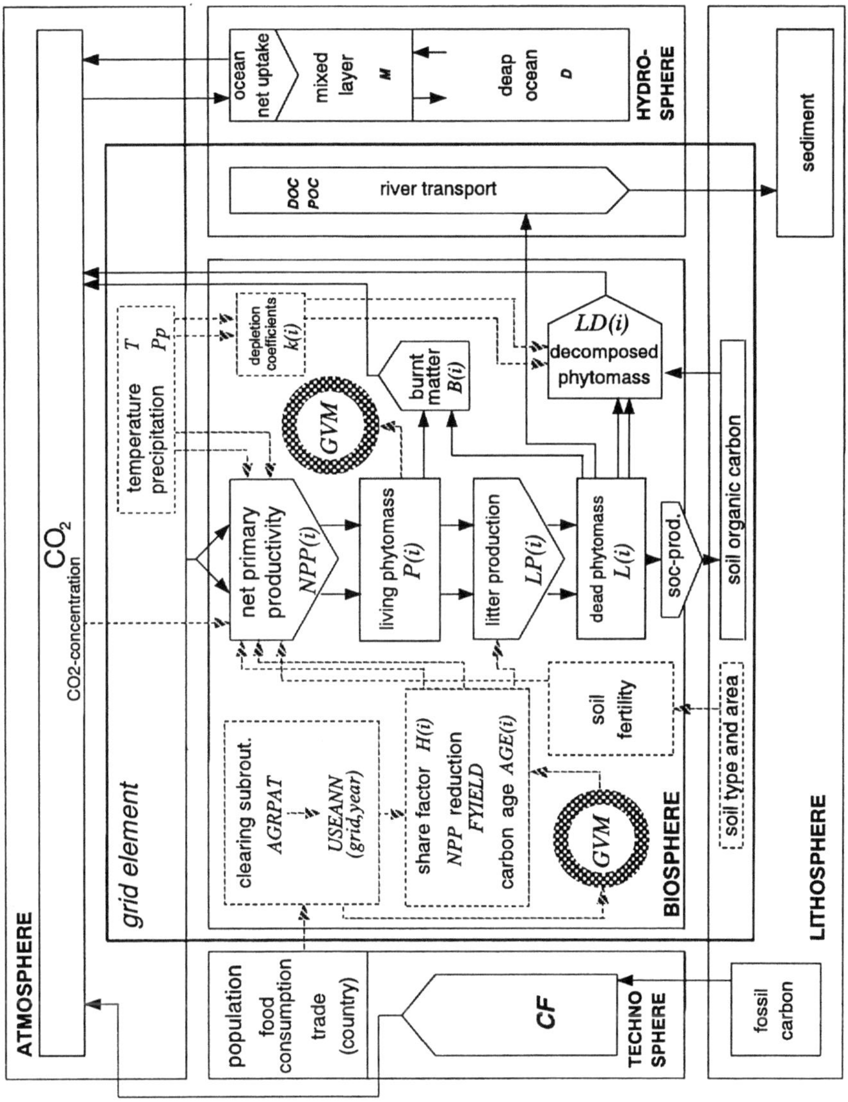

Figure 1: Simplified structure of the Osnabrück Biosphere Model. The global grid size is 2.5 degrees latitude and longitude. Carbon pools are represented by rectangles, carbon fluxes by pentagons, driving forces by dashed rectangles, mass relations by solid arrows, control relations by dashed arrows. *GVM* indicates the exchange with a coupled global vegetation model (PRENTICE et al., 1992). From ESSER (1986), modified.

Table 2: Principal biospheric variables of the Osnabrück Biosphere Model, their computation, and references for further information.

biospheric variable	calculated from	references
net primary productivity	temperature, precipitation, soil fertility, agricultural yield, conversion factors yield → productivity, land-use areas, CO_2-fertilization	ESSER 1991
land-use changes 1860–1981	statistical data on country basis, remote sensing	ESSER 1989, RICHARDS et al. 1983
land-use changes after 1981	scenarios considering: land-use density in 1970, natural productivity, soil fertility, increase of agricultural areas 1950–1980; options: preservation of natural vegetation, minimum-fertility for economic uses	ESSER 1989
cleared phytomass	land-use changes, natural phytomass, crop phytomass	ESSER 1987
soil fertility	empirical function for major 37 out of 106 soil units of Soil Map (FAO 1974 ff.)	ESSER 1984, ESSER et al. 1982
conversion factor yield → productivity	empirical factor for major field crops	ASELMANN, LIETH 1983, ESSER 1991
CO_2-fertilization	atmosph. CO_2-conc., soil fertility	ESSER 1991
phytomass	net primary productivity, mean stand age	ESSER 1984, 1987
litter production	net primary productivity, phytomass change	ESSER 1987
litter pool	litter production minus depletion	ESSER, LIETH 1989, ESSER et al. 1982
litter depletion	depletion coefficient, litter pool	ESSER, LIETH 1989, ESSER et al. 1982

continued next page

Table 2 continued

biospheric variable	calculated from	references
litter depletion coefficient	temperature, precipitation, material depleted	ESSER 1991, ESSER, LIETH 1989, ESSER et al. 1982
soil organic carbon production	litter production, lignin content	ESSER 1990
soil organic carbon	production minus depletion	ESSER 1990
soil organic carbon depletion	temperature, precipitation, soil organic carbon pool	ESSER 1990
leaching of dissolved and particulate org. C	precipitation	ESSER, KOHLMAIER 1991
leaching of minerals due to land–use changes	soil type, precipitation, element concentration in phytomass, cleared amount of phytomass	ESSER, LIETH, CLÜSENER–GODT, 1989
atmosphere	balanced by: fossil emissions, ocean exchange, net primary production, depletion fluxes of litter and soil organic carbon, burnt biomass	
ocean	box diffusion ocean, 1 mixed layer, 43 deep sea boxes	OESCHGER et al. 1975

are computed using a fixed atmospheric CO_2 concentration. The atmosphere then acts as an "unlimited" carbon source to fill the pools. This "pre–run" procedure may need considerable computing time ($\sim 2{,}000$ model years) to get the large soil pools stable and to prevent model drift in the consecutive model run.

Some of the model fluxes of Table 2 are particularly important for the biosphere–atmosphere exchange of CO_2 or give rise to irreversible carbon transport into final sinks. Those fluxes shall be discussed more in detail.

2.1 Net primary productivity and the CO_2 fertilization.

The total flux net primary productivity is calculated as an annually integrated two–dimensional array according to equation (1). The productivity of the potential natural vegetation is provided by the equations (2) and (3) which represent the original MIAMI model developed by LIETH (1975). The result is then modified by the soil fertility in the respective grid element (second line of equation (1)). The soil factor array $F(m)$ is the weighted mean (see (4)) of the individual soil factors of each soil type found in the grid element.

$$NPP(j,m) = \Big(\min[NPP_T(j,m), NPP_{Pp}(j,m)] \cdot F(m) \cdot \frac{AV(j,m)}{AG(m)} + yield(j,m) \cdot FA(m) \cdot \frac{AA(j,m)}{AG(m)}\Big) \cdot FCO_2(j) \quad (1)$$

$$NPP_T(j,m) = 3000/\{1+\exp[1.315 - 0.119 \cdot T(j,m)]\} \quad (2)$$

$$NPP_{Pp}(j,m) = 3000 \cdot \{1 - \exp[-0.000664 \cdot Pp(j,m)]\} \quad (3)$$

$$F(m) = \frac{\sum_{o=1}^{O(m)} f(o) \cdot AS(m,o)}{\sum_{o=1}^{O(m)} AS(m,o)} \quad (4)$$

$$FCO_2(j,m) = A \cdot [1 - \exp\{-R \cdot (CO_2(j) - 80)\}] \quad (5)$$

$$A = 1 + \frac{F(m)}{4}$$

$$R = -\ln\left(\frac{A-1}{A}\right) \cdot \frac{1}{240}$$

The individual soil factors are empirical correction factors which relate the productivity measured on a given soil type to the productivity given by the climate–driven model (ESSER 1986). The values of $f(o)$ for the major FAO–Unesco soil units (FAO–Unesco 1974 ff.) are found in Table 3.

In the third line of equation (1) the productivity of agriculturally used areas of the grid element is provided. Yields of the main field crops in the grid element are recalculated to total productivity by use of conversion factors (factor array $FA(m)$). The variables *yield* and FA are again weighted means of the crop yields and their factors $fa(c)$ in the grid element. For the factor array $fa(c)$ see Table 4. The variable *yield* (yield fresh weight of major field crops) was derived from the FAO production yearbooks on a country basis (FAO 1980 ff.) so that the grid elements under agricultural use of a given country get the similar yields.

The term in line 4 of equation (1) is the CO_2 fertilization effect. The OBM uses an empirical approach. The productivity is modified by a factor which is calculated from the actual atmospheric CO_2 concentration and the soil fertility according to equation (5) (ESSER 1991).

From an ecophysiological point of view, the CO_2 fertilization effect is not a single process but represents a complex of different processes all contributing to the experimentally observed CO_2 fertilization.

The principal processes contributing to the CO_2 fertilization effect on plant growth include:

- The rise of the quantum–yield of CO_2 assimilation in plants. The activity of the enzyme Ribulose–bisphosphate–carboxylase (Rubisco) which catalyzes the first reaction of CO_2 with ribulose–bisphosphate as the first step of the Calvin–cycle, depends on the partial pressure of CO_2 in the cells of the leaf mesophyll. If the partial pressure is low, an oxygenase reaction with O_2 is prevailing rather than carboxylation, thus decomposing ribulose–bisphosphate to CO_2. As a consequence, the over–all quantum yield, *kph*, of assimilation, i.e., the net amount of CO_2 bound per mol light quanta absorbed by the leaf, goes down. This occurs at high levels of photosynthetically active radiation (PAR) since the consumption of CO_2 is high and the flux into the leaf is limited by the diffusion resistance

Table 3: Soil factors $f(o)$ of the OBM which characterize the fertility of the main soil units of the world. Names of soil units according to FAO–Unesco (1974 ff.). From ESSER, 1991.

soil unit o	$f(o)$	soil unit o	$f(o)$
Gleyic Acrisol	0.87	Chromic Luvisol	1.04
Humic Acrisol	0.22	Ferric Luvisol	1.65
Orthic Acrisol	0.70	Gleyic Luvisol	2.78
other Acrisol	0.60	Orthic Luvisol	0.85
Dystric Cambisol	0.94	Dystric Histosol	1.39
Eutric Cambisol	1.69	Humic Podzol	0.56
Humic Cambisol	1.58	Orthic Podzol	0.61
Gelic Cambisol	0.76	other Podzols	0.55
Luvic Chernozem	0.99	Calcaric Regosol	1.61
Dystric Podzoluvisol	0.83	Eutric Regosol	1.14
Xanthic Ferralsol	0.55	Gelic Regosol	0.91
Humic Gleysol	0.47	other Regosol	1.20
Gelic Gleysol	0.57	Orthic Solonetz	0.59
other Gleysol	0.50	Vitric Andosol	1.65
Lithosol	0.52	Haplic Xerosol	0.42
Lithosol–Yermosol	1.14	Yermosol	0.30
Fluvisol	0.49	Haplic Yermosol	0.66
Eutric Fluvisol	0.61	Luvic Yermosol	0.23
other Fluvisol	0.55	Takyric Yermosol	0.09
Haplic Kastanozem	1.96	Orthic Solonchak	0.44
Luvic Kastanozem	1.61	Takyric Solonchak	0.03
other Kastanozem	1.80	other Solonchak	0.20
Albic Luvisol	0.34		

Table 4: Factors $fa(c)$ for the calculation of NPP (dry weight) of some agricultural crops from the yields (fresh weight). From ASELMANN & LIETH (1983).

crop	$fa(c)$	crop	$fa(c)$
wheat	2.15[1]	cotton	5.00
barley	2.12[5]	sugar cane	0.44
oats	3.44[1]	mandioka (cassava)	0.64[4]
rye	2.65[5]	potato	0.54[2]
maize	2.46[1]	batate	0.55[3]
sorghum	3.44[1]	soybeans	2.46[1]
pearl millet	3.44[1]	Phaseolus beans	2.86[1]
paddy rice	2.86[1]	rape	2.81[5]
sugar beets	0.32[6]		

dry matter content in yield:
[1] 86%, [2] 32.5 %, [3] 30%, [4] 35%, [5] 87%, [6] 23%

of the leaf (Figure 2), thus leading to low CO_2 concentrations in the leaf mesophyll cells. Diffusion is enhanced by higher atmospheric CO_2 concentration thus allowing for higher *kph* at a given PAR level.

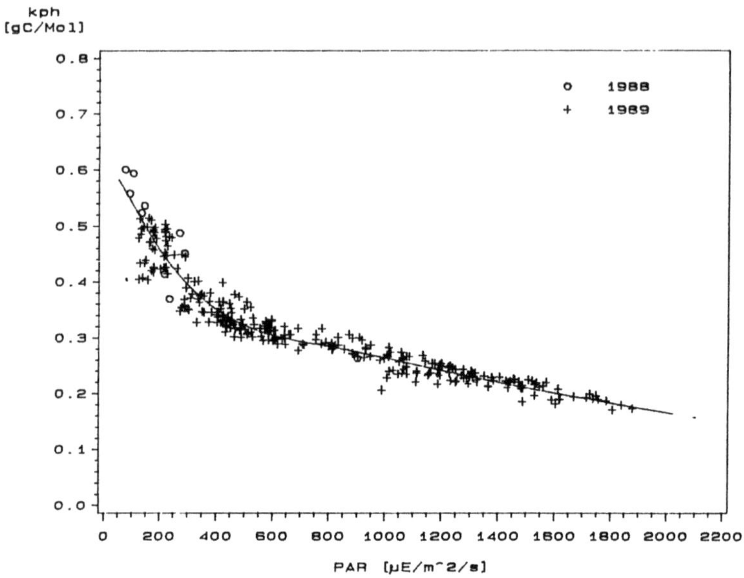

Figure 2: Over–all quantum yield of CO_2 assimilation, *kph*, as a function of irradiation of photosynthetically active radiation (PAR) of upper leaves of wheat (*Triticum aestivum* L.). Soil water was not limiting. The experiments were carried out throughout the vegetation period. The influence of temperature is low and included in the deviation of the points from the spline function. From ESSER (unpublished data).

- Plants are capable of regulating the partial pressure of CO_2 in the cells within certain limits by modifying the leaf (stomatal) resistance. This results in a better water–use efficiency of plant productivity at high atmospheric CO_2, i.e., the amount of water transpired per amount of CO_2 fixed. This effect may be most important, since the majority of sites is water limited (Figure 3).

- Although the photosynthesis may be enhanced by elevated atmospheric CO_2, the allocation of the assimilates as plant structural material ("growth") may be limited by the lack of nutrients (N,P,S, and others). Therefore, the nutrient supply and the soil fertility play an important role in the over all fertilization effect of CO_2 on the short run. Since a distinct, although low, nutrient input is observed in practically all vegetation units of the world, it is unlikely that the nutrient limitation plays a major part on the long run. The fertilization effect is probably only retarded.

- Normally a distinction is made between C3 and C4 pathways of carbon fixation by plants. C4 plants basically use the same processes of final carbon fixation in the Calvin cycle, but in addition have a sort of "CO_2–pump" to increase the CO_2 concentration in leaf mesophyll cells. Therefore, C4 plants are capable of living under low atmospheric CO_2 concentrations. Since "pumping" CO_2 against a gradient is an energy consuming process C4 plants are at present only competitive where water supply is critical, i.e., in the tropical and subtropical lowlands. C4 metabolism is therefore most probably restricted to herbs and shrubs of open tropical and arid to semiarid habitats (OSMOND et al., 1981). C4 metabolism may be an comparatively recent adaptation to the low atmospheric CO_2 concentrations in

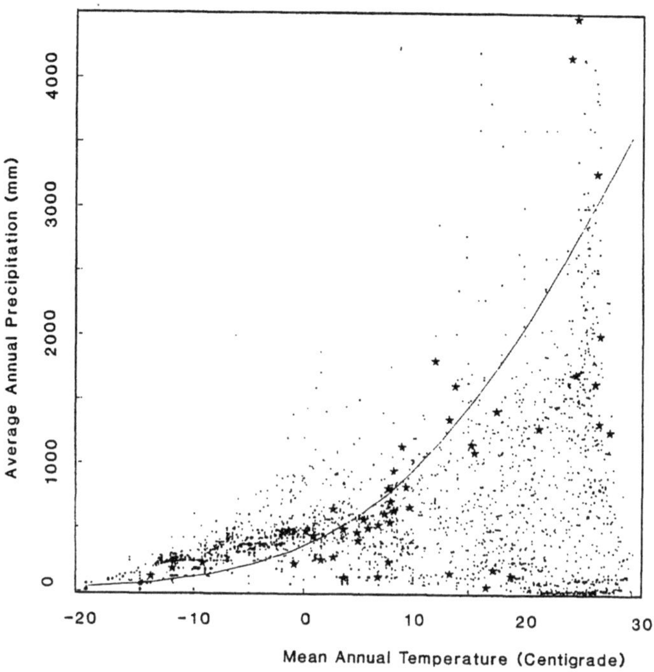

Figure 3: Limitation of net primary productivity of vegetation in the 2433 grid elements of the OBM in the temperature–precipitation field of the model. The function is the trajectory where the equations (2) and (3) give the same result. Data points below the trajectory are thus limited by precipitation, which is obviously the case for most of the grid elements having mean annual temperatures above 5 °C. The stars indicate the values used to calibrate equations (2,3).

glacial periods. This speculation is supported by the observation, that C4 plants are members of genetically active groups with short reproductive cycles (*Poaceae, Chenopodiaceae, Amaranthaceae*, and some members of the genera *Atriplex* and *Heliotropium*).

Equation (5) was calibrated by use of the results of such experiments and represents an effort to parameterize the related processes globally. The shortcoming is that most of the experiments which provided the data were carried out under controlled conditions using mainly field–crops as objects. Therefore, the reliability of equation (5) for global use is obviously questionable, although I believe it is a "good fit" of present knowledge. A three–dimensional graph of equation (5) is given in Figure 4.

The FCO_2 values used in the OBM may be compared with β–factors used by other model concepts. The FCO_2 corresponds to β values which decrease with increasing CO_2 concentration. In Table 5 the β factors corresponding to FCO_2 calculated from equation (5) are listed for several combinations of soil quality and CO_2 concentrations.

The factors given in Table 5 may be compared with the β values used in other models. GOUDRIAAN and KETNER (1984) used $\beta = 0.5$ for all CO_2 concentrations and recently GOUDRIAAN (1990) suggested $\beta = 0.45$. $\beta = 0.45$ corresponds to $FCO_2 = 1.047$ which value is given by equation (5) for 351 $\mu l \cdot l^{-1}$ CO_2 and $F(m) = 1.0$, i.e. the conditions of 1985 on a medium quality soil.

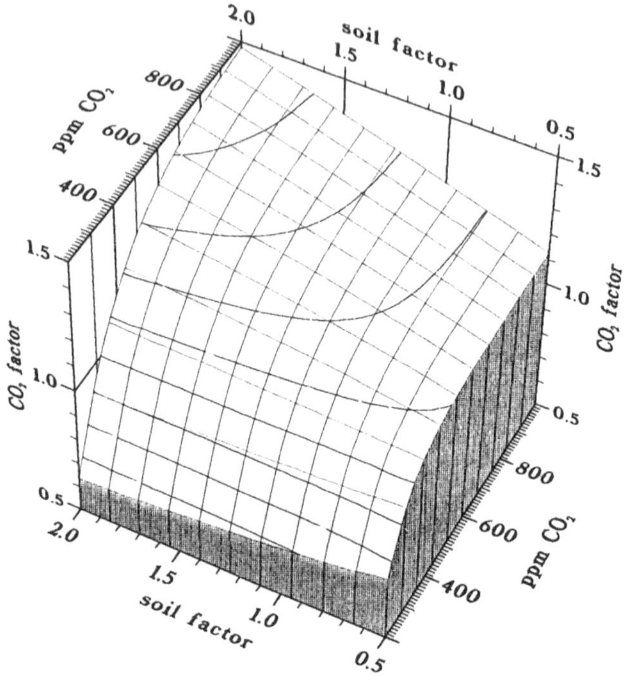

Figure 4: Plot of function (5), $FCO_2(j,m) = f[CO_2(j), F(m)]$, to calculate the CO_2 fertilization factor on net primary productivity. The CO_2 compensation concentration for entire plants is assumed to be 80 $\mu l \cdot l^{-1}$ ($FCO_2 = 0$ irrespective of F(m)). For 320 $\mu l \cdot l^{-1}$ CO_2 concentration $FCO_2 = 1.0$, since it was assumed that the data used to calibrate the functions (2) and (3) were measured mainly during the 1960s when CO_2 was about 320 $\mu l \cdot l^{-1}$. The function achieves saturation at 1.125 for $F(m) = 0.5$, at 1.250 for $F(m) = 1.0$, and at 1.375 for $F(m) = 1.5$. For the present atmospheric CO_2 of 355 $\mu l \cdot l^{-1}$ the function yields the factors 1.034 for $F(m) = 0.5$, 1.052 for $F(m) = 1.0$, and 1.065 for $F(m) = 1.5$. From ESSER 1991.

Table 5: β factors calculated from the FCO_2 given by equation (5) for certain combinations of soil quality $F(m)$ and CO_2 concentration.

CO_2	soil factor ($F(m)$)		
$\mu l \cdot l^{-1}$	0.5	1.0	1.5
285	0.408	0.551	0.675
330	0.355	0.527	0.642
600	0.184	0.337	0.466

2.2 Small "irreversible" fluxes

Transport of organic compounds in rivers (DOC, POC). The transport of dissolved and particulate organic carbon (DOC, POC) in rivers gave rise to the speculation that the amount of carbon transported may increase due to human activities, mainly through land–use practises.

DOC and POC loads for many of the larger river systems in the world have been determinated in the past years, as well as further information about drainage and mineral composition (Degens 1982, Degens et al. 1983, 1985). Esser and Kohlmaier (1991) analyzed the data using linear correlation techniques with driving functions, ignoring cross–correlation. There was no fundamental difference between the behavior of DOC and POC. As a consequence, both showed a positive correlation to each other (correlation coefficient 0.82). The strongest correlation existed between the watershed *discharge* and the DOC and POC values (0.94 and 0.92 respectively). This result strongly suggests that leaching and soil surface erosion processes may be the most important mechanisms for DOC and POC formation. Since discharge mainly depends on precipitation there is a strong correlation of DOC and POC values with precipitation too (0.81 and 0.66 respectively). The correlation with the *litter* pool is weak. The reason for that may be that the annual export through dissolved and particulate organic carbon rarely exceeds 0.1% of the source pool *litter* and 0.2% of the biospheric input. The global total transport was 226 $Tg \cdot yr^{-1}$ DOC-carbon and 78 $Tg \cdot yr^{-1}$ POC-carbon, in total about 300 $Tg \cdot yr^{-1}$ or 0.3 $Gt \cdot yr^{-1}$ carbon.

Since, due to their study, river freights seem not to be changed by human activities, DOC and POC are not contributing to a biogenic carbon sink. Nevertheless, if the transport is considered a normal feature of the global carbon cycle, a compensation flux from the ocean to the atmosphere that has already existed since preindustrial and preagricultural times must be postulated. The ultimate fate of DOC and POC in the ocean is probably oxidation, the pathways and rates of which are still a matter of research. Until they are established, the source/sink characteristics of riverine export of organic carbon from land to sea with regard to atmospheric CO_2 cannot be quantified (Spitzy and Leenheer 1991).

Charcoal formation during burning. Burning of grasslands and savannas, forest fires, and burning in the shifting cultivation cycle release charcoal on the topsoil which is later incorporated in the soil. Although the depletion coefficient of charcoal is unknown it is obviously much lower than that of organic material. Carbon deposited as charcoal is eliminated from the cycling for decades or probably even centuries.

Although mechanistic approaches to burning, on a global scale, are still under development, a crude estimate of charcoal formation, based on the OBM, may outline the importance of this problem. Grasslands and coniferous forests are two important biomes with respect to fires. Moreover, areas limited by precipitation and by temperature may be separated, since either have different fire frequencies.

About 1.5×10^6 km^2 of grasslands or $\sim$ 1 Gt of phytomass carbon of grasslands is limited by temperature, while 17×10^6 km^2 or 13 Gt C of grassland phytomass are limited by precipitation, on average (Esser 1992). Thus, grassland phytomass amounts to only 2% of the global phytomass and most is limited by precipitation. We assume that half the grassland phytomass is above ground and therefore subject to burning; further that grasslands limited by humidity are annually burnt, which results in about 7 Gt of grassland phytomass burnt annually; finally that 1% of the burnt phytomass is transformed to charcoal.

The charcoal production then amounts to $\sim$ 0.07 $Gt \cdot yr^{-1}$ or about 8 Gt C since 1860.

Using a similar approach we find that 15×10^6 km^2 of coniferous forests are temperature limited, while only 2×10^6 km^2 are limited by precipitation. The related phytomass is 47 Gt C and 18 Gt C, respectively. Coniferous forests provide 11% of the global phytomass, $\sim$ 40% being limited by precipitation. The following assumptions are made: 80% of the phytomass is above ground; the fire frequency is once in 20 years for precipitation limited sites and once in 100 years for temperature limited sites, which leads to about 1 Gt of coniferous phytomass being burnt annually on average; 1% of phytomass is converted to charcoal while burning.

The fires in coniferous forests then produce 0.01 Gt·yr^{-1} of charcoal–carbon or 1.5 Gt C since 1860.

Carbon storage in organic soils (Histosols). According to the Soil Map of the World (FAO–Unesco 1974 ff.) Histosols cover about 2.1×10^6 km^2 or 1.6% of the global land area. Since, in most cases, Histosols are oligotrophic and occur mainly in boreal, subpolar, and mountainous regions, the net primary productivity of vegetation on Histosols will not exceed 2% of the global value, that is 0.8 Gt·yr^{-1}. Most probably it is much lower. Therefore, Histosols may not be responsible for a large biospheric sink in northern latitudes, even if part of the productivity is stored as organic material.

Anthropogenic nitrogen import. The input of anthropogenic nitrogen compounds into natural and agricultural systems increased during the last decades. In the Federal Republic of Germany the emissions of NO_x reached 3×10^6 t in the 1980s (Umweltbundesamt 1988).

The nitrogen import potentially enhances the net primary productivity in nitrogen limited systems. As a consequence, the phytomass may increase and additional carbon may be sequestered hereby. On the other side, the sites with high nitrogen immissions are the places of forest decline. The net effects may eventually be small, therefore. Additional soil effects, i.e., carbon fixation by higher nitrogen availability in soils, can not be excluded. At present any quantification of the nitrogen effects must be very uncertain. Mechanistic models which resolve the relevant processes are urgently required.

3 Regionalization of the biosphere–atmosphere exchange

A regionalized global view of the carbon exchange between atmosphere and biosphere is required to investigate the question whether a temperate sink of the magnitude of 1–3 Gt·yr^{-1}, as postulated by calculations based on the ocean uptake, might exist. If the fossil emissions and the ocean uptake are considered in atmospheric transport models, a supplementary sink in the northern hemisphere is required to meet the measured north–south gradient of the atmospheric CO_2. Due to the fractionation of stable carbon isotopes this postulated sink is considered to be biospheric. KEELING et al. (1989) came to the conclusion that an additional sink of only 1 Gt·yr^{-1} is required, while TANS et al. (1990) concluded that the sink must be 2.0–2.7 Gt·yr^{-1}.

Using the OBM we can calculate the regionalized biospheric sinks based on biological processes and independently of any ocean data. We assume that the atmosphere is well mixed in latitudinal direction and, therefore, calculate the source/sink function of the land surface for 10° latitudinal belts. We can separate effects of CO_2 fertilization, land–use changes, deforestation and reforestation, and secondary effects on the litter and soil organic carbon pools.

In Table 6 the results are summarized. Column (a) is the amount of carbon additionally bound in the phytomass by the fertilization effect in the year 1980. It is highest in the humid tropics where the productivity is high. A second maximum is found in the northern temperate zone. In total, about 1.25 Gt·yr^{-1} are bound. In column (b) the numbers mean fertilization effect plus phytomass–changes due to deforestation, reforestation, and land–use changes. From (a) and (b), the net values for deforestation, reforestation, and land–use changes can be calculated $(b - a)$. The total global balance of those direct human impacts is -0.8 Gt·yr^{-1} of carbon. The tropical deforestation amounts to -0.9 Gt·yr^{-1}, the global net deforestation is -1.1 Gt·yr^{-1}. In an recent paper HOUGHTON (1991) gives a range for the flux from changes in land use of 0.6 to 2.6 Gt C for 1980, although he stated in a personal communication that he subjectively considers 1.0 to 2.0 Gt C more likely. The OBM result is on the low side of this range. 0.2

Table 6: Zonal integrals of the fertilization effect (a), the phytomass change due to fertilization effect and land–use changes together (b), and the total biospheric balance including phytomass, litter, and soil (c). The difference $(b - a)$ is the net effect of deforestation, reforestation, and land–use changes alone. It is not directly calculated by the model and, therefore, must be derived from subtraction of columns b and a. Values are in 10^6 tons of carbon per year for zones of 10° latitude. Negative values are losses from the biosphere. The reference year is 1980. From ESSER 1991.

hemisphere	latid. zone	fertilization (a)	phytomass change (b)	deforestation, reforestation $(b-a)$	biosphere total (c)
north	70–80	1.1	1.0	−0.1	5.6
	60–70	40.4	34.9	−5.5	64.2
	50–60	80.7	62.6	−18.1	10.6
	40–50	102.5	171.1	68.6	143.9
	30–40	67.5	178.1	110.6	116.9
	20–30	80.5	−113.6	−194.1	−50.8
	10–20	82.3	−95.4	−177.7	−71.5
	0–10	233.9	79.5	−154.4	144.7
south	10–0	295.0	180.5	−114.5	215.0
	20–10	147.7	67.5	−80.2	115.4
	30–20	90.5	−102.3	−192.8	−57.3
	40–30	19.4	−20.9	−40.3	−31.0
	50–40	9.3	1.8	−7.5	1.6
	60–50	1.1	0.9	−0.2	1.9
total net biosph. sink		1251.9	445.7		608.2
total net biosph. source				−806.2	

$Gt \cdot yr^{-1}$ are bound by reforestation in the northern 30–50° belt. Therefore, the model suggests that the global phytomass could have been a sink for carbon of 0.45 $Gt \cdot yr^{-1}$ in 1980. In column (c) secondary effects of the impacts of columns (a) and (b) on litter and soil organic carbon are included. By comparing (c) and (b) the net effects on soil and litter can be assessed. Obviously the fertilization effect is responsible for the fixation of carbon in those pools of dead material, too. Losses were only observed in the 30–60° northern belt, where we presumed an increase of phytomass due to reforestation. In immature stands the phytomass increase lowers the litter production flux, which gives rise to losses of carbon from litter and soils.

Column (c) is relevant for the net atmosphere–biosphere exchange. The extratropical northern hemisphere (> 20° N) is a net sink of about 0.3 Gt C annually, while the similar southern latitudes are a source of about −0.08 Gt C. This northern hemispheric sink given by the OBM is considerably lower than the respective value estimated by TANS et al. (1990) of 2.0–2.7 $Gt \cdot yr^{-1}$. It can not be excluded at present, that the northern hemispheric sink might be larger than calculated by the OBM since this model does not yet consider the possible fertilization effect of anthropogenic emissions to the air besides CO_2, i.e. NO_x compounds. On the other side it is also possible that in the preindustrial atmosphere there was already a north–south gradient of the CO_2 concentration. Then the sink in the northern temperate zone could be smaller than suggested by KEELING at al. (1989) respective TANS et al. (1990) and eventually be in the range suggested by the OBM.

4 Conclusions

For the year 1980 used as a reference year the Osnabrück Biosphere Model (OBM) results suggest the following conclusions.

- Global carbon cycle models as the OBM do not yet resolve all the relevant processes in the biosphere mechanistically. There are still many fluxes in the models calculated with empirical functions. This introduces uncertainties into the models, particularly under changing (transient) boundary conditions.
- The fertilization effect of raised atmospheric CO_2 and the deforestation, reforestation, and land–use complex are the most important features influencing the biosphere–atmosphere exchange.
- Small "irreversible" fluxes, including transport of dissolved and particulate organic compounds in rivers, charcoal formation during burning, carbon storage in organic soils (Histosols), and the consequences of anthropogenic nitrogen input into natural and agricultural systems, are considered to be of minor importance. At least it is unlikely that those processes give rise to a large temperate sink.
- The OBM calculates a temperate sink of about 0.3 $Gt{\cdot}yr^{-1}$. This is much lower than the respective values calculated from the atmospheric north–south gradient of CO_2 considering the fossil source/ocean sink. According to different authors, these values span from 1–2.7 $Gt{\cdot}yr^{-1}$. At present, there is no reasonable way to increase the temperate sink in the OBM.
- In the tropical belt, the carbon fixation due to the fertilization effect of CO_2 overcompensates the losses from deforestation.

Acknowledgment

I cordially thank J. Goudriaan for carful review of and helpful comments on the manuscript.

References

ASELMANN I, LIETH H (1983) The implementation of agricultural productivity into existing global models of primary productivity.
In: Degens, Kempe, Soliman (eds.) Transport of carbon and minerals in major world rivers, Part 2, Mitt. Geolog. Paläontolog. Inst. Univ. Hamburg, SCOPE/UNEP Sonderband **55**: 107–118.

DEGENS E T (ed.) (1982) Transport of carbon and minerals in the major world rivers, Part 1. Mitt. Geol.–Paläont. Inst. Univ. Hamburg, SCOPE/UNEP Sonderband **52**.

DEGENS E T, KEMPE S, SOLIMAN H (eds.) (1983) Transport of carbon and minerals in major world rivers, Part 2.
Mitt. Geol.–Paläont. Inst. Univ. Hamburg, SCOPE/UNEP Sonderband **55**.

DEGENS E T, KEMPE S, HERRERA R (eds.) (1985) Transport of carbon and minerals in major world rivers, Part 3.
Mitt. Geol.–Paläont. Inst. Univ. Hamburg, SCOPE/UNEP Sonderband **58**.

ESSER G (1984) The significance of biospheric carbon pools and fluxes for the atmospheric CO_2: A proposed model structure.
Progress in Biometeorology **3**: 253–294.

ESSER G (1986) The carbon budget of the biosphere — structure and preliminary results of the Osnabrück Biosphere Model (in German with extended English summary).
Veröff. Naturf. Ges. zu Emden von 1814, New Series Vol. **7**, 160 pp. and 27 Figures.

ESSER G (1987) Sensitivity of global carbon pools and fluxes to human and potential climatic impacts.
Tellus **39**B: 245–260.

ESSER G (1989) Global land use changes from 1860 to 1980 and future projections to 2500.
Ecological Modelling **44**: 307–316.

ESSER G (1990) Modelling global terrestrial sources and sinks of CO_2 with special reference to soil organic matter.
In: Bouwman, A. F. (ed.) Soils and the Greenhouse Effect, Chapter 10: 247–261, John Wiley & Sons, Chichester · New York · Brisbane · Toronto · Singapore.

ESSER G (1991) Osnabrück–Biosphere–Model: structure, construction, results.
In: Esser G, Overdieck D (eds.) Modern Ecology: Basic and Applied Aspects. Elsevier Sci. Publ. Amsterdam, Chapter 31: 679–709.

ESSER G (1992) Implications of climate change for production and decomposition in grasslands and coniferous forests.
Invited paper for a SCOPE workshop in Woods Hole, Ma., April 1989, on "Ecosystem response to climate change: The effects of climate change on production and decomposition in coniferous forests and grasslands"; Ecological Applications **2** (1): 47–54.

ESSER G, ASELMANN I, LIETH H (1982) Modelling the Carbon Reservoir in the System Compartment "Litter".
Mitt. Geolog.–Paläontolog. Inst. Univ. Hamburg **52**: 39–58.
SCOPE/UNEP Sonderband.

ESSER G, KOHLMAIER G H (1991) Modelling terrestrial sources of nitrogen, phosphorus, sulphur, and organic carbon to rivers.
In: Degens E T, Kempe S, Richey J E (eds.) Biogeochemistry of major world rivers, SCOPE 42: 297–322, Wiley & Sons Chichester.

ESSER G, LIETH H (1989) Decomposition in tropical rain forests compared with other parts of the world.
In: LIETH, WERGER (eds.) Tropical Rain Forest Ecosystems. Ecosystems of the World Vol. 14B: 571–580, Elsevier Science Publ. Amsterdam.

ESSER G, LIETH H, CLÜSENER GODT M (1989) Assessment of P, K, Ca dynamics during land use changes.
In: Ittekkot et al. (eds.) Facets of Modern Biogeochemistry, Chapter 10: 102–115, Springer Verlag, Berlin, Heidelberg.

FAO–Unesco (1974 ff.) Soil Map of the World, 1:5,000,000.
Vol. I–X Unesco Paris.

FAO–Unesco (1980 ff.) Production Yearbook 1979 ff., Vol. 33 ff., FAO–Statistics Series No. 28, Rome.

FRIEDLI H, LÖTSCHER H, OESCHGER H, SIEGENTHALER U, STAUFFER B (1986) Ice core record of the $^{13}C/^{12}C$ ratio of atmospheric CO_2 in the past two centuries.
Nature **324**: 237–238.

GOUDRIAAN J (1990) Atmospheric CO_2, global carbon fluxes and the biosphere.
In: Rabbinge R, Goudriaan J, Keulen H van, Penning de Vries F W T, Laar H H van (eds.), Theoretical Production Ecology: reflections and prospects. pp 17–40. Pudoc Wageningen.

GOUDRIAAN J, KETNER P (1984) A simulation study for the global carbon cycle, including man's impact on the biosphere.
Climatic Change **6**: 167–192.

HANSEN J, FUNG I, LACIS A, RIND D, LEBEDEFF S, RUEDY R, RUSSELL G, STONE P (1988) Global climate changes as forecast by Goddard Institute for Space Studies three-dimensional model.
J. Geophys. Research **93**: 9341–9364.

HOUGHTON R A (1991) Tropical deforestation and atmospheric carbon dioxide.
Climatic Change (in press).

KEELING C D (1986) Atmospheric CO_2 concentrations — Mauna Loa Observatory, Hawaii 1958–1986.
NDP–001/R1 Carbon Dioxide Information Centre, Oak Ridge, Tennessee (regularly updated).

KEELING C D, PIPER S C, HEIMANN M (1989) A three-dimensional model of atmospheric CO_2 transport based on observed winds: 4. Mean annual gradients and interannual variations.
In: Peterson D H (ed.), Aspects of climate variability in the Pacific and the western Americas.
Geophysical Monograph **55**: 305–363, American Geophysical Union.

LIETH H (1975) Modeling the primary productivity of the world.
In: Lieth, Whittaker (eds.) Primary productivity of the biosphere, Ecological Studies **14**: 237–283, Springer–Verlag, New York, Heidelberg, Berlin.

MAIER–REIMER E, HASSELMANN K (1987) Transport and storage of CO_2 in the ocean — an inorganic ocean–circulation carbon cycle model.
Climate Dynamics **2**: 63–90.

MARLAND G, ROTTY R M (1983) Carbon dioxide emissions from fossil fuels: A procedure for estimation and results for 1950–1981.
Report DOE/NBB–0036 for U.S. Dept. of Energy.

OESCHGER H, SIEGENTHALER U, SCHOTTERER U, GUGELMANN A (1975) A box diffusion model to study the carbon dioxide exchange in nature.
Tellus **27**: 168–192.

OSMOND C B, WINTER K, ZIEGLER H (1981) Functional significance of different pathways of CO_2 fixation in photosynthesis.
In: Encyclopedia of Plant Physiology. New Series, Vol. 12B, Pison A, Zimmermann M H (eds.) pp 480–547, Springer Berlin Heidelberg.

PRENTICE C I, CRAMER W, HARRISON S P, LEEMANS R, MONSERUD R A, SOLOMON A M (1992) A global biome model based on plant physiology and dominance, soil properties and climate.
Submitted to Journal of Vegetation Science.

RICHARDS J F, OLSON J S, ROTTY R M (1983) Development of a data base for carbon dioxide releases resulting from conversion of land to agricultural uses.
Institute for Energy Analysis, Oak Ridge Ass. Universities, ORAU/IEA–82–10(M); ORNL/TM–8801.

SPITZY A, LEENHEER J (1991) Dissolved organic carbon in rivers.
In: Degens E T, Kempe S, Richey J E (eds.) Biogeochemistry of major world rivers, SCOPE 42: 213–232, Wiley & Sons Chichester.

TANS P P, FUNG I Y, TAKAHASHI T (1990) Observational constraints on the global atmospheric CO_2 budget.
Science **247**: 1431–1438.

Umweltbundesamt (UBA) (1988) Materialien zum vierten Immissionsschutzbericht der Bundesregierung.
Berlin

ORGANIC CARBON IN SOIL AND THE GLOBAL CARBON CYCLE †

Wilfred M. Post III
Environmental Sciences Division
Oak Ridge National Laboratory
P. O. Box 2008, Bldg. 1000
Oak Ridge, Tennessee 37831–6335
USA

1. INTRODUCTION

Soil organic matter is, simultaneously, the most inert carbon cycle component of terrestrial ecosystems, and the most dynamic component of terrestrial geologic systems placing it in a pivotal position in the biogeochemistry of carbon. The large size and potentially long residence time of the soil organic matter pool make it an important component of the global carbon cycle (Schlesinger 1977, Post et al. 1982, 1985). Figure 1 shows the relationship of the soil organic matter pool to other components of global terrestrial ecosystems with both significantly longer and shorter turnover times. Net terrestrial primary production of about 60 Pg C·yr^{-1} is, over a several-year period of time, balanced by an equivalent flux of litter production and subsequent decomposition of detritus and soil organic matter (Post et al. 1990). Using estimates in Figure 1, the turnover time, T_t, of all organic matter (litter and soil of Figure 1) globally is

$$T_t = \frac{litter + soil}{litterfall} = \frac{1372}{62} = 22\ years \qquad (1)$$

However, the input rates and decomposition rates for different terrestrial ecosystems vary over several orders of magnitude resulting in widely different amounts and turnover rates of soil organic matter. The amounts of carbon stored in soils and the rates of exchange of soil carbon with the atmosphere depend on many factors related to the chemistry, biology, and physics of soil and soil organic matter.

2. PATTERNS OF SOIL ORGANIC MATTER STORAGE AND DYNAMICS

The amount of carbon stored in soil is determined by the balance of two biotic processes — productivity of terrestrial vegetation and decomposition of organic matter. Each of

†Research sponsored by the U.S. Department of Energy, Carbon Dioxide Research Program, Atmospheric and Climate Research Division, Office of Health and Environmental Research, under contract DE–AC05–84OR21400 with Martin Marietta Energy Systems, Inc. Publication No. 3876, Environmental Science Division, Oak Ridge National Laboratory.

NATO ASI Series, Vol. I 15
The Global Carbon Cycle
Edited by M. Heimann

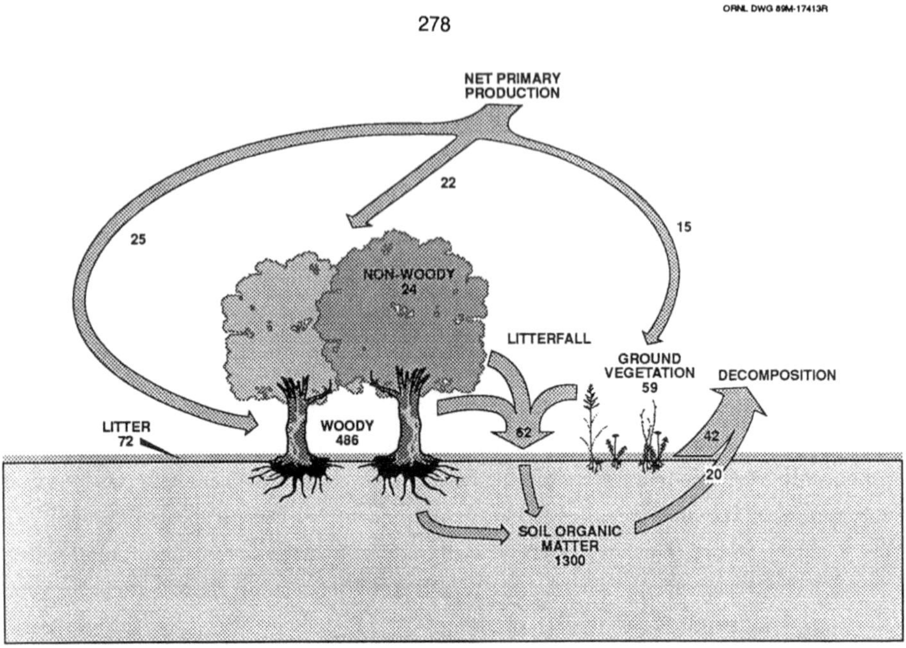

Figure 1. Terrestrial component of the global carbon cycle showing central importance of the soil organic matter pool. Compartment sizes are Pg C, fluxes are net amounts in units of Pg C$\cdot$yr^{-1}.

these processes has strong physical and biological controlling factors. These include climate; soil chemical, physical, and biological properties; and vegetation composition. Interactions among these controlling factors are of particular importance.

2.1 Climate.

Temperature and precipitation show striking relationships with soil carbon content (Figure 2). Climate affects soil carbon storage by influencing rates of vegetation production and rates of organic matter decomposition. Soil carbon concentration generally increases with increasing precipitation, and for any particular level of precipitation there is an increase in soil carbon with decreasing temperature. The highest concentrations of soil carbon occur in cold-wet regions, particularly wet tundra, where soil carbon concentration averages 36 kg$\cdot$m^{-3}. Here plant production rates are low, but decomposition rates are extremely low as a result of low soil temperatures and water saturated conditions. Moist and wet tropical soils also contain a high concentration of carbon, 14-18 kg$\cdot$m^{-3}, as a result of high plant production rates. In desert regions with low precipitation, soil

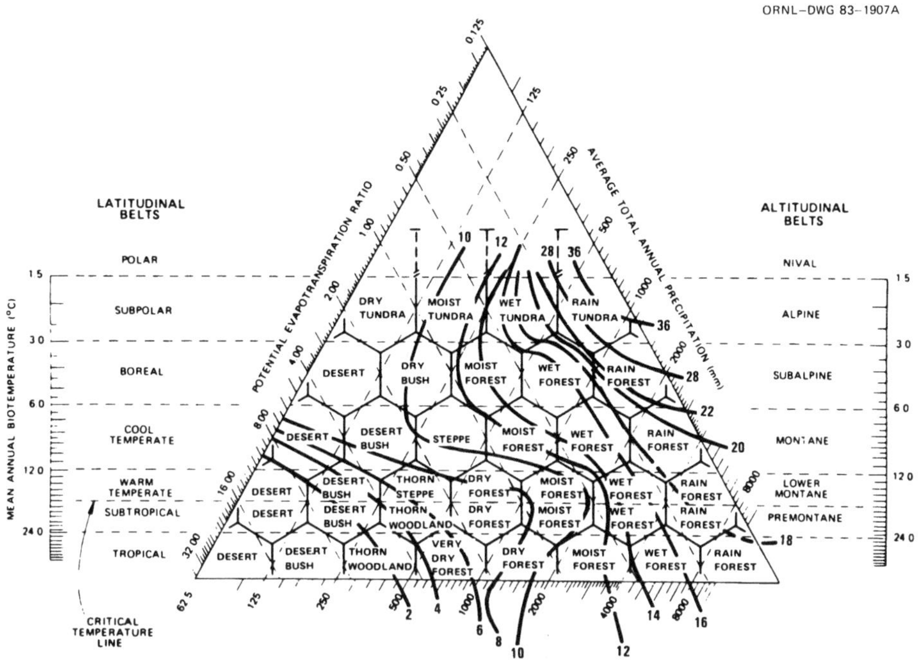

Figure 2. Soil carbon concentration (kg·m^{-3}) as a function of Holdridge Life Zone (from Post et al. 1985 and reprinted by permission from Nature vol. 317 p. 614 Copyright © 1985 Macmillan Magazines Ltd.).

carbon storage is low and mostly in inorganic forms (Schlesinger 1985). Warm deserts average less than 2 kg·m^{-3} and cool deserts reach 6-8 kg·m^{-3}. Decomposition rates are low in desert regions, but plant production rates are extremely low.

The ratio of potential evapotranspiration (PET) to precipitation is the upper left axis of the triangular Holdridge life zone chart (Figure 2). The separation between dry and moist life zones for all latitudinal belts occurs where PET is approximately equal to precipitation or the PET ratio is one. Along this line, traversing the center of the chart from upper left to bottom center, soil carbon concentrations are approximately 10 kg·m^{-3}. The life zones along this line are those that support most of the world's agriculture. Thus 10 kg·m^{-3} is a common carbon concentration for soils brought under cultivation. Cultivation generally reduces soil carbon concentration which I shall discuss later.

2.2 Soil properties

Wetland and coastal marsh ecosystems have soils with very high carbon concentrations. Bogs can have concentrations as high as 177 kg·m^{-3} (Zinke et al. 1984). Marsh and swamp soils that support some trees or shrubs may contain 24 $kg \cdot m^{-3}$. These ecosystems, under undisturbed conditions, accumulate carbon in their soils as a result of excess production over decomposition. Decomposition is inhibited by partial or complete waterlogging resulting from a combination of topographic position and local ground water hydrology.

Boreal and subarctic peatlands contain 455 Pg C (Gorham 1991) that is not included in the global soil carbon pool estimate of 1300 Pg C given in Figure 1. This estimate of peat is higher than previous estimates (Sjörs 1980, Armentano and Menges 1986, Oechel 1989) largely because a greater average peat depth of 2.3 meters rather than 1 meter was used in this calculation. Boreal and subarctic peatlands are located largely in the USSR, Canada, the USA, and Fennoscandian countries. All of this carbon has accumulated during the postglacial period at an average net rate of approximately 0.01 $Pg \cdot yr^{-1}$.

The physical and chemical nature of the underlying parent material can also influence the amount of soil organic matter stored. Walker and Adams (1958) hypothesized that the level of available phosphorus during the course soil development is the primary determinant of terrestrial net primary production. Numerous workers have examined this hypothesis. Tiessen et al. (1984) and Roberts et al. (1985) found that available phosphorus explained about one-forth of the variance in soil organic matter in many different soil orders. The relationship between phosphorus and carbon is strongest during the aggrading stage of vegetation-soil system development (Anderson 1988). Initially, the production of acidic products by pioneer vegetation promotes the release of phosphorus by weathering of parent material. Organic matter builds up in the soil, increasing the storage of phosphorus in decomposing organic compounds. Nitrogen-fixing bacteria populations, which depend on a supply of organic carbon and available phosphorus, can grow to meet ecosystem demands for nitrogen. Plant growth is enhanced by this increasing nitrogen and phosphorus cycling, resulting in increased rates of weathering. This process continues until the vegetation is constrained by other factors affecting phosphorus availability (leaching losses become larger than the weathering inputs (Jenny 1980), or an increasing fraction of the phosphorus becomes unavailable by adsorption or precipitation with secondary minerals (Walker and Syers 1976)) or affecting nitrogen availability (denitrification or leaching reaching or exceeding nitrogen inputs and fixation (Schlesinger 1991a). In mature soils, net primary production is more likely to be limited by nitrogen. Availability of other nutrients that are largely derived from parent materials, such as most base cations, may also influence soil organic matter accumulation during early soil development. Soils derived from base cation rich volcanic

Table 1. Soil carbon concentration in relationship to parent material (modified from Zinke et al. 1984).

Parent Material	Mean Carbon Density ($kg \cdot m^{-3}$)
Acid intrusive	10.9
Basic intrusive	12.2
Ultrabasic	10.6
Acid extrusive	15.1
Basic extrusive	21.8
Sedimentary (consolidated)	12.6
Metamorphic	11.2
Sedimentary (weakly consolidated)	14.9

parent materials (basic extrusive) have much higher carbon contents on average than soils from other parent materials (Zinke et al. 1984, Table 1).

2.3 Vegetation type

Vegetation production depends not only on climate, nutrient supply from geochemical weathering and other abiotic factors but also on biotic factors, in particular, plant species composition. Decomposition is also to some degree controlled by species composition. Each terrestrial plant species produces and sheds a characteristic blend of carbon compounds, each of varying decomposability. This range of decomposability may be summarized by the lignin and nitrogen content of the organic material (Aber and Melillo 1982, Meentenmeyer 1978). Nitrogen is made available to plants during the decomposition process. Nitrogen is a limiting element for productivity in most terrestrial ecosystems so the rate at which it is released during decomposition is an important factor in ecosystem production. Thus the interactions between processes regulating plant populations and their productivity, and microbial processes regulating nitrogen availability result in some of the observed variation in soil carbon and nitrogen storage and thus the rate of carbon exchange between the soil and atmosphere.

Computer simulations of forest succession reflect the importance of these interactions in terrestrial ecosystems (Pastor and Post 1986). Two different successional sequences after forest harvest, one with early stages dominated by pin cherry, the other with early and mid-stages dominated by aspen, occur in the Northeastern United States. Forest simulations for this region were run using 4 different initial organic matter contents of 25, 50, 75 and 100 $Mg \cdot ha^{-1}$ for both species successions. For both successions, soil organic matter concentration declines initially, then rises as productivity increases (Figure 3). The initial decline is greater when initial organic matter concentration is high and negligible when organic matter is low. It is less during aspen succession than

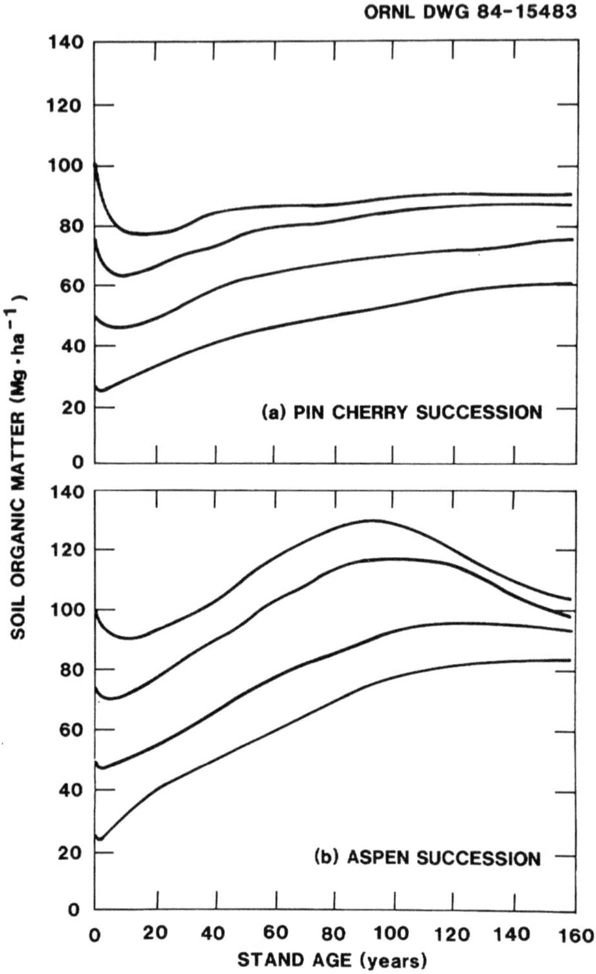

Figure 3. Soil carbon from Pin cherry vs. Aspen succession simulations (from Pastor and Post 1985, and reprinted by permission from Biogeochemistry vol. 2 p. 23 Copyright © 1985 Kluwer Academic Publishers).

during pin cherry succession. Aspen succession maintained higher levels of soil organic matter than pin cherry succession. Under aspen succession, the soil organic matter rises to a peak at about 90–100 years, then declines back to about 100 Mg·ha^{-1}. In all simulations, aspen succession returns 1.5 to 2 times more woody litter than does pin cherry succession by virtue of its longer life, sustained high rate of productivity, and a larger fraction of production allocated to wood. After 120 years, both successions

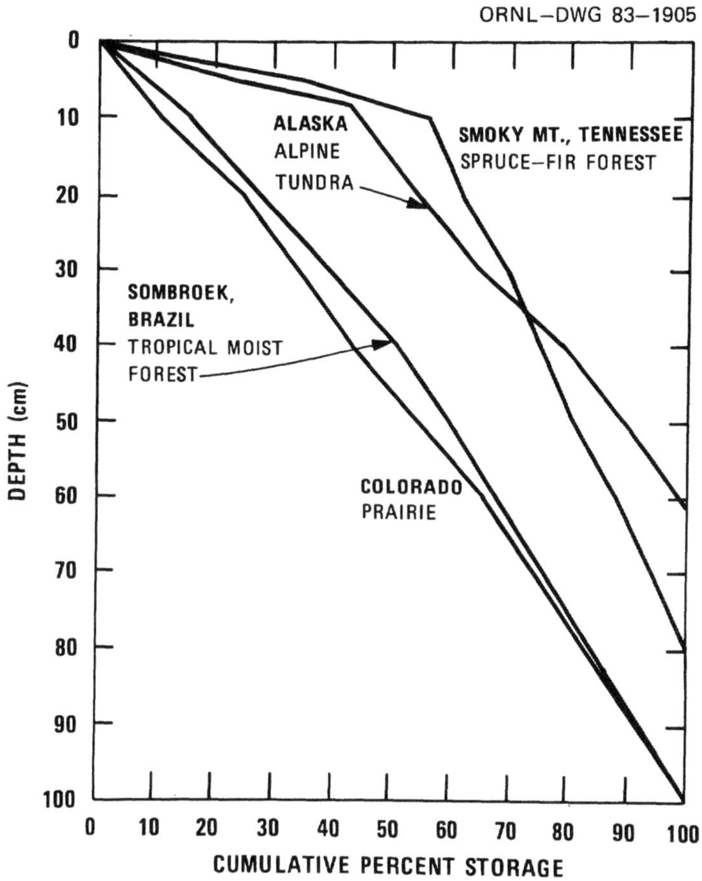

Figure 4. Relative cumulative carbon distribution with depth in the soil for several contrasting soil profiles (from Zinke et al. 1984)

result in similar northern hardwood forest compositions, and soil organic matter levels converge to 60–90 Mg·ha^{-1}.

The deeper that fresh detritus is placed in the soil, the slower it decomposes. This is a result of declining decomposer activity and increased protection from oxidation with depth in the soil. Prairies have a somewhat lower productivity than forests, and produce no slowly decomposing woody material. Nevertheless, prairies have a very high soil organic matter content because prairie grasses allocate twice as much production to belowground roots and tillers than to aboveground leaves (Sims and Coupland 1979). The result is high soil organic matter contents with a uniform distribution in the upper 1 meter of soil (Figure 4). In contrast, a spruce-fir forest contains 50% of its soil organic matter in the top 10 cm. There are interesting exceptions to the rule that above-/below-ground plant allocation determines soil organic matter distribution patterns in soil. Tropical moist forest soils show a uniform depth distribution similar to the depth distribution of temperate grasslands, however, in tropical forests this is largely due to a long-term accumulation of recalcitrant organic materials at lower depths in the soil rather than increased allocation to roots. Alpine tundra soils support a largely herbaceous flora but show a similar depth distribution as forest soils because of inhibition of surface litter decomposition by low temperatures and high water saturation.

The physical and biological factors that control patterns of soil organic matter described above give an indication of the potential influence of this pool on the global carbon cycle. In general, climate, soil physical factors of water holding capacity, and fertility form constraints within which ecosystem processes of production and decomposition interact to determine the dynamics of soil organic matter at a local scale (Pastor et al. 1984, Pastor and Post 1986).

3. CHEMICAL ASPECTS OF DECOMPOSITION AND STABILIZATION

Soil organic matter consists of a diversity of chemical compounds with widely varying turnover times. Decomposition of organic matter is a continual transformation of humic substances into more resistant substrates and CO_2. In most soils a portion of the soil organic matter consists of undecayed organic residues, mostly carbohydrates, with a smaller amount of proteinaceous compounds and lipids. This organic matter fraction is usually measured as organic fragments obtained by flotation on heavy liquids of density 1.6–2.0 $g \cdot cm^{-3}$ (Spycher et al. 1983). The turnover time of this soil fraction is less than 25 years. The remaining soil organic matter consists of humus. Humus may be further characterized chemically by sequential extraction of organic matter by alkali and acid to give the operationally defined and chemically heterogeneous fractions of humic acid, fulvic acid, and insoluble humin (Stevenson 1982, see Figure 5). Turnover times of these refractory soil organic matter constituents are on the order of 500–1200 years (Stout et al. 1981).

3.1 Chemical stabilization

The structure of organic materials will influence the time that they are retained in the soil. The decomposition rates of a wide range of organic compounds labeled with ^{14}C or ^{15}N have been reviewed (Oades and Ladd 1977; Paul and Van Veen 1978). In general, simple monomers from carbohydrates, proteins, and polyphenolic materials decompose within weeks. Polymers are decomposed more slowly and decomposition rate decreases with chemical complexity (Oades 1988). The percent of original material remaining

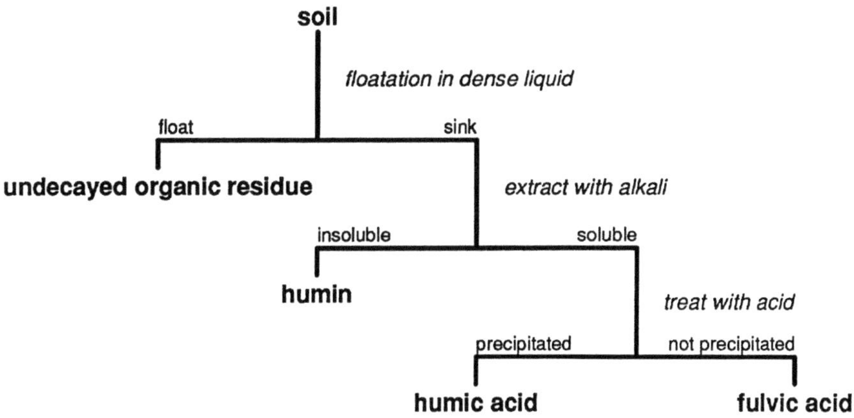

Figure 5. Traditional chemical fractionation of soil organic matter (modified from Kononova 1966).

as a function of time is well described with a negative exponential curve with data of different chemicals having a range of decay constants (Haider and Martin 1975). The time scale of the decomposition of most simple chemical substances in soil is on the order of weeks.

Decomposition of plant materials yields curves similar to those for individual chemical compounds except the time scale for the bulk material is on the order of years (Olson 1963, Jenkinson and Rayner 1977, Ladd et al. 1985). The initial rapid decline in organic matter remaining is due to decomposition of readily mineralized materials. After that, a succession of organisms utilize more resistant plant components, dead organisms, and metabolic products as sources of energy and nutrients.

As decomposition progresses, soil humus is formed, a process which has been extensively reviewed (Swain 1963; Swaby and Ladd 1962, 1966, (Stout et al. 1981). The complex structure of soil humic substances contributes to their stability. Humus formation (Stevenson 1986) is a multistage process beginning with the decomposition of all plant components, including lignin into simple monomers and concurrent polymerization of reactive monomers into high-molecular-weight brown colored polymers (Stout et al. 1981). The importance of lignin was demonstrated by tracer work (Martin et al. 1974) which showed that 34% of the added tracer incorporated into humus originated from lignin while only 6% was from cellulose. A description of the principle steps in the formation of humic substances from lignin and products of microbial synthesis can be found in Flaig (1975), Flaig et al. (1975).

3.2 Physical stabilization

As organic matter is fragmented into smaller sizes and converted into humic molecules, the physical structure of soil particles becomes an important factor in the rate of decomposition. The formation of complexes between organic compounds and clay particles plays a central role in stabilizing organic matter in soil (Anderson 1979). Newly formed humic acids have a relatively high aliphatic component, much of it as long side chains. This structure and high molecular weight results in a tendency for humic acids to be adsorbed by fine mineral colloids (Greenland 1965). Soil clay fraction includes a diversity of minerals with different surface areas and properties, and different arrangements of individual clay particles. The formation of soil particle aggregates through polycation bridging, organism glues and mucilages, or organic-inorganic bonds, results in micropore structure that can entrap organic matter and protect it from organisms and their extracellular enzymes.

Using tracer techniques, McGill and Paul (1976) found that much recently formed humic material was associated with mineral colloids $<0.04\mu m$ in size. In an eight year long decomposition experiment the loss of carbon due to decomposition was strongly related to clay content (Ladd et al. 1985). In four experimental plots kept free of vegetation, the content of soil organic matter was followed. The soils were all calcareous and influenced by the same climate. Soils with 5%, 12%, 20%, and 42% clay lost 40%, 25%, 15% and 0% organic carbon respectively. In most models of soil organic matter usually around 50% of the soil humus is assumed to be a physically stabilized fraction (Paul and van Veen 1978, Jenkinson and Rayner 1977, Parton et al. 1987).

Kilburtus (1980) reports that bacteria within soil aggregates only exist in pores at least three times their own diameter. The greater the clay content, the greater the proportion of small pore spaces, and therefore less pore space is accessible to bacteria. As much as 90% of the pore space in clayey soils may harbor organic matter from which bacteria and other organisms are physically excluded (Oades 1988).

Interactions between organic matter and inorganic minerals, involving various sized mineral particles and organic colloids with the formation of aggregates, can result in the physical protection of organic matter and increase its retention in soil (Oades 1988). Cultivation breaks these natural aggregates through mechanical stress. The decrease in physical protection of organic matter by such disturbances then can lead to an increase in decomposition rates.

4. QUANTIFYING SOIL ORGANIC MATTER DYNAMICS

Early models described the rate of organic matter input to soils as constant and the rate of loss through decomposition of total mass as a negative exponential using a single first order differential equation (Jenny et al. 1949, Olson 1963). Generalization of this approach became necessary in order to account for different components of soil organic matter with varying turnover times resulting from diverse biological, chemical, and physical processes. Additional differential equations were introduced later for each soil organic matter component, each containing terms expressing the transformation of one type into another or into CO_2. Many of these are reviewed by Wieder and Lang (1982). Additional refinements include models that describe soil organic matter along a continuum of decomposability (Carpenter 1981, Bosatta and Ågren 1985) and/or depth in the soil (O'Brien and Stout 1978, Nakane 1978).

There have been numerous attempts to estimate parameters and compartment sizes for these models using tracer experiments in which radiotracers, usually ^{14}C or ^{15}N, are introduced intentionally, or unintentionally as with nuclear weapons testing. These studies indicate that freshly applied material is rapidly incorporated into all soil organic matter components regardless of the procedures by which they are identified. The conclusion to be drawn from these results is that current fractionation techniques do not adequately separate soil organic matter into meaningful biological or chemical components such that they contain organic matter constituents with turnover times sufficiently uniform for modeling (Paul 1984, Trumbore 1988).

The models that have the greatest potential of being general enough for incorporation into global carbon cycle models have taken a more phenomenological approach that does not attempt to model all the transformations that occur during decomposition. There are two different approaches. The most common approach is simulation of a numerical model in which organic materials are partitioned into a minimum number of conceptual components based on whether biological, chemical, or physical processes dominate their decomposition rate. The other approach views decomposition as a microbial process which modifies organic matter into a continuous distribution of amounts distributed across a quality spectrum ranging from fresh, easily decomposed material to refractory humus substances.

4.1 Numerical simulation models

Most numerical simulation models of soil organic matter share, essentially, the same five-component structure shown in Figure 6. Plant residues are divided into two components, one with simple compounds subject to rapid uptake, transformation and mineralization

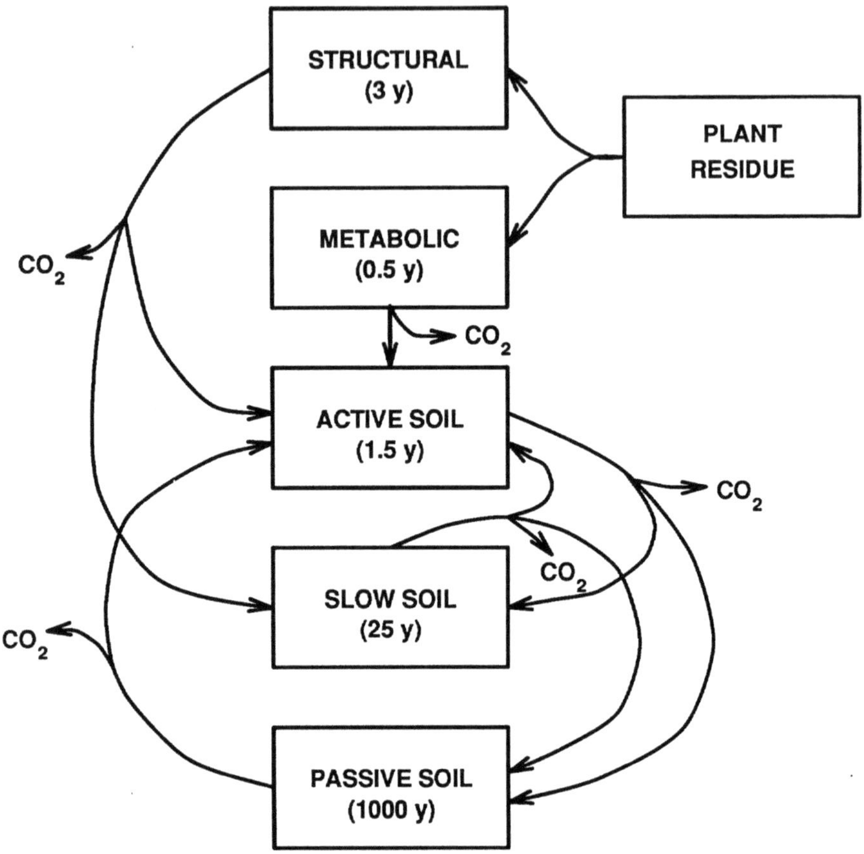

Figure 6. Soil organic matter components and flows in the CENTURY model (modified from Parton et al. 1987).

by decomposers (METABOLIC), and the other contains plant components such as lignin which are not readily attacked (STRUCTURAL). The soil humus is divided into three components with turnover times differing by an order of magnitude between each one. The ACTIVE SOIL is mostly live microbes, and microbial products such as extracellular enzymes. The SLOW SOIL component consists of organic compounds that are physically protected or in chemical forms with biological resistance to decomposition. The PASSIVE SOIL component consists of chemically refractory forms with the longest turnover times.

Various formulations of this basic model structure differ in the number and magnitudes of the interactions between the components and the turnover times of the components. They also differ somewhat in the descriptions of components which are heuristically defined, but may bear close correspondence to operationally defined soil organic matter fractions (Anderson 1979). For example, Paul and Juma (1981) found that the relative size and radiocarbon age of four operationally defined components of a Chernozemic soil were useful for analyzing predictions from a simulation model calibrated from independent data. Paul and Juma's (1981) operational identification, estimated turnover time, and approximate correspondence to the components described above and shown in Figure 6 are presented in Table 2.

Table 2. Paul and Juma's (1981) method of determination, turnover times, and correspondence with soil organic matter pools identified in Figure 7.

Organic matter fraction	Method of determination	Turnover time (years)	Approximate correspondence with Figure 7 components
Biomass	Fumigation incubation	0.5	ACTIVE
Active non-biomass	Isotope dilution	1.5	STRUCTURAL +METABOLIC +part of ACTIVE
Stabilized	By difference	22.0	SLOW
Old	Carbon dating	600.0	PASSIVE

The earliest model of this type was developed by Jenkinson and Rayner (1977) to describe turnover of soil organic matter in the classical field experiments employing different cultivation systems at the Rothamsted Experimental Station in England. Similar models have been used to simulate soil organic matter development (Paul and Van Veen 1978, Van Veen and Paul 1981) and incorporate the effect of soil erosion (Voroney et al. 1981). In these studies, the magnitudes of the fluxes between compartments were estimated by data fitting procedures, making their application restricted to single locations.

Recently, two models (CENTURY and the Rothamsted turnover model) have been developed that generalize this basic model structure by formulating the component fluxes as functions of a small set of easily obtained environmental factors or driving variables such as climate, soil texture, and lignin:N content of plant litter. This allows application of these models to broad geographic regions and perhaps incorporation into global carbon cycle models that have a spatially distributed terrestrial component.

The CENTURY model has been used to describe the geographic pattern of soil organic carbon and nitrogen storage in the Great Plains of the U.S. (Parton et al. 1987). The Century model has also been used for simulating large-scale and long-term consequences of climate and management changes on organic matter dynamics in native grasslands and agroecosystems (Parton et al. 1988). Critical parameters in the CENTURY model are the turnover times for the SLOW SOIL and PASSIVE SOIL pools. Estimates of these rates are based on long-term incubation studies (5 years with the first year discarded). It is not known how much these rates differ between different ecosystems or soil conditions so general application to global scales is not currently feasible.

The Rothamsted turnover model (Jenkinson 1990) simulates soil organic matter over the years-to-centuries timespan. Parameters for the model are derived, largely, by fitting data from long-term field experiments on Rothamsted and Woburn experimental farms. This model has been used to predict the decomposition of ^{14}C-labeled plant material in a range of climates and soils: 19 locations in the humid tropics, 6 in the temperate zone, 4 in the warm temperate forest zone, and 2 in the cool temperate steppe zone. Experiments on the decomposition of labeled plant material under field conditions are particularly valuable in model testing, as they give reliable measures of C retention over much longer decomposition periods than incubations. Unfortunately, only a few types of labeled plant material have been used (annual crop plants and grasses).

Similarity of the Rothamsted turnover model with CENTURY is striking. For example, both use experimental data from Sørensen's (1981) incubation of cellulose in soils with different textures to incorporate the effect of soil texture. Both models use similar methods of applying these data fits to modify the efficiency of converting ACTIVE SOIL organic matter into SLOW SOIL organic matter. However, CENTURY also uses this information to modify the decomposition rate of the ACTIVE SOIL compartment. In both models, the dominant parameters governing the equilibrium content of soil organic matter are the rate constants that determine the turnover rate of the SLOW SOIL and PASSIVE SOIL compartments. Current methods of fitting these parameters in either of these or similar models, however, is not direct and introduces considerable uncertainty for predicting decade to century long effects from these models.

A major difference between the two models has to do with the PASSIVE SOIL compartment. In the CENTURY model this compartment receives very small fluxes of organic matter from both the ACTIVE SOIL and SLOW SOIL compartments and has a very long turnover time. In the Rothamsted turnover model, the PASSIVE SOIL compartment is represented as completely inert, with no inputs or outputs. An earlier version of the Rothamsted turnover model (Jenkinson et al. 1987) did postulate that a portion of the incoming material entered a highly resistant (but not inert) compartment. But this more elegant formulation was abandoned for practical reasons – an inability to reconcile a model containing a dynamic PASSIVE SOIL organic matter compartment with empirical data.

Both models have been used to examine soil organic matter dynamics under a warmer climate (Schimel et al. 1988, Jenkinson et al. 1991). Not surprising, both obtain similar general results. With an increase in temperature of 2° to 4° C, soil organic matter storage decreases around 5% in the Rothamsted turnover model and 20% in the CENTURY model. The larger decline results from a feedback with vegetation production. In the CENTURY model, lower soil organic matter levels reduce nitrogen availability and result in lower plant productivity and lower organic matter inputs into the soil. The Rothamsted model does not include a plant production feedback and therefore predicts smaller declines in soil organic matter.

Further development and application of models like CENTURY and the Rothamsted turnover model to ecosystems in addition to the temperate zone forests and grasslands currently modeled may increase the confidence in using this modeling framework for understanding the dynamics of soil organic matter in global carbon cycle studies.

4.2 Microbial transformation models

Another approach to modeling soil organic matter dynamics, rather than defining phenomenologically convenient compartments, is to treat soil organic matter as a heterogeneous substrate in which the heterogeneity is described by a continuously varying variable, q, commonly referred to as litter/soil organic matter "quality" (Ågren and Bosatta 1991). For convenience, let $q = 0$ be the most refractory material and $q = 1$ be the most easily decomposed material. Decomposition can be though of as a process that moves organic matter along the q-axis under the effect of several driving forces that transform material into other qualities and result in losses to CO_2. Mathematically, this can be expressed as a continuity equation

$$\frac{\partial \rho(q,t)}{\partial t} = -S(q,t) + \frac{\partial \Phi(q,t)}{\partial q}, \tag{2}$$

where $\rho(q,t)dq$ is the amount of carbon with qualities between q and $q + \mathrm{d}q$, $\Phi(q,t)$ describes the transformation of carbon of quality q into other qualities, and $S(q,t)$ is a sink term related to losses of carbon during the decomposition process (Figure 7). Microbial activity is the mechanism behind decomposition. Making some reasonable assumptions it is possible to express Eq. (2) in terms of two microbial properties,

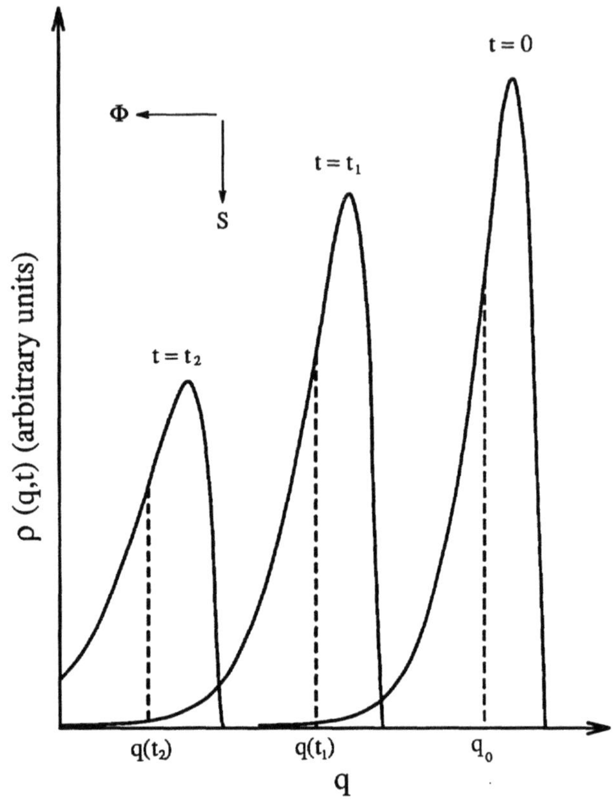

Figure 7. Hypothetical development of the distribution of organic matter quality at three different times resulting from the effects of losses of carbon, S, and transformation of carbon to other qualities, Φ (Modified from Bosatta and Ågren 1985).

efficiency of converting a unit carbon contained in substrate of quality q into a unit of microbial biomass, $e(q)$, and the rate of substrate utilization by microbes, $u(q)$, as

$$\frac{\partial \rho(q,t)}{\partial t} = -\frac{1-e(q)}{e(q)} f_c u(q) \rho(q,t) + \epsilon f_c \frac{\partial}{\partial q}[u(q)\rho(q,t)], \tag{3}$$

where f_c is the mean concentration of carbon in the microbial biomass, and ϵ is a scale factor (Bosatta and Ågren 1985). With appropriate initial conditions and boundary conditions Eq. (3) can be solved exactly using Laplace transformation for arbitrary functions $e(q)$ and $u(q)$. Solution for the general form of Eq. 3 as well as solutions for specific forms of $e(q)$ and $u(q)$ are given in Bosatta and Ågren (1985). Some qualitative results of the general form are of interest concerning long-term sequestering of carbon in soils.

It is possible that $u(q)$ has such a form that there exists a $t = t_{max}$ such that q will go to 0 at t_{max}. For later times there will be no further changes in substrate quality and decomposition will stop. On the other hand, $u(q)$ may have a form such that q will not go to zero in finite time and the decomposition can proceed for an infinitely long time. Thus two qualitatively different outcomes are possible.

A second result is related to the behavior of the function $e(q)$. If the efficiency with which the microorganisms can utilize the substrate goes to zero more rapidly than q goes to zero during the decomposition process there will remain some substrate which will never decompose (unless some other, more efficient microorganism takes over at this stage). If efficiency does not decrease faster than q, then all substrate will eventually be consumed. The four consequences of different combinations of $e(q)$'s and $u(q)$'s are schematically illustrated in Figure 8. Although plausible forms of the functions $u(q)$ and $e(q)$ have been proposed and compared to data (Bosatta and Ågren 1985, 1991), not enough is known about microbial dynamics in relation to organic matter quality to rule out any of the possibilities depicted in Figure 8.

4.3 Observational methods for soil carbon dynamics

During soil development, organic matter often shows an initial period of rapid increase, for up to 3,000 years, followed by a period with a lower rate of increase that may continue for many more millennia. This lower rate of accumulation can be viewed as accretion of recalcitrant organic matter into the PASSIVE SOIL pool. Schlesinger (1991b) compiled data from chronosequences in order to compute the average long-term rate of organic carbon accumulation for 3,000 to 10,000-year-old soils. In upland ecosystems (those with well drained soils), the long-term rate of organic carbon storage ranges from 0.2 g C $m^{-2}yr^{-1}$ in polar deserts to over 10 g C $m^{-2}yr^{-1}$ in some temperate and boreal forests. The average for 16 soils was 2.4 g C $m^{-2}yr^{-1}$.

After long periods of time ($\sim$ 6,000 years) accretion must be balanced by losses to erosion. Erosion causes some of this organic matter to be transported to the oceans. If one assumes that 2.4 g C $m^{-2}yr^{-1}$ is an average rate of erosion and transport and multiplies this by the upland area under natural vegetation (133 x 10^6 km^2, Schlesinger 1977) then 0.32 Pg C yr^{-1} would be contributed to oceans from upland soils. This corresponds well with estimates of global transport of organic carbon in rivers (0.4 Pg C yr^{-1}, Schlesinger and Melack 1981, Maybeck 1982) which get additional inputs from wetlands and fresh litter directly.

If we can use chronosequences to determine the long-term accretion rate of soil organic matter (which may be related to the rate of increase of the PASSIVE SOIL pool), why

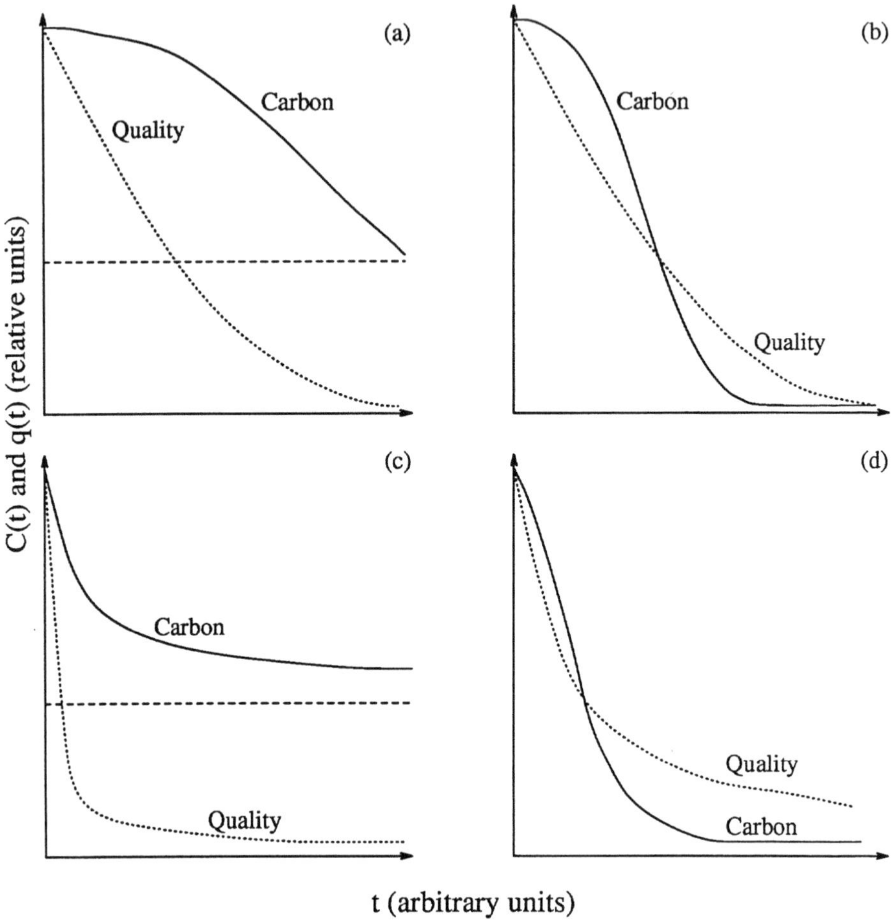

Figure 8. Quality and quantity of a substrate decomposing according to Eq. 3 for 4 different combinations of efficiency functions, $e(q)$, and utilization rates, $u(q)$. Solid curve $= C(t)$, dashed curve = quality, q. (a) Decomposition stops in finite time with finite amount of C remaining. (b) Decomposition stops in finite time with all C mineralized. (c) Decomposition process proceeds to infinity but a finite amount of C will remain. (d) Decomposition proceeds to infinity and all C is mineralized (Modified from Bosatta and Ågren 1985).

can't we use chronosequences to determine the sizes and turnover times of other soil organic pools? In theory, it should be possible to go out to a chronosequence, dig up samples from each exposure period, bring it back into the laboratory and radiocarbon date various soil organic matter fractions. In the 1950's and early 1960's atmospheric nuclear weapons testing produced ^{14}C in amounts much greater than natural ^{14}C abundances in the atmosphere. Enrichment of soil carbon with bomb-^{14}C enabled studies on the uptake and movement of carbon into the soil (Rafter and Stout 1970, Stout and O'Brien 1973). However, the amount of bomb-^{14}C that has been taken up by plants and soil in a particular place is not generally known. Therefore, in addition to current samples, a time series of ^{14}C data in soil C is needed, covering the period of the rise in atmospheric bomb-^{14}C. It is generally impossible now to obtain such data. Some information, however, can be obtained using a simple model outlined by O'Brien (1984) and measuring bomb-^{14}C penetration, of carbon input rate, decomposition time, and downward diffusion in the soil profile. This model represents material that would be called SLOW SOIL carbon in the CENTURY model or the Rothamsted turnover model as a simple first order process. This method also assumes that "old" carbon (equivalent to PASSIVE SOIL carbon) is uniformly distributed through the soil profile and that the overwhelming amount of fresh organic matter enters the soil from the surface. While these assumptions appear to be reasonable for the soils studied by O'Brien (1984), it is not known whether these assumptions are generally applicable to other soils.

Through experimental manipulation at Rothamsted, England of several fields for 150 years and the periodic sampling and storage of soil samples (which, before 1950, are uncontaminated with bomb-^{14}C) it has been possible to derive parameters for a model of soil organic matter turnover. This approach has several advantages over chronosequence studies. The different treatments alter the kinds and amounts of organic matter entering the soil. Such perturbations from a control or presumed steady-state soil can reveal interesting aspects of soil organic matter dynamics. Additionally, the stored samples provide valuable direct information on amounts of soil organic matter especially in the absence of bomb-^{14}C contamination. In order to further demonstrate how valuable long-term experimental plots are to understanding soil organic matter dynamics, I present additional results from a study by Balesdent et al. (1988).

Experimental plots were established at the Sanborn Field, University of Missouri, in 1888 to compare crop rotations with continuous cropping, with regard to effects on soil fertility. Historical soil samples were collected from the plots in 1915, 1928, 1938, 1962, and 1975, air-dried, and stored in sealed glass bottles. The native prairie on which these plots were established consisted of a mixture of predominantly C-4 grasses with some C-3 broadleaf plants also present. Coarse debris separated from the prairie topsoil had a δ^{13}C of -19°/₀₀. The topsoil organic matter had a δ^{13}C of -18.6°/₀₀, suggesting that a slight enrichment occurred during humification of the prairie vegetation. Analysis of crop samples collected from Sanborn Field yielded δ^{13}C values as follows: wheat -27.9°/₀₀; timothy -27.8°/₀₀; both of which are C-3 crops and corn, a C-4 crop, -11.6°/₀₀. These measured values were adjusted to account for the slight enrichment during humification by using -27.0°/₀₀ for C-3 crops, and -11.0°/₀₀ for C-4 corn, and -18.6°/₀₀ for bulk native prairie in all calculations.

Using the $\delta^{13}C$ information, it is possible, for each historical soil sample, to determine the total amount of soil organic matter as well as the proportion of soil organic matter that was derived from native prairie or introduced crop plants. The main results appear in Figure 9. Soil organic carbon levels decline rapidly during the first thirty years of cultivation toward new equilibria that depends on each crop and soil management regime. Levels of soil organic matter are lower under wheat cultivation than under timothy cultivation. The wheat field requires annual plowing and sowing. The timothy field, since timothy is a perennial crop, is plowed only once at establishment and then is harvested with mowing. A larger fraction of the native-prairie derived soil organic carbon was lost under wheat cultivation. After 1950 the wheat plot received larger fertilizer applications and surface residues were left and incorporated into the soil. This approximately doubled the annual C input and soil organic C.

Continued turnover of soil organic matter after 1915 or 1928 was primarily that originating from the recently cultivated crop with little change in $\delta^{13}C$. The data show a very slow mineralization of the remaining stable soil organic matter. The mean slope of the curve after 1938 is -0.0032±0.0078 mg C g^{-1} yr^{-1} (complete turnover requiring over 600 years) for wheat and 0.0012±0.0086 mg C g^{-1} yr^{-1} (complete turnover requiring over 1400 years) for timothy.

After 100 years, the carbon of prairie origin was 61% of total carbon under timothy and 49% of the carbon under wheat. The higher amount under timothy suggests that some of the original prairie soil carbon is chemically labile but physically protected and that annual cultivation results in the loss of some physical protection.

On some plots that had been cultivated to C-3 crops corn was cultivated continuously starting in 1950. In Figure 9c we see the rapid decline of the labile C-3 crop-derived organic matter that was no longer being added each year and being more or less replaced by organic matter derived from corn residues. A mean decay constant for the C-3 carbon is 0.067 yr^{-1} indicating a half-life of 10 years.

5. CONCLUSIONS

Soil organic matter is a large and heterogeneous collection of carbon-containing compounds that interact with both active and relatively inactive components of the global carbon cycle. The turnover of carbon in soils is controlled mainly by water regimes and

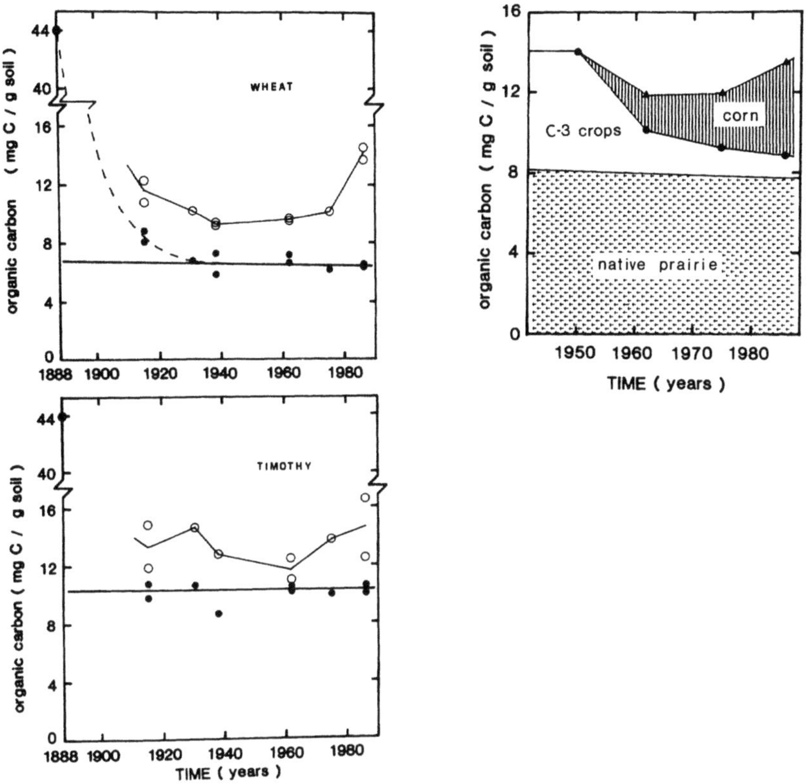

Figure 9. Changes in amount and origin of soil organic C as a result of cultivation on formerly virgin prairie soil. Open circles denote total C, and solid circles represent C of prairie origin with upper and lower points at different dates for the 0 to 10 and 10 to 20 cm depth samples, respectively. The straight, solid line indicates the level of PASSIVE SOIL C (from Balesdent et al. (1988) and reprinted by permission from Soil Science Society of America Journal vol. 52 p. 121 Copyright © 1988 Soil Science Society of America, Inc.).

temperature but is modified by a large number of processes involving the physiochemical properties of organic-matter additions, distribution in the soil matrix, decomposer population dynamics, and interactions with mineral particles.

It is useful to divide the soil organic matter into several fractions of various degrees of decomposability and therefore turnover time. Some progress has been made in identifying the types of organic compounds associated with different operationally defined fractions. Soil organic matter fractions can be characterized chemically, physically, and biologically. Classical descriptions of soil organic matter are based on chemical extraction. More recently, approaches that also consider soil as a physical and biological system have been used. These approaches are better aligned with biological and biochemical processes that occur during organic matter decomposition and humus formation. However, it is still difficult to arrive at a separation of soil organic matter into components of distinctly uniform turnover times that is satisfactory in modeling soil carbon dynamics.

Various ecosystem disturbances alter the balances between production of organic matter and its decomposition and therefore change the amount of carbon in soil. The most severe perturbation is conversion of natural vegetation to cultivated crops. This conversion results in a lowered input of slowly decomposing material which causes a reduction in overall carbon levels. Disruption of soil matrix structure by cultivation leads to lowered physical protection of organic matter resulting in an increased net mineralization rate of soil carbon resulting in further reductions in soil carbon content. Climate change is another perturbation that affects the amount and composition of plant production, litter inputs, and decomposition regimes but does not affect soil structure directly. Nevertheless, large changes in soil carbon storage are possible with anticipated CO_2 induced climate change, particularly in northern latitudes where anticipated climate change will be greatest (MacCracken and Luther 1985) and large amounts of soil organic matter are found (Post 1990).

It is impossible, given the current state of knowledge of soil organic matter processes and transformations, to develop detailed process models of soil carbon dynamics. Phenomenological models or ones that are based on empirically defined components and transformation rates appear to be developing into general tools for understanding the role of soil organic matter in the global carbon cycle. In particular, these models will be useful in quantifying soil carbon changes due to human land-use and to anticipated global climate and vegetation changes.

REFERENCES

Aber JD, Melillo JM (1982) Nitrogen immobilization in decaying hardwood leaf litter as a function of initial nitrogen and lignin content. Canadian Journal of Botany 58:416–421

Ågren GI, Bosatta E (1991) Dynamics of carbon and nitrogen in the organic matter of the soil: A generic theory. American Naturalist 138:227–245

Anderson DW (1979) Processes of humus formation and transformation in soils of the Canadian Great Plains. Journal of Soil Science 30:77–84

Anderson DW (1988) The effect of parent material and soil development on nutrient cycling in temperate ecosystems. Biogeochemistry 5:71–97

Armentano TV, Menges ES (1986) Patterns of change in the carbon balance of organic-soil wetlands of the temperate zone. Journal of Ecology 74:755–557

Balesdent J, Wagner GH, Mariotti A (1988) Soil organic matter turnover in long-term field experiments as revealed by carbon-13 natural abundance. Soil Science Society of America Journal 52:118–124

Bosatta E, Ågren GI (1985) Theoretical analysis of decomposition of heterogeneous substrates. Soil Biology and Biochemistry 17:601–610

Carpenter S (1981) Decay of heterogeneous detritus a general model. Journal of Theoretical Biology 89:539–547

Flaig, W (1975) Biochemistry of soil organic matter. FAO Soils Bulletin 27:31–35

Flaig W, Buetelspacher H, Reitz E (1975) Chemical composition and physical properties of humic substances. In: Gieseking JE (ed) Soil Components, Volume 1: Organic Components. Springer, Berlin Heidelberg New York

Gorham E (1991) Northern peatlands: role in the carbon cycle and probable responses to climatic warming. Ecological Applications 1:182–195

Greenland DJ (1965) Interactions between clays and organic compounds in soils. Soils Fertil. 28:415–25 and 521–32

Haider K, Martin JP (1975) Decomposition of specifically carbon-14 labelled benzoic and cinnamic acid derivatives in soil. Soil Science Society of America, Proceedings 39:657–662

Jenkinson DS (1990) The turnover of organic carbon and nitrogen in soil. Phil. Trans. R. Soc. Lond. B 329:361–369

Jenkinson DS, Rayner JH (1977) The turnover of soil organic matter in some of the Rothamsted classical experiments. Soil Science 123:298–305

Jenkinson DS, Hart PBS, Rayner JH, Parry LC (1987) Modelling the turnover of organic matter in long-term experiments at Rothamsted. Intecol. Bull. 15:1–8

Jenkinson DS, Adams DE, Wild A (1991) Global warming and soil organic matter. Nature 351:304–306

Jenny H (1980) The Soil Resource. Springer, Berlin Heidelberg New York

Jenny H, Gessel SP, Bingham T (1949) Comparative study on the decomposition rates of organic matter in temperate and tropical regions. Soil Science 68:419–432

Kilburtus G (1980) Etudes des microhabitats contenus dans les agregats du sol, leur relation avec la biomasse bacterienne et la taille des procaryotes presents. Rev. Ecol. Biol. Sol 17:545–557

Kononova MM (1966) Soil Organic Matter. Pergamon Press, Oxford

Ladd JN, Amato M, Oades JM (1985) Decomposition of plant material in Australian soils. III. Residual organic matter decomposing under field conditions. Australian Journal of Soil Research 23:603-611

MacCracken MC, Luther FM (eds) (1985) Detecting the climatic effects of increasing carbon dioxide. US Department of Energy, Washington, DC (DOE/ER-0235)

Martin JP, Haider K, Farmer WJ, Fustec-Mathon E (1974) Decomposition and distribution of residual activity of some ^{14}C-microbial polysaccharides and cells, glucose, cellulose, and wheat straw in soil. Soil Biology and Biochemistry 6:221–230

Maybeck M (1982) Carbon, nitrogen and phosphorus transport by world rivers. American Journal of Science 282:401–450

McGill WB, Paul EA (1976) Fractionation of soil and ^{15}N nitrogen to separate the organic and clay interactions of immobilized N. Canadian Journal of Science 56:203–212

Meentemeyer V (1978) Macroclimate and lignin control of litter decomposition rates. Ecology 59:465–472

Nakane K (1978) A mathematical model of the behavior and vertical distribution of organic carbon in forest soils. II. A revised model taking the supply of root litter into consideration. Japanese Journal of Ecology 28:169–178

Oades JM (1988) The retention of organic matter in soils. Biogeochemistry 5:35–70

Oades JM, Ladd JN (1977) Biogeochemical properties: carbon and nitrogen metabolism. In: Russell JS, Greacen EL (eds) Soil Factors in Crop Production in a Semi-arid Environment. University of Queensland Press, St. Lucia, Queensland

O'Brien BJ (1984) Soil organic carbon fluxes and turnover rates estimated from radiocarbon enrichments. Soil Biology and Biochemistry 16:115–120

O'Brien BJ, Stout JD (1978) Movement and turnover of soil organic matter as indicated by carbon isotope measurements. Soil Biology and Biochemistry 10:309–317

Oechel WC (1989) Nutrient and water flux in a small arctic watershed: an overview. Holarctic Ecology 12:229–237

Olson JS (1963) Energy storage and the balance of producers and decomposers in ecological systems. Ecology 44:322–331

Parton WJ, Schimel DS, Cole CV, Ojima DS (1987) Analysis of factors controlling soil organic matter levels in Great Plains grasslands. Soil Science Society of America Journal 51:1173–1179

Parton WJ, Stewart JWB, Cole CV (1988) Dynamics of C, N, P, and S in grassland soils: a model. Biogeochemistry 5:109–131

Pastor J, Aber JD, McClaugherty CA, Melillo JM (1984) Aboveground production and N and P cycling along a nitrogen mineralization gradient on Blackhawk Island, Wisconsin. Ecology 65:256–268

Pastor J, Post WM (1986) Influence of climate, soil moisture, and succession on forest carbon and nitrogen cycles. Biogeochemistry 2:3–27

Pastor J, Post WM (1988) Response of northern forests to CO_2-induced climate change. Nature 334:55–58

Paul EA (1984) Dynamics of organic matter in soils. Plant and Soil 76:275–285

Paul EA, Juma NG (1981) Mineralization and immobilization of soil nitrogen by microorganisms. In: Clark FE, Rosswall T, (eds) Terrestrial Nitrogen Cycles: Processes, Ecosystems Strategies and Management Impacts. Ecological Bulletin (Stockholm) 33:179–204

Paul EA, Van Veen H (1978) The use of tracers to determine the dynamic nature of organic matter. Transaction of the 11th Congress of the International Society of Soil Science, Edmonton 3:61–102

Post WM (ed) (1990) Report of a Workshop on Climate Feedbacks and the Role of Peatlands, Tundra, and Boreal Ecosystems in the Global Carbon Cycle. Oak Ridge National Laboratory, Oak Ridge, TN (ORNL/TM-11457)

Post WM, Emanuel WR, Zinke PJ, Stangenberger AG (1982) Soil carbon pools and world life zones. Nature 298:156–159

Post WM, Pastor J, Zinke PJ, Stangenberger AG (1985) Global patterns of soil nitrogen storage. Nature 317:613–616

Post WM, Peng T-H, Emanuel WR, King AW, Dale VH, DeAngelis DL (1990) The Global Carbon Cycle. American Scientist 78:310–326

Rafter TA, Stout JD (1970) Radiocarbon measurements as an index of the rate of turnover of organic matter in forest and grassland ecosystems in New Zealand. In:

Olsson IU (ed), Radiocarbon Variations and Absolute Chronology. Almquist and Wiksell, Stockholm, pp. 401–418

Roberts TL, Stewart JWB, Bettany JR (1985) The influence of topography on the distribution of organic and inorganic soil phosphorus across a narrow environmental gradient. Canadian Journal of Soil Science 65:651–665

Schimel DS, Parton WJ, Kittel TGF, Ojima DS, Cole CV (1988) Grassland biogeochemistry. Links to Atmospheric Processes. Climatic Change 17:13–25

Schlesinger WH (1977) Carbon balance in terrestrial detritus. Annual Reviews of Ecology and Systematics 8:51–81

Schlesinger WH (1985) The formation of caliche in soils of the Mojave Desert, California. Geochim. Cosmochim. Acta 49:57–66

Schlesinger WH (1991a) Biogeochemistry: An Analysis of Global Change. Academic Press, New York

Schlesinger WH (1991b) Evidence from chronosequence studies for a low carbon-storage potential of soils. Nature 348:232–234

Schlesinger WH, Melack JM (1981) Transport of organic carbon in the world's rivers. Tellus 33:172–187

Sims PL, Coupland RT (1979) Producers. In: Coupland RT (ed.) Grassland Ecosystems of the World: Analysis of Grasslands and Their Uses, Chapter 5. Cambridge University Press, Cambridge

Sjörs H (1980) Peat on earth: multiple use or conservation? Ambio 9:303–380

Sørensen LH (1981) Carbon-nitrogen relationships during the humification of cellulose in soils containing different amounts of clay. Soil Biol. Biochem. 13:313–321

Spycher G, Sollins P, Rose SL (1983) Carbon and nitrogen in the light fraction of a forest soil: vertical distribution and seasonal patterns. Soil Science 135:79–87

Stout JD, O'Brien BJ (1973) Factors affecting radiocarbon enrichment in soil and the turnover of organic matter in forest and grassland ecosystems in New Zealand. Proceedings of the 8th International Conference on Radiocarbon Dating, Vol 2. Wellington, New Zealand, pp. 394–407

Stevenson FJ (1982) Humus Chemistry: Genesis, Composition, Reactions. Wiley-Interscience, New York

Stevenson FJ (1986) Cycles of Soil Carbon, Nitrogen, Phosphorus, Sulfur, Micronutrients. John Wiley and Sons, New York

Stout JD, Goh KM, Rafter TA (1981) Chemistry and turnover of naturally occurring resistant organic compounds in soil. In: EA Paul, JN Ladd, (eds) Soil Biochemistry, Volume 5. Marcel Dekker, New York

Swaby RJ, Ladd JN (1962) Chemical nature, microbial resistance, and origin of soil humus. In: Neale GJ (ed) Transactions of Joint Meeting of Commissions IV and V. International Society of Soil Science, Palmerston North, New Zealand

Swaby RJ, Ladd JN (1966) Stability and origin of soil humus. In: The use of Isotopes in Soil Organic Matter Studies. International Energy Agency, Vienna

Swain FM (1963) Geochemistry of humus. In: Bregen IA (ed) Organic Geochemistry. Pergamon, New York

Tiessen HJ, Stewart WB, Cole CV (1984) Pathways of phosphorus transformations in soils of differing pedogenesis. Soil Science Society of America Journal 48:853–858

Trumbore SE (1988) Carbon cycling and gas exchange in soils. PhD thesis, Columbia University

Van Veen JA, Paul EA (1981) Organic carbon dynamics in grassland soils. I. Background information and computer simulation. Canadian Journal of Soil Science 61:185-201

Voroney RP, Van Veen JA, Paul EA (1981) Organic C dynamics in grassland soils. II. Model validation and simulation of the long-term effects of cultivation and rainfall erosion. Canadian Journal of Soil Science 61:211-224

Walker TW, Adams AFR (1958) Studies on soil organic matter: I. Influence of phosphorus content of parent materials on accumulations of carbon, nitrogen, sulfur, and organic phosphorus in grassland soils. Soil Science 85:307–318

Walker TW, Syers JK (1976) The fate of phosphorus during pedogenesis. Geoderma 15:1–19

Wieder RK, Lang GE (1982) A critique of the analytical methods used in examining decomposition data obtained from litter bags. Ecology 63:1636–1642

Zinke PJ, Stangenberger AG, Post WM, Emanuel, WR, Olson JS (1984) Worldwide Organic Soil Carbon and Nitrogen Data. Oak Ridge National Laboratory, Oak Ridge, Tennessee (ORNL/TM–8857)

Is the Magnitude of the Deep Outflow from the Atlantic Ocean Actually Governed by Southern Hemisphere Winds?

J. R. Toggweiler and B. Samuels
U.S. Department of Commerce/NOAA
Geophysical Fluid Dynamics Laboratory
P.O. Box 308, Princeton University
Princeton, New Jersey 08542 USA

Abstract

The large-scale overturning in the Atlantic Ocean and its export of salty water to the other basins of the ocean is usually thought of as a thermohaline process driven by the formation of dense bottom water in the isolated basins of the North Atlantic. In this paper the output from several different runs of a global ocean GCM is used to show that the inflow of upper kilometer water in the South Atlantic and the outflow of deep water varies in direct proportion to the westerly wind stress in the circumpolar region of the southern hemisphere. According to the results presented here, the production of dense bottom water in the North Atlantic makes it possible for an Atlantic overturning to exist, but southern hemisphere winds appear to determine the magnitude of the inflow and outflow. The connection between southern hemisphere winds and the Atlantic overturning is due to a unique dynamic constraint which operates in the latitude band of Drake Passage. This constraint suggests the possibility of a very simple relationship between the magnitude of the northward wind drift at the latitude of the tip of South America (i.e. $\int(\tau_x/f)dx_{SA}$) and the magnitude of the inflow and outflow from the Atlantic basin.

NATO ASI Series, Vol. I 15
The Global Carbon Cycle
Edited by M. Heimann

Introduction

The dominant mode of large-scale overturning in the ocean is thought to consist of a widely distributed upwelling of abyssal water that is balanced by the formation of bottom water in a few confined high-latitude basins (Stommel, 1958; Warren, 1981). The overturning is usually assumed to be a thermohaline, or density-driven, process induced by the slow warming of the ocean's interior by the downward diffusion of heat from the surface. According to this model, diffusive warming makes the interior of the ocean less dense, i.e. it adds buoyancy, which must be balanced at steady state by the upwelling and continuous replacement of cold bottom waters. The formation of exceptionally dense bottom waters is known to occur in the Norwegian and Greenland Seas north of Iceland and in the Weddell and Ross Seas adjacent to Antarctica. Low- and high-latitude thermohaline processes are assumed to be coupled by an equator-to-pole overturning of the ocean's waters in which cold bottom waters flow away from the high-latitude source regions while warm upper ocean waters flow poleward.

The thermohaline overturning in the Atlantic Ocean is a special case in that most of the upwelling associated with deep water formation in the Atlantic occurs in other basins. The Atlantic exports deep water to the Indian and Pacific Oceans where the exported deep water is thought to upwell and return to the Atlantic as part of the Agulhas current south of Africa (Gordon, 1986) or through Drake Passage as part of the Antarctic Circumpolar Current (Rintoul, 1991). This process has been referred to as the ocean's global "conveyor belt" (Broecker, 1987). Broecker and Denton (1989) argue that the driving force for the conveyor is the removal of fresh water from the Atlantic Ocean by an excess of evaporation over precipitation and a net outflow of water vapor from the Atlantic and its drainage basin. The excess evaporation and the outflow of water vapor removes buoyancy from the Atlantic Ocean's surface waters making them saltier. According to this argument, when the atmosphere extracts heat from the salty surface waters of the Norwegian and Greenland Seas an exceptionally dense water mass is produced that flows southward out of the Atlantic Ocean.

The idea of a dominant density-driven overturning in the ocean is not without its inconsistencies. The rate of formation of deep water in the North Atlantic, 15-20x10^6 $m^3 \cdot s^{-1}$ (Roemmich and Wunsch, 1985; Broecker et al., 1991; Rintoul, 1991), greatly exceeds the rate of formation of deep water around Antarctica, 2-5x10^6 $m^3 \cdot s^{-1}$ in the Weddell Sea (Carmack and Foster, 1975; Weiss et al., 1979), yet the bottom water formed around Antarctica is denser. If the density of the bottom water is a deterministic factor governing the ocean's overturning, one has to wonder why the Atlantic overturning is so large in relation to the overturning induced in the Antarctic. Furthermore, a simple density-driven overturning does not allow for the effects of the earth's rotation (Warren, 1981). On a completely water covered earth the creation of north-south density gradients as described above should give rise to purely *zonal* flows. It is only the existence of boundaries along the eastern and western margins of the ocean's basins that permit geostrophically balanced *meridional* flows of the type that can sustain the thermohaline circulation. In the end, the inflow and outflow from the Atlantic basin is mainly a function of the east-west pressure gradients along the length of the basin.

The Geostrophic Constraint in the Latitude Band of Drake Passage

In a series of general circulation model experiments with idealized geometry, Cox (1989) showed that the water masses of an ocean with no gap between South America and Antarctica are completely dominated by the bottom water produced around Antarctica. When a deep gap is opened between South America and Antarctica, the influence of Antarctic bottom water is greatly weakened with respect to intermediate and deep water masses produced in other parts of the ocean. Our purpose in the present paper is to illustrate how the same constraint which limits bottom water production in the Antarctic might also be invoked to account for the magnitude of overturning in the Atlantic Ocean.

In figure 1, meridional overturning diagrams from Cox's closed-Drake Passage case (hereafter referred to as Experiment 1) and his open-Drake Passage case (Experiment 2) have been reproduced. Surface temperatures and salinities in all of Cox's idealized model runs are restored toward simple one-dimensional fields of the observed, zonally-averaged sea surface temperature and salinity. There is no wind stress applied in Experiments 1 and 2. When Drake Passage is closed (fig. 1a), the extraction of heat and the addition of salt by the high-latitude boundary conditions in the southern hemisphere support a huge overturning of about 50x$10^6$$m^3 \cdot s^{-1}$. When a gap is opened between South America and Antarctica (fig. 1b) the total overturning is reduced to about

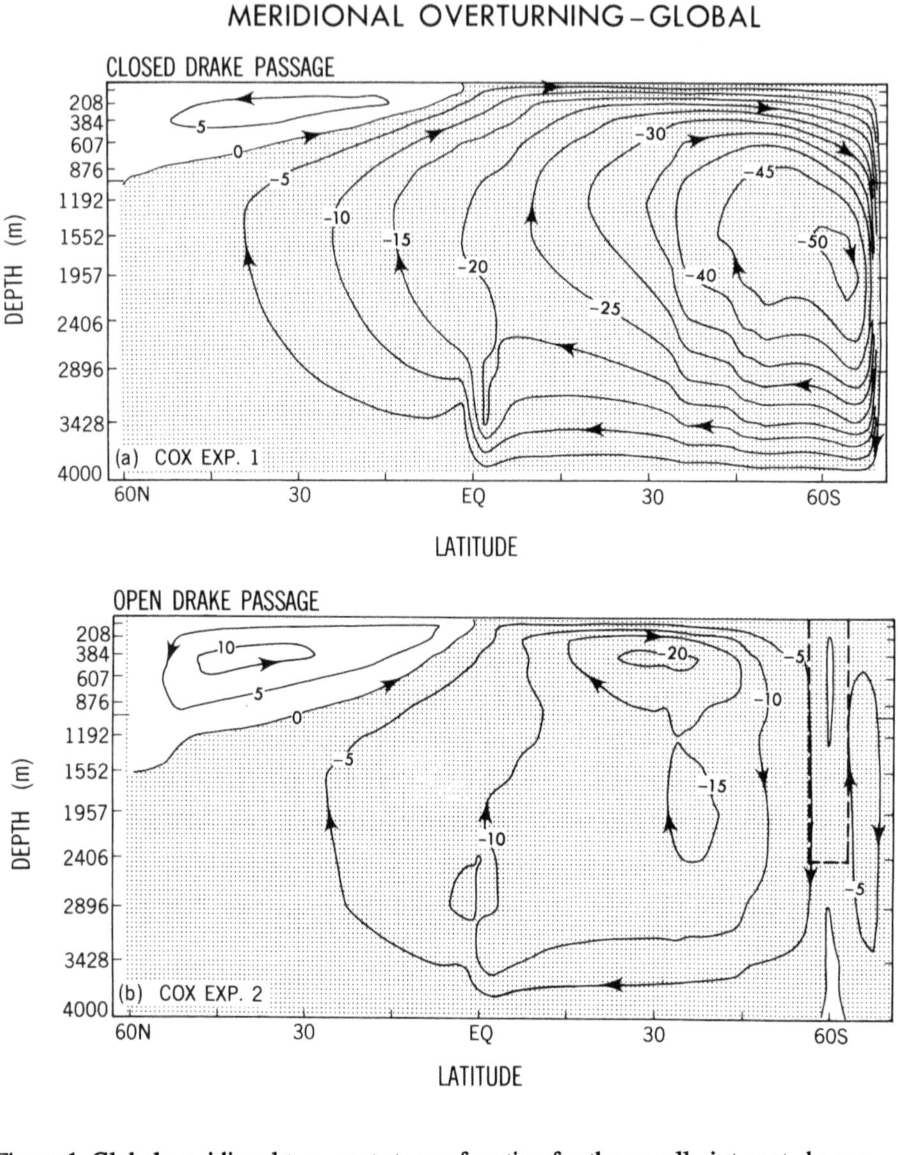

Figure 1 Global meridional transport stream function for the zonally integrated overturning in Experiments 1 (top) and 2 (bottom) described by Cox (1989). A land barrier between South America and Antarctica closes up Drake Passage in Experiment 1. It is removed in Experiment 2. Heavy dashed lines in the bottom panel show the location of the open Drake Passage. There is no wind forcing in these model runs.

$15 \times 10^6\ m^3 \cdot s^{-1}$. Most of the sinking in this case occurs *north* of Drake Passage. A $5 \times 10^6\ m^3 \cdot s^{-1}$ overturning cell south of Drake Passage circulates dense water into the deep ocean near Antarctica, but the Antarctic circulation cell is quite separate from the main overturning cell north of Drake Passage.

Cox's idealized ocean model was constructed on a 2° (latitude) x 2.5° (longitude) model grid with idealized continents. The model has its northern and southern boundaries at 60°N and 70°S, respectively. It has 12 vertical levels and a flat bottom at 4000 m. There are explicit elevated sills connecting South America and Antarctica, Australia/New Guinea and New Zealand, and Australia/New Guinea and S.E. Asia.

The overturning diagrams in figure 1 (and in other figures to follow) are stream functions calculated by constructing two-dimensional fields of $\int v \cdot dx$, where v represents the meridional component of velocity at each model grid point, and x the zonal direction. The value of the stream function is the cumulative sum of $\int v \cdot dx$ at each latitude starting at the surface and summing downward. Since the total northward flow across every latitude circle must be compensated by a southward flow of equal magnitude, the cumulative sum at the bottom is always zero. An overturning diagram like figure 1 is a simple contour plot of the cumulative sum. The zonal limit of the $\int v \cdot dx$ integration can be the whole globe or a single ocean basin.

Cox's results resurrect an argument, first expressed by Gill and Bryan (1971), that a poleward flow of warm, upper ocean water into the Southern Ocean is dynamically constrained to be very weak because there can be little direct flow across the latitude band of Drake Passage to complete a thermohaline circulation. Drake Passage is the one place in the world ocean where no north-south continental barrier exists to support a meridional geostrophic flow in the upper ocean, i.e. there can be no net meridional geostrophic flow across the latitude band of Drake Passage because east-west pressure gradients cannot exist to balance the forces of the earth's rotation. At depths below the sill connecting South America and Antarctica, or at depths below the tops of mid-ocean ridges spanning the latitude band of Drake Passage, east-west pressure gradients can exist. Therefore, geostrophically balanced meridional flows are possible across the latitude band of Drake Passage, but they must be deep. Water mass distributions in the Southern Ocean clearly support the idea of a deep poleward flow and upwelling from depth in the circumpolar belt (Deacon, 1937; Sverdrup et al., 1942; Warren, 1990).

Ageostrophic meridional flows are possible in regions with no continental barriers but may not be important unless they are directly wind forced. Warren (1981) pointed out that a poleward flow in the upper ocean is conceivable on a completely water covered earth as part of a surface/bottom pair of frictionally balanced Ekman layers. In Warren's example, a poleward flow at the surface and an equatorward flow at the bottom are generated as a result of the bottom drag on a zonal geostrophic current. Using a scaling argument, he showed that such flows should be quite inconsequential. Significant meridional transports can be generated in the surface Ekman layer by the stress due to zonal winds. In the latitude band containing Drake Passage, strong winds blow from west to east and give rise to a northward drift at the surface which is spread around the circumpolar belt.

Figure 2a shows the results of Cox's Experiment 3 in which a simple, latitudinally varying wind stress is applied to the model. The zonal wind stress reaches a maximum between 45° and 50°S that is reflected in the overturning diagram by a $>20x10^6\ m^3 \cdot s^{-1}$ surface transport. The surface wind drift is not geostrophic and is not affected by the existence of Drake Passage. The shear in the wind stress to the north of the wind-stress maximum produces a convergence in the surface transport and downwelling. The shear south of the wind-stress maximum produces a divergence in the surface transport and upwelling. The combination of the geostrophic constraint, the northward wind drift, and the upwelling require, in some sense, that deep water be "pulled" poleward into the latitude band of Drake Passage. In figure 2a one sees a shallow return flow north of Drake Passage, where a continental barrier (South America) exists to sustain a poleward geostrophic flow. But within the latitude belt containing Drake Passage, the return flow shifts to much greater depths.

In the first row of model grid points containing Drake Passage, i.e. the first row south of the tip of South America, there is a northward surface drift of $14x10^6\ m^3 \cdot s^{-1}$ imposed by the wind which must be compensated by $14x10^6\ m^3 \cdot s^{-1}$ of poleward flow. Only $8x10^6\ m^3 \cdot s^{-1}$ of the total poleward return flow in Cox's Experiment 3 actually takes place below the depth of the sill in the model's Drake Passage (2406 m), indicating that the poleward return flow in this particular model is not entirely geostrophic. Low-resolution models like these require large amounts of parameterized eddy viscosity to assure smooth numerical solutions. The braking force applied to the Antarctic

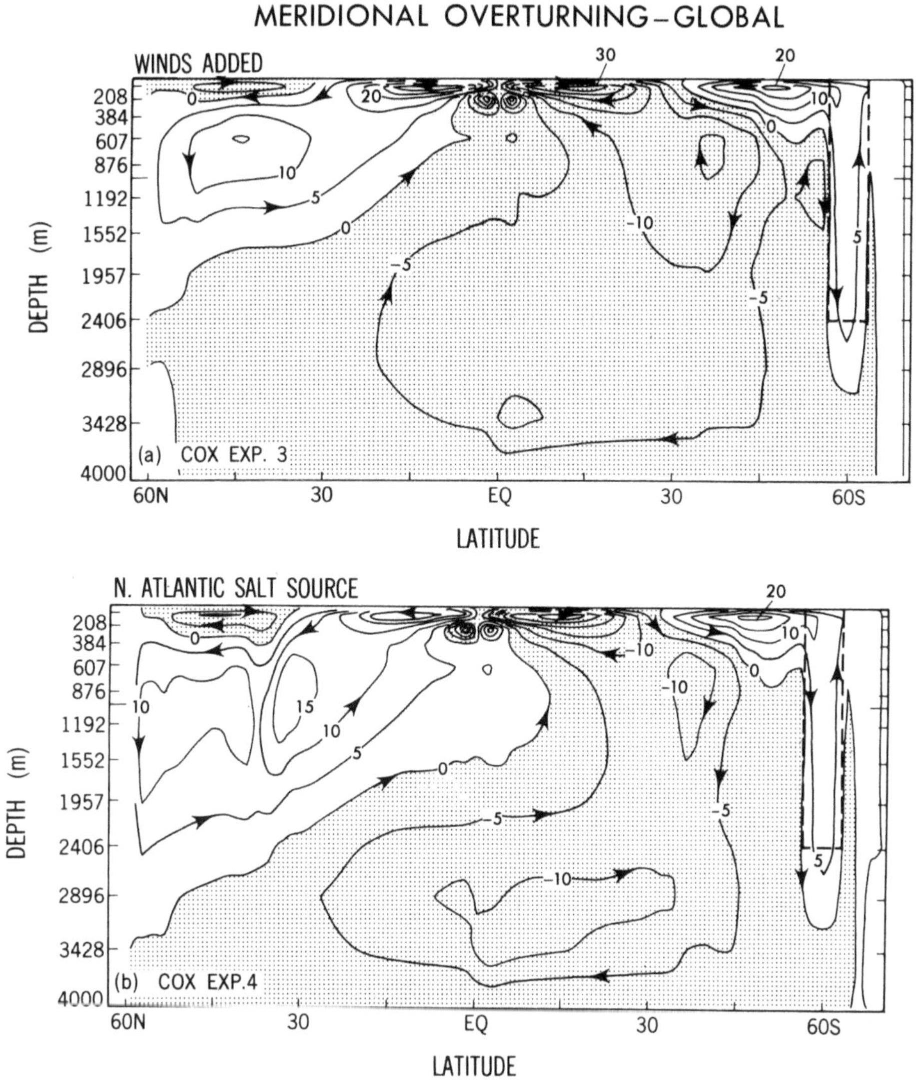

Figure 2 Global meridional transport stream function for the overturning in Experiments 3 (top) and 4 (bottom) described by Cox (1989). Experiments 3 and 4 are identical to Cox's Experiment 2 (figure 1b) except that a zonally uniform wind stress has been added to the forcing. In Experiment 4 a salt source is added to the North Atlantic basin (see text).

Circumpolar Current by lateral friction appears to substitute for the missing pressure gradient force and thereby sustains some net southward flow above the sill (Gill and Bryan, 1971). High-resolution ocean models do not seem to share this feature.

Figure 3a shows the overturning in Cox's Experiment 3 for the Atlantic basin alone. There is no outflow from the Atlantic Ocean in Experiment 3. Because the simplified surface temperature and salinity fields used to force Cox's experiments are uniform around every latitude circle there is no provision for the formation of a cold and salty North Atlantic water mass. In Cox's Experiment 4 a salt source is added to the North Atlantic along the basin's eastern boundary, the region of the Mediterranean outflow. The eastern salt source makes the whole basin saltier including the deep water formation area along the northern boundary. The Atlantic overturning for Experiment 4 is shown in figure 3b. (The global overturning is shown in figure 2b.) By making Atlantic deep water denser Cox induces an outflow of nearly $6 \times 10^6\ m^3 \cdot s^{-1}$ across 30°S. The $6 \times 10^6\ m^3 \cdot s^{-1}$ of Atlantic outflow is very similar in magnitude to the $8 \times 10^6\ m^3 \cdot s^{-1}$ of poleward flow below the depth of the sill in Drake Passage.

There is actually a total of $8 \times 10^6\ m^3 \cdot s^{-1}$ of outflow from the Atlantic in figure 3b, but $2 \times 10^6\ m^3 \cdot s^{-1}$ of the total outflow comes into the Atlantic as bottom water from the south. When the phrase "Atlantic outflow" is used hereafter, we will be referring only to that part of the outflow which has a North Atlantic origin.

New Results

Equilibrium Solutions for Models with Fixed Boundary Conditions

The addition of salt to the North Atlantic in Cox's Experiment 4 makes the deep water formed in the North Atlantic dense enough to flow out of the Atlantic basin. We noted earlier that the northward wind drift in the circumpolar belt coupled with the geostrophic constraint in the latitude band of Drake Passage act to pull deep water poleward into the latitude band of Drake Passage. Cox did not run an experiment in which an Atlantic salt source was added *without* winds, so it is not clear what effect the winds and Drake Passage might have by themselves when a provision exists for the formation of dense water in the North Atlantic. We have taken a second global ocean model,

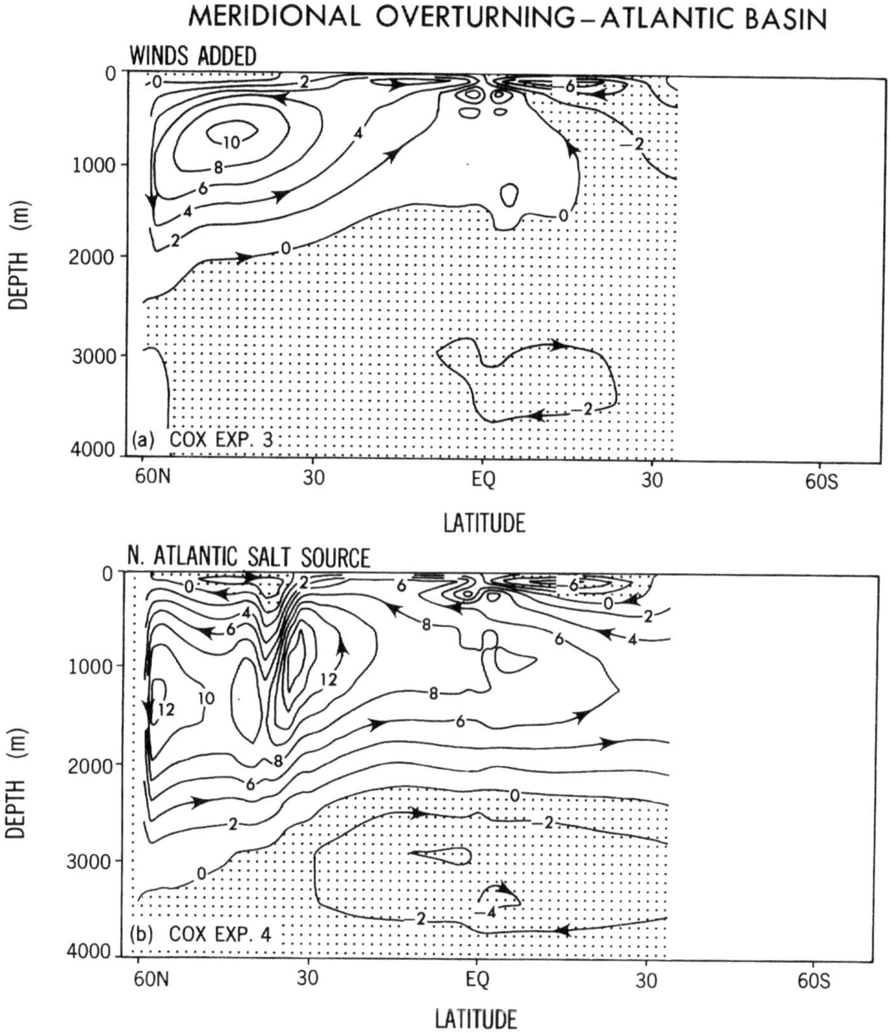

Figure 3 Meridional overturning in the Atlantic basin from Experiments 3 and 4 of Cox (1989).

governed by the same basic equations as the models of Cox (1989), which contains quasi-realistic continental outlines and bottom topography, plus fully two-dimensional, annual-mean surface forcing fields for temperature, salinity (Levitus, 1982), and wind stress (Hellerman and Rosenstein, 1983). In this set of experiments the zonal component of the applied wind stress south of 30°S, τ_{x30S}, is multiplied by factors of 0.5, 1.0, and 1.5 to create variable amounts of northward surface transport in the circumpolar belt. The average zonal wind stress in each row of grid boxes is plotted as a function of latitude in figure 4.

This model will be referred to hereafter as the "realistic geometry model". It is nearly identical to the coarse-resolution (4.5° latitude x 3.75° longitude) model described in Toggweiler et al. (1989a) and is similar in terms of resolution and topography to the ocean model used in the coupled ocean-atmosphere model experiments of

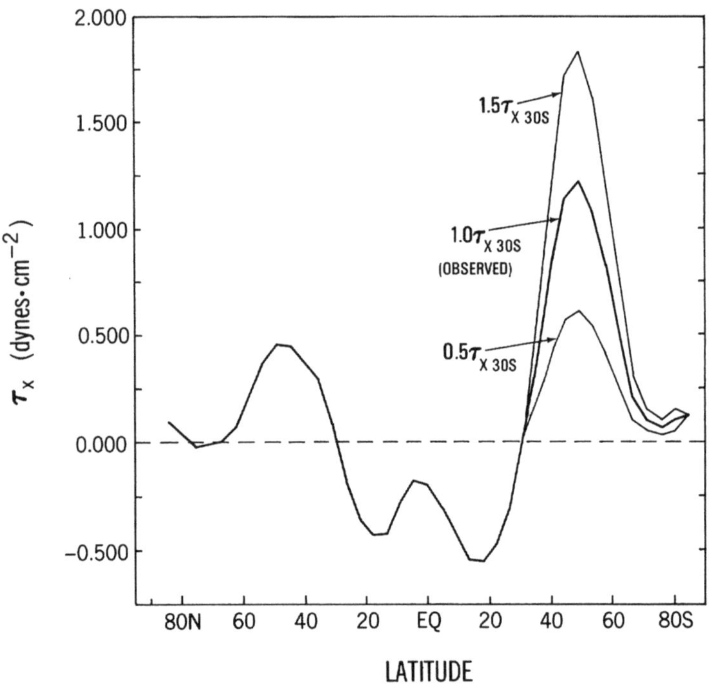

Figure 4 Average zonal component of the wind stress used in the three realistic geometry model experiments described in the text. The zonal component of the gridded wind stress data in Hellerman and Rosenstein (1983) has been multiplied by factors of 0.5, 1.0, and 1.5 in the region south of 30°S.

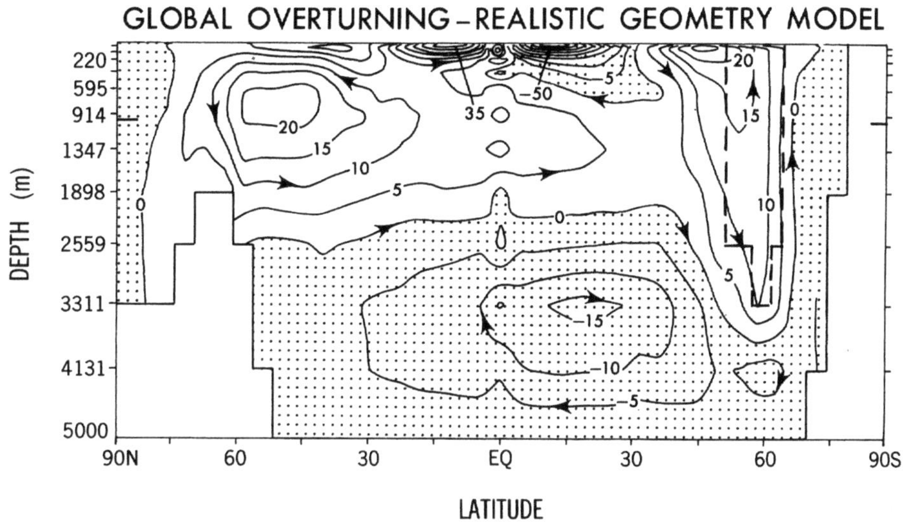

Figure 5 Meridional transport stream function for the global overturning in the standard 1.0·τ_{x30S} "realistic geometry" model. The dashed lines near 60°S show the depth and latitudinal extent of the gap between South America and Antarctica in Drake Passage.

Manabe and Stouffer (1988) and Stouffer et al. (1989). A more detailed account of the differences between the standard model in this paper and the standard model in Toggweiler et al. (1989a) is given in Toggweiler and Samuels (this volume).

Figure 5 shows a diagram of the global overturning in the standard realistic geometry model (1.0·τ_{x30S}) which can be compared with the results from Cox's model. Drake Passage is wider in the coarser realistic geometry model, and the poleward flow at the sill depth is substantially larger, $>10x10^6\ m^3{\cdot}s^{-1}$ vs. $8x10^6\ m^3{\cdot}s^{-1}$ in Cox's Experiments 3 and 4. The Atlantic outflow is also substantially larger ($11.5x10^6\ m^3{\cdot}s^{-1}$ vs. $6x10^6\ m^3{\cdot}s^{-1}$). The larger poleward flow below the sill in Drake Passage and the larger Atlantic outflow are accounted for by the different amounts of wind stress applied to each model at the latitude the tip of South America (see below). In the global stream function in figure 5, the greater level of overturning in the Atlantic basin negates part of the thermohaline overturning in the Indian and Pacific basins and changes the sign of the global stream function integral with respect to Cox's models. The deep closed cell along the bottom will be referred to later as the "southern hemisphere bottom water cell". The realistic geometry model resolves a weak, $5x10^6\, m^3 \cdot s^{-1}$, northward flow of Antarctic bottom water across the latitude band of Drake Passage.

Figure 6 shows overturning diagrams for the overturning within the Atlantic basin for each of the three realistic geometry runs. The inflow and outflow through the South Atlantic amounts to $11.7x10^6\, m^3 \cdot s^{-1}$ in the standard $1.0 \cdot \tau_{x30S}$ case. In the $0.5 \cdot \tau_{xHR}$ and the $1.5 \cdot \tau_{xHR}$ cases the outflow is $6.5x10^6\, m^3 \cdot s^{-1}$ and $13.9x10^6\, m^3 \cdot s^{-1}$, respectively. The relative magnitudes of the overturning reveal that the inflow and outflow in the South Atlantic is positively correlated with the imposed circumpolar wind stress.

The maximum overturning in the North Atlantic is larger in each case than the outflow in the South Atlantic but it also increases when stronger circumpolar winds are imposed. It is noteworthy, perhaps, that the difference between the maximum North Atlantic overturning and the Atlantic outflow remains roughly $10x10^6\, m^3 \cdot s^{-1}$. Are we seeing here a superposition of two meridional overturning cells: one confined to the North Atlantic that doesn't care about southern hemisphere winds, and a full-Atlantic cell which is dominated by southern hemisphere winds?

In figure 7 the Atlantic outflow at 30°S has been plotted as a function of the northward surface transport in the first row of model grid boxes below the tip of South America, i.e. $\int(\tau_x/f)dx_{SA}$, where f is the Coriolis parameter, and the integral refers to a simple sum of τ_x/f around the appropriate latitude circle. Results are included from the $0.5 \cdot \tau_{x30S}$, $1.0 \cdot \tau_{x30S}$, and $1.5 \cdot \tau_{x30S}$ realistic geometry model runs, Cox's Experiments 3 and 4, and both the "sinking" and "non-sinking" solutions from Manabe and Stouffer (1988). The sinking and non-sinking solutions of Manabe and Stouffer represent an identical pair of coupled ocean-atmosphere model solutions started from different initial conditions. The coupled solution initiated with an active overturning in the Atlantic continues to form deep water in the North Atlantic. The solution initiated with an ocean at rest does acquires a low-salinity cap north of 40°N in the North Atlantic and does not overturn.

With the exception of Cox's Experiment 3 and the non-sinking solution of Manabe and Stouffer, the results from all three models follow a linear trend which intercepts the y axis close to the origin. A regression line has been drawn on figure 7 which passes through all the results with an Atlantic outflow. It intercepts the y axis at an overturning

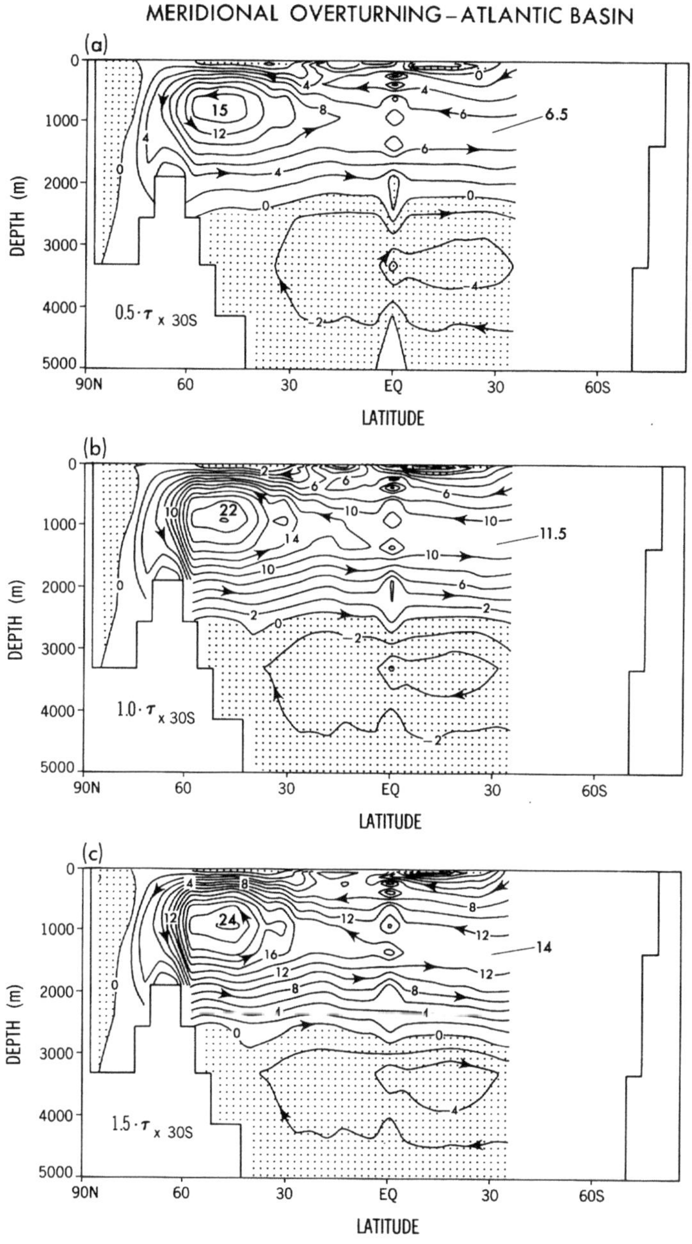

Figure 6 Meridional overturning in the Atlantic basin from the three realistic geometry models forced by the 0.5x, 1.0x, and 1.5x wind stress fields shown in figure 5.

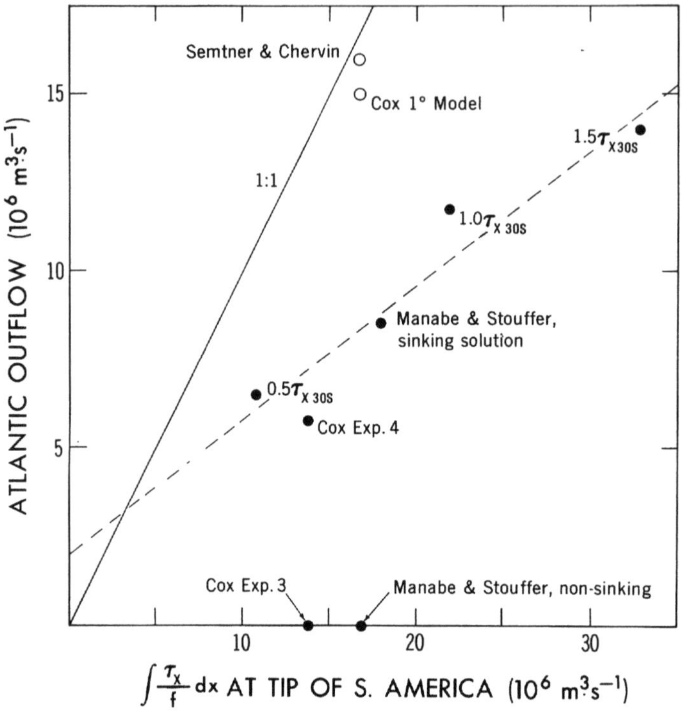

Figure 7 Plot of Atlantic Overturning as a function of the zonally integrated wind drift in the first grid box south of the tip of South America (both in units of 10^6 $m^3 \cdot s^{-1}$ or Sverdrups) in different model simulations. The results labeled '$0.5 \cdot \tau_{x30S}$', '$1.0 \cdot \tau_{x30S}$', and '$1.5 \cdot \tau_{x30S}$', and the results labeled 'Cox Exp. 3' and 'Cox Exp. 4' are described here. The results labeled 'Manabe and Stouffer, sinking' and 'Manabe and Stouffer, non-sinking' represent experiments I and II, respectively from Manabe and Stouffer (1988). The results labeled 'Cox 1° Model' are from an unpublished model run produced by M. Cox at GFDL. The results labeled 'Semtner and Chervin' are from Semtner and Chervin (1992).

of only 2×10^6 $m^3 \cdot s^{-1}$. While the regression line should not be taken too seriously, the trend is clear. This result suggests that an ocean model with no southern hemisphere westerlies would have a very weak outflow from the Atlantic, even though it was able to form dense bottom water in the North Atlantic.

A simple way to think about these results is to consider that the ocean is confronted with a problem of how to compensate for the northward wind drift in the latitude band of Drake Passage when only a deep poleward return flow is allowed. The North Atlantic turns out to be the most convenient place to link up the shallow northward flow with a deep poleward flow because it is the only place where sufficiently dense water is formed anywhere north of Drake Passage. According to this hypothesis, the ocean builds the appropriate east-west pressure gradients along the length of the Atlantic basin, not to maintain a thermohaline circulation, but to maintain this wind-driven overturning.

The Atlantic outflow in the standard $1.0{\cdot}\tau_{x30S}$ model is about twice as large as the outflow in Cox's Experiment 4 and is about the 50% larger than the outflow in Manabe and Stouffer's sinking solution. This is because Experiment 4 and Manabe and Stouffer's models might be considered wind sensitivity experiments in their own right. The topographic resolution in the coarser "realistic geometry" model actually puts the tip of South America at 51°S (vs. 55°S in reality, and 56°S in Cox's idealized 2° model). The tip of South America is thus much closer to the latitude of the wind-stress maximum (47°S). The northward wind drift in the first row of model grid boxes south of the tip of South America is therefore substantially larger, $22 \times 10^6\ m^3{\cdot}s^{-1}$, vs. $14 \times 10^6\ m^3{\cdot}s^{-1}$ in Cox's model. Manabe and Stouffer's ocean model is part of a coupled ocean-atmosphere model that generates its own wind stresses. The coupled model winds are weaker than the winds used to force ocean-only models, and the Atlantic outflow is also weaker.

One should note that the poleward flow below the Drake Passage sill in the standard realistic geometry model (figure 4) is substantially less than the northward drift at the surface. Only about $11 \times 10^6\ m^3{\cdot}s^{-1}$ out of $22 \times 10^6\ m^3{\cdot}s^{-1}$ flows poleward below sill depth. Roughly $11 \times 10^6\ m^3{\cdot}s^{-1}$ flows poleward above the sill. As before, we interpret the poleward flow above the sill to be a consequence of the parameterized viscosity introduced into the model to maintain smooth numerical solutions.

This same analysis has been performed on two high-resolution ocean models, one a 1/2° x 1/2° model described by Semtner and Chervin (1988), the other an unpublished 1° (latitude) x 1.2°(longitude) model run out by M. Cox. High-resolution ocean models operate with much lower levels of parameterized viscosity. The horizontal viscosity coefficient used by Semtner and Chervin is 40x smaller than the value used in Cox's (1989) 2° model. It is 250x smaller than the coefficient used in the 4° wind-sensitivity experiments presented here.

The Atlantic outflow at 30°S and the northward wind drift in the first row of model grid boxes below the tip of South America is plotted for the Semtner and Chervin (1988) and Cox 1° models on figure 7. In both models the Atlantic outflow is very close in magnitude to the wind-induced surface transport. Thus, the high-resolution results plot nearly on the 1:1 line passing through the origin. We do not have the benefit of testing the sensitivity of these models to variations in the wind stress. However, in light of the low-resolution model results in figure 7, the close agreement between the Atlantic outflow and the northward wind drift in Drake Passage suggests that the simple relation-ship between these properties is a fundamental property of ocean models, if not of the real ocean.

The Semtner and Chervin model and the Cox 1° model are very different. The Semtner and Chervin model is forced by annual-mean boundary conditions and is largely diagnostic below 700 m, i.e. the model's density field below 700 m is continuously restored toward the observed density field. The model's pressure gradients in the South Atlantic, which are dictated by the slope of observed density surfaces, force a larger amount of water into and out of the South Atlantic, $16x10^6 m^3 \cdot s^{-1}$, than actually sinks as deep water in the North Atlantic, $8x10^6 m^3 \cdot s^{-1}$. Cox's 1° model, on the other hand, was forced with monthly mean surface boundary conditions and is fully prognostic, i.e. the interior density field is predicted by the model. Cox's model produces over $20x10^6 m^3 \cdot s^{-1}$ of sinking in the North Atlantic while $15x10^6 m^3 \cdot s^{-1}$ flow out in the south. Thus the magnitude of the Atlantic outflow in these two models remains close to the magnitude of the wind drift in Drake Passage in spite of very different amounts of deep water produced in the North Atlantic.

The $16x10^6 m^3 \cdot s^{-1}$ stream line tracing the deep poleward flow in Drake Passage in the Semtner and Chervin model is found at about 2,000 m (Semtner and Chervin, 1992). This is shallower than the actual sill in Drake Passage (~3,000 m) but easily below the depths of other topographic barriers that should be resolved by a 1/2° model (e.g. the Scotia Arc and Kerguelen Plateau). This suggests that the geostrophic constraint pointed out by Gill and Bryan (1971) is being followed very closely in this particular model. To the extent that this is true in the real ocean, one should be able to equate the deep poleward flow below topography in the latitude band of Drake Passage with both the northward wind drift at the surface and the outflow from the Atlantic. Of course, the outflow from the Atlantic and the deep poleward flow in the latitude band of Drake Passage need not be the same water masses. The Atlantic outflow moves eastward after passing south of Africa and becomes part of the Antarctic Circumpolar Current. During its eastward flow it mixes with outflow from the Indian and Pacific before moving poleward.

Solutions for Models with a Flux Boundary Condition for Salinity

Manabe and Stouffer (1988) concluded that the inflow and outflow from the Atlantic is a self-sustaining process. The large-scale overturning makes it possible for dense water to form in the North Atlantic by transporting salty, low-latitude water into a high-latitude region where it can be cooled. With no overturning, the input of fresh water in high latitudes makes high-latitude surface waters too fresh to sink no matter how much they are cooled. The model solutions illustrated thus far represent equilibrium solutions for models run with fixed boundary conditions. With fixed boundary conditions, changes in the large-scale overturning produced by variable winds are not allowed to have any impact on surface salinities. In the next series of experiments the boundary condition for salinity has been changed to a flux condition in order to allow a feedback between high-latitude salinities and the level of overturning.

The restoring of model salinities toward the Levitus (1982) observations in a fixed-boundary-condition model is computed as $\gamma\cdot(S^* - S)$, where γ is an arbitrary constant (in this case 1/120 days^{-1}), S^* is the observed salinity and S is the model's predicted salinity at the beginning of a time step. Multiplying $\gamma\cdot(S^*-S)$ by the depth of the surface layer (Δz_1) and dividing by 35 (the average salt content of seawater in g/kg) produces an implied water flux into or out of each surface grid box which is the model equivalent of the net water flux in the real world (evaporation - precipitation - runoff).

During a special 100-year run of the standard $1.0\cdot\tau_{x30S}$ model the two-dimensional field of $\gamma\cdot(S^* - S)$ was averaged and saved as a field of numerical constants. The salinity restoring term was then removed from the tracer equation and the field of constants was inserted in its place. The model was then run for an additional 1000 years. This model will be referred to as "$1.0\cdot\tau_{x30S}$ (water flux)". Figure 8 shows a map of the equivalent water flux which was applied to the model during the 1000-year integration.

Although the water flux version of $1.0 \cdot \tau_{x30S}$ was no longer tied to the original Levitus (1982) boundary conditions, the circulation remained the same. Two additional experiments were then initiated from the equilibrium state of $1.0 \cdot \tau_{x30S}$ (water flux) in which the southern hemisphere winds south of 30°S were altered by factors of 1.5x and 0.5x as before (figure 4). The water flux remained the same. The new $1.5 \cdot \tau_{x30S}$ (water flux) model was run for 1500 years, while the $0.5 \cdot \tau_{x30S}$ (water flux) model was run for 3000 years.

Figure 9 shows the time variation of the maximum North Atlantic overturning and the Atlantic outflow in the two models. Both the North Atlantic overturning and the Atlantic outflow increase initially in the $1.5 \cdot \tau_{x30S}$ (water flux) model and then remain very steady. Neither the North Atlantic overturning or the Atlantic outflow reach the levels of the $1.5 \cdot \tau_{x30S}$ (fixed B.C.) model within 1500 years but they move in this direction. Surface salinities increase around Antarctica because the water fluxes from the $1.0 \cdot \tau_{x30S}$ (water flux) model are too small to balance the larger Antarctic upwelling induced by stronger circumpolar winds.

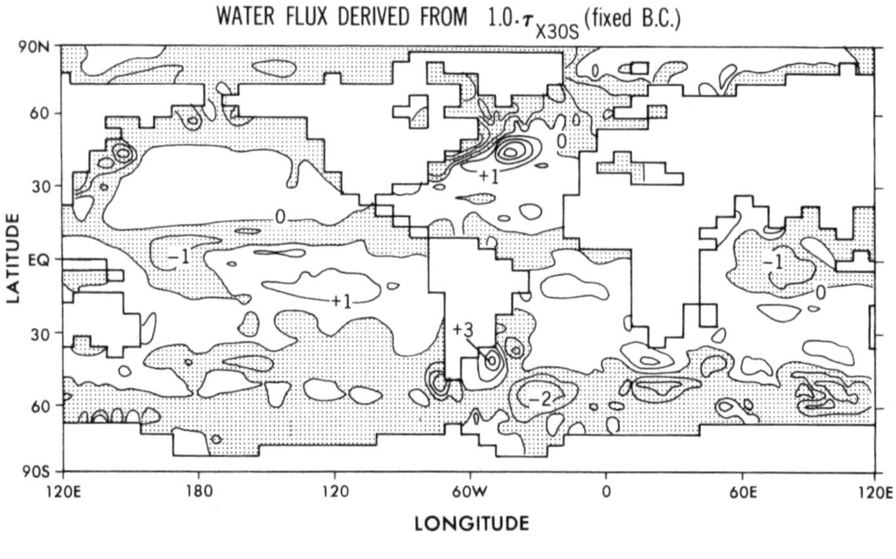

Figure 8 Map of the implied fresh water flux across the ocean's surface as derived from the $1.0 \cdot \tau_{x30S}$ model from the expression , $\gamma \cdot (S^*-S) \cdot \Delta z_1/35$, (units - m/yr). Water flux is into the ocean in stippled (negative) areas.

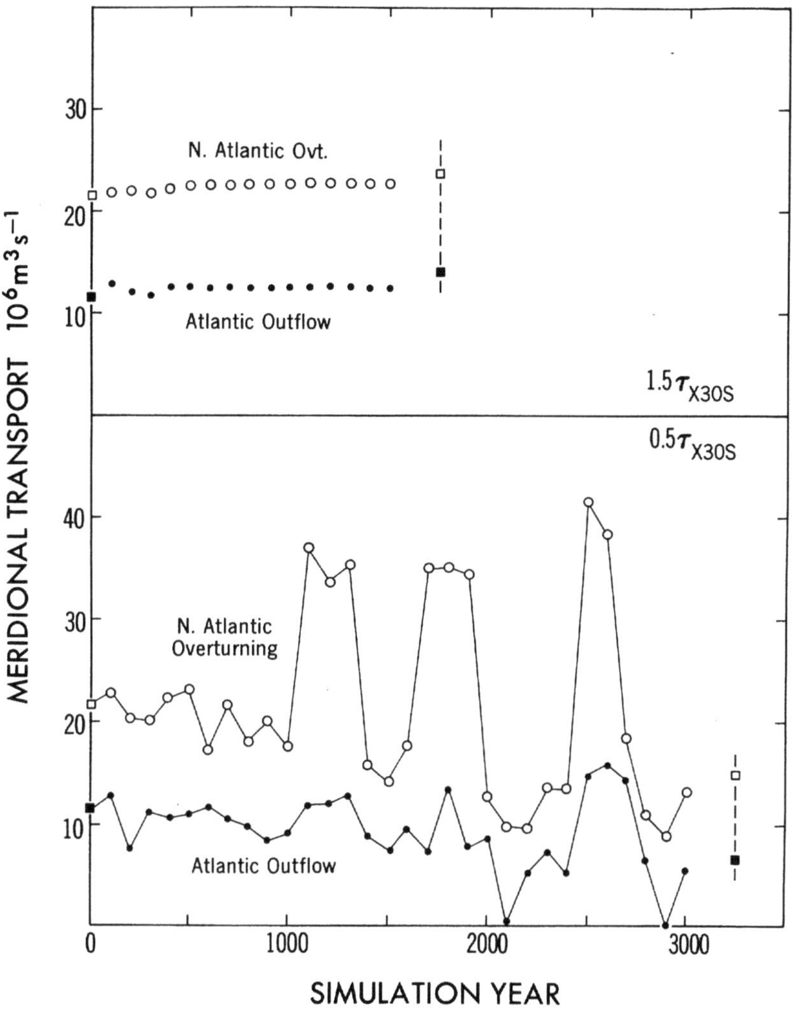

Figure 9 Magnitude of the maximum North Atlantic overturning (open circles) and Atlantic outflow (closed circles) as a function of time in two models run with flux boundary conditions for surface salinity. Results from the $1.5 \cdot \tau_{x30S}$ (water flux) model are shown in the top panel. Results from the $0.5 \cdot \tau_{x30S}$ (water flux) model are shown in the bottom panel. Both models were run with the implied water fluxes shown in figure 8. The square symbols at the beginning of each time series denote the overturning in the $1.0 \cdot \tau_{x30S}$ (water flux) model from which each run was initiated. The square symbols at the end of each run denote the equilibrium overturning in the $1.5 \cdot \tau_{x30S}$ and $0.5 \cdot \tau_{x30S}$ models with restoring boundary conditions for salinity.

The model run with weaker winds, $0.5 \cdot \tau_{x30S}$ (water flux), begins a series of large oscillations in which the North Atlantic overturning jumps up to levels of 30-40x10^6 $m^3 \cdot s^{-1}$ on three separate occasions. The North Atlantic overturning remains at these levels for about 300 years before collapsing back to lower levels. In all three cases, the large excursions in North Atlantic overturning coincide with increases in North Atlantic salinity and increases in the density of North Atlantic Deep Water, but the increases in overturning are mostly confined to the region north of 30°N. The outflow in the South Atlantic seems to fluctuate in phase with the North Atlantic overturning but with a much smaller amplitude. At one point near the end of the run the Atlantic outflow stopped completely and restarted. Over the course of the 3000 year run the average level of overturning and outflow seem to drift downward toward the levels of the $0.5 \cdot \tau_{x30S}$ (fixed B.C.) model. Very large thermohaline fluctuations also develop in the southern hemisphere which seemed to fluctuate independently of the fluctuations in the Atlantic. Tables 1 and 2 list the output of the two water flux experiments at 100 year intervals. Tables 1 and 2 include the information plotted in figure 9 as well as information on the magnitude of overturning in the southern hemisphere bottom water cell and the transport of the Antarctic Circumpolar Current through Drake Passage.

Bryan (1986) showed that models with mixed surface boundary conditions (i.e. flux boundary conditions for salt and fixed boundary conditions for temperature) are subject to a spontaneous "halocline catastrophe" in which the North Atlantic overturning can completely shut down due to an accumulation of fresh water in the high latitudes of the North Atlantic. Bryan's results were generated in a sector model of a single idealized ocean basin. The results in figure 9 show that a halocline catastrophe like that of Bryan (1986) does not occur in a realistic geometry model which includes Drake Passage and a westerly wind stress in that latitude band. A complete shutdown of North Atlantic overturning would probably require a substantial additional fresh water input to the North Atlantic in order to overcome the northward salt transport driven by the winds in the south.

Mikolajewicz and Maier-Reimer (1990), using a global ocean model with mixed boundary conditions, described changes in North Atlantic overturning which are very similar to those generated by the $0.5 \cdot \tau_{x30S}$ (water flux) model. Mikolajewicz and Maier-Reimer applied a variable water flux (randomly variable in space, "white" in

Table 1. Transports in $1.5 \cdot \tau_{x30S}$ (water flux) model
all units $10^6\ m^3 \cdot s^{-1}$

Year	North Atlantic Overturning Cell*	Atlantic Outflow†	Southern Hemisphere B.W. Cell‡	Antarctic Circumpolar Current◊
Initial Value♦	21.5	11.4	-16.6	181
100	21.8	13.0	-20.7	237
200	22.0	12.0	-25.2	240
300	21.8	11.7	-21.6	224
400	22.1	12.5	-19.6	239
500	22.4	12.5	-19.9	241
600	22.5	12.4	-19.3	240
700	22.6	12.4	-18.6	239
800	22.7	12.4	-18.0	240
900	22.7	12.3	-18.2	240
1000	22.8	12.5	-17.8	238
1100	22.9	12.5	-17.6	238
1200	22.8	12.5	-17.6	237
1300	22.8	12.5	-17.3	237
1400	22.7	12.4	-17.3	237
1500	22.7	12.4	-17.3	238

Notes for Tables 1 and 2

* Maximum value of meridional stream function in North Atlantic, ~ 50°N, e.g. figure 6.

† Maximum value of meridional stream function at ~ 1200 m in South Atlantic averaged between the equator and 31°S, e.g. figure 6.

‡ Maximum value of meridional stream function in the bottom water cell in the southern hemisphere centered at ~3500 m. Limit of integration in this case is the whole globe, e.g. figure 4.

◊ Eastward transport through Drake Passage.

♦ Intial transports from $1.0 \cdot \tau_{x30S}$ (water flux) model.

Table 2. Transports in $0.5 \cdot \tau_{x30S}$ (water flux) model
all units $10^6\ m^3 \cdot s^{-1}$

Year	North Atlantic Overturning Cell*	Atlantic Outflow†	Southern Hemisphere B.W. Cell‡	Antarctic Circumpolar Current◊
Initial Value ♦	21.5	11.4	-16.6	181
100	22.9	12.8	-11.7	137
200	20.3	7.5	-14.5	210
300	20.1	11.1	-7.8	135
400	22.2	10.7	-4.7	150
500	23.0	10.9	-3.8	152
600	17.3	11.5	-14.6	193
700	21.4	10.7	-8.4	151
800	18.0	9.8	-10.0	163
900	19.9	8.2	-12.4	151
1000	17.6	9.0	-16.0	152
1100	36.9	11.9	-24.8	188
1200	33.9	12.0	-16.5	205
1300	35.3	12.9	-7.0	135
1400	15.8	8.8	-15.5	203
1500	14.2	7.3	-12.9	182
1600	17.6	9.5	-5.4	146
1700	34.9	7.2	-33.3	185
1800	35.2	13.6	-8.6	144
1900	34.4	7.7	-24.0	178
2000	12.6	8.7	-15.1	159
2100	9.9	0.2	-28.7	164
2200	9.4	5.2	-11.5	199
2300	13.4	7.4	-6.6	167
2400	13.4	5.1	-26.4	168
2500	41.2	14.9	-7.3	149
2600	38.3	15.9	-6.3	155
2700	18.3	14.4	-15.7	169
2800	10.9	6.6	-25.9	179
2900	8.9	< 0	-41.2	181
3000	13.1	5.7	-7.3	161

time) to a realistic geometry GCM and generated a ~300-year thermohaline oscillation which appears to be very similar to the fluctuations in figure 9. They concluded that organized, low-frequency thermohaline fluctuations in response to a high-frequency variation in the water flux are a characteristic feature of the ocean's circulation. The results in figure 9 suggest that the tendency to oscillate exists without any variability in the water flux when $\int(\tau_x/f)dx_{SA}$ is weak. Stronger levels of wind forcing seem to stabilize the circulation against fluctuations. There seems to be a threshold in terms of $\int(\tau_x/f)dx_{SA}$ at which the model circulation changes from an unsteady thermohaline-dominated mode to a steady wind-dominated mode. The threshold value could very well be model dependent.

Mikolajewicz and Maier-Reimer (1990) run the Hamburg ocean GCM. It is based on a different grid and differencing scheme, and uses a different set of approximations to the Navier-Stokes equation than the models described here. The Mikolajewicz and Maier-Reimer model is also a low-resolution model, but the model grid puts the tip of South America much closer to its actual position. Therefore, $\int(\tau_x/f)dx_{SA}$ should be substantially less strong in the Hamburg GCM for a given wind stress field. In this case both models are forced by the Hellerman and Rosenstein winds. We estimate the magnitude of $\int(\tau_x/f)dx_{SA}$ to be about $14 \times 10^6\ m^3 \cdot s^{-1}$ in Mikolajewicz and Maier-Reimer (1990), vs. $22 \times 10^6\ m^3 \cdot s^{-1}$ in $1.0 \cdot \tau_{x30S}$ and $11 \times 10^6\ m^3 \cdot s^{-1}$ in $0.5 \cdot \tau_{x30S}$. Thus $\int(\tau_x/f)dx_{SA}$ in the Mikolajewicz and Maier-Reimer experiments is closer in magnitude to the weak-wind experiment in figure 9.

Is an ocean with weak circumpolar winds inherently unstable with respect to thermohaline fluctuations? A second version of the $0.5 \cdot \tau_{x30S}$ (water flux) model was initiated from the equilibrium state of the $0.5 \cdot \tau_{x30S}$ (fixed B.C.) run. This time the water flux field was saved from the $0.5 \cdot \tau_{x30S}$ (fixed B.C.) run, and the model's surface salinities were initially in balance with respect to the circulation and the water flux field. After 1000 years, the model circulation had *not* begun to fluctuate. While this result indicates that the equilibrium state of this particular model is not inherently unstable with regard to mixed boundary conditions, it is conceivable that fluctuations might develop if the model were run out longer. A second sensitivity experiment was run in which the implied water fluxes derived from $0.5 \cdot \tau_{x30S}$ (fixed B.C.) were applied to the $1.0 \cdot \tau_{x30S}$ (water flux) model. After 1000 years, the overturning generated by this model was different from the original, but the solution did not fluctuate. Thus the application of a water flux field which is out of equilibrium with the initial condition of the model is not enough to produce fluctuations when the wind forcing is stronger.

Discussion

The simple experiment performed by Cox (1989) (figure 1) shows why rather small quantities of Antarctic Bottom Water are being formed in the modern ocean. A vigorous thermohaline circulation in the southern hemisphere is prohibited by the fact that a direct poleward flow of upper ocean water cannot exist across the latitude band of Drake Passage. As a result, warm upper ocean water cannot be transported very easily down to Antarctica where the cooling of surface water to the freezing point is possible. The actual poleward flow associated with bottom water formation consists instead of a relatively small flow of ~2°C water that flows poleward beneath the ridges and sills that span this latitude band, where east-west pressure gradients can exist to balance the force of the earth's rotation.

The Drake Passage constraint also helps explain why large quantities of North Atlantic Deep Water are being formed in the modern ocean. The existence of Drake Passage not only inhibits poleward flows in the upper ocean, but the prevailing wind stress in this latitude band pushes ~20×10^6 $m^3 \cdot s^{-1}$ in the other direction. The northward wind drift is fed by water which is pumped up to the surface by the shear in the wind stress field. According to the Drake Passage constraint, the shear-induced upwelling can only be fed by a deep poleward flow below the ridges and sills spanning the latitude band of Drake Passage. Models faced with this constraint tie the northward wind drift in the circumpolar belt with cooling and sinking in the North Atlantic. Given that the northward flow at the surface and the poleward flow at depth cannot be joined without the northward flow becoming more dense, one can understand the model's "logic" in predicting such a closure. The pertinent question in the real world is whether the wind's efforts to push South Atlantic water northward can actually cause more cooling and deep water formation to take place in the North Atlantic. In the model results described here, stronger circumpolar winds cause more deep water to form because they force warm thermocline water to encounter cooler fixed boundary conditions in the North

Atlantic. It is not obvious that this has to happen in the real world.

For a given level of $\int(\tau_x/f)dx_{SA}$, there is much less inflow and outflow through the Atlantic in the low-resolution models in figure 7 than in the two high-resolution models. It is interesting to ask what happens to the volume of water associated with the northward wind drift which does not flow northward into the Atlantic. Basically, the fraction of $\int(\tau_x/f)dx_{SA}$ which does not flow northward downwells within the circumpolar belt and completes a local short-circuit back into the zone of upwelling. This feature is readily identified in the stream function diagram in figure 4 by the closed 5, 10, and $15 \times 10^6\ m^3 \cdot s^{-1}$ contours trending downward and poleward between 35° and 60°S. If, in fact, all the water raised to the surface south of the tip of South America were to move northward into the Atlantic Ocean, then the number of contours defining the closed overturning cell would be reduced. This is largely what one sees in the Semtner and Chervin model stream function. Flow contours tracking the water upwelled by the wind in the latitude band of Drake Passage, i.e. the $0\text{-}16 \times 10^6 m^3 \cdot s^{-1}$ contours, all turn northward in the region north of 40°S, while only the contours tracking the wind-induced flow initiated north of Drake Passage, i.e. the $18\text{-}30 \times 10^6\ m^3 \cdot s^{-1}$ contours, close locally (Semtner and Chervin, 1992).

In low-resolution models the downwelling north of the wind-stress maximum carries water with subantarctic surface properties below 1000 m as it tries to connect the equatorward wind drift with the deep poleward flow. This makes intermediate-depth properties in this region anomalously warm and salty (Toggweiler et al., 1989b) and makes model isotherms stand nearly vertical in meridional section. Some of the flow associated with the downwelling crosses isotherms and isopycnals indicating that heat is probably being removed from the downward branch of the overturning cell and is being added to the upward branch by horizontal mixing, which in this environment is operating in a nearly diapycnal mode. If, on the other hand, the entire mass of $\int(\tau_x/f)dx_{SA}$ were to move northward into the Atlantic, as in the high-resolution models, then the removal of heat from the downward branch of the overturning would take place in the North Atlantic, where the process is more physically acceptable.

The conclusion presented here is based largely on empirical model results. If the inflow and outflow from the Atlantic Ocean is functionally related to $\int(\tau_x/f)dx_{SA}$ in the real world, this relationship holds because the poleward flow crossing the latitude band of Drake Passage is forced to be deep by the geostrophic constraint operating in this latitude band. Even in a high-resolution configuration, ocean models are known to have serious problems with regard to the dynamics of the Antarctic Circumpolar Current and the flow through Drake Passage. Given the complex dynamics of this region and the importance of eddies in the overall dynamic balance, it is by no means certain that the poleward flow in the real world is geostrophically balanced to the degree that it is in the Semtner and Chervin and the Cox 1° models. This needs to be looked at more carefully. After all, the atmosphere generates significant meridional overturning cells (e.g. the Hadley Cells) without the benefit of a strict geostrophic balance.

The conclusions of this paper must also be tempered by the fact that the actual magnitude of the time mean wind stress at the tip of South America is not known very well. The Hellerman and Rosenstein (1983) data set used to force all the models referred to in this paper (except for those of Manabe and Stouffer, 1988) puts the annual-mean, zonally averaged wind stress at the latitude of the tip of South America at 0.95 $dynes \cdot cm^{-2}$, which forces a northward wind drift of $16.8 x 10^6\ m^3 \cdot s^{-1}$ when integrated around the world. This number happens to be in very good agreement with the magnitude of outflow from the South Atlantic Ocean (Broecker et al., 1991; Rintoul, 1991), but the agreement could be a coincidence. More recent wind data sets compiled by Trenberth et al. (1990) and Chelton et al. (1990) suggest that the wind stress at this latitude should be substantially higher, $30 x 10^6\ m^3 \cdot s^{-1}$ according to Trenberth et al. The results from the water flux experiments in figure 9 indicate that the magnitude of $\int(\tau_x/f)dx_{SA}$ could be rather critical in terms whether or not a particular model's circulation is dominated by an unsteady thermohaline mode or a steady wind-dominated mode.

Conclusions

The model results presented in this paper suggest that the strength of the inflow and outflow through the South Atlantic is determined more by the winds in the southern hemisphere than by the production of dense water in the North Atlantic. According to this interpretation, the inflow and outflow from the Atlantic represents an attempt by the ocean to balance the northward wind drift in the latitude band containing Drake Passage with the requirement for a deep poleward return flow below the topography in this latitude band. The combination of wind forcing in the southern hemisphere and the constraints on geostrophy imposed by Drake Passage provide a very simple and economical explanation for why there is so much more North Atlantic Deep Water formed in the modern ocean than Antarctic Bottom Water, even though AABW is more dense. In ocean models with fixed boundary conditions, we find that the inflow and outflow from the Atlantic Ocean scales with the magnitude of $\int(\tau_x/f)dx$ at the latitude of the first model grid box below the tip of South America. In models with mixed boundary conditions, we find that low levels of $\int(\tau_x/f)dx_{SA}$ lead to the development of large thermohaline fluctuations. Mixed-boundary-condition models forced with higher levels of $\int(\tau_x/f)dx_{SA}$, on the other hand, seem particularly stable against fluctuations. Low-resolution models seem to divert a significant amount of their potential Atlantic overturning into a spurious downwelling within the circumpolar region.

Acknowledgements

This paper could not have been written without the pioneering work of Michael Cox and the illustrations provided by his idealized model experiments (Cox, 1989). Cox's 1989 paper is the last paper that he wrote, as he died in 1989 after a long bout with cancer. Before he died, Cox was planning to write a second paper on the circulation in his idealized model experiments which would have undoubtedly included the figures we have made use of here (figures 1, 2, and 3). Cox's work was clearly moving in a direction that would have caused him to confront the same issues that we have raised in this paper. In following up on the work in his final paper, we have clearly benefited from his insights and his model experience.

We would like to acknowledge Kirk Bryan, not only for his review of this manuscript, but also for his support and his encouragement as head of the GFDL Ocean group. Suki Manabe deserves special mention for his suggestion that we run the water flux experiments. We would like to thank Bruce Warren for a particularly thorough unsolicited review, and would also like to acknowledge the suggestions of Arnold Gordon, Isidoro Orlanski, Jorge Sarmiento, and Jerry Mahlman. We are also pleased to acknowledge the assistance of Phil Tunison, Jeff Varanyak, Cathy Raphael, and Wendy Marshall, who helped in preparing the camera-ready copy figures and tables.

References

Broecker W S (1987) The biggest chill. Natural History 97:74-82.

Broecker W S, and G H Denton (1989) The role of ocean-atmosphere reorganizations in glacial cycles. Geochim Cosmochim Acta 53:2465-2501.

Broecker W S, A Virgilio, and T H Peng (1991) Radiocarbon age of waters in the deep Atlantic revisited. Geophys Res Lett 18:1-3.

Bryan F (1986) High latitude salinity effects and interhemispheric thermohaline circulations. Nature 323:301-304.

Carmack E C, and T D Foster (1975) On the flow of water out of the Weddell Sea. Deep-Sea Res 22:711-724.

Chelton D B, A M Mestas-Nuñez, and M H Freilich (1990) Global wind stress and Sverdrup circulation from the Seasat Scatterometer. J Phys Oceanogr 20:1175-1205.

Cox M D (1989) An idealized model of the world ocean Part 1: the global-scale water masses. J Phys Oceanogr 19:1730-1752.

Deacon G E R (1937) The hydrology of the Southern Ocean. Discovery Rep 15:1-124, plates I-XLIV.

Gill A E, and K Bryan (1971) Effects of geometry on the circulation of a three-dimensional southern hemisphere ocean model. Deep-Sea Res 18:685-721.

Gordon A L (1986) Interocean exchange of thermocline water. J Geophys Res 91:5037-5046.

Hellerman S, and M Rosenstein (1983) Normal monthly wind stress over the world ocean with error estimates. J Phys Oceanogr 13:1093-1104.

Levitus S (1982) Climatological Atlas of the World Ocean. NOAA Prof Paper 13 US Government Printing Office, Washington DC.

Manabe S, and R J Stouffer (1988) Two stable equilibria of a coupled ocean-atmosphere model. J Climate 1:841-866.

Mikolajewicz U, and E Maier-Reimer (1990) Internal secular variability in an ocean general circulation model. Max-Plank-Institut für Meteorologie Report No. 50 Hamburg FRG 20 pp.

Rintoul S R (1991) South Atlantic interbasin exchange. J Geophys Res 96:2675-2692.

Roemmich D, and C Wunsch (1985) Two transatlantic sections: meridional circulation and heat flux in the subtropical North Atlantic Ocean. Deep-Sea Res 32A:619-664.

Semtner A J, and R M Chervin (1988) A simulation of the global ocean circulation with resolved eddies. J Geophys Res 93:15502-15522.

Semtner A J, and R M Chervin (1992) Ocean general circulation from a global eddy-resolving model, J Geophys Res 97:5493-5550.

Stommel H (1958) The abyssal circulation. Deep-Sea Res 5:80-82.

Sverdrup H U, M W Johnson, and R H Fleming (1942) The Oceans. Prentice Hall Englewood Cliffs NJ 1087 pp.

Stouffer R J, S Manabe, and K Bryan (1989) Interhemispheric asymmetry in climate response to a gradual increase of atmospheric CO_2. Nature 342:660-662.

Toggweiler J R, K Dixon, and K Bryan (1989a) Simulations of radiocarbon in a coarse-resolution world ocean model 1. Steady state Prebomb distributions. J Geophys Res 94:8217-8242.

Toggweiler J R, K Dixon, and K Bryan (1989b) Simulations of radiocarbon in a coarse-resolution world ocean model 2. Distributions of bomb-produced carbon 14. J Geophys Res 94:8243-8264.

Toggweiler J R, and B. Samuels (1992) New radiocarbon constraints on the upwelling of abyssal water to the ocean's surface. This volume.

Trenberth K E, W G Large, and J G Olson (1990) The mean annual cycle in global ocean wind stress. J Phys Oceanogr 20:1742-1760.

Warren B A (1981) Deep circulation of the world ocean. In: B A Warren and C Wunsch (eds) Evolution of Physical Oceanography Scientific Surveys in Honor of Henry Stommel. MIT Press, Cambridge, MA, pp 6-41.

Warren B A (1990) Suppression of deep oxygen concentrations by Drake Passage. Deep-Sea Res 37A:1899-1907.

Weiss R F, H G Ostlund, and H Craig (1979) Geochemical studies of the Weddell Sea. Deep-Sea Res 26A:1093-1120.

New Radiocarbon Constraints on the Upwelling of Abyssal Water to the Ocean's Surface

J. R. Toggweiler and B. Samuels
U.S. Department of Commerce/NOAA
Geophysical Fluid Dynamics Laboratory
P.O. Box 308, Princeton University
Princeton, New Jersey 08542 USA

Abstract

The output from seven different ocean model simulations is compared on the basis of the $\Delta^{14}C$ difference between North Pacific deep water and Antarctic surface water. This set of models produces a range of North Pacific-Antarctic $\Delta^{14}C$ differences between -173‰ and -108‰, all but the smallest of which are substantially larger than the actual pre-bomb difference, -80 to -110‰. Predicted values are highly correlated with the quantity of mid-depth water which flows out of the Pacific to the south. A circulation in which most of the Antarctic bottom water flows back out of the basin at mid-depth produces the smallest North Pacific-Antarctic $\Delta^{14}C$ differences, whereas a circulation in which all the inflow of bottom water upwells through the thermocline produces the largest and least realistic differences. According to the models, upwelled abyssal water becomes entrained into the wind-driven convergence of thermocline water toward the equator. When it reaches the surface it spreads to the north and south, producing a $\Delta^{14}C$ minimum along the equator. A detailed analysis of both pre-bomb and post-bomb $\Delta^{14}C$ data indicates that the oldest water in the tropical Pacific is actually found south of the equator and is associated with the upwelling off Peru, not

NATO ASI Series, Vol. I 15
The Global Carbon Cycle
Edited by M. Heimann

the upwelling along the equator. Toggweiler et al. (1991) trace the low-$\Delta^{14}C$ signal in the Peru upwelling to deep water raised to the surface around Antarctica which is pushed northward into the thermocline by circumpolar winds. According to the models, even a small amount of abyssal water upwelling through the thermocline ($\sim 3 \times 10^6 \; m^3 \cdot s^{-1}$) leaves a characteristic signal in the surface $\Delta^{14}C$ distribution which is not observed. One is left with the general conclusion that there is very little upwelling associated with a top-to-bottom thermohaline circulation in the world ocean. Virtually all the upwelling of abyssal water to the ocean's surface occurs around Antarctica where it is mainly wind-forced. The implications of this conclusion for the carbon cycle are discussed.

Introduction

The deep ocean is usually thought to overturn as cold, dense bottom waters formed in high latitudes flow equatorward along the bottom and displace water in the interior which has become warmer and lighter due the downward diffusion of heat in low latitudes. According to the conventional view, e.g. Stommel and Arons (1960), diffusively warmed deep water upwells through the thermocline and then flows poleward in the upper ocean. In 1977 Valentine Worthington published a paper entitled, "The case for near-zero production of Antarctic Bottom Water" (Worthington, 1977), in which he pointed out a contradiction between the distribution of silica in the Pacific and the thermohaline circulation that is generally assumed to characterize the Pacific's overturning.

It is quite obvious that the deepest parts of the Pacific Ocean are filled with bottom water of Antarctic origin. Yet, silica-rich Antarctic Bottom Water cannot upwell through the thermocline in the Pacific Ocean and then flow poleward as silica-poor upper ocean water without leaving behind an enormous accumulation of silicious sediment. The numbers speak for themselves. Worthington calculated that roughly 30×10^{14}g of silica should be deposited on the sea floor north of 28°S in the Pacific basin given the circulation predicted by the Stommel and Arons model. Only $3\text{-}6 \times 10^{14}$g of silica are deposited in marine sediments worldwide, and of this only 0.7×10^{14}g are deposited north of the Antarctic zone in the Pacific basin (DeMaster, 1981).

Worthington also pointed out that silica-poor upper ocean water cannot be converted into silica-rich Antarctic Bottom Water without a substantial addition of silica. Worthington concluded that the silica budget could only be reconciled with a top-to-bottom thermohaline circulation if very little Antarctic Bottom Water is forming at the present time.

At the time Worthington's paper was published, oceanographers were just beginning to realize that most of the northward flow of Antarctic Bottom Water in the Pacific Ocean actually returns poleward as a mid-depth flow (Reid, 1973; Cox, 1975; Fiadeiro, 1982; Wunsch et al., 1983). Some of the earliest results from numerical ocean models predicted the existence of a mid-depth outflow (Cox, 1975; reviewed by Bryan, 1991). Wunsch et al. (1983), using an inverse method to solve for the flow across 28° and 43°S, concluded that virtually all the northward transport in the bottom water returns to the south between 1000 and 3000 m.

A mid-depth outflow provides a very appealing solution for the silica problem: if all the northward flowing bottom water returns poleward below the thermocline, then silica in the bottom water will not come into contact with phytoplankton living at the surface and will not end up as silicious sediment. The strong shear in the circumpolar westerlies south of the Antarctic Circumpolar Current (ACC) is known to raise large quantities of deep water to the surface. This wind-forced upwelling raises water to the surface from the depths of the mid-depth outflow and accounts for the high silica levels in Antarctic surface waters.

While Worthington's enigma seems to have been solved by the discovery of the mid-depth outflow, two problems remain. First, the discovery of the mid-depth outflow does not explain why the mid depth outflow exists. If the ocean's deep circulation is primarily driven by thermohaline processes, the question still remains, "why doesn't Antarctic Bottom Water upwell through the thermocline?" Second, the message contained in Worthington's original paper does not seem to have taken root among oceanographers. Recent schematic constructions of the ocean's general circulation still include the upwelling of abyssal water in the Indian and Pacific Oceans (e.g. Gordon, 1986; Broecker, 1991). Ocean modellers generally seem quite content to see their models upwell large quantities of abyssal water into the thermocline (Maier-Reimer et al., 1990; Stocker and Wright, 1991; Weaver and Sarachik, 1991).

In this paper GEOSECS and coral radiocarbon data will be used along with the output from several different ocean general circulation models to buttress the conclusion that very little abyssal water upwells through the thermocline in the Pacific Ocean. The surface distribution of $\Delta^{14}C$ will be used to show that low $\Delta^{14}C$ levels in the tropical Pacific arise from a subantarctic thermocline source, not a deep water source. Low-$\Delta^{14}C$ levels in the tropical Pacific and the mid-depth outflow from the Pacific can be understood as part of a global wind-driven circulation which extends into the deep ocean due to a unique dynamic constraint which operates in the latitude band of Drake Passage (Toggweiler and Samuels, this volume).

This paper will make a number of references to the pre-bomb and post-bomb radiocarbon simulations described by Toggweiler, Dixon, and Bryan (1989a, b). In order to avoid confusion in referencing this and related works by the same authors, the 1989 papers describing the set up of the $\Delta^{14}C$ model and the pre-bomb simulations will be referred to as "TDBa". The paper describing the post-bomb simulations will be referred to as "TDBb". The Toggweiler and Samuels paper appearing in this volume will be referred to as TSa.

The North Pacific - Antarctic $\Delta^{14}C$ Difference

Figure 1 shows a north-south section of the observed radiocarbon distribution in the Pacific Ocean. The distribution is characterized by a prominent mid-depth minimum in the North Pacific in which the dissolved CO_2 has about 24% less ^{14}C than a sample of atmospheric CO_2 containing the same amount of ^{12}C. At face value, the deficit of ^{14}C in the minimum implies that the water in the minimum has been isolated from the atmosphere for over 2,200 years. In reality, North Pacific deep waters are not nearly so old. Newly formed Antarctic deep waters start out with an initial $\Delta^{14}C$ level that is already quite depleted with respect to atmospheric $\Delta^{14}C$. The evolution of North Pacific $\Delta^{14}C$ from a depleted initial value implies that the ventilation time for North Pacific deep water is actually much shorter. Depending on the initial value one chooses, the ventilation time for the North Pacific $\Delta^{14}C$ minimum is about 700-1000 years.

The deep $\Delta^{14}C$ minimum in the North Pacific would appear to be an important target for ocean general circulation models. It turns out, however, that ocean circulation

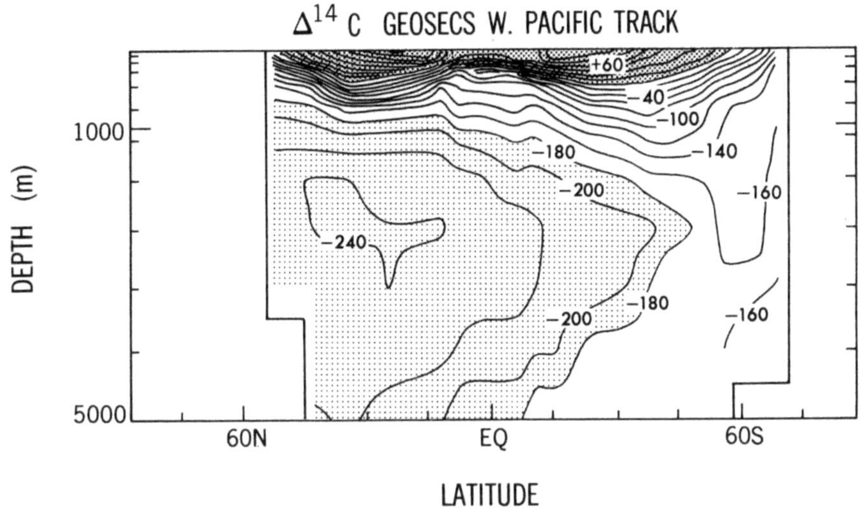

Figure 1 Latitude vs. depth section of $\Delta^{14}C$ (‰) along the western Pacific GEOSECS track (data source - Ostlund et al., 1987). The upper kilometer of the section includes bomb ^{14}C.

models generally have little trouble generating the $\Delta^{14}C$ minimum in the North Pacific. Virtually every model simulation described in the literature which has included $\Delta^{14}C$ as a tracer succeeds in producing a mid-depth $\Delta^{14}C$ minimum with approximately the right magnitude (Fiadeiro, 1982; Maier-Reimer and Hasselmann, 1987; TDBa; Bacastow and Maier-Reimer, 1990). The really challenging problem for ocean models is reproducing the initial value of Antarctic surface waters and the $\Delta^{14}C$ difference between Antarctic surface waters and North Pacific deep waters. Here, ocean models can diverge greatly from each other.

The property of $\Delta^{14}C$ which makes the Antarctic surface value so fundamentally important stems from the fact $^{14}C/^{12}C$ isotope ratios in oceanic surface waters equilibrate with the $^{14}C/^{12}C$ ratio in the atmosphere very slowly. About ten years are required for full equilibration (Broecker and Peng, 1974). This means that the $\Delta^{14}C$ content of Antarctic surface water is far more sensitive to the underlying circulation than it is to the surface forcing. This property makes $\Delta^{14}C$ quite unlike other tracers, especially temperature or the simple gaseous tracers like oxygen or the CFC's. Surface concentrations of these tracers are altered by interactions with the atmosphere on time scales of about one month and reflect the surface forcing far more than the circulation.

In Table 1 we have compiled output from seven different global model simulations run out at GFDL since the appearance of TDBa,b. Each of the models listed in Table 1 is an equilibrium solution derived from a 2000-year run under steady forcing. Each model generates a mid-depth North Pacific $\Delta^{14}C$ minimum ranging from -291 to -252‰. The first column of Table 1 lists the minimum $\Delta^{14}C$ in a zonally averaged section through the basin. For comparison, the oldest $\Delta^{14}C$ values in the GEOSECS data set are about -240 to -250‰. The second column lists the average surface $\Delta^{14}C$ at a latitude of 78°S, which includes only the high latitude regions of the Weddell and Ross Seas. The Antarctic surface values from the seven simulations range from -111 to -155‰. The third column lists the difference between columns 1 and 2. The model values range from -173‰ to -108‰. The model with the oldest North Pacific deep water ($0.5 \cdot \tau_{x30S}$) has the youngest Antarctic surface water, while the model with the oldest Antarctic surface water (Antarctic Salt) is among the models with the youngest

Table 1. North Pacific - Antarctic $\Delta^{14}C$ difference

	‰			$10^6\ m^3 \cdot s^{-1}$		
	$\Delta^{14}C$ minimum	$\Delta^{14}C$		Bottom Water	Upwelling	Mid-Depth
Model	N. Pacific	Ant. Surface	Difference	Inflow	across 2000 m	Outflow
$1.0 \cdot \tau_{x30S}$	-269	-132	-137	7.90	3.85	4.05
3 Ridges	-268	-131	-137	8.20	4.00	4.20
Antarctic Salt	-263	-155	-108	10.25	3.10	7.15
DP 56	-274	-129	-145	8.50	5.30	3.20
$0.5 \cdot \tau_{x30S}$	-291	-118	-173	7.45	6.00	1.45
$1.5 \cdot \tau_{x30S}$	-255	-111	-144	8.90	4.60	4.30
ECMWF	-252	-115	-137	7.90	3.75	4.15

North Pacific deep water. This tendency accounts for the relatively large spread in the North Pacific-Antarctic $\Delta^{14}C$ differences. For the moment, the reader should not be concerned with the descriptive labels applied to each model or with the reasons why these particular models are different.

Pre-bomb $\Delta^{14}C$ of Antarctic Surface Water

The nuclear bomb tests of the 1950s and early 1960s left most of the upper ocean at the time of GEOSECS heavily contaminated with bomb ^{14}C. In order to recover the steady state $\Delta^{14}C$ difference between deep North Pacific water and Antarctic surface water, one is faced with the necessity of reconstructing the pre-bomb $\Delta^{14}C$ level of Antarctic surface water from sparse pre-bomb data and information contained in the deep distribution of $\Delta^{14}C$. The GEOSECS data show that $\Delta^{14}C$ levels in the deep water around Antarctica were fairly uniform, about -158±6‰ (Stuiver et al., 1983). These deep water samples were free of bomb tritium, indicating that there had been no penetration of bomb ^{14}C into Antarctic deep water at this time. Given the long equilibration time for carbon isotopes in surface waters and the relatively rapid replacement of Antarctic surface waters with upwelled deep water (Gordon et al., 1977), one could almost assume that Antarctic surface water in the Weddell and Ross Seas had a pre-bomb $\Delta^{14}C$ equal to that of the deep water, i.e. -158‰. Weiss et al. (1979) determined that newly formed Weddell Sea Bottom Water is only 6% re-equilibrated with atmospheric ^{14}C. They estimated its initial pre-bomb $\Delta^{14}C$ to be -148‰.

A small number of seawater samples were collected for radiocarbon analyses in the South Atlantic and South Pacific during 1957 and 1958. Figure 2 shows the distribution of pre-bomb surface $\Delta^{14}C$ plotted against surface temperature. $\Delta^{14}C$ drops off very strongly with decreasing temperature in water colder than 14°C due to the upwelling of abyssal water south of the ACC. Of all the observations in figure 2 only two samples can be said to have been collected from truly Antarctic waters, i.e. from water with temperatures <2°C. These samples come from 66.5°S in the sector of the Antarctic containing the Ross Sea (Rafter, 1968) and 57.1°S in the Weddell Sea sector (Broecker et al., 1960). These locations are at least 1000 kilometers north of the areas where deep waters form in the Weddell and Ross Seas. Both samples have $\Delta^{14}C$ levels of about -125‰. Figure 2 also includes the surface $\Delta^{14}C$ output from the TDBa model

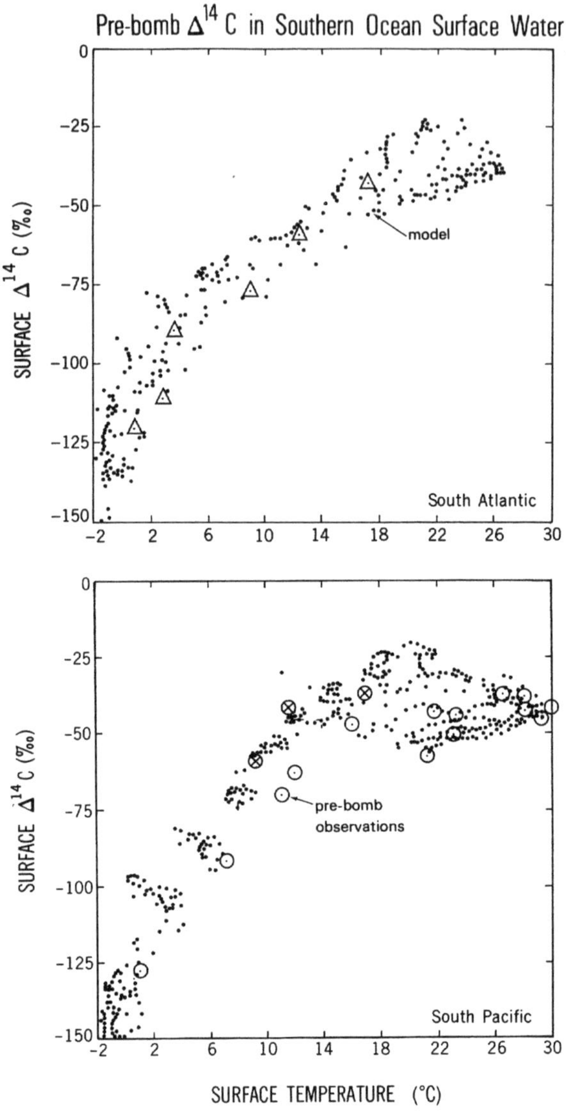

Figure 2 Pre-bomb $\Delta^{14}C$ in South Atlantic (top) and South Pacific (bottom) surface water plotted against surface temperature. Data sources are Broecker et al. (1960), open triangles in top panel; Rafter (1968), open circles, and Bien et al. (1965), open circles with 'x', in bottom panel. Small solid dots represent predicted pre-bomb surface $\Delta^{14}C$ from TDBa. Surface $\Delta^{14}C$ falls off rapidly with latitude and decreasing temperature poleward of the subtropical convergence, ~40°S and 14°C.

plotted against temperature. In general, $\Delta^{14}C$ levels in the TDBa model are slightly higher than the observed $\Delta^{14}C$ levels at a given temperature, but they reproduce the basic downward trend. Model $\Delta^{14}C$ levels go down as low as -150‰ in the highest latitudes of the Weddell and Ross Seas.

Broecker and Olsen (1961) measured the radiocarbon content of the right rear flipper from a seal killed in 1959 in McMurdo Sound (77°S) in the Ross Sea. They found that the seal's flesh had a $\Delta^{14}C$ content of -149‰. This is the only evidence which documents the pre-bomb $\Delta^{14}C$ of surface water in the Ross Sea. While we can offer no proof that the seal's flesh was derived entirely from food eaten at the latitude of McMurdo Sound, its radiocarbon content is consistent with the location where it was killed.

All the information at our disposal yields a fairly consistent picture regarding the pre-bomb $\Delta^{14}C$ of Antarctic surface water in both the Weddell and Ross Seas. We will assume an average value of -150±10‰. Thus, the pre-bomb North Pacific-Antarctic $\Delta^{14}C$ difference should have been about -80 to -110‰. This is substantially smaller than the $\Delta^{14}C$ difference for most of the models in Table 1. Only the results from the Antarctic Salt model fall within the range of observed values. At -150‰, Antarctic surface water can be considered a 50:50 mixture of old, decay-depleted Pacific and Indian Ocean outflow water (~ -200‰) and relatively young North Atlantic Deep Water (~ -100‰).

Patterns of Deep Water Flow and the North Pacific - Antarctic $\Delta^{14}C$ Difference

The right hand side of Table 1 lists the magnitude of the bottom water inflow to the Pacific basin for each of the models along with the amount of the bottom inflow which upwells across 1898 m (the bottom of level 8 in the model). The difference between the total inflow and the upwelling across 1898 m is the mid-depth outflow through the South Pacific which is given in the last column. The total bottom water inflow varies between a high of $10.25 \times 10^6\ m^3 \cdot s^{-1}$ and a low of $7.45 \times 10^6\ m^3 \cdot s^{-1}$. The quantity of upwelling varies between 3.10 and $6.00 \times 10^6\ m^3 \cdot s^{-1}$. The model with the largest inflow at the bottom (Antarctic Salt) has the weakest upwelling and the strongest mid-depth outflow. The model with the weakest inflow ($0.5 \cdot \tau_{x30S}$) has the largest upwelling. None of the models in Table 1 has any flow from the Pacific to the Indian

Ocean through the Indonesian Straits. This restriction is not a realistic feature from an oceanographic point of view, but at least one can be certain that the variation among the models in Table 1 is not due to any variation in this flowthrough.

In figure 3 the North Pacific-Antarctic $\Delta^{14}C$ difference has been plotted as a function of the mid-depth outflow. The results of all seven GFDL models fall along a linear trend which suggests that the North Pacific-Antarctic $\Delta^{14}C$ difference is functionally

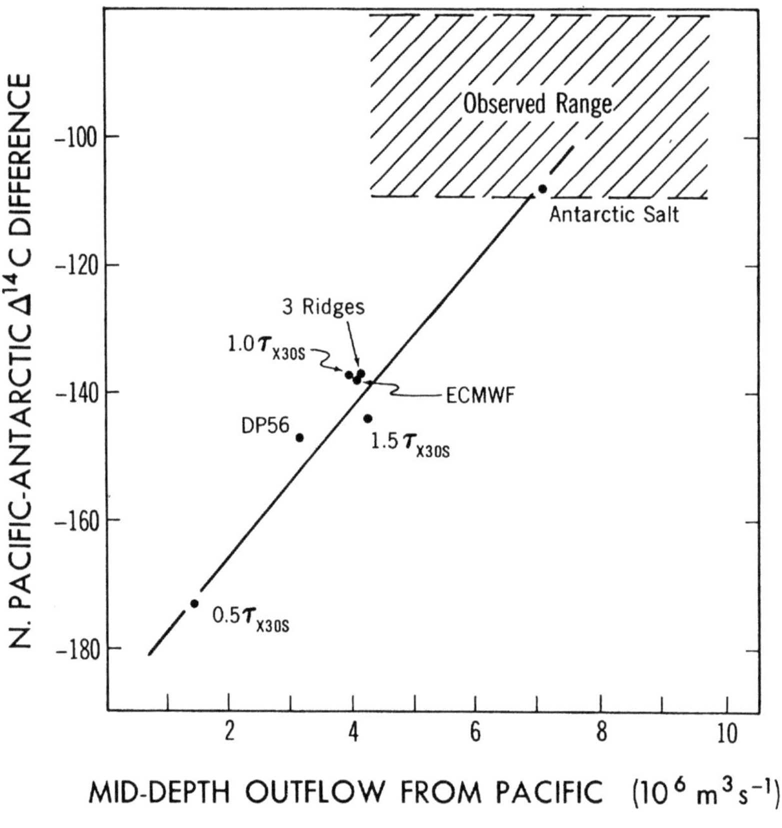

Figure 3 Variation of the $\Delta^{14}C$ difference (‰) between deep North Pacific water and Antarctic surface water as a function of the mid-depth poleward outflow from the South Pacific (10^6 $m^3 \cdot s^{-1}$ or Sverdrups) as predicted by the seven model simulations in Table 1. The actual pre-bomb North Pacific - Antarctic $\Delta^{14}C$ difference was between -80 and -110‰. A description of each of the models is given in Table 2.

tied to the magnitude of the outflow. The model with the strongest mid-depth outflow (Antarctic Salt) has the smallest, and most realistic, $\Delta^{14}C$ difference. This relationship is not hard to understand. A large mid-depth outflow delivers decay-depleted water from the interior of the Pacific to the circumpolar region, where it is picked up by upwelling south of the ACC and is raised to the surface. Water in the mid-depth return flow does not come into contact with the atmosphere until it upwells south of the ACC.

The model in which most of the bottom water inflow upwells into the thermocline ($0.5 \cdot \tau_{x30S}$) delivers its decay-depleted deep water into low-latitude surface layers. The surface return flow remains in continuous contact with high $\Delta^{14}C$ levels in the atmosphere as it flows poleward. By the time the return flow reaches the areas of deep water formation its $\Delta^{14}C$ is much higher. Most importantly, there is much less direct infusion of -200‰ deep Pacific water into the Antarctic upwelling.

The subantarctic zone is also a zone of net fresh water input. A near-surface return flow acquires a lower salinity while it picks up ^{14}C from the atmosphere. The low salinities characteristic of a near-surface return flow are carried into the deep water formation areas. Thus, the $0.5 \cdot \tau_{x30S}$ model has the freshest, and least realistic, bottom water salinities of any of the models. While a direct return flow of upper ocean water across the latitude band of Drake Passage is inhibited by the constraint discussed in TSa, tracer quantities like $\Delta^{14}C$ and salinity can be mixed across this latitude band and can influence surface properties in the Weddell and Ross Seas. A deep return flow, on the other hand, misses the zone of freshening entirely until it upwells to the surface around Antarctica.

Figure 4 shows the zonally averaged distributions of $\Delta^{14}C$ through the Pacific in the Antarctic Salt and $0.5 \cdot \tau_{x30S}$ models. Superficially, the patterns in the two $\Delta^{14}C$ distributions are not that different. Each has a mid-depth minimum in the North Pacific, but the $0.5 \cdot \tau_{x30S}$ model is about 30‰ older. The south-pointing tongue of low-$\Delta^{14}C$ water is more distinct in Antarctic Salt reflecting the stronger mid-depth outflow. Most importantly, surface $\Delta^{14}C$ levels are about 50‰ lower next to Antarctica and about 20‰ lower at the latitude of the tip of South America. The Antarctic salt model has a much more homogenized water column adjacent to Antarctica.

Figure 5 diagrams the zonally integrated overturning in the Pacific basin which produced the radiocarbon distributions in figure 4. The deep circulation in $0.5 \cdot \tau_{x30S}$

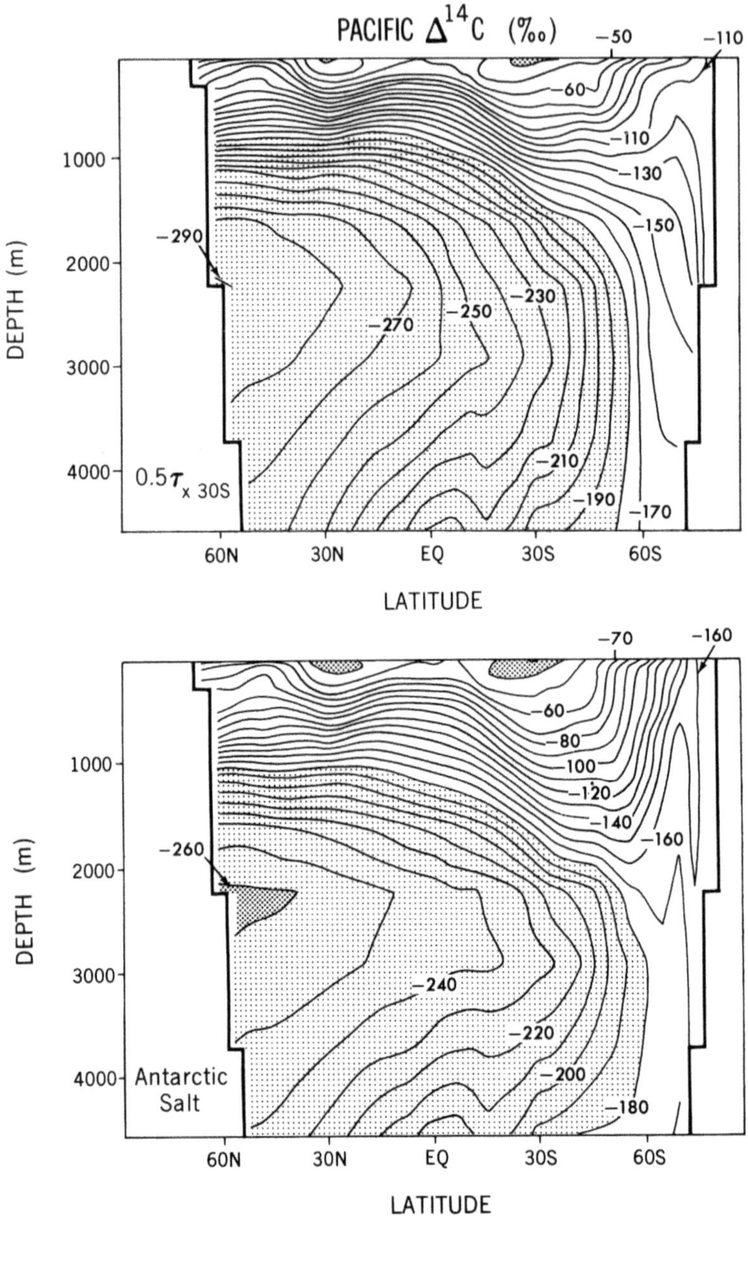

Figure 4 Latitude vs. depth section of the zonally averaged $\Delta^{14}C$ (pre-bomb) in the Pacific basin as predicted by the $0.5{\cdot}\tau_{x30S}$ (top) and Antarctic Salt (bottom) models.

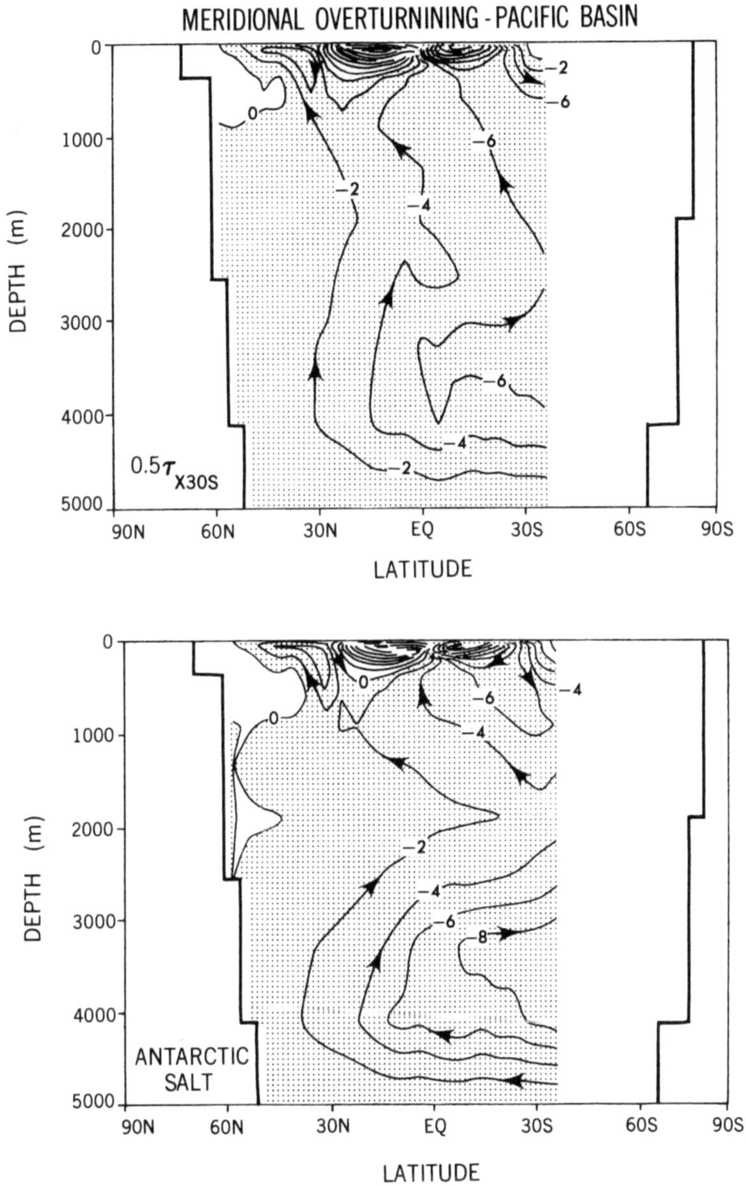

Figure 5 Meridional overturning stream function for the Pacific basin in the $0.5{\cdot}\tau_{x30S}$ (top) and Antarctic Salt (bottom) models. The circulation in the $0.5{\cdot}\tau_{x30S}$ model tends toward a single top-to-bottom overturning, while the circulation in the Antarctic Salt model has a two-cell structure.

consists of a single top-to-bottom circulation cell in which the quantity of water flowing in at the bottom is nearly the same as the amount upwelling into the thermocline and the amount flowing poleward out of the basin in the upper 600 m. The Antarctic salt model shows a clear tendency toward a two-cell deep circulation in which most of the water flowing in at the bottom flows out at mid-depth. Only about $3x10^6\ m^3{\cdot}s^{-1}$ of the nearly $10x10^6\ m^3{\cdot}s^{-1}$ flowing in at the bottom upwells across the thermocline and flows poleward near the surface A second northward inflow of about $3x10^6\ m^3{\cdot}s^{-1}$ develops above 2000 m, which when combined with the $3x10^6\ m^3{\cdot}s^{-1}$of water upwelling from below produces a combined upwelling into the upper ocean of $6x10^6\ m^3{\cdot}s^{-1}$. Because the inflowing water in the upper cell is younger than the water upwelling from below, tropical surface water is about 10‰ younger in Antarctic Salt than its counterpart in $0.5{\cdot}\tau_{x30S}$.

Origin of the Variation in North Pacific - Antarctic $\Delta^{14}C$ Differences

Table 2 gives a brief description of the changes which produced each of the model results in Table 1. Each model is a direct descendent of the "prognostic" model of TDBa, b and the $1.0{\cdot}\tau_{x30S}$ model described in TSa.[1]

The model "DP 56" has an opening between South America and Antarctica which has been shifted south by one grid box so that the tip of S. America is at 56°S instead of 51°S as in the standard model. The opening remains three grid boxes wide. With regard to the hypothesis presented in TSa, the poleward extension of South America in DP 56 should be roughly equivalent to a northward shift in the latitude of the maximum westerlies.

The models "3 Ridges" and "Antarctic Salt" have a bottom topography that includes three mid-ocean ridges in the latitude band of Drake Passage that are meant to simulate the effects of the Scotia Arc, Kerguelen Plateau, and the Macquarie Rise on the deep poleward flow. These ridges are normally not resolved by the bottom topography in a four-degree ocean model. The top of the inserted Scotia Arc and Kerguelen Plateau is at the base of level 7 (1350 m) while the top of the inserted Macquarie Rise is at the base of level 8 (1900 m). As such, the three inserted ridges displace the sill between South America and Antarctica as the highest topography in the latitude band containing Drake Passage and allow a shallower poleward flow than the standard model. The three ridges by themselves appear to have very little effect on radiocarbon (Table 1), but do

Table 2. Description of Models

New model	Source	Changes from source model
$1.0 \cdot \tau_{x30S}$ (standard)	"prognostic" model (Toggweiler et al. 1989a,b)	Indonesian straits closed. Surface salinity restoring constant weakened to 1/120 $days^{-1}$.
$0.5 \cdot \tau_{x30S}$	$1.0 \cdot \tau_{x30S}$	Westerly wind stress south of 30°S weakened by 0.5x.
$1.5 \cdot \tau_{x30S}$	$1.0 \cdot \tau_{x30S}$	Westerly wind stress south of 30°S strengthened by 1.5x.
ECMWF	$1.0 \cdot \tau_{x30S}$	Hellerman and Rosenstein (1983) wind-stress field replaced by ECMWF analysis wind-stress field (Trenberth, 1990).
DP56	$1.0 \cdot \tau_{x30S}$	Opening between S. America and Antarctica shifted south by one grid box.
3 Ridges	$1.0 \cdot \tau_{x30S}$	Three north-south ridges raised in latitude band of Drake Passage to simulate effect of Scotia Arc, Kerguelen Plateau, and Macquarie Rise.
Antarctic Salt	3 Ridges	Observed salinities (S*) in restoring boundary condition increased to 35.0 psu for grid rows at 73°S and 78°S (Weddell and Ross Seas).

enhance the poleward transport of salt from the Atlantic Ocean. The Antarctic Salt model also includes a modification of the salinity boundary condition in which surface salinities in the two southernmost rows of grid boxes are restored toward a value of 35.0‰, making the bottom water formed in the Antarctic more dense.

The two-cell circulation in Antarctic Salt (figure 5) first appears when the wind stress is changed in going from $0.5\cdot\tau_{x30S}$ to $1.0\cdot\tau_{x30S}$. The $1.0\cdot\tau_{x30S}$ model has a stronger mid-depth outflow and weaker upwelling across 2000 m than $0.5\cdot\tau_{x30S}$ even though the bottom water inflow in the two models is roughly the same (Table 1). As pointed out in TSa, any poleward flow feeding the wind-forced upwelling south of the

---

1. All the models described here are different from the standard "prognostic" or "P" models described in TDBa in two respects. A longer time constant is used for restoring surface salinities toward the Levitus observations, 1/120 days^{-1} vs. 1/30 days^{-1} in TDBa. This has a relatively minor effect on $\Delta^{14}C$. Of more importance is the fact that all the models described here have their Indonesian Straits filled with "land" so that there is no flow between the Pacific and Indian Oceans.

The models in TDBa, b have a 1900 m-deep passage through the Indonesian archipelago, but the flow through the passage is governed by an arbitrary condition that the net flow through the strait is zero. This is a common feature in low-resolution GFDL models of this vintage (see also Manabe and Stouffer, 1988). A net flow set to zero, or any specified value, means that Australia is not treated as a true island, in which case the flow around the island and through the strait would be explicitly calculated. In the zero-net-flow mode, the TDBa,b model produces a southward flow in its upper levels through the strait of about $15x10^6$ $m^3\cdot s^{-1}$, which is not unrealistic. But $15x10^6$ $m^3\cdot s^{-1}$ also move northward in the lower layers of the strait. This bi-directional flow acts like a pump which moves young, warm, and salty thermocline water from the Indian Ocean into the deep North Pacific, distorting the radiocarbon field.

Shortly after the results of TDBa,b were published, the model was modified in order to resolve a net flow of water through the Indonesian straits. The modified model failed to produce any westward flow from the Indian to the Atlantic Ocean as in Gordon (1986), or as in the model of Cox (1975), and simply recirculated $25x10^6$ $m^3\cdot s^{-1}$ through the Indian Ocean and into the Circumpolar Current. The explicit calculation also doubled the computational cost of running the model. Both the zero-net-flow calculation and the explicit calculation were deemed unsatisfactory. Simply filling the passage with land afforded the most economical and straightforward way of dealing with this problem.

tip of South America must be below the topography in the latitude band containing Drake Passage. Thus, the enhanced mid-depth outflow in $1.0 \cdot \tau_{x30S}$ appears to be caused by the "pulling" effect of a stronger wind-induced upwelling south of the tip of South America. The effect of making Antarctic bottom water denser in Antarctic Salt intensifies the vertical overturning near Antarctica and intensifies the flow in the deep cell. Roughly the same amount water upwells across 2000 m in Antarctic Salt as in $1.0 \cdot \tau_{x30S}$.

Model streamlines in figure 5 show that water upwelling across 1000 m becomes entrained into the convergent subsurface flow of the wind-driven cells on either side of the equator. The -2, -4, and -6 stream lines between 200 and 1000 m are virtually identical in both $0.5 \cdot \tau_{x30S}$ and Antarctic Salt, even though the source of the upwelled water is different. This suggests that the same dynamic balance is obtained in the upper kilo-meter of both models. This is most likely a balance upwelling and vertical mixing in conformance with classic thermocline theory.

The stronger circumpolar winds forcing the model in Antarctic Salt shift the source of the upwelled water from the bottom to the south in the depth range between 1000 and 2000 m and produce the upper cell of the two-cell structure. As such, the two-cell overturning pattern in the Antarctic Salt model bears some resemblance to the meridional circulation pattern in the South Pacific identified by Wunsch et al. (1983). We are tempted to interpret the northward flow between 1000 and 2000 m as an attempt by the model to generate Antarctic Intermediate Water. The upper level flow in Wunsch et al. (1983) is entirely above 1000 m. This is the depth range one normally associates with the water masses which appear to be moving northward in the real ocean (e.g. Subantarctic Mode Water and Antarctic Intermediate Water). If the northward flow between 1000 and 2000 m in Antarctic Salt is, in fact, a counterpart to the northward flow of intermediate-depth water in the real ocean, it is certainly quite crude. The 1000-2000 m northward flow in the models with the two-cell structure does not carry any sort of low-salinity signature.

It is important to point out that the tendency toward a two-cell circulation is not enhanced when the circumpolar wind stress is increased from $1.0 \cdot \tau_{x30S}$ to $1.5 \cdot \tau_{x30S}$. In the $1.5 \cdot \tau_{x30S}$ model, a $1 \times 10^6 m^3 \cdot s^{-1}$ increase in bottom water inflow results in a 1×10^6 $m^3 \cdot s^{-1}$ increase in abyssal upwelling and almost no increase in mid-depth outflow. The same effect is produced when the stronger ECMWF wind stresses are applied to the model. Increases in circumpolar wind stresses beyond a certain level appear to result mainly in the intensification of a persistent downward flow centered near 45°S. This downward flow combines with the wind-driven upwelling south of the ACC to generate a closed overturning cell within the circumpolar region in global stream function diagrams (see fig. 4 in TSa). In our opinion, the intensification of the local overturning in the $1.5 \cdot \tau_{x30S}$ and ECMWF models at the expense of the mid-depth outflow illustrates a very undesirable tendency in low-resolution models.

The Surface $\Delta^{14}C$ Distribution in the Tropical Pacific

Most of the evidence offered in support of the idea that abyssal water upwells through the thermocline is derived from theoretical considerations or models, not from oceanographic data. There is, however, one substantial piece of evidence which has been invoked to demonstrate that an upwelling of abyssal water actually takes place. Surface waters of the tropical Pacific are known to have lower levels of $\Delta^{14}C$ than the adjacent subtropical areas, and it has been suggested that the low-$\Delta^{14}C$ signal is an expression of deep water upwelling through the tropical thermocline (Broecker et al., 1985). The ocean models in Table 1, all of which include some upwelling of abyssal water, produce a $\Delta^{14}C$ minimum along the equator which, at first glance, seems consistent with the $\Delta^{14}C$ observations. Our purpose in this section is to show that the Pacific $\Delta^{14}C$ observations, when looked at in detail, are actually tracking a very different kind of upwelling.

The data set used by Broecker et al. comes from water samples collected in 1973-74 during the GEOSECS expedition. The GEOSECS samples were collected about ten to fifteen years after the atmospheric nuclear weapons tests of the late 1950s and early 1960s. According to Broecker et al. (1985), the bomb ^{14}C coming into the tropical Pacific from the atmosphere is flushed out the equatorial zone by upwelled deep water and is transported laterally to the subtropical gyres where it accumulates in a downwelling environment. There is no doubt that bomb ^{14}C was flushed out of the equatorial zone. At the time of GEOSECS, the region 10°S and 10°N typically held only half the amount of bomb ^{14}C that should have entered the region from the atmosphere (Broecker et al., 1985).

Thermocline water in the low latitudes of the Pacific Ocean converges toward the equator as part of the shallow overturning driven by low-latitude winds (see figure 5). According to the models, any deep water upwelling into the thermocline between 30°S and 30°N is picked up by this convergent flow and reaches the surface as part of the equatorial divergence. Upon upwelling at the equator, the deep water is carried away from the equator to the north and south spreading its low-$\Delta^{14}C$ signal more-or-less equally between the hemispheres.

Following is a detailed comparison between the $\Delta^{14}C$ distributions predicted by the TDBa, b model and the actual observations as published in Toggweiler et al. (1991). Although the TDBa,b model is not one of the models in Table 1, its radiocarbon distributions are similar to all the models examined here. The TDBa surface $\Delta^{14}C$ distribution in the tropical Pacific is virtually identical to that of Antarctic Salt. In fact, the spatial patterns predicted are very similar for all the models: only the magnitude of the $\Delta^{14}C$ minimum varies with the quantity of upwelled abyssal water. The oldest equatorial surface water in $0.5 \cdot \tau_{x30S}$ is -70‰, compared to -55‰ in Antarctic Salt and in TDBa. The analysis in Toggweiler et al. (1991) begins with the pre-bomb $\Delta^{14}C$ distribution which is assumed to be steady in time.

Figure 6 shows the pre-bomb surface distribution of $\Delta^{14}C$ as revealed by six corals and one shell that grew during the late 1940s and early 1950s. Each data point represents an average of several $\Delta^{14}C$ measurements covering multiple growth bands. It is evident from the data in figure 6 that the lowest pre-bomb $\Delta^{14}C$ levels come from the eastern equatorial Pacific as recorded by the Galapagos coral (-72‰, Druffel, 1981). $\Delta^{14}C$ levels south of the equator (Fiji, -60‰; Tahiti, -62‰) were somewhat lower than those north of the equator (Fanning, -59‰). This distinction, while quite small, is noteworthy because the Fiji and Tahiti locations are quite distant from any upwelling zone, while Fanning Is. is largely within the equatorial up-welling zone. One expects that surface water drifting slowly toward Fiji and Tahiti picks up some ^{14}C from the atmosphere. Thus, the nearly equivalent $\Delta^{14}C$ levels at Fiji, Tahiti, and Fanning actually imply that the water reaching Fiji and Tahiti began its journey with a significantly lower $\Delta^{14}C$ content than the water upwelling along the equator.

$\Delta^{14}C$ levels during the early 1950s, even though they pre-date the major bomb tests, are not directly comparable with the results of a steady-state ocean model.

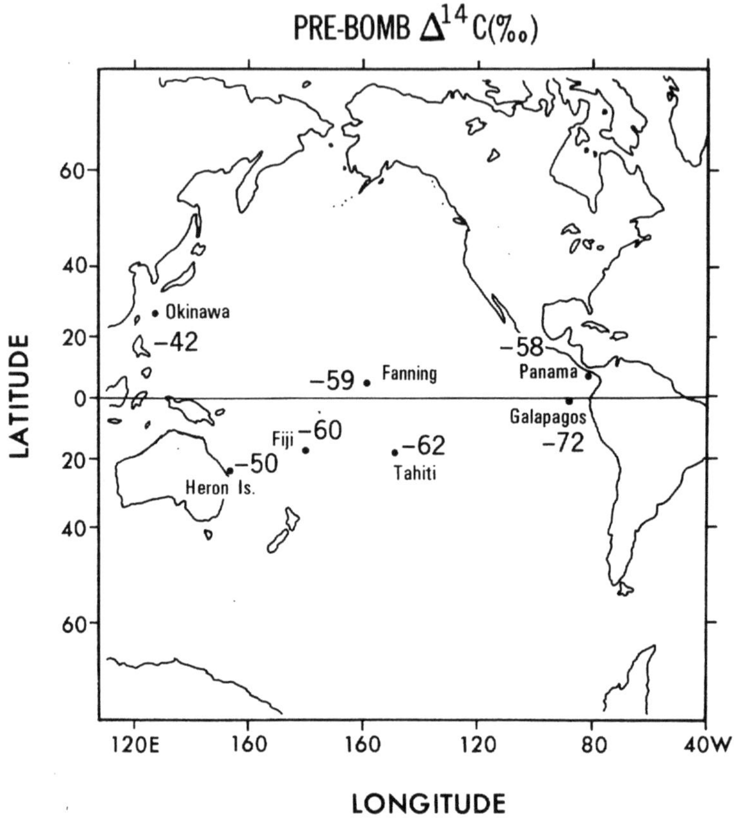

Figure 6 Location and pre-bomb (1950) $\Delta^{14}C$ levels in six Pacific corals and shells. Data sources - Okinawa, Konishi et al., 1982; Fanning and Panama, Druffel, 1987; Galapagos, Druffel, 1981; Heron Is., Druffel (personal comm., 1990); Fiji, Toggweiler et al. 1991; Tahiti, Broecker and Olson, 1961.

Oceanic and atmospheric $\Delta^{14}C$ levels were changing all through the first half of the 20th century due to the burning of fossil fuels. Fossil fuel CO_2 contributes plenty of $^{12}CO_2$ to the atmosphere, but no $^{14}CO_2$. In order to find ocean data representative of a more nearly steady state system one must go further back in time. Four of the records in figure 6 extend back before 1920 when the dilution of oceanic $\Delta^{14}C$ with fossil fuel CO_2 was much less significant. Using these records, Toggweiler et al. (1991) compile a pre-1920 coral and shell data set which is representative of the ocean-atmosphere $\Delta^{14}C$ gradient at that time and is directly comparable with the output of the TDBa model.

In the top panel of figure 7, the pre-1920 coral and shell results from Toggweiler et al. (1991) are laid on top of the predicted TDBa surface $\Delta^{14}C$ distribution. The bottom panel shows data-model differences at the sites of the coral and shell data. The model produces its oldest surface water along the equator between 105 and 135°W. There is a clear $\Delta^{14}C$ minimum over the whole tropical band which is nearly symmetric about the equator. The pre-1920 data are too sparse to indicate whether an equatorial minimum existed or not. They do show, however, that the main difference between the model and the observations is a clear north-south tendency in which the symmetrical model $\Delta^{14}C$ distribution is systematically too old in the northern hemisphere and too young in the southern hemisphere.

If one now looks at the GEOSECS era $\Delta^{14}C$ data in light of the pre-1920 distribution, the same pattern emerges. Figure 8 shows a pair of diagrams like those in figure 7 in which seawater and coral $\Delta^{14}C$ data for the year 1974 are laid over model results. Again, the bottom panel of figure 8 shows data-model differences. There are far more data points and better spatial coverage in 1974 than in the earlier period. The model predicts the existence of a symmetric minimum and puts its oldest values (+20‰) against the eastern margin. Again, the model is systematically too old north of the equator and too young south of the equator. The observations show evidence of a tropical $\Delta^{14}C$ minimum centered on the equator in the central and western Pacific, but not in the east. The most tell-tale observations are from two GEOSECS samples collected along 125°W at 4.5°S (+41‰) and at 14°S (+45‰). $\Delta^{14}C$ levels at these locations were much older than along the equator (+73‰). While the model predicts that the lowest values should be found both north and south of the equator along the Central and South American coasts, the observations show that the 1974 $\Delta^{14}C$ level off Panama at 8°N (+67‰) was much higher than in the Galapagos coral at 1°S (+25‰).

If one looks at another geochemical upwelling tracer with much better spatial coverage, like nitrate for example, one sees the actual source of the low-$\Delta^{14}C$ water in the tropical Pacific labeled quite clearly. Figure 9 shows a map of the annual mean distribution of surface nitrate (Levitus et al. submitted). Maximum nitrate values are found off the coast of Peru, centered between 5° and 7°S at 90°W. A distinct nitrate maximum spreads westward south of the equator until it merges with the equatorial nitrate maximum somewhere near 150°W. The coastal region off the coast of Peru is a

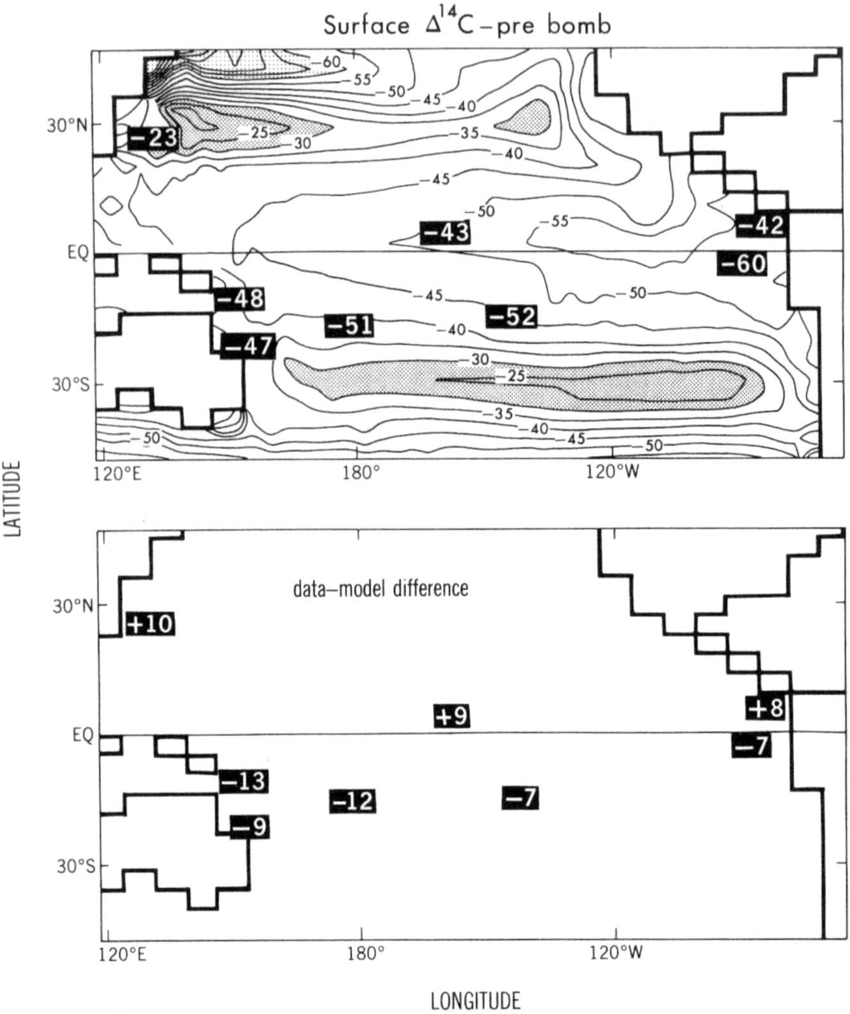

Figure 7 (top panel) Pre-bomb surface $\Delta^{14}C$ distribution (‰) in the tropical Pacific as predicted by the TDBa model overlaid with $\Delta^{14}C$ observations from the coral sites in figure 6. The coral data from figure 6 have been modified to reflect the ocean-atmosphere $\Delta^{14}C$ gradient between 1850 and 1920 (see Table 1 in Toggweiler et al., 1991).

(bottom panel) $\Delta^{14}C$ difference between coral data and the model at the coral sites. The observations are systematically younger than the model in the northern hemisphere and are older than the model in the southern hemisphere. Copied from Toggweiler et al. (1991).

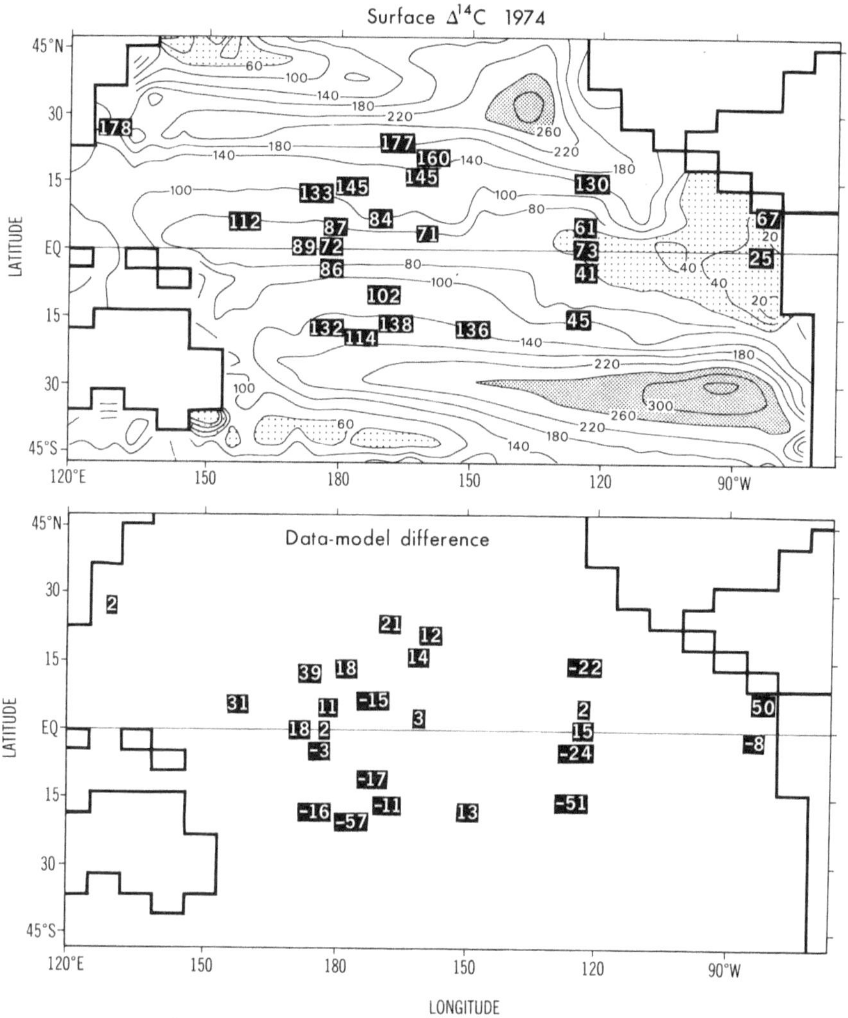

Figure 8 (top panel) Surface $\Delta^{14}C$ distribution (‰) in the tropical Pacific in the year 1974 as predicted by the TDBb model overlaid with GEOSECS seawater measurements (Ostlund et al., 1987) and coral data. Sources for coral data are given in Toggweiler et al. (1991). The model $\Delta^{14}C$ simulation was forced by a 1950-1990 time history of atmospheric $\Delta^{14}C$ measurements which was partitioned into three latitude bands.

(bottom panel) $\Delta^{14}C$ difference between the GEOSECS and coral data and the TDBb model. Copied from Toggweiler et al. (1991).

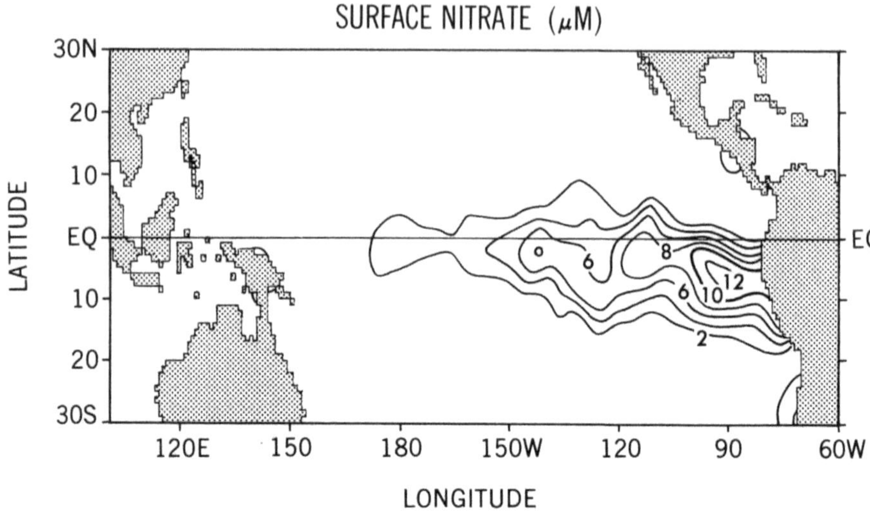

Figure 9 Surface distribution of nitrate (μM) in the tropical Pacific as mapped from data collected at all times of the year. (data source - Levitus et al., submitted).

well known source of unusually cold, nutrient-rich upwelled water.

High-resolution ocean models, which resolve vertical motions close to the equator much better than the low-resolution models discussed here, show that the shallowest parts of the equatorial upwelling system recirculate downwelled surface water back into the equatorial divergence in the central Pacific between 3° and 10°S and 3° and 10°N (Philander et al., 1987). The impression one gets from the complete set of observations in figures 7, 8, and 9 is that the equatorial $\Delta^{14}C$ minimum in the central and western Pacific is best explained by the lateral spreading of water upwelled off Peru and the *subduction* of Peruvian water into the shallow upwelling system in the central Pacific.

Besides its low $\Delta^{14}C$ content, the water upwelled off Peru is characterized by its high salinity (~35 psu). Deep water along the coast of South America, while low in $\Delta^{14}C$ and high in nutrients, is significantly fresher than the upwelled water off Peru and can-not be the source of the Peru upwelling. Toggweiler et al. (1991) show that the salty, high-nutrient water upwelling off Peru has its origin in 11-14°C water from the lower layers of the Equatorial Undercurrent.

The water in the lower part of the Equatorial Undercurrent has a distinctly southern hemisphere source (Tsuchiya et al., 1989). On the basis of its particular $\Delta^{14}C$ content, the 11-14°C water in the lowest part of the Undercurrent and the water upwelled off Peru, are shown to have a high-latitude southern hemisphere source. Toggweiler et al. (1991) determine that these water masses are probably injected into the thermocline as ~8°C Subantarctic Mode Water (SAMW) in the western South Pacific. SAMW is an actively ventilated layer of the lower thermocline which is formed along the northern edge of the Antarctic Circumpolar Current. Here, the wind stress field forces surface water downward, preconditioning the water column for deep convection during the winter.

The water forced downward into the thermocline north of the ACC carries a low-$\Delta^{14}C$ Antarctic signal into the South Pacific thermocline. The low-$\Delta^{14}C$ content of this water is a consequence of the fact that the wind-induced upwelling south of the ACC raises decay-depleted deep water from the interior of the Pacific to the surface. Ironically, the tropical $\Delta^{14}C$ signal comes from the same source, whether it upwells directly from the deep Pacific, as in a thermohaline circulation, or whether it upwells around Antarctica and reaches the surface in the tropical Pacific through the thermocline as part of a wind-dominated circulation.

Constraints on the Pacific Ocean's deep circulation derived from 1.) Worthington's (1977) analysis of the silica budget, 2.) Wunsch et al.'s (1983) analysis of the flow through the South Pacific, and 3.) the analysis presented here of North Pacific-Antarctic $\Delta^{14}C$ differences, all suggest that a minimal amount of abyssal water upwells into the thermocline of the Pacific Ocean. According to the models examined here, even a small amount of abyssal upwelling, when entrained into the equatorial divergence, produces a distinctive symmetrical tropical $\Delta^{14}C$ surface distribution. A detailed examination of equatorial Pacific $\Delta^{14}C$ data does not support the notion that low-$\Delta^{14}C$ water upwells to the surface in this way.

None of the models examined in this paper is capable of transporting subantarctic water through the thermocline into the Peru upwelling. A major stumbling block is the fact that low-resolution models cannot generate an important feature of this flow path, namely a vigorous Equatorial Undercurrent. Therefore, we cannot say with certainty that the two circulation features discussed in this paper are related, i.e. we do not have a model which demonstrates that a wind-dominated circulation with a large mid-depth outflow is an indispensable component of the Peru upwelling. On the other hand, we have shown that the mid-depth outflow in the Antarctic Salt model is part of a two-cell structure in the Pacific's meridional circulation (figure 5). The two-cell structure features a net northward flow of relatively shallow water above the mid-depth outflow. The upper cell develops when the circumpolar wind stress is stronger.

Discussion - Reconfiguring the Global Conveyor

Broecker's (1987, 1991) diagram of the "Global Conveyor" embodies the conventional view about the way in which the deep circulation of the ocean is organized and driven. One of the major components of the Global Conveyor is the upwelling of abyssal water through the thermocline in the Indian and Pacific Oceans. It is almost axiomatic among oceanographers that the Conveyor is a thermohaline heat engine driven by formation of dense water in the North Atlantic and the upwelling of diffusively warmed deep water through the thermoclines in the other ocean basins. The evidence documented in this paper does not support the idea that there is a substantial amount of abyssal water upwelling through the thermocline of the Pacific Ocean. The same arguments might also apply to the Indian Ocean. If these arguments are correct, how does one characterize the global circulation?

One might argue that the *Global* Conveyor is really an *Atlantic* Conveyor in which the upwelling balancing NADW formation occurs around Antarctica and not in the Indian and Pacific basins. This modification of the conveyor belt concept allows for the upward branch of the Conveyor to be wind driven, not buoyancy driven. Toggweiler and Samuels (TSa) go further and show evidence that the sinking in the North Atlantic might itself have a critical wind-forced component. They suggest that the overturning in the Atlantic Ocean may represent the closure of a wind-driven circulation initiated in the latitude band of Drake Passage.

The existence of a deep gap between South America and Antarctica presents the ocean with a special dynamical problem. Net north-south geostrophic flows across the latitude band of Drake Passage are prohibited by the lack of east-west pressure gradients to balance the effect of the earth's rotation. Thus there can be no net poleward flows of upper ocean water across the latitude band of Drake Passage to sustain a strong thermohaline circulation in the southern hemisphere (Gill and Bryan, 1971). Meanwhile, the winds in this latitude belt push the surface water northward. The shear in the circumpolar westerlies south of the wind stress maximum (~50°S) pumps up about $30\times10^6\ m^3\cdot s^{-1}$ of deep water. Part of this upwelling occurs in the latitude band containing Drake Passage (55° to 63°S).

Because the northward wind drift and upwelling arising within this latitude belt are not geostrophically balanced, they are not affected by the existence of Drake Passage. However, any subsurface return flows feeding the upwelling and compensating the surface wind drift are subject to the geostrophic constraint. Ocean models deal with this problem by generating a deep poleward return flow which is below the depth of the highest topography in this latitude belt, i.e. where east-west pressure gradients can exist. This explains why the upwelling within the latitude band of Drake Passage generally comes from depths below 2000 m.

The surface water pushed north of Drake Passage is warmed and freshened by the atmosphere. Even though it is forced downward by the wind shear north of the maximum wind stress it is too buoyant to join up with the deep return flow. According to TSa, the surface wind drift is linked up to the deep poleward return flow by a long-range overturning cell whose downward branch is the sinking associated with the formation of NADW. TSa claim that the magnitude of the Atlantic outflow is determined by $\int(\tau_x/f)dx_{SA}$, the magnitude of the zonally integrated wind drift at the latitude of the tip of South America. The formation of dense water in the North Atlantic makes it possible for this closure to exist, but it does not account for the magnitude of the Atlantic overturning.

Most of the northward wind drift at the latitude of the tip of South America occurs in the Indian and Pacific Oceans. If this flow is to be quantitatively linked up with the overturning in the Atlantic Ocean, this northward flow must find its way back to the Atlantic Ocean in some way. We would interpret the flow of Pacific water through the Indonesian Straits, and the westward flow south of Africa, not as a return flow associated with upwelled deep water, but as a return flow for upper ocean water forced northward into the thermocline by circumpolar winds.

Figure 10 shows a schematic of a reconfigured conveyor belt which is driven by the winds in the south. In this view, the ocean's deep circulation can be divided into two cells, one of which might be labelled a "wind-driven" cell and one a "thermohaline" cell. The thermohaline cell embodies the formation of Antarctic Bottom Water and its northward distribution to the three ocean basins. The return flow of the thermohaline cell is constrained to be deep because of the geostrophic constraint in Drake Passage. The wind-driven cell is driven by $\int(\tau_x/f)dx_{SA}$. Closure is provided by sinking in the North Atlantic and outflow from the Atlantic at depth. Intermediate-depth water pushed northward into the thermoclines of the Indian and Pacific Oceans returns to the Atlantic around the tip of Africa (Gordon, 1986) or through Drake Passage (Rintoul, 1991).

One should not view the diagram in figure 10 and the labelling of the Atlantic overturning as a wind-driven cell any more literally than one views the Global

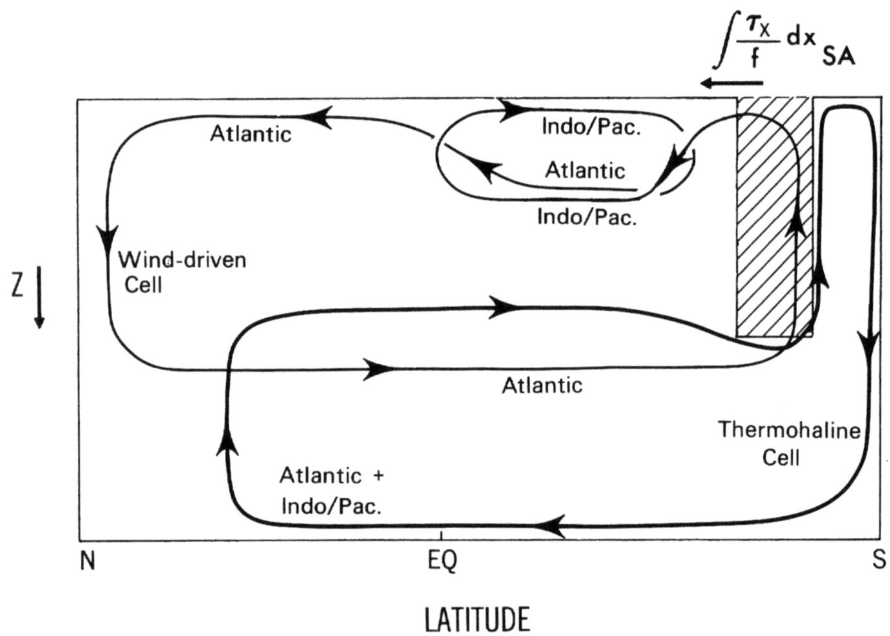

Figure 10 A reconfigured global circulation in which the ocean's deep overturning is divided between a purely thermohaline bottom water circulation cell and a mainly wind-driven deep circulation cell in the upper 3000 m.

Conveyor schematic in Broecker (1987, 1991). The reconfigured circulation schematic in figure 10 is offered as an alternative paradigm to a purely thermohaline deep circulation. Surely both wind and thermohaline processes are important in the real ocean. The real circulation is an intricate balance between the two kinds of forcing. However, the conventional view, in presupposing a dominant thermohaline circulation, is grossly weighted to the one point of view.

Importance of a Reconfigured Global Conveyor for the Carbon Cycle

The warming and cooling of oceanic surface water by the ocean's meridional circulation has a big effect on the ocean's pCO_2 distribution because of the sensitivity of the CO_2 system to changes in temperature. As with $\Delta^{14}C$, the way the circulation is configured can make a big difference in the distribution of CO_2 sources and sinks. Bottom water, with an initial temperature of ~0°C, upwelling through the thermocline in the Indian and Pacific Oceans will take up a great deal of heat from the atmosphere and give off a tremendous amount of CO_2. When it flows poleward toward Antarctica it will give up that heat and take up the CO_2 again as it cools down. Without the mid-depth outflow, all the sinking around Antarctica has to be derived from mid-latitude upper ocean water. The cooling of mid-latitude, upper ocean water as it flows poleward will make the southern ocean a sink for CO_2, much like the North Atlantic. It will give the latitudinal pCO_2 distribution a symmetric structure with minima in the high latitude regions of both hemispheres.

If, on the other hand, the poleward flow of abyssal water in the southern hemisphere occurs at depth, very little warming or cooling takes place when this water upwells to the surface around Antarctica. Some CO_2 may actually be lost to the atmosphere as the upwelled water is pushed northward by the wind and is warmed up. On this basis alone, the southern ocean should have a higher pCO_2 than the North Atlantic. Of course, the wind-forced transport of subantarctic water to the tropical Pacific and Indian Oceans would be expected to create a tropical pCO_2 maximum as in the abyssal upwelling case, but the poleward flow of this water should be confined to latitudes north of the tip of Africa or the tip of South America. The cooling of these water masses as they flow poleward will indeed result in an uptake of CO_2, but the pCO_2 drawdown will be limited by the fact that these water masses are only cooled off to ~13°C for flow around Africa (Gordon, 1986) and ~5°C for flow around South America. None of the poleward flow is cooled to 0°C as in the thermohaline model.

In principle, the global oceanic uptake of fossil fuel CO_2 is not affected by the operation of the natural carbon cycle or the pre-industrial distribution of pCO_2. One expects the uptake of fossil fuel CO_2 per unit area to be relatively constant over the ocean. However, measured fluxes of CO_2 across the air-sea interface are a function of both the pre-industrial carbon cycle and the anthropogenic perturbation. Recent papers by Tans et al. (1990) and Keeling et al. (1989) show that anthropogenic releases of CO_2 (mostly in the northern hemisphere) which are balanced by large oceanic sinks in the southern hemisphere (where there is more ocean) are inconsistent with the CO_2 transports derived from atmospheric CO_2 transport models which assume a relatively constant oceanic uptake of fossil fuel CO_2 per unit area. Tans et al. conclude that the main sink for fossil fuel CO_2 must be in the northern hemisphere.

If, however, there were a pre-existing natural CO_2 cycle in which CO_2 outgassed from the southern ocean and was taken up in the North Atlantic, the northern hemisphere today would appear to be a larger fossil fuel CO_2 sink than the southern hemisphere. Under such a regime, the southern hemisphere would still be taking up large quantities of fossil fuel CO_2, but it would be partially masked by the outgassing of CO_2 in the natural cycle. The atmospheric transport of fossil fuel CO_2 across the equator would also be partially offset by a northward natural transport of CO_2 such that the total transport across the equator would be smaller and more in agreement with atmospheric models. Thus the wind-driven upwelling of deep water around Antarctica linked with a large natural sink for CO_2 in the North Atlantic helps accommodate the constraint imposed by atmospheric circulation models with a substantial southern hemisphere sink for fossil fuel CO_2.

Conclusions

According to the classic recipe for the ocean's deep circulation, abyssal water upwells through the thermocline and flows poleward in the upper ocean. However, recent evidence from the Pacific Ocean indicates that the poleward flow which balances the northward flow of Antarctic Bottom Water moves poleward at mid-depth, not in the upper ocean. Any tendency to shift the Pacific's poleward flow from mid-depth to the upper ocean replaces a silica-rich, low-$\Delta^{14}C$ mid-water flow with a low-silica, high-$\Delta^{14}C$ upper ocean flow. Low levels of $\Delta^{14}C$ observed in pre-bomb Antarctic surface water can only be derived from a dominant mid-depth outflow of deep water. We have shown that any upwelling of abyssal water across the Pacific thermocline produces a symmetric $\Delta^{14}C$ minimum all along the equator, which is not observed. Existing observations define the Peru upwelling as the only clear source of low-$\Delta^{14}C$ surface waters in the tropical Pacific. The Peru upwelling has a southern hemisphere thermocline source, not a deep water source. We find no clear evidence for any abyssal upwelling into or across the thermocline in the Pacific Ocean. Only our preconceived notions about the dominance of a buoyancy-driven deep circulation suggest otherwise. Radiocarbon observations are consistent with a circulation scheme in which virtually all the upwelling of abyssal water occurs around Antarctica, where it is mainly wind-forced.

Acknowledgements

We would like to acknowledge three very helpful reviews from Kirk Bryan, Rick Murnane, and Bruce Warren. We are also pleased to acknowledge the assistance of Wendy Marshall, Phil Tunison, Jeff Varanyak, and Cathy Raphael, who helped in preparing the camera-ready copy, figures and tables.

References

Bacastow R, and E Maier-Reimer (1990) Ocean-circulation model of the carbon cycle. Climate Dynamics 4:95-125.

Bien G S, N W Rakestraw, and H E Suess (1965) Radiocarbon in the Pacific and Indian Oceans and its relation to deep water movements. Limnol Oceanogr 10:Suppl R25-R37.

Broecker W S (1987) The biggest chill. Natural History 97:74-82.

Broecker W S (1991) The Great Global Conveyor. Oceanography, 2:79-89.

Broecker W S, R Gerard, M Ewing, and B C Heezen (1960) Natural radiocarbon in the Atlantic Ocean. J Geophys Res 65:2903-2931.

Broecker W S, and E A Olson (1961) Lamont radiocarbon measurements VIII. Radiocarbon 3:176-204.

Broecker W S, and T H Peng (1974) Gas exchange rates between air and sea. Tellus 26:21-35.

Broecker W S, T H Peng, H G Ostlund, and M Stuiver (1985) The distribution of bomb radiocarbon in the ocean. J Geophys Res 90:6953-6970.

Bryan K (1991) Michael Cox (1941-1989): his pioneering contributions to ocean circulation modeling. J Phys Oceanogr 21:1259-1270.

Cox M D (1975) A baroclinic numerical model of the world ocean: preliminary results. In: Numerical Models of Ocean Circulation. National Academy of Sciences Washington DC 107-118.

DeMaster D J (1981) The supply of and accumulation of silica in the marine environment. Geochim Cosmochim Acta 45:1715-1732.

Druffel E M (1981) Radiocarbon in annual coral rings from the eastern tropical Pacific Ocean. Geophys Res Lett 8:59-62.

Druffel E R M (1987) Bomb radiocarbon in the Pacific: Annual and seasonal timescale variations. J Mar Res 45:667-698.

Fiadeiro M E (1982) Three-dimensional modeling of tracers in the deep Pacific Ocean II. Radiocarbon and the circulation. J Mar Res 40:537-550.

Gill A E, and K Bryan (1971) Effects of geometry on the circulation of a three-dimensional southern hemisphere ocean model. Deep-Sea Res 18:685-721.

Gordon A L, H W Taylor, and D T Georgi (1977) Antarctic oceanographic zonation. In: M J Dunbar (ed)Polar Oceans. Arctic Institute of North America Calgary 45-76.

Gordon A L (1986) Interocean exchange of thermocline water. J Geophys Res 91:5037-5046.

Hellerman S, and M Rosenstein (1983) Normal monthly wind stress over the world ocean with error estimates. J Phys Oceanogr 13:1093-1104.

Keeling C D, S C Piper, and M Heimann (1989) A three-dimensional model of atmospheric CO_2 transport based on observed winds: 4. Mean annual gradients and interannual variations. In: D H Peterson (ed) Aspects of Climate Variability in the Pacific and the Western Americas. Geophysical Monograph 55 American Geophysical Union Washington DC 305-363.

Konishi K, T Tanaka, and M Sakanoue (1982) Secular variation of radiocarbon concentration in sea water: Sclerochronological approach. In: E D Gomez (ed) Proceedings of the Fourth International Coral Reef Symposium Manila. Marine Science Center University of the Philippines 1:181-185.

Levitus S, J L Reid, M E Conkright, and R Najjar (to be published) Distribution of nitrate, phosphate, and silicate in the world oceans. Prog Oceanogr.

Maier-Reimer E, and K Hasselmann (1987) Transport and storage of CO_2 in the ocean - An inorganic ocean-circulation carbon cycle model. Clim Dyn 2:63-90.

Maier-Reimer E, U Mikolajewicz, and T. Crowley (1990) Ocean general circulation model sensitivity experiment with an open Central American isthmus. Paleoceanography 5:349-366.

Manabe S, and R J Stouffer (1988) Two stable equilibria of a coupled ocean-atmosphere model. J Climate 1:841-866.

Ostlund H G, H Craig, W S Broecker, and D Spencer (1987) GEOSECS Atlantic, Pacific, and Indian Ocean Expeditions. Shorebased Data and Graphics vol 7 U S Government Printing Office Washington D C.

Philander S G H, W J Hurlin, and A D Siegel (1987) Simulation of the seasonal cycle of the tropical Pacific Ocean. J Phys Oceanogr 17:1986-2002.

Rafter T A (1968) Carbon-14 variations in nature, 3, ^{14}C measurements in the South Pacific and Antarctic Oceans. N Z J Science 11:551-589.

Reid J L (1973) Transpacific hydrographic sections at Lats. 43°S and 28°S: the SCORPIO Expedition - III. Upper water and a note on southward flow at mid-depth. Deep-Sea Res 20:39-49.

Rintoul S R (1991) South Atlantic interbasin exchange. J Geophys Res 96:2675-2692.

Stocker T F, and D G Wright (1991) Rapid transitions of the ocean's deep circulation induced by changes in surface water fluxes. *Nature 351*:729-732.

Stommel H, and A B Arons (1960) On the abyssal circulation of the world ocean - II. An idealized model of the circulation pattern and amplitude in ocean basins. Deep-Sea Res 6:217-233.

Stuiver M, P D Quay, and H G Ostlund (1983) Abyssal water carbon-14 distribution and the age of the world oceans. Science 219:849-851.

Tans P P, I Y Fung, and T Takahashi (1990) Observational constraints on the global atmospheric CO_2 budget. Science 247:1431-1438.

Toggweiler J R, K Dixon, and K Bryan (1989a) Simulations of radiocarbon in a coarse-resolution world ocean model, 1, Steady state Prebomb distributions. J Geophys Res 94:8217-8242.

Toggweiler J R, K Dixon, and K Bryan (1989b) Simulations of radiocarbon in a coarse-resolution world ocean model, 2, Distributions of bomb-produced carbon 14. J Geophys Res 94:8243-8264.

Toggweiler J R, K Dixon, and W S Broecker (1991) The Peru upwelling and the ventilation of the South Pacific thermocline. J Geophys Res 96:20467-20497.

Toggweiler J R, and B Samuels (1992) Is the magnitude of the deep outflow from the Atlantic Ocean actually governed by southern hemisphere winds? This volume.

Trenberth K E, W G Large, and J G Olson (1990) The mean annual cycle in global ocean wind stress. J Phys Oceanogr 20:1742-1760.

Tsuchiya M, R Lukas, R A Fine, E Firing, and E Lindstrom (1989) Source waters of the Pacific Equatorial Undercurrent. Prog Oceanogr 23:101-147.

Weaver A., and E S Sarachik (1991) The role of mixed boundary conditions in numerical models of the ocean's climate. J Phys Oceanogr 21:1470-1493.

Weiss R F, H G Ostlund, and H Craig (1979) Geochemical studies of the Weddell Sea. Deep-Sea Res 26A:1093-1120.

Worthington L V (1977) The case for near-zero production of Antarctic Bottom Water. Geochim Cosmochim Acta 41:1001-1006.

Wunsch C, D X Hu, and B Grant (1983) Mass, heat, salt, and nutrient fluxes in the South Pacific Ocean. J Phys Oceanogr 13:725-753.

MODELLING THE PRESENT-DAY OCEANIC CARBON CYCLE

Ulrich Siegenthaler
Physics Institute
University of Bern
Sidlerstrasse 5
3012 Bern
Switzerland

THE ANTHROPOGENIC CO_2 INCREASE

Carbon dioxide is an important greenhouse gas, and its atmospheric increase causes concern because of a potential impact on the global climate. A prerequisite for assessing possible future climatic changes is a means to estimate the future development of the atmospheric CO_2 concentration, given some energy consumption scenarios. For this purpose, models have been developed which represent the global carbon cycle and the processes which are important for the redistribution of anthropogenic CO_2. This paper describes the carbon cycle mechanisms which are relevant in this context, the way in which they are integrated in models and some model results.

The atmospheric CO_2 concentration has grown by about 25% in the last 200 years, and it continues to grow from year to year (Figure 1). The reason for this are emissions from fossil fuel combustion and from deforestation (Figure 2), which together amounted to about 345 GtC from 1860 to 1989 according to the numbers given by Keeling (1991), Marland and Boden (1991) and Houghton (1991). As Figure 2 shows, the year-to-year increase in the atmosphere has been roughly half the estimated input; more precisely, the airborne fraction (defined here as the ratio of atmospheric increase to emissions from fossil fuels plus deforestation) is 40% for the period 1860-1989 and 45±7% for 1980-1989. Where has the non-airborne CO_2 gone? Two reservoirs can have acted as sinks for anthropogenic carbon, the ocean and the land biota. Table 1 gives a budget of the atmospheric perturbations for the decade 1980 to 1989. Estimated emissions exceed the accumulation in atmosphere plus ocean by about 2 GtC yr^{1}. The question therefore is if the numbers in Table 1 are

NATO ASI Series, Vol. I 15
The Global Carbon Cycle
Edited by M. Heimann

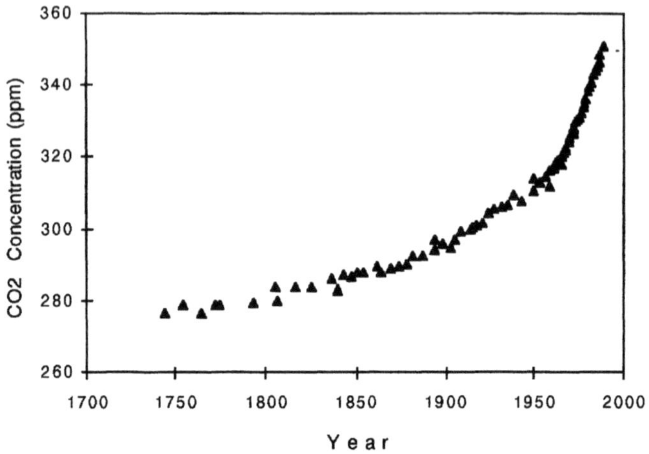

Figure 1 Atmospheric CO_2 increase in the past two centuries, as recorded by measurements at Mauna Loa (Keeling et al., 1989a) and in air trapped in the ice core from Siple station, Antarctica (Neftel et al., 1985, Friedli et al., 1986).

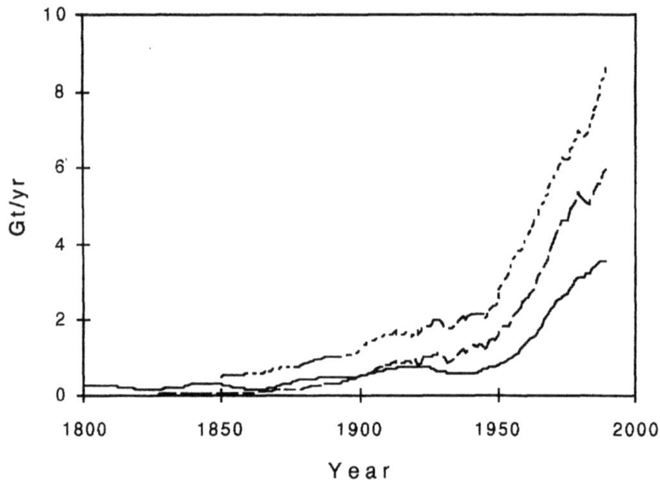

Figure 2 Anthropogenic CO_2 emissions from fossil fuel combustion (long-dashed line, Keeling 1991, Marland and Boden 1991), from fossil fuels plus deforestation (Houghton, 1991) together (short-dashed line), and year-to-year increase of atmospheric CO_2 (solid line) derived from the combined Mauna Loa-Siple record of Figure 1.

wrong or if there is some important sink missing there. In this paper I discuss how the estimate for the oceanic uptake is obtained and why there are good reasons to assume that this uptake is known to ±25%. If the numbers of Table 1 are accepted, there must be a significant sink in the land biota, and it has been suggested that this is due to stimulation of biological activity on land by the increased atmospheric CO_2 concentration or by anthropogenic nitrogen compounds.

Table 1: Budget of anthropogenic CO_2 perturbations for the decade 1980-1989. The table is revised from the IPCC assessment (Watson et al. 1990). Numbers are in GtC yr^{-1}. A range instead of an error is given for deforestation, because estimates from different authors cannot just be averaged to get a best value.

Emissions	
Fossil fuels (Marland and Boden 1991)	5.4 ± 0.5
Deforestation (Watson et al. 1990, 1992)	0.6 to 2.5
Accumulation	
Atmosphere	3.4 ±0.2
Ocean	2.0 ± 0.5
Net imbalance	1.6 ± 1.2

Before discussing in detail the processes that have to be represented in carbon cycle models, we briefly consider the global carbon cycle (Figure 3). CO_2 is continuously exchanged between atmosphere and ocean by gas transfer. In pre-industrial time, there was an approximate steady state between gross flux into the ocean (including a minor flow via vegetation - soils - rivers) and the back flux to the atmosphere. This is indicated by analyses on an ice core from the South Pole which show that atmospheric CO_2 levels between 1000 and 1800 A.D.were constant within 5 to 10 ppm (Siegenthaler et al., 1988). As the atmospheric CO_2 concentration grows, also the flux to the ocean increases. In the ocean, CO_2 is transported to depth by water motion. Biological processes lead to a continuous rain of dead organic particles to depth and to the production of dissolved

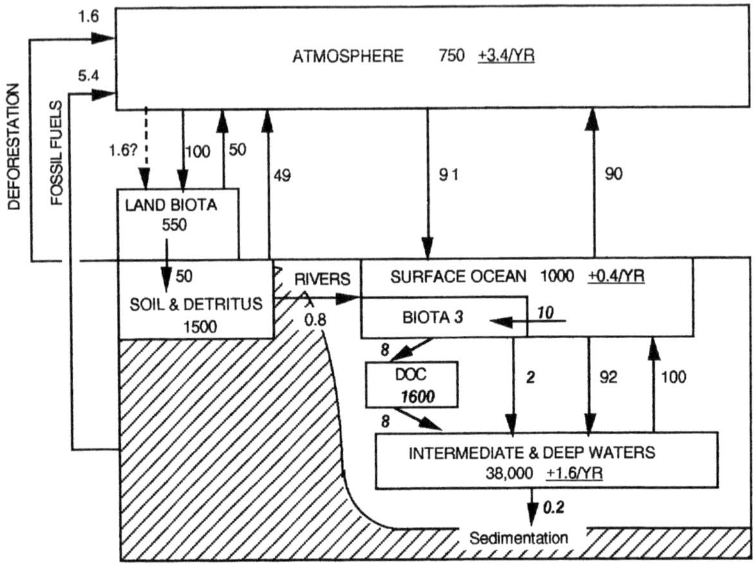

Figure 3 Major global carbon reservoirs and fluxes for present-day conditions (after Watson et al. 1990, revised). Numbers are in GtC (gigatons carbon, 1 GtC = 10^{12} kg C) for reservoirs and GtC yr^{-1} for fluxes. Underlined numbers indicate annual increases, numbers in bold-face italics quantities not affected by the anthropogenic perturbation. The dashed arrow from atmosphere to land biota represents the biospheric CO_2 sink inferred from the balance of atmospheric perturbations (Table 1).

organic carbon. Particulate as well as dissolved organic carbon are both remineralized at depth to inorganic carbon and are therefore important for the natural cycling of carbon. Marine biology does, however, not sequester anthropogenic carbon, because - according to current knowledge - biological production in the ocean is not influenced by changes in pCO_2, but is controlled by other factors like light, nutrients (notably nitrogen and phosphorus) and grazing by zooplankton. Therefore, marine biology is often omitted from the models designed for simulating the oceanic CO_2 uptake. Carbon cycle modellers have always used the assumption that the natural transfers of carbon were in a steady state before industrialization began and that this is still the case. The task of modelling the fate of anthropogenic CO_2 can then be considerably simplified by using perturbation models in which only perturbations (i.e. the differences between the present and the pre-industrial pools and fluxes) are considered, instead of including the whole natural carbon cycle.

The assumption that the natural carbon cycle was in a steady state can be justified by the constancy of the pre-industrial atmospheric concentration mentioned above, which indicates that natural control mechanisms are at work which stabilize atmospheric carbon dioxide as long as there are no external (i.e. anthropogenic) perturbations. On long time scales, CO_2 varied in parallel to the glacial-interglacial climatic cycle, but this is beyond the subject of this paper.

INGREDIENTS OF OCEANIC CARBON CYCLE MODELS

Carbonate chemistry

The ocean is the most important carbon reservoir; it contains about 65 times as much carbon as the pre-industrial atmosphere. Its equilibrium capacity for taking up excess CO_2 is, however, only about 6.5 times as large as that of the atmosphere. Inorganic carbon is present in the ocean mainly in ionic form; less than 1% of the total dissolved inorganic carbon (ΣCO_2) is dissolved CO_2 gas, ca. 90% is HCO_3^-, ca. 10% $CO_3^=$. With the invasion of additional CO_2, the chemical equilibria are shifted such that the relative increase of the concentration of ΣCO_2 is considerably smaller than of dissolved CO_2 gas ($CO_{2,aq}$) (e.g. Broecker and Peng 1982, Peng et al. 1987). More precisely, the relative changes are related by the buffer factor $\xi \approx 10$ as follows:

$$\xi = \frac{pCO_2 - pCO_{20}}{\Sigma CO_2 - \Sigma CO_{20}} \cdot \frac{\Sigma CO_{20}}{pCO_2}$$

pCO_2 is the equilibrium partial pressure of CO_2 in the surface sea water, defined as

$$pCO_2 = [CO_{2,aq}]/s$$

where s = solubility of CO_2, and the index 0 indicates a reference state.

This means that when ocean and atmosphere have come to a new equilibrium after cessation of anthropogenic CO_2 emissions (which will take a long time, ~10^3 years), the ocean will have taken up $65/\xi \approx 6.5$ parts of the excess CO_2 and the atmosphere 1 part. In other words, 14% of the perturbation stays in the atmosphere. On a still longer time scale (several 1000 years), dissolution of carbonate sediments sets in which can reduce this by another factor of 2. The buffer factor ξ depends on the chemical composition of the water, on its salinity and temperature. It increases with growing ΣCO_2, i.e. the oceanic uptake capacity diminishes with growing CO_2 concentration. ξ is higher at high latitudes, that is in the outcrop regions of deep water than for low-latitude surface waters.

The anthropogenic carbon emissions have grown roughly exponentially, with a time scale (e-folding time) of about 30 years. In this time span, only waters near the surface have taken up excess CO_2. In order to get quantitative estimates of the amount of excess CO_2 in the ocean, models have to take into account the two processes controlling the rate of uptake, that is gas air-sea gas transfer and downward transport of surface water to depth, i.e. ocean circulation.

Air-sea gas exchange

The net air-to-sea flux can be written as (index s for surface water):

$$F_{net} = k_g \; (pCO_{2atm}-pCO_{2s})$$

The gas exchange coefficient k_g depends on the physical conditions, in particular on wind speed (Liss and Merlivat, 1986; Watson, this volume). There is a discrepancy between the global mean value of the exchange flux, and therefore of k_g, as determined by means of two independent methods. The first method is based on the oceanic distribution of ^{14}C (see Siegenthaler 1989 for a review of ^{14}C in the oceans.) The gross exchange flux is, according to this method, determined from the following schematic balance:

$$F_{ex}(R_{atm}-R_s) = \frac{d}{dt}(\text{oceanic } ^{14}C \text{ inventory}) - {}^{14}C \text{ decay in the ocean}$$

where R_{atm} and R_s are $^{14}C/C$ ratios in atmosphere and ocean surface water. The distribution of natural ^{14}C was approximately stationary, so that the first right-hand-side term can be neglected. The result for the natural isotope balance is not very precise, because it is sensitive to the difference $R_{atm}-R_s$ which is not very well known. For bomb-produced ^{14}C, it is the first r.h.s. term which is important, while radioactive decay can be neglected. Best estimates of the exchange flux based on bomb-produced and of natural ^{14}C are in the range 15 to 18 mol $m^{-2}yr^{-1}$ for the pre-industrial atmospheric CO_2 concentration of 280 ppm, or k_g = 0.054 to 0.064 mol $m^{-2}yr^{-1}ppm^{-1}$; this corresponds to an atmospheric residence time of CO_2 with respect to air-sea exchange of about 8 years. On the other hand, laboratory and field determinations of the gas exchange coefficient (mostly carried out with other gases than CO_2, however) yield lower global mean values, typically 9 to 11 mol $m^{-2}yr^{-1}$ (Etcheto et al. 1991). The reason for this difference is not clear. Since the ^{14}C method provides a globally integrated value, the ^{14}C-derived value is appropriate for global simulations of the oceanic CO_2 uptake.

Fortunately, the model-calculated flux of anthropogenic CO_2 to the ocean is not strongly dependent on the assumed gas exchange coefficient (Oeschger et al., 1975). Sarmiento et al. (1992) found in a 3-dimensional model study that a factor of 2.1 decrease in k_g resulted in a decrease of the oceanic uptake by only 11%. The reason for this small sensitivity is that the uptake is mainly controlled by the transport of CO_2-laden surface water to deeper layers. According to model calculations, vertical transport by ocean circulation has been responsible for about 80 % of the total transport resistance, in other words, average sea surface water has a degree of equilibration of ca. 80% with the atmospheric excess CO_2 (Siegenthaler, 1983).

Vertical transport by water circulation

A schematic concentration profile of anthropogenic CO_2 - or a time-dependent (transient) tracer like $^{14}CO_2$ or CFCs - is shown in Figure 4. The two resistances to the penetration of the perturbation into the deep ocean - gas exchange and vertical circulation - are reflected in two quantities which can, for tracers like bomb-produced ^{14}C, be determined from observations: the degree of equilibrium of surface water with the atmosphere and the penetration depth into the water column. The penetration depth $\bar{z}$ is defined as the ratio of the oceanic inventory of the tracer to the surface concentration (Broecker et al. 1980):

$$\bar{z} = \frac{\int c(z)\, dz}{c(z=0)}$$

where c = tracer concentration. The faster the vertical circulation, the larger the penetration depth. The degree of equilibrium of surface water depends on both, gas exchange and

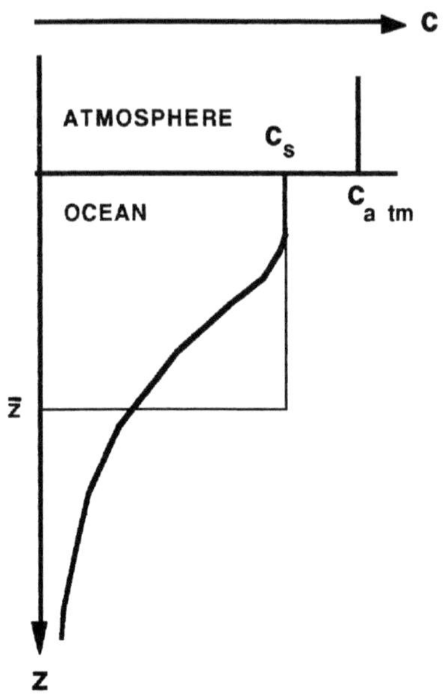

Figure 4 Schematic vertical profile of anthropogenic CO_2 (or of another perturbation) in atmosphere and ocean. The Figure illustrates the concept of the penetration depth $\bar{z}$ which is defined as

$$\bar{z} = \frac{\int c(z)\,dz}{c_s},$$

where c(z) is the concentration at depth z and c_s the concentration in the water at the surface.

vertical circulation; it is near 1 for high gas exchange or for low vertical circulation rate. Both quantities also depend on the rate of change with time of the perturbation: a signal with a small time scale will penetrate less deep into the ocean than a slowly changing perturbation. Penetration depths for ^{14}C produced by the thermonuclear bomb test of 1961/62 were typically a few 100 m (Broecker et al., 1985) at the time of the GEOSECS observations 1972-1978 (time since injection into the atmosphere: 10 to 16 years). For anthropogenic CO_2, a larger mean penetration depth is expected, because its characteristic time is longer, ca. 30 years (obtained from fitting an exponential curve to the anthropogenic emissions); a model calculation yields a penetration depth of 650 m for 1990 (Siegenthaler and Joos, 1992). Models designed for studying the invasion of man-made CO_2 or other short-lived tracers into the ocean must therefore have a sufficient depth resolution in the upper ocean to resolve these penetration depth, i.e. a grid spacing of the order of 100 m or less. Coarser models, e.g. two-box models, can at best be fitted such as to produce reasonable oceanic inventories for one fixed time scale, but will fail if the time scale changes, as is e.g. the case for the invasion of bomb-^{14}C with time.

Eddy diffusion

The crucial process which has to be simulated by carbon cycle models designed for studying the anthropogenic CO_2 increase is the vertical exchange between surface and deep waters. In "simple" models, this exchange is represented in a parameterized way, by turbulent (eddy) diffusion or by exchange between boxes. For eddy diffusion, the flux is assumed proportional to the gradient of the concentration c:

$$F = - K \frac{dc}{dz}$$

A tracer which increases with a time scale τ penetrates a mean distance $\overline{z} = \sqrt{K\tau}$ into a medium with constant diffusivity K. The assumption that vertical transport can be parameterized in this way thus involves that the penetration depth varies as the square root of the time scale of a perturbation. Bryan et al. (1984) studied, using a 3-D ocean general circulation model, the penetration of a heat perturbation into the ocean by prescribing a step-function temperature change at the surface, and they found that the heat flux decreased approximately as $\tau^{-0.5}$ (τ = time since onset of the perturbation), as would be predicted for a diffusive ocean. The eddy diffusion concept thus seems reasonable for simulating the large-scale penetration of a tracer into the ocean.

Eddy diffusion is formally analogous to molecular diffusion, which, however, plays no role except on a microscopic scale; eddy diffusivities are many orders of magnitudes larger than molecular diffusivities. Turbulent diffusion occurs on very different space scales; the distinction between advective and turbulent (eddy diffusive) flux depends on the scale considered. Local vertical mixing, as e.g. important for the dispersion of a dye patch, can be caused by breaking of internal waves or by double diffusion (see review by Garrett 1979). In a global model, eddy diffusion represents the mixing due to motions which are not resolved in the model, i. e. sub-grid scale processes. In a 3-D model, the horizontal grid size is typically several hundred km. In simple global models, the ocean is often very coarsely or not at all subdivided horizontally, so that here, eddy diffusion represents mixing by eddies up to the largest oceanic circulation cells. It is clear that this is something different than local mixing. Typical eddy diffusivities for large-scale models are of order 10^{-4} m^2s^{-1}, while for local mixing in the upper ocean (but below the mixed layer), values of 10^{-6} to 10^{-5} m^2s^{-1} have been found. Theoretical studies show that the local eddy diffusivity depends on the vertical stability of the water column, with lower values for higher stability (e.g. Welander 1968, Garrett 1982), and 1-dimensional modelling of ^{222}Rn and ^{228}Ra depth profiles has confirmed such a dependency (Sarmiento et al. 1976). Thus, one might expect small K values in the thermocline (strong stratification) and low values in the deep ocean. However, this is not appropriate for simple diffusive global ocean models, as discussed below.

OCEAN CARBON CYCLE MODELS

The box-diffusion model (BD model) of Oeschger et al. (1975) is based on the assumption that below a wind-mixed surface layer of typically 75 m thickness, vertical transport occurs by eddy diffusion. Oeschger et al. found that in this way, the distribution of natural ^{14}C (governed by the turnover time of ca. 1000 years for the whole ocean) and of bomb-^{14}C can both be simulated with approximately the same value for the eddy diffusivity, while two-box ocean models had to assume rather different exchange coefficients for the two tracers. Later studies, with a larger bomb-^{14}C data set available, showed that for simulating bomb-^{14}C, a larger eddy diffusivity is necessary than for natural ^{14}C (Siegenthaler, 1983). This makes sense, since the wind-driven and the thermohaline circulation are more vigorous near the surface than at depth (Siegenthaler and Joos, 1992). Thus, the dependence of local K values on stability is overridden by the large circulation cells which are, in such a model, sub-grid scale features.

In diffusion models, K must be determined from distributions of tracers like natural or bomb-produced ^{14}C. Siegenthaler (1983) calibrated the box diffusion with both and found K=4005 m^2yr^{-1} and 7685 m^2yr^{-1} (1.3 and 2.4 cm^2s^{-1}), respectively. (Siegenthaler and Oeschger (1987) slighly revised the model, e.g. by using exact instead of linearized equations for isotopic perturbations; a recalibration with bomb-^{14}C yielded 7100 m^2yr^{-1} = 2.2 cm^2s^{-1}). Bomb-^{14}C, which has invaded essentially only the upper ocean, is a better analogue for anthropogenic CO_2 than natural ^{14}C; the distribution of natural ^{14}C is strongly influenced by the deep ocean with a turnover time of many centuries. Therefore, the higher values are preferable. Li et al. (1984)

found 1.7 cm^2s^{-1} from GEOSECS observations of tritium, about comparable to the bomb-^{14}C result.

The way how simple models are calibrated is essential for their value, since the eddy diffusivity (or similar parameters) cannot be determined from basic physical considerations. When calibrated using bomb-^{14}C, a model has to reproduce the surface water bomb-^{14}C concentration as well as the oceanic inventory (or equivalently the penetration depth). This gives two degrees of freedom, which in the BD model correspond to the gas exchange coefficient and the eddy diffusivity. Fitting only one of these quantities is not sufficient, since there is then still one degree of freedom left open. Enting (1990) related oceanic CO_2 uptake to the surface concentration of bomb-^{14}C and found that the surface ^{14}C concentration does not strongly constrain the CO_2 uptake; this is not surprising, since he did not use the full ^{14}C information by ignoring the observed penetration depth. Enting concluded that a very low oceanic CO_2 uptake, as postulated by Tans et al. (1990), can be reconciled with the oceanic ^{14}C observations, but this is incorrect; rather, the surface concentration and the inventory of bomb-^{14}C (or equivalently the degree of equilibrium between surface water and atmosphere and the penetration depth) together put rather strong constraints on the net air-sea flux of CO_2.

The role of high latitudes

Purely 1-dimensional models like the box-diffusion model do not take into account the fact that vertical exchange in the ocean is rather non-uniform, in particular that the cold deep waters are ventilated exclusively from the high-latitude surface. The exchange there is much faster than in low and mid-latitudes, where it is sluggish due to the stability of the thermocline. This is witnessed e.g. by low values of ^{14}C in high latitude surface water and high concentrations of ΣCO_2, phosphate and other properties that are enriched in deep water. In order to take this into account, Siegenthaler (1983) designed an outcrop-diffusion model, which is the BD model plus a direct air-sea exchange for deep waters representing in a crude way the outcrop in high latitudes. Since deep water is partly ventilated directly from the atmosphere, the eddy diffusivity is smaller here than in the purely one-dimensional BD model, namely 5180 m^2yr^{-1} for calibration with bomb-^{14}C (2224 m^2yr^{-1} for calibration with natural ^{14}C) for an outcrop covering 10% of the ocean

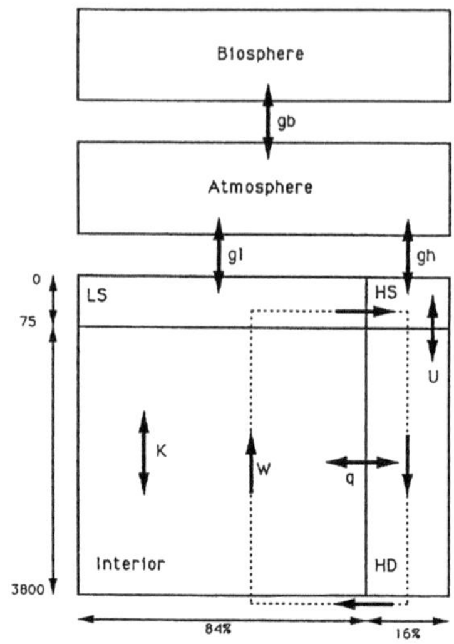

Figure 5 Structure of the HILDA model (High-latitude exchange/interior diffusion-advection) of Shaffer and Sarmiento (1992) as used by Joos et al. (1991 a, b) and Siegenthaler and Joos (1992). Oceanic reservoirs are low- and high-latitude surface box (LS, HS), HD is the high-latitude deep box; in the Interior reservoir vertical transport occurs by eddy diffusion (diffusivity K). w is an advective circulation, upwelling in the Interior reservoir; the latter exchanges laterally (q) with the deep box HD. u denotes exchange between HS and HD. gl, gh and gb are gas exchange fluxes. A biosphere box is included for studying isotopic perturbations.

surface. The outcrop-diffusion model was designed to study the sensitivity of the oceanic uptake of anthropogenic CO_2 to the ventilation of the deep waters through their surface outcrop. If both are calibrated in a consistent way with bomb-^{14}C, the outcrop-diffusion ocean takes up 24% more CO_2 than the BD model in the period 1959-1983 (Siegenthaler and Oeschger 1987), with airborne fractions (= ratio of atmospheric increase to net input into the atmosphere) of 58% and 48%. Since the assumption of infinitely fast mixing between high-latitude surface and deep ocean is obviously not realistic, the result of the model with outcrop should be considered only as an upper limit for the oceanic uptake of anthropogenic CO_2, rather than as a valid estimate.

Shaffer and Sarmiento (1992; also Shaffer, this volume) devised a model in which the high latitudes are included in a still simple way, but more realistically than in the outcrop-diffusion model, with the main purpose to simulate the distributions of natural oceanic properties. This HILDA (High-latitude exchange/interior diffusion-advection) model was adapted by Joos et al. (1991 a, b) and Siegenthaler and Joos (1992) for simulating anthropogenic perturbations of the carbon cycle (Figure 5). The high-latitude ocean is represented by two well-mixed boxes. Besides eddy diffusion, also vertical advection, with an upwelling velocity w, takes place in the low-latitude deep waters (Interior reservoir). Advective-diffusive models have often been used for representing oceanic distributions of tracers (e.g. Munk 1966, Craig 1969).

Three-dimensional models

In 3-D general circulation models of the ocean, the oceanic circulation is not prescribed by adjusting some parameters to observations, as is the case for the simple models. Rather, it is computed, using the laws of dynamics, as the result of external wind forcing at the surface and of horizontal pressure gradients arising from density differences related to temperature and salinity (thermohaline circulation). 3-D models yield simulations of the regional pattern of CO_2 or tracer distributions, but they also allow to study the response of the ocean to changes in external forcing, e.g. due to climatic change, while it is difficult to estimate how the transport parameters of simple models should be adjusted such as to represent different climatic conditions. Vertical transport is accomplished in 3-D models by advection, by convection (usually represented simply by mixing of two vertically adjacent grid boxes whenever the density gradient becomes unstable) and in some models also by explicit eddy diffusion.

Studies of the anthropogenic perturbations of the carbon cycle have been performed by Maier-Reimer and Hasselmann (1987) and Bacastow and Maier-Reimer (1990), both using the Hamburg 3-D ocean model, and by Sarmiento et al. (1992) who used the time-averaged transport field of the Princeton model in the version of Toggweiler et al. (1989a,b). A good way to characterize a model is by considering its response to a pulse input of CO_2 into the atmosphere (Siegenthaler and Oeschger 1978). A comparison of the box-diffusion and the HILDA model with 3-D models shows that the latter simulates the pulse response functions of the ocean general circulation models relatively well. Thus, these simple models, if approproately calibrated, yield very similar results for invasion of anthropogenic CO_2 into the ocean on very different time scales as 3-D ocean models (Maier-Reimer and Hasselmann 1987, Sarmiento et al. 1992, Siegenthaler and Joos 1992).

RESULTS FOR THE OCEANIC UPTAKE OF ANTHROPOGENIC CO_2

The oceanic response to the anthropogenic perturbation can be studied in two ways, either by prescribing the emissions of CO_2 into the atmosphere as a function of time or by deconvolving the atmospheric CO_2 increase in the past 200 years, i.e. by prescribing the atmospheric concentration history and calculating the resulting flux into the ocean. In a deconvolution, the net flux into the atmosphere-ocean system is obtained by adding the annual increase in the atmosphere (prescribed) and in the ocean (computed) (Siegenthaler and Oeschger 1987). This net flux is equal to the emissions from

Figure 6 Result of a deconvolution of the atmospheric CO_2 record of Figure 1 using the HILDA model (Siegenthaler and Joos, 1992). The total production is determined as the sum of observed atmospheric increase and model-computed oceanic uptake.

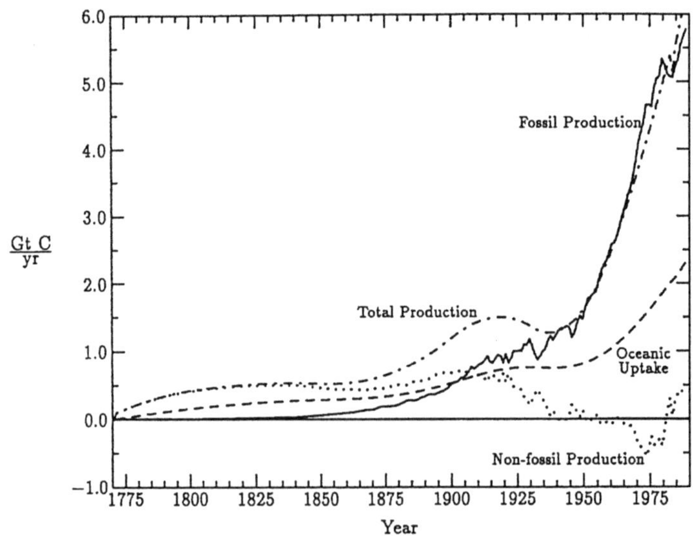

fossil fuels and deforestation minus additional uptake by the land biota , e.g. due to CO_2 fertilization, plus/minus any other sources and sinks. The atmospheric concentration since about 1750 is quite well known from ice core analyses and atmospheric monitoring, while the net non-fossil flux is poorly known, since there are no reliable means to estimate the effect of CO_2 fertilization and also the release from deforestations is quite uncertain. Therefore, a deconvolution is preferable to a simulation where emissions are prescribed, because the model input data are better known. Figure 6 shows the result of a deconvolution using the HILDA model. The computed net biospheric flux (= computed total net input minus fossil input) is near zero in the past several decades, indicating that a sink must exist that happens to approximately compensate the deforestation source. This "missing" sink, tentatively assigned to CO_2 fertilization of the land biota, increased roughly parallel to atmospheric CO_2 (Houghton, this volume).

Table 2 shows results from different models for the mean annual CO_2 uptake by the ocean for the decade 1980-1989. Only such results are included that were obtained by a deconvolution of the Siple-Mauna Loa record of Figure 1, and only from models which were either calibrated or tested with a time-dependent tracer, so that quantitative information is available on their time-dependent behaviour.

Table 2: Oceanic CO_2 uptake rate (GtC/yr), average for 1980-89, obtained by deconvolving the atmospheric CO_2 record

Model	*Oceanic uptake*	*Flux from biota*[1)]
Box-diffusion model, $K=7100\cdot m^2yr^{-1}$ (Siegenthaler and Oeschger 1987)	2.4	0.3
HILDA (Siegenthaler and Joos 1992)	2.1	0.1
Princeton 3-D model (Sarmiento et al. 1992)	1.9	-0.2
Hamburg 3-D model (E. Maier-Reimer, personal communication)	1.5	-0.6
Outcrop-diffusion model (Siegenthaler and Oeschger 1987) [2)]	(3.6)	(1.5)

1) Column 3 gives the net biospheric source (or sink, if negative) required to balance the difference between fossil-fuel CO_2 emissions and oceanic uptake.

2) The result from the outcrop-diffusion model is not considered as realistic, because the model assumes an infinitely fast mixing between the deep sea and its high-latitude surface outcrop.

Of the models on Table 2, BD and HILDA are calibrated with bomb-^{14}C. The 3-D models are not calibrated, since they produce their own water circulation from dynamical laws and do not simulate it with adjustable parameters (except for eddy diffusivity). Toggweiler et al. (1989b) found that the Princeton model (their version P) yields a smaller oceanic inventory of bomb-^{14}C than observed, while the simulated average surface ^{14}C concentration is well simulated. Toggweiler et al. point out that a higher ^{14}C inventory could be obtained by either increasing the gas exchange rate or the vertical mixing in the upper ocean. However, the two processes affect the surface concentration in opposite ways (Broecker et al., 1980). Higher gas exchange rate will increase the air-to-sea transfer of bomb ^{14}C, but also lead to higher surface concentrations. Stronger oceanic mixing will lead to more rapid transport to depth and therefore to larger inventories but lower surface concentrations. In order to increase the inventory but not to change the average surface concentration in the Princeton model, both, gas exchange and upper ocean mixing would therefore have to be increased. Based on this, Sarmiento et al. (1992) conclude that their result (1.9 GtC yr^{-1} on average for 1980-1989) probably represents a lower limit for the actual CO_2 uptake. Thus, it seems that the Hamburg model, yielding 1.5 GtC yr^{-1}, underpredicts the CO_2 uptake (and therefore probably also that of bomb-^{14}C).

Summarizing the results of Table 2, I conclude that the best estimate for the CO_2 flux to the ocean in the decade 1980-89 is 2.0 GtC yr^{-1}, with an estimated error of ±0.5 GtC yr^{-1}. In the IPCC assessment (Watson et al. 1990), a conservative, larger error was indicated under the impression of the very low estimate of oceanic CO_2 uptake by Tans et al. (1990). However, the estimate of Tans et al. is most probably significantly too low, because the oceanic pCO_2 observations are insufficient to strongly constrain the northern-hemispheric uptake and because there must have been a non-zero natural interhemispheric CO_2 concentration difference which was not taken into account by Tans et al.; see Keeling et al. (1989a), Broecker and Peng (1992) and Sarmiento and Sundquist (1992).

Obviously, the flux into the ocean changes with time as atmospheric CO_2 increases. This flux, and therefore the airborne fraction of the anthropogenic emissions, also depends on the history of atmospheric CO_2 concentration (or equivalently, of the emissions). Oeschger et al. (1975) discussed this for an exponentially increasing emission rate. The shorter the e-folding time, the steeper the resulting vertical gradient of excess CO_2 in the ocean and therefore the larger the oceanic uptake. Also, a substantial CO_2 input into the atmosphere in the past, such as reconstructed for the 19th century from deconvolutions of the CO_2 record (Figure 6), can affect the flux to the ocean in later time by providing a decreasing atmospheric "baseline" (Siegenthaler et al. 1978, Oeschger and Heimann 1983). The assumption that the rate of atmospheric increase is a constant fraction of the current emissions, which has sometimes been made, is therefore not a good approximation. This is most clearly illustrated by a hypothetical stop of CO_2 emissions: after such a stop, the atmospheric concentration will decline because of the oceanic sink, while a "constant airborne fraction" concept would predict that the concentration remains constant.

With the model-based estimate of the oceanic uptake, a "missing", probably biotic, sink of 1-2 GtC yr^{-1} must exist (Table 1). For predicting future CO_2 levels based on emission scenarios, the change of this ill-defined sink with time is important: will it remain constant in future, or will it increase approximately in proportion to the atmospheric CO_2 concentration, as suggested for the past by model deconvolutions? Obviously, the computed future greenhouse forcing depends on the assumptions about this (Wigley 1992). It is therefore important to determine the magnitude and nature of this biotic sink. Unfortunately, this is a very difficult task in view of the extreme heterogeneity of the terrestrial vegetation and soils.

Regional distribution of CO_2 uptake and oceanic inventories

Regions of strong uptake of anthropogenic CO_2 do not necessarily coincide with the regions where the inventory is large, since ocean currents can redistribute oceanic properties. The net air-sea flux, given by $k_g(pCO_{2atm}-pCO_{2s})$, depends on wind speed (via gas exchange coefficient k_g) and on ocean circulation (influencing pCO_{2s}). Sarmiento et al. (1992) showed that the flux is particularly large in high latitudes and in low-latitude upwelling zones such as the equatorial Pacific (Figure 7a). In the high latitudes, strong winds cause a large gas exchange coefficient and at the same time, the rapid oceanic circulation exports anthropogenic CO_2 rapidly to depth and to lower

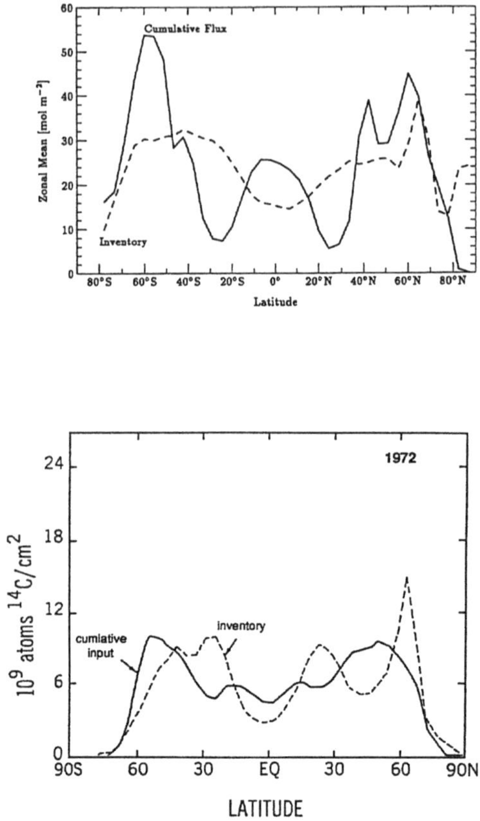

Figure 7 Results of simulations of the oceanic uptake of anthropogenic CO_2 (top; Sarmiento et al. 1992) and of bomb-^{14}C (bottom; Toggweiler et al. 1989b) obtained using the 3-dimensional Bryan-Cox ocean general circulation model. Shown are zonal means of the cumulative air-sea flux of CO_2 (till 1986) and ^{14}C (till 1972) as well as the corresponding column inventories.

latitudes. The water which upwells in coastal regions or near the equator has been separated from the atmosphere for some time and is therefore relatively little contaminated by excess CO_2. It is interesting that the equatorial Pacific, where pCO_2 of surface water is generally much higher than in the atmosphere because of rapid advection of carbon-rich water from deeper layers, is an efficient sink for man-made CO_2. This apparent paradox can be understood such that the outgassing flux of carbon dioxide from equatorial waters is smaller at present than in pre-industrial time, the difference corresponding to a net uptake of excess CO_2. As Figure 7a shows, the column inventory of excess CO_2 varies much less with latitude than the input. The circulation transports the perturbation from equatorial and high latitudes towards the mid-latitude subtropical gyres, where Ekman-induced sinking - and therefore convergence - takes place. This transport pattern creates at the same time a relatively large air-sea pCO_2 difference in the regions of divergence, which leads to large net invasion fluxes.

It is of interest to compare the pattern of input and inventory for the perturbations of CO_2 and of ^{14}C (Figures 7a and 7b). For bomb-^{14}C, the input varies much less with latitude than for CO_2. This is so because surface water is equilibrated more rapidly with respect to CO_2 than with respect to ^{14}C, due to the chemical buffer factor which only affects CO_2 but not the $^{14}C/C$ ratio. For both tracers, the flux can be represented by

$$F = k_g\ (c_{atm}-c_s) = k_g\ c_{atm}\ (1-e_s)$$

where c_{atm} and c_s represent either pCO_2 or the $^{14}C/C$ ratio in air and in sea water, and $(1-e_s) = (1-c_s/c_{atm})$ is the relative concentration deficit (or degree of disequilibrium with the atmosphere) which drives the net flux. The relative concentration deficit is small for pCO_2, but large for bomb-^{14}C as can be verified from the GEOSECS observations. Consequently, a relatively small change of c_s causes a significant change of $(1-e_s)$ and therefore of the flux for CO_2, but not for ^{14}C.

ISOTOPIC PERTURBATIONS: BOMB-PRODUCED ^{14}C AND THE SUESS EFFECT

Carbon of different origin is labelled by its isotopic composition, i.e. by its isotope ratios $^{14}C/C$ and $^{13}C/^{12}C$ (usually indicated as $\delta^{13}C$*). These isotope ratios are valuable tracers

---

* $\delta^{13}C$ denotes the relative deviation of the $^{13}C/^{12}C$ ratio in a sample from that of a standard (usually PDB). $^{14}C/C$ ratios are often indicated by the $\Delta^{14}C$ value, that is the relative deviation of the ^{13}C-normalized $^{14}C/C$ ratio from that of the internationally used standard which corresponds to 100% modern. $\delta^{13}C$ and $\Delta^{14}C$ are both indicated in permil.

for studying the carbon cycle. Fossil fuels bear no ^{14}C, and fossil as well as biospheric carbon has a lower $\delta^{13}C$ (ca.-25‰) than atmospheric CO_2 (ca.-7‰). Therefore, addition of anthropogenic CO_2 to the atmosphere has resulted in observable isotopic signals. In addition, the atmospheric ^{14}C concentration was strongly perturbed by bomb-produced radiocarbon from the early 1950s on.

When considering the increase of the atmospheric CO_2 concentration, a relevant question is which fraction of the anthropogenic carbon resides in which reservoir. For the isotopes, the analogous question, which fraction of an isotopic perturbation resides in the different carbon reservoirs at a given time, is not clearly posed and has no simple answer. An isotopic ratio is the ratio of two concentrations, and correspondingly, $\delta^{13}C$ or $\Delta^{14}C$ of atmospheric CO_2 can change either because the concentration of CO_2 changes or because the concentration (atoms per kg of air) of ^{13}C or ^{14}C changes. This makes the interpretation of isotopic changes more complicated than of CO_2 concentration changes. For bomb-produced ^{14}C, the above question can approximately be answered, because the CO_2 increase between the large thermonuclear tests (1961/62) and the GEOSECS survey, which provided bomb-^{14}C inventories for the major oceanic basins, was small. According to a model calculation using the box-diffusion ocean model and a four-box biosphere model (Siegenthaler and Oeschger 1987), the perturbations of $\Delta^{14}C$ in atmosphere, land biota and whole ocean were 435‰, 122‰ and 13.9‰, respectively, in 1974 (approximate mid-time of the GEOSECS survey). With the appropriate reservoir sizes, this yields the following rough partitioning of bomb-^{14}C in 1974: atmosphere 30%, ocean 45%, land biota 25%. In reality, the amounts of carbon in the different pools did change somewhat, so that even for a constant $^{14}C/C$ ratio, the atmospheric amount of $^{14}CO_2$ would have increased by about 4% from 1962 to 1974. However,this is small compared to the addition of bomb-produced ^{14}C.

The situation is different when biospheric or fossil-fuel CO_2 is added to the atmosphere (or withdrawn from it). The seasonal exchange of CO_2 between the atmosphere and the land biota occurs fast enough that attenuation of the signal due to exchange with the ocean can be neglected. This exchange creates a signal in concentration as well as in $\delta^{13}C$, as demonstrated in Figure 8. A plot of $\delta^{13}C$ vs. CO_2 concentration (Fig. 8c) shows that the data points are clustered along a line corresponding to addition or

Figure 8 Seasonal variations in CO_2 concentration and $\delta^{13}C$ of atmospheric CO_2 at Mauna Loa (Mook et al., 1983). Time scale: 81 denotes 1 January, 1981. Bottom: $\delta^{13}C$ versus concentration for the same samples. The arrow in the indicates the effect of the addition of CO_2 with $\delta^{13}C$=-25‰ for a sample of arbitrarily chosen initial composition.

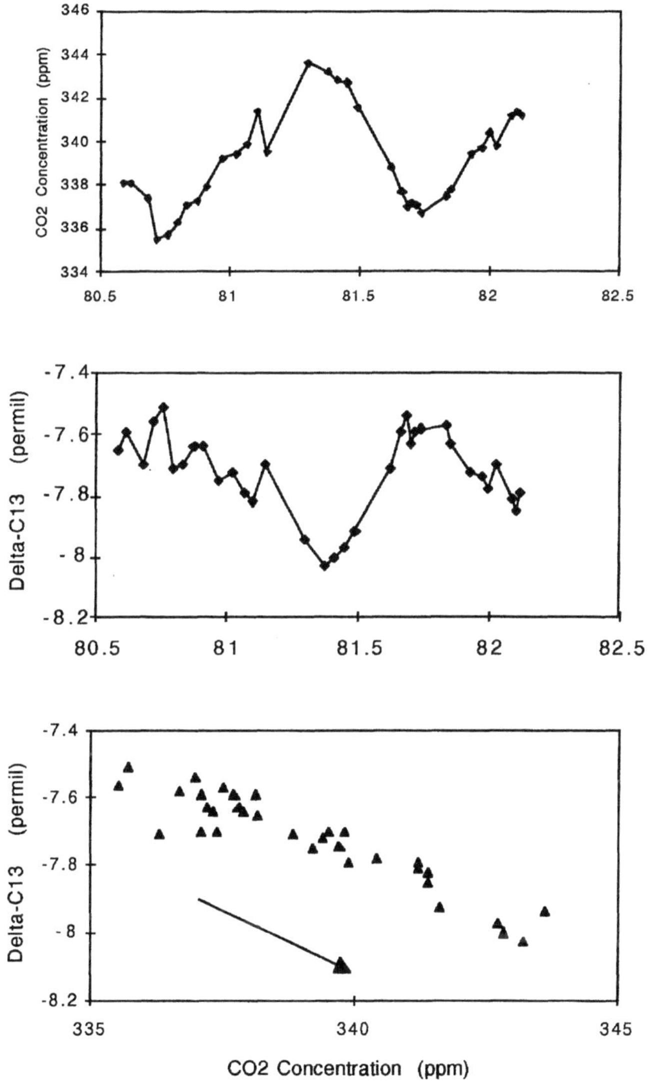

withdrawal of carbon with $\delta^{13}C \approx -25‰$. This correlation, which corresponds to a two-component mixture, is found for rapid perturbations, but not for variations with a longer time scale.

The anthropogenic CO_2 increase has been accompanied by a decrease of $\delta^{13}C$ and, till the first thermonuclear tests, of $\Delta^{14}C$ (Suess effect). The atmospheric CO_2 increase until 1950 was about 11%, of which ca. 5% were due to fossil fuel burning according to model calculations, while the $^{14}C/C$ ratio decreased only by ca. 1.7% up to that date (if corrected for a small trend due to the varying solar magnetic field; Stuiver and Quay 1981). If no exchange with the ocean and terrestrial biota had taken place, the relative changes of the $^{14}C/C$ ratio and of the CO_2 concentration due to fossil fuels would simply be inversely related, as is the case for the seasonal variations. While the atmospheric $^{14}C/C$ ratio had decreased by 1.7% due to the input of ^{14}C-free carbon, the number of $^{14}CO_2$ molecules in the atmosphere had increased by about 9% till 1950. The reason is that ocean and biosphere do not only take up CO_2 by gas exchange and by photosynthesis, but they also release CO_2 with the corresponding $^{14}C/C$ ratio back to the atmosphere. The isotopic composition of the carbon released from ocean and biosphere is less perturbed than of the carbon taken up. In this way, all exchanges between atmosphere and other reservoirs help to dilute an isotopic perturbation, even if these exchanges do not provide a net carbon transport and therefore do not influence the atmospheric CO_2 concentration. The land biota take up bomb-^{14}C as well as the isotopic signals of the ^{14}C and ^{13}C Suess effect. As to the exchange with the ocean, the buffer factor reduces the oceanic uptake capacity for CO_2 by about a factor 10, compared to the amount of carbon it contains, while the whole exchanged amount of carbon dilutes the isotopic perturbations.

Figure 9 shows the time trends of CO_2 concentration and $\delta^{13}C$ in the atmosphere, as obtained from direct observations and analyses of CO_2 in air bubbles trapped in a polar ice core. In Figure 9c, the correlation $\delta^{13}C$-CO_2 for these data is plotted. While this correlation is approximately linear, the slope deviates considerably from that for an addition of CO_2 with $\delta^{13}C=-25‰$ (arrow), because the isotopic signal is attenuated more strongly than the concentration increase. This is illustrated by the different atmospheric responses to a concentration pulse and an isotopic pulse (Figure 10a). In the first few years after the perturbation, the slope of the $\delta^{13}C$-CO_2 curve (Fig. 10b) is as for a simple two-component mixture,

Figure 9 CO_2 concentration and $\delta^{13}C$ in the atmosphere, as obtained from direct atmospheric observations (Keeling et al. 1979; Mook et al 1983) and from analyses of CO_2 extracted from air trapped in an ice core from Siple station, Antarctica (Neftel et al., 1985, Friedli et al. 1986). Bottom: $\delta^{13}C$ versus concentration for the same samples. The arrow indicates the effect of the addition of CO_2 with $\delta^{13}C$=-25‰ for a sample of 280 ppm and -6.4‰.

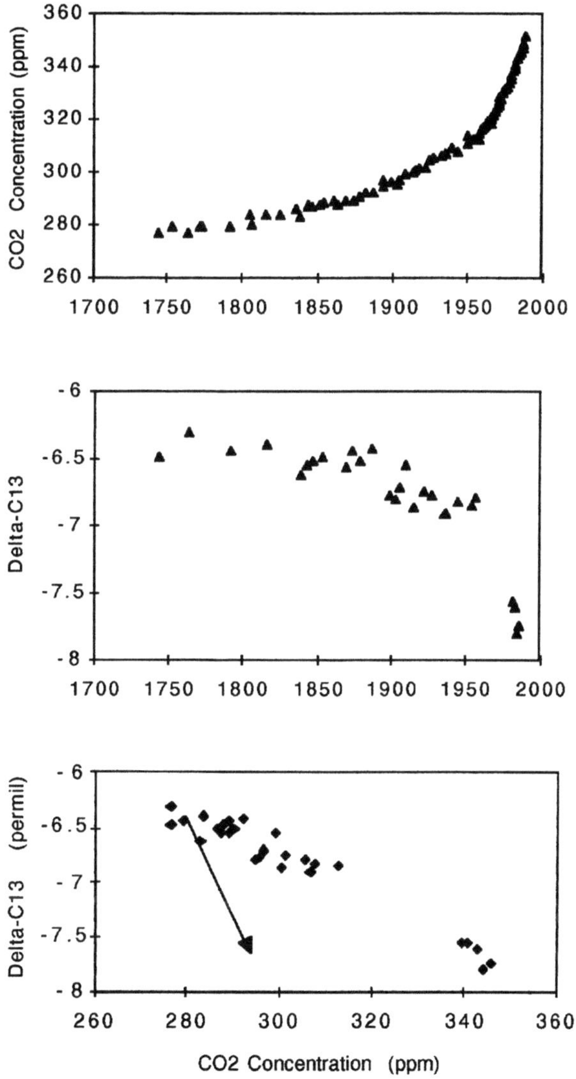

but then the curve becomes flatter because of exchange with the ocean and the biota.

Siegenthaler and Oeschger (1987) calculated the isotopic changes due to the anthropogenic carbon emissions, using the box-diffusion model with a four-box model for vegetation and soils. They obtained a $\delta^{13}C$ decrease of 1.19‰ until 1980, compared to an observed shift of 1.14‰ (Friedli et al. 1986). For $\Delta^{14}C$, the observed decrease until 1950 is 17‰, and the model result is 19‰. The model results depend on the model ocean as well as on the biospheric submodel. The influence of the biosphere can be tested in the model by setting the exchange flux atmosphere-biophere equal zero. Then, instead of the above figures, the following changes are obtained: $\delta^{13}C$ decrease until 1980: 1.49‰, $\Delta^{14}C$ decrease until 1950: 25‰. While the oceanic transport parameters can be calibrated using bomb-^{14}C, there is no similar means to calibrate the biosphere model with tracers, which limits the accuracy of the model results for the isotopic perturbations.

An interesting application of $\delta^{13}C$ measurements stems from Quay et al. (1992) who estimated the oceanic uptake of anthropogenic CO_2 based on a combination of balances for the perturbations of total carbon and of ^{13}C. The two balances yield two equations, from which two unknowns - the oceanic CO_2 uptake and the net change of global biomass - can be determined. The balance for carbon between the two times t_0 and $t_1 = t_0 + \Delta t$ can be written as

$$\Delta N_a = (F_{ff} + F_{br} - F_{bu})\Delta t - A_{oc}\int (c_1 - c_0)\,dz \qquad (1)$$

where
- N_a amount of CO_2 in the atmosphere
- F_{ff} emission rate of fossil-fuel derived CO_2
- F_{br} biospheric release by respiration plus deforestation
- F_{bu} biospheric uptake by photosynthesis
- A_{oc} ocean surface area
- c_0, c_1 oceanic ΣCO_2 concentration at times t_0 and t_1.

The fluxes F_{ff}, F_{br} and F_{bu} are time averages over the period t_0 to t_1. The last term in eq. (1) gives the CO_2 uptake by the ocean. The balance for ^{13}C reads, with $R = {}^{13}C/{}^{12}C$ ratio:

$$\Delta (N_a R_a) = (F_{ff} R_{ff} + F_{br} R_{br} - F_{bu} R_{bu})\Delta t - A_{oc}\int (c_1 R_1 - c_0 R_0)\,dz \qquad (2)$$

The relation to the oceanic CO_2 uptake and the rate of deforestation can be made by decomposing two terms of eq. (2) as follows:

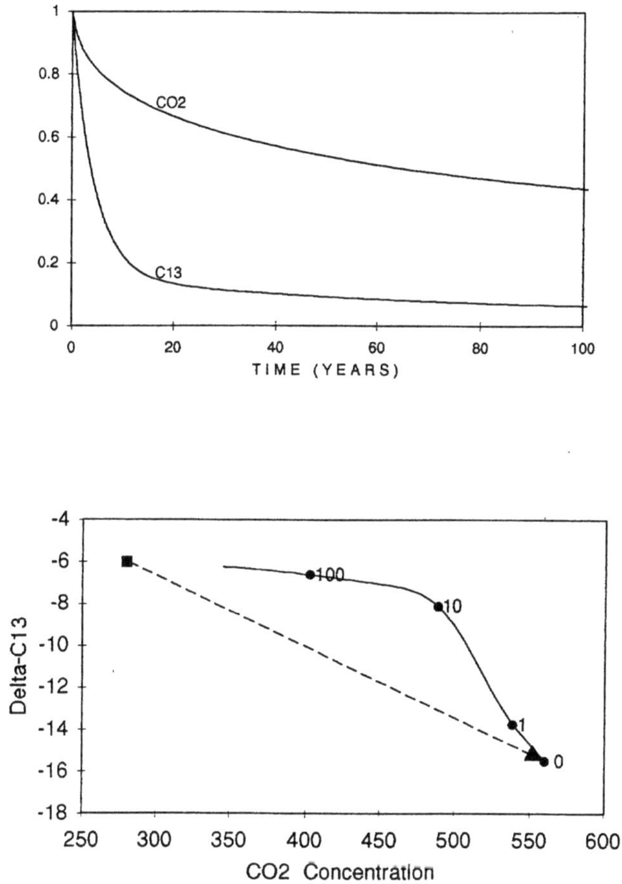

Figure 10 Top: Response of the concentration and the $\delta^{13}C$ value of CO_2 in the atmosphere to a pulse input of isotopically labelled CO_2 ($\delta^{13}C$=-25‰), yielding an initial doubling of the concentration. In order to compare the two curves, they are plotted on a normalized scale, with 0 corresponding to the original composition (280 ppm, -6‰) and 1 to the values immediately after the pulse injection (560 ppm, -15.5‰). Bottom: Plot of $\delta^{13}C$ vs. CO_2 concentration. The square denotes the original composition, circles mark the times 0, 1, 10 and 100 years.

$$F_{br}R_{br} - F_{bu}R_{bu} = (\mathbf{F_{br}} - \mathbf{F_{bu}})R_{br} + F_{bu}(R_{br} - R_{bu}) \tag{3}$$

$$A_{oc}\int(c_1R_1 - c_0R_0)\,dz = R_0\mathbf{A_{oc}}\int\mathbf{(c_1 - c_0)\,dz} + A_{oc}c_1\int(R_1 - R_0)\,dz \tag{4}$$

Eq. (4) means that the change in oceanic amount of ^{13}C can be considered as the sum of increase in ΣCO_2 times an appropriate average oceanic $^{13}C/^{12}C$ ratio at time t_0 plus an average ΣCO_2 concentration at time t_1 times the depth-integrated change in the $^{13}C/^{12}C$ ratio; eq. (3) can be interpreted in a similar way. By combining eqs. (1)-(4), the oceanic uptake, $A_{oc}\int(c_1 - c_0)dz$, and the rate of deforestation, $(F_{br} - F_{bu})$, can be calculated, provided all other terms are known. Atmospheric CO_2 amount, emission rates and isotopic composition of fossil fuels, oceanic ΣCO_2 and photosynthesized carbon are well known, and the isotopic changes in atmosphere and ocean have been measured. The global net primary productivity (F_{bu}) is reasonably well known, but the isotopic difference between CO_2 taken up by the biosphere, $R_{br} - R_{bu}$, must be estimated. In this way, Quay et al. (1992) estimated that the average oceanic CO_2 uptake rate between 1970 and 1990 was 2.1±0.8 GtC yr^{-1}, the corresponding net biospheric release -0.1 GtC yr^{-1}. For the decade 1980-90, model calculations yield about 13% higher oceanic uptake than for 1970-90, so that Quay et al.'s result corresponds to 2.4 GtC yr^{-1} for this period, which is within the range of carbon cycle model estimates (Table 2). The method is not precise, as indicated by the large estimated error (1 standard deviation). One reason is that the atmospheric and oceanic $\delta^{13}C$ measurements are not yet sufficient to precisely determine the isotopic changes Thus, Quay et al. estimated a $\delta^{13}C$ change of -0.39‰ for average surface water between 1970 and 1990, compared to an atmospheric change of -0.40‰. This near-equality is in contrast to expectations from models, according to which the change in surface water should be clearly smaller than in the atmosphere: the box-diffusion model yields changes of -0.29‰ and -0.55‰, respectively, for this time period. A second problem is that the $\delta^{13}C$ difference between respired and photosynthesized CO_2 is difficult to assess. It cannot be measured directly, and estimates based on simple biota models are uncertain. Thus, Quay et al. have exchanged the uncertainties of other methods used to estimate the oceanic uptake for possibly larger uncertainties connected to isotopic changes. Nevertheless, their approach is valuable, since it approaches the question about the role of the ocean as a CO_2 sink in a novel and very stimulating way.

References

Bacastow R, Maier-Reimer E (1990) Ocean-circulation model of the carbon cycle. Climate Dynamics 4: 95-126

Broecker WS, Peng T-H, Eng R (1980) Modeling the carbon system. Radiocarbon 22: 565-580

Broecker WS, Peng T-H (1982) Tracers in the Sea. Eldigio Press Lamont-Doherty Geological Observatory Palisades NY

Broecker WS, Peng T-H (1992) Interhemispheric transport of carbon dioxide by ocean circulation. Nature 356: 587-589

Broecker WS, Peng T-H, Ostlund G, Stuiver M (1985) The distribution of bomb radiocarbon in the ocean. J Geophys Res 90: 6953-6970

Bryan K, Komro FG, Rooth C (1984) The ocean's transient response to global surface temperature anomalies. In: Hansen JE, Takahashi T (eds) Climate Processes and Climate Sensitivity. AGU Washigton DC Geophys Monograph 29: 29-38

Craig H (1969) Abyssal carbon and radiocarbon in the Pacific. J Geophys Res 74: 5491-5506

Enting IG (1990) Ambiguities in the calibration of carbon cycle models. Inverse Problems 6: L39-L46

Etcheto J, Boutin J, Merlivat L (1991) Seasonal variation of the CO_2 exchange coefficient over the global ocean using satellite wind speed measurements. Tellus 43B: 247-255

Friedli H, Loetscher H, Oeschger H, Siegenthaler U, Stauffer B (1986) Ice core record of the $^{13}C/^{12}C$ ratio of atmospheric carbon dioxide in the past two centuries. Nature 324: 237-238

Garrett C (1979) Mixing in the ocean interior. Dyn Atmos Oceans 3: 239-265

Garrett C (1982) On the parameterization of diapycnal fluxes due to double-diffusive intrusions. J Phys Oceanography 12: 952-959

Houghton RS (1991) The role of forests in affecting the greenhouse gas composition of the atmosphere. In: Wyman RL (ed) Global Climate Change and Life on Earth: 43-55 Chapman and Hall New York

Joos F, Sarmiento JL, Siegenthaler U (1991a) Estimates of the effect of Southern Ocean iron fertilization on atmospheric CO_2 concentrations. Nature 349: 772-775

Joos F, Siegenthaler U, Sarmiento JL (1991b) Possible effects of iron fertilization in the Southern Ocean on atmospheric CO_2 concentration. Global Biogeochem Cycles: in press

Keeling CD (1991) In: Boden TA, Sepanski RJ, Stoss FW (eds) Trends '91 - A Compendium of Global Change: 382-385. CDIAC Carbon Dioxide Information Analysis Center Oak Ridge National Laboratory Oak Ridge TN

Keeling CD, Mook WG, Tans PP (1979) Recent trends in the $^{13}C/^{12}C$ ratio of atmospheric carbon dioxide. Nature 277: 121-123

Keeling CD, Bacastow RB, Carter AF, Piper SC, Whorf TP, Heimann M, Mook WG, Roeloffzen H (1989a) A three dimensional model of atmospheric CO_2 transport based on observed winds: 1. Analysis of observational data. Geophysical Monograph 55: 165-236

Keeling CD, Piper SC, Heimann M (1989b) A three dimensional model of atmospheric CO_2 transport based on observed winds:

4. Mean annual gradients and interannual variations. Geophysical Monograph 55: 305-363
Li Y-H, Peng T-H, Broecker WS, Östlund G (1984) The average vertical mixing coefficient for the oceanic thermocline. Tellus 36B: 212-217
Liss PS, Merlivat L (1986) In: Buat-Ménard P (ed) The Role of Air-Sea Gas Exchange in Geochemical Cycling: 113-127 Reidel
Maier-Reimer E, Hasselmann K (1987) Transport and storage of CO_2 in the ocean - an inorganic ocean-ciculation carbon cycle model. Climate Dynamics 2: 63-90
Marland G, Boden T (1991) In: Boden TA, Sepanski RJ, Stoss FW (eds) Trends '91 - A Compendium of Global Change: 386-389 CDIAC Carbon Dioxide Information Analysis Center Oak Ridge National Laboratory Oak Ridge TN
Mook WG, Koopmans M, Carter AF, Keeling CD (1983) Seasonal, latitudinal and secular variations in the abundance and isotopic composition of atmospheric carbon dioxide. J Geophys Res 88: 10915-10933
Munk WH (1966) Abyssal recipes. Deep-Sea Res 13: 707-730
Neftel A, Moor E, Oeschger H, Stauffer B (1985) Evidence from polar ice cores for the increase in atmospheric CO_2 in the past two centuries. Nature 315: 45-47
Oeschger H, Siegenthaler U, Schotterer U, Gugelmann A (1975) A box diffusion model to study the carbon dioxide exchange in nature. Tellus 27: 168-192
Oeschger H, Heimann M (1983) Uncertainties in predictions of future atmospheric CO_2 concentrations. J Geophys Res 88: 1258-1262
Peng T-H, Takahashi T, Broecker WS (1987) Seasonal variability of carbon dioxide, nutrients and oxygen in the North Atlantic surface water: observations and a model. Tellus 39B: 439-458
Quay PD, Tilbrook B, Wong CS (1992) Oceanic uptake of fossil fuel CO_2: carbon-13 evidence. Science 256: 74-79
Sarmiento JL, Feely HW, Moore WS, Bainbridge AE, Broecker WS (1976) The relationship between vertical eddy diffusion and buoyancy gradient in the deep sea. Earth Planet Sci Lett 32: 357-370
Sarmiento JL, Sundquist ET (1992) Revised budget for the oceanic uptake of anthropogenic carbon dioxide. Nature 356: 589-593
Sarmiento JL, Orr JC, Siegenthaler U (1992) A perturbation simulation of CO_2 uptake in an Ocean General Circulation Model. J Geophys Res 97: 3621-3645
Shaffer G, Sarmiento JL (1992) Biogeochemical cycling in the global ocean 1. A new, analytical model with continuous vertical resolution and high latitude dynamics. J Geophys Res: in press
Siegenthaler U (1983) Uptake of excess CO_2 calculated by an outcrop diffusion model of the ocean. J Geophys Res 88: 3599-3608
Siegenthaler U (1986) Carbon dioxide: its natural cycle and anthropogenic perturbation. In: Buat-Ménard P (ed) The Role of Air-Sea Exchange in Geochemical Cycling: 209-247 Reidel
Siegenthaler U (1989) Carbon-14 in the oceans. In: Fritz P, Fontes J-C (eds) Handbook of Environmental Isotope Geochemistry vol 3, the Marine Environment: 75-137 Elsevier A Amsterdam

Siegenthaler U, Oeschger H (1978) Predicting future atmospheric carbon dioxide levels. Science 199: 388-395
Siegenthaler U, Oeschger H (1987) Biospheric CO_2 emissions during the past 200 years reconstructed by deconvolution of ice core data. Tellus 39B: 140 - 154
Siegenthaler U, Joos F (1992) Studying anthropogenic CO_2 perturbations and oceanic tracers using a diffusive-advective ocean model (HILDA). Tellus: in press
Siegenthaler U, Heimann M, Oeschger H (1978) Model responses of the atmospheric CO_2 level and $^{13}C/^{12}C$ ratio to biogenic CO_2 input. In: Buat-Ménard P (ed) Carbon Dioxide, Climate and Society: 79-87 Pergamon Press Oxford
Siegenthaler U, Friedli H, Loetscher H, Moor E, Neftel A, Oeschger H, Stauffer B (1988) Stable-isotope ratios and concentration of CO_2 in air from polar ice cores. Annals of Glaciology 10: 1-6
Stuiver M, Quay P (1981) Atmospheric ^{14}C changes resulting from fossil fuel CO_2 release and cosmic ray flux variability. Earth Planet Sci Lett 53: 349-362
Stuiver M, Östlund HG, McConnaughey TA (1981) GEOSECS Atlantic and Pacific results. In Bolin B (ed) Carbon Cycle Modelling SCOPE 16 Wily: 201-222
Tans PP, Fung IY, Takahashi T (1990) Observational constraints on the global atmospheric CO_2 budget. Science 247: 1431-1438
Toggweiler JR, Dixon K, Bryan K (1989a) Simulations of radiocarbon in a coarse resolution world ocean model 1. Steady state prebomb distributions. J Geophys Res 94: 8217-8242
Toggweiler JR, Dixon K, Bryan K (1989b) Simulations of radiocarbon in a coarse resolution world ocean model 2. Distributions of bomb-produced carbon 14. J Geophys Res 94: 8243-8264
Watson RT, Rodhe H, Oeschger H, Siegenthaler U (1990) Greenhouse gases and aerosols. In: Houghton JT, Jenkins GJ, Ephraums JJ (eds) Climate Change - The IPCC Scientific Assessment: 1-40 Cambridge University Press
Watson RT, Meira Filho LG, Sanhueza E, Janetos A (1992) Greenhouse gases: sources and sinks. In: Houghton JT, Callander BA, Varney SK (eds) Climate Change 1992 - The Supplementary Report to the IPCC Scientific Assessment: 25-46 Cambridge University Press
Welander P (1968) Theoretical forms for the vertical exchange coefficient in a stratified fluid with application to lakes and seas. Acta Reg Soc Sci Litt Gothoburgensis Geophys 1:1-27
Wigley TML (1992) Balancing the carbon budget: implications for projections of future carbon dioxide concentration changes. Tellus: submitted

AIR-SEA GAS EXCHANGE AND CARBON DIOXIDE

By Andrew Watson
Plymouth Marine Laboratory
Prospect Place, West Hoe, Plymouth PL1 3DH UK

INTRODUCTION

An understanding of air-sea exchange, and the ability to calculate fluxes of interesting atmospheric gases into and out of the oceans, is an important tool in global biogeochemistry. For some gases, (dimethylsulphide for example) the process of gas exchange is itself a limiting factor controlling the flux: this gas is generated in excess in surface water, and the amount getting into the atmosphere is limited by the rate of air-sea exchange.

For the net uptake of carbon dioxide by the ocean, gas exchange is not however a very important limiting factor. For the relatively slow increase of CO_2, the main limitation to ocean uptake seems to be the rate of vertical mixing of the ocean. The surface layer has a rather limited capacity to absorb atmospheric increases, so, for example, doubling the gas transfer rate gives only a small (of order 10%) increase in the rate of uptake. (Sarmiento, Orr and Siegenthaler, 1991). However, the rate of exchange with the ocean is nevertheless of great interest in carbon cycle studies for two reasons. First, given adequate observations of surface ocean pCO_2 we ought to be able to calculate the rate of ocean uptake by integrating the gas transfer rate over the world ocean, thus providing an independent check on the rate of uptake derived from ocean models. Second, not only the magnitude but also the distribution of the fluxes to and from the ocean are required to constrain atmospheric inversion techniques, which attempt to infer the distribution of the net sources and sinks on land and sea. At present, these calculations have shown that the northern hemisphere sink exceeds that in the southern hemisphere by about 2 Gt C yr^{-1} (Heimann, this volume). How much of this is asymmetry is due to land sinks and how much to the ocean depends on how the gas exchange calculation is performed.

NATO ASI Series, Vol. I 15
The Global Carbon Cycle
Edited by M. Heimann

BACKGROUND

The sea surface forms a barrier which slows down vertical transport. Away from the boundary in both the ocean and the atmosphere, molecules can be transported rapidly in the vertical by eddy motions. But at the boundary between the two phases these vertical motions are suppressed by the presence of the horizontal surface. Very close to the surface convective motions become so slow that molecular diffusion alone is left to transport the molecules the last few tens of microns. A measurable concentration gradient therefore tends to build up across the barrier for gases with significant fluxes into or out of the oceans. With a suitable model of the gas exchange process, observations of this gradient can be used to calculate the flux.

The basic parameter used to describe the rate of gas transfer is the transfer velocity K.. K is *defined* by the equation

$$F=K.\Delta C \qquad \text{(i)}$$

where F is the flux across the surface and

$$\Delta C = C_W - \alpha C_{air}$$

is the difference in concentration across the water-side boundary layer limiting the exchange. In this equation a is the Ostwald solubility constant of the gas in question, so that αC_{air} is the concentration of gas expected at the top of the boundary layer, in contact with, and at equilibrium with, the air. For CO_2, equation (i) is more usually expressed in terms of partial pressure differences, writing $C_W = S pCO_2$, where pCO_2 is the partial pressure of CO_2 which would be in equilibrium with the bulk water and S is the solubility of CO_2 per unit of pressure, we can alternatively write

$$F = K'.\Delta pCO_2 \qquad \text{(ii)}$$

where $K' = KS$ is called the CO_2 gas transfer coefficient, and $\Delta pCO_2 = pCO_{2w} - pCO_{2a}$ is the difference in partial pressures of CO_2 between air and sea. The basic equation contains little real physical information. The units of K in (i) turn out to be those of velocity, but it is not the "real" velocity of the molecules, or the waves, or anything else. The equation implies that the flux is linear in concentration difference and equal to zero when ΔC is zero. While this should be true when the surface is not broken up by whitecaps, we shall see that for the real ocean these assumptions are probably not correct.

The theory and experiments which go to make up our understanding of gas transfer processes have been reviewed by Liss and Merlivat (1986) and Liss (1983). The present paper is intended to stand by itself and therefore repeats a little of this material by way of introduction, but it is not intended to be a comprehensive review of the subject of air-sea exchange. Reference should be made to the earlier literature for a more detailed treatment.

Transport is inhibited in the boundary layers on both the air and water sides of the interface. As described by i.e. Liss (1983), the overall transfer can be modelled as two series resistances to transport. However, for CO_2 and all less soluble gases, the resistance of the air-side boundary layer is much less than that of the water side, so that for all practical purposes the air side can be neglected. In common with previous treatments, that given here concentrates exclusively on the water-side boundary layer.

Models

At low and moderate winds, transfer occurs exclusively by molecular transport across the "viscous sub-layer" -- the region so close to the boundary that vertical eddy motions are supressed and molecular diffusion is dominant. Three kinds of models for this transport have been proposed: The *stagnant film* model envisages that the surface is permanently covered by a film of specified thickness within which there is no transfer except by molecular diffusion. The model is physically unrealistic but because of its simplicity has been widely used. The model gives for the gas transfer velocity:

$$K = \frac{D}{z} \qquad \text{(iii)}$$

where D is the diffusivity of the gas in water and z is a free parameter corresponding to the thickness of the film.

In *surface renewal* models (i.e Danckwerts, 1951) the surface film is periodically renewed from the bulk of the material with frequency of renewal f. The gas transfer velocity is given by

$$K = \text{Const} \times \sqrt{D.f} \qquad \text{(iv)}$$

The most sohisticated models are *turbulent boundary layer* treatments. That of

Deacon (1977) is applicable to a smooth water surface under very calm conditions and predicts that K is proporttonal to $D^{2/3}$. The analysis of Ledwell (1982) is applicable to a roughened (but not broken) surface. Ledwell derives

$$K=\frac{2}{\sqrt{10.\pi.a}}\cdot\sqrt{\frac{D}{\nu}}\cdot U^*_w \qquad \text{(v)}$$

where a is a dimensionless constant of order 2.5 and ν is the viscosity of water. U^*_w is the friction velocity in the water (friction velocity is the scale velocity generated at the surface by the action of the wind on the water), calculated as

$$U^*_w=\sqrt{C_d}\cdot\sqrt{\frac{\sigma_a}{\sigma_w}}\cdot U \qquad \text{(vi)}$$

where U is the wind speed measured at a reference height, normally 10m, C_d is an effective coefficient of drag, and σ_a, σ_w are the densities of air and water respectively. The coefficient of drag is frequently taken to be a constant of order 1.5×10^{-3}, though actually it is apparent that it too will be a function of the wind speed, roughness of the sea and the thermal stability of the lowermost atmosphere. For example, Trenburth *et al.*, (1989) suggest that for a neutral atmosphere, C_d may be parameterised as

$$C_d = \begin{cases} (0.62 + 1.56U) \times 10^{-3} & (U < 3\,\mathrm{m\,s^{-1}}) \\ 1.14 \times 10^{-3} & (3 < U < 10\,\mathrm{m\,s^{-1}}) \\ (0.49 + 0.065U) \times 10^{-3} & (U > 10\,\mathrm{m\,s^{-1}}) \end{cases} \qquad \text{(vii)}$$

The analysis of Ledwell combined with such a representation of the friction velocity thus makes rather specific predictions about the gas transfer velocity and its variation with wind speeds over the ocean. The absolute value cannot be defined by even this fairly sophisticated treatment without reference to an empirically determined constant, (i.e the parameter a in equation (v)), but this can be determined with reasonable accuracy by comparison with experiments and field measurements. Figure 1 shows the predicted variation of K with wind speed obtained by combining equations (v) and (vii) above, compared with some measurements obtained at sea by the radon deficit and dual tracer techniques and with the "synthesis" of Liss and Merlivat (1986), all of which are discussed further below. For moderate wind speeds, the theoretically derived relationship fits the data almost as well as the empircal Liss-Merlivat fit.

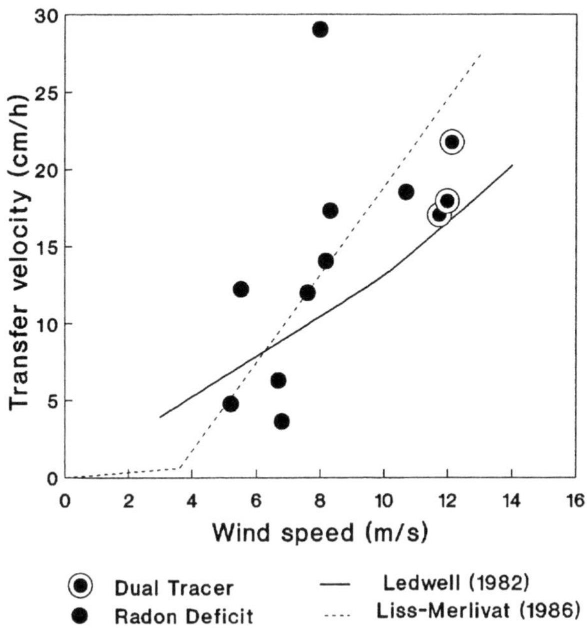

Figure 1. Solid line: the gas-exchange-wind speed curve obtained from the theoretical relationship derived by Ledwell (1982, see text) for gas transfer rate as a function of friction velocity, combined with the representation of Trenburth *et al.* (1989) for drag coefficient in a neutral atmosphere. Dotted line: the Liss-Merlivat (1986) curve. The points are gas exchange measurements made at sea.

The theoretical prediction that, at a rough surface, gas transfer should vary with the square root of molecular diffusivity (or as it is more usefully expressed, with $Sc^{-1/2}$ where Sc is the Schmidt number, ν/D) has been confirmed by experiment (Ledwell, 1982; Jahne *et al.*, 1984). Jahne *et al.* also confirmed the $Sc^{-2/3}$ dependence at a smooth surface. These results have the important practical use that exchange rates measured for one gas can be scaled to apply to any other providing the diffusivity is known. The scaling is useful because most gas transfer measurements in the environment have been made with inert gases such as radon, sulphur hexafluoride and helium, rather than with CO_2. It is common practice now to scale all such measurements to be appropriate to CO_2 at 20°C. Some care in using the scaling is advisable however, because the at high wind speeds, where effects of spray and bubbles will be important, the models so far discussed are not adequate and the $Sc^{-1/2}$ scaling may not apply.

Large regions of the ocean are subject to frequent high winds and rough seas under which the unbroken surface gas exchange will be augmented by other processes occuring at the surface of bubbles and jets. Though a number of modelling studies have been made and are in progress, (i.e Merlivat and Memery, 1983, Memery and Merlivat, 1985, Thorpe, 1982, Woolf and Thorpe, 1991) the complexity of the processes involved means that the number of ill-defined parameters is increased and a resort to relatively crude parameterisation is in practice necessary for purposes of calculation. As an example, Monahan and Spillane (1984) split the transfer velocity into separate unbroken surface and a turbulent whitecap components:

$$K_{tot} = (1\text{-}W).K_u + W.K_w.$$

where W is the proportion of the surface covered by whitecaps. This proportion is itself a function of wind speed (Monahan and Muircheartaigh, 1980, fit a power law curve to whitecap coverage data, obtaining a best fit with $W \sim U^{3.4}$). Monahan and Spillane assumed that K_u and K_w were constants, and attempted to use field measurements of transfer velocities combined with (non-simultaneous) observations of whitecap coverage to derive empirical values for them. Though the statistics of the fit are too poor to make it very useful as a predictive tool, the treatment is mentioned here as the emphasis on the role of whitecaps highlights the area where our present ignorance is greatest.

Though gas transfer velocity is normally expressed as a function of sea surface temperature and wind speed, there is every reason to suppose that parameterisation solely in terms of these variables is insufficient to determine it adequately. It can be expected to vary also as a function of fetch, atmospheric stability (Erickson, 1992), and surface films (Broecker *et al.*,1978). The historical concentration on wind speed and sea temperature has arisen because these variables are easily measured and global data sets are available for them. In laboratory experiments, gas transfer is actually more closely related to RMS wave slope than to the wind speed itself (Jahne *et al.*, 1987). Wanninkhof and Bliven (1989) have pointed out that, since the satellite scatterometry signal used to derive wind speeds is itself dependent on the wave slopes, it is probable that that exchange coefficients are better correlated with radar backscatter than with wind speeds, and have made first attempts to measure this correlation in wind-wave tunnels.

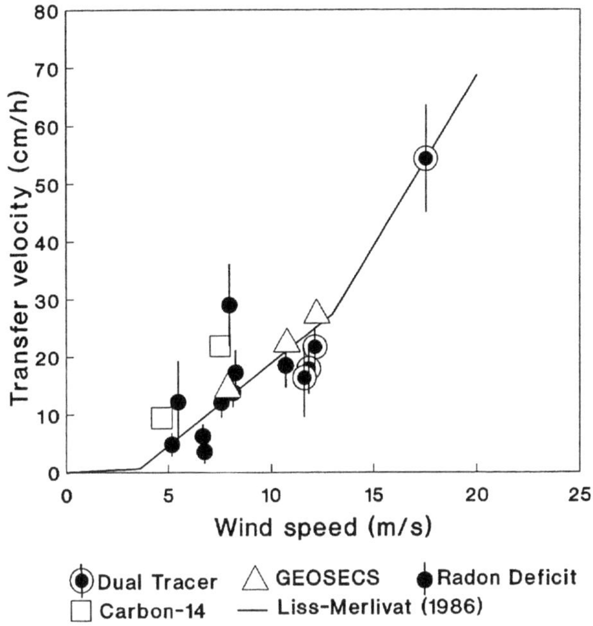

Figure 2. Gas exchange measurements made at sea as a function of wind speed. The solid line is the Liss-Merlivat curve. The points are: open squares, mean bomb-^{14}C transfer rate plotted at global mean wind speed (Broecker *et al.*, 1985, upper square) and from the Red Sea (Cember, 1989, lower square); closed circles, radon-deficit measurements from site-specific studies (Roether and Kromer, 1984, Emerson, *et al.*,1991); triangles, GEOSECS radon-deficit averages plotted against climatological mean winds; double circles, dual tracer experiments (Watson *et al.*, 1991).

MEASUREMENTS IN THE FIELD

Figure 2 summarises many of the measurements made at sea. The most useful constraint on global CO_2 atmosphere-ocean exchange has come from studies of the $^{14}C/^{12}C$ burden of the ocean. Prior to the atmospheric nuclear testing of the late fifties and early sixties, ^{14}C was formed by natural cosmic ray activity in the atmosphere from where it transferred into the ocean. Analysis of this natural ^{14}C budget allowed Broecker *et al.* (1980) to derive a value for the global mean air-sea transfer velocity. The injection of a large ^{14}C transient by the nuclear tests upset the natural budget, but allowed another independent assessment of the rate of transfer as the response of the ocean $^{14}C/^{12}C$ ratio to the injection was measured. Broecker et al (1985) describe this calculation, which is the more accurate of the two methods. The global mean transfer rate derived the bomb ^{14}C technique and corrected to CO_2 at 20^oC is $21{,}9 \pm 3.3$ cm h^{-1}.This is plotted at the global mean wind speed as the upper square in Fig. 2.

The global ^{14}C rate tells us nothing about how the rate of gas exchange may differ in various locations around the world. However, a similar bomb-carbon budgeting technique has been used successfully in the nearly-enclosed Red Sea by Cember (1989). Mean winds over the Red Sea are significantly lower than the global average, so the transfer velocity he derives can be used as an independent constraint on the long-term average transfer velocities derived from ^{14}C budgeting. The value derived by Cember is plotted as the lower square on Fig. 2. During the GEOSECS and TTO expeditions, numerous measurements were made by the radon deficit technique, which uses the naturally-generated concentration of ^{222}Rn in surface ocean waters to obtain estimates of gas transfer rates. A single radon deficit measurement averages over an ill-defined region of the ocean, and the difficulties of interpretation mean that the derived gas transfer velocity is subject to a great deal of uncertainty. However, by averaging the GEOSECS and TTO data over ocean regions and plotting against climatological mean winds a strong correlation with the wind emerges. These data are plotted as triangles without error bars in Fig. 2. Alternatively, several campaigns of intensive measurements made at fixed locations over a period of weeks have also defined useful averages (Roether and Kromer, 1984), and these are plotted with error bars.

In recent years, many measurements in the literature have been made by one of two techniques which make use of releases of small quantities of SF_6, an inert substance which can be measured at very low concentrations. Wanninkhof *et al.*(1985, 1987) and Upstill-Goddard *et al.* (1990) used releases of SF_6 in lakes. The experiments have clearly demonstrated the dependence of gas exchange on wind speeds in the range 2-12 m s^{-1}, consistent with the theory discussed above. Larger wind speeds have proved to be difficult to study by this technique because, though they are common over important regions of the ocean (the North Atlantic in winter and the Southern Ocean all year round), they are rare over lakes. The method cannot be applied to the open sea because it relies on being able to construct an accurate budget for the released SF_6, which is possible in an enclosed lake but not in open water.

Most recently, a dual tracer method has been used to circumvent these problems and obtain measurements at sea at wind speeds up to 17 ms^{-1} (Watson *et al.*, 1991). The technique involves release of both SF_6 and ^3He in a controlled ratio, and the experiment is best conducted in coastal waters which are well mixed to the bottom and of constant depth. It can be shown that under these circumstances, the ratio R = $[^3He]/[SF_6]$ changes with time according to

$$\frac{1}{R} \cdot \frac{dR}{dt} = \frac{(K_{SF6} - K_{3He})}{H}$$

where H is the depth of the water and the Ks are the gas transfer velocities of ^3He and SF_6 respectively. Monitoring the time rate of change of the ratio therefore enables the difference between these two transfer rates to be calculated. Absolute values for the transfer rates, and conversion to apply to CO_2, can be found if the relation $K \sim Sc^{-0.5}$ is assumed.

An advantage of this technique is that, like the SF_6 budgeting method used on lakes, it gives quite accurate values of transfer velocity over relatively short periods of time (24-72 hours) during which the wind and sea surface conditions may be relatively constant and therefore well-defined. A disadvantage is that an elaborate field campaign is necessary, so that large numbers of determinations are unlikely to be made. An added complication is that, at high wind speeds, the $K \sim Sc^{-1/2}$ scaling may break down. Investigations using wind-wave tunnels (Wanninkhof and Asher, pers. comm) suggest that for these two tracers, K may vary with a lower power of the Schmidt number when extensive whitecaps are present, in which case it may be necessary to revise upwards the determinations made at the highest wind speeds. Fortunately this correction can be made at a later date if experiments bear out these doubts about the Schmidt number scaling.

Comparison of ^{14}C and other techniques

The SF_6-based techniques give a consistent picture which is broadly summarised by the "synthesis" of Liss and Merlivat (1986), who suggested a piecewise linear relationship for gas exchange as a function of wind speed. Though advanced in 1986 and based on only the first published SF_6 data, subsequent measurements even at the highest wind speeds have generally lain close to this curve, which is also plotted in Fig 2. The SF_6- based values are substantially lower at a given wind speed than the ^{14}C measurements: for example the value of the Liss-Merlivat curve at the global mean wind of 7.47 m s^{-1} is 11.6 cm hr^{-1}, just over half the ^{14}C value. A part at least of the reason for the difference is that the two are based on measurements over different time periods, The ^{14}C measurements being long-term averages while the SF_6 is over a day or two only. Since the "instantaneous" gas exchange rate increases in a super-linear fashion with increasing wind speeds, averages over long time periods will give transfer rates which are higher than that corresponding to the average wind. Wanninkhof (1991) has used a technique of numerical deconvolution using an assumed frequency distribution for wind speeds over the ocean and a $K \sim U^2$ dependency for the transfer velocity to suggest that this effect accounts for a discrepancy of about 30% between short-term and long term averages. Heimann and Monfray, (1991) used a similar wind frequency distribution and the Liss Merlivat relationship, combined with a monthly mean climatology to calculate exchange coefficients over the global ocean. Their more detailed treatment gives a mean global transfer velocity for CO_2 which is a factor of 1.7 lower than the $^{14}CO_2$-derived mean. At present therefore there appears still to be a substantial discrepancy between the ^{14}C and SF_6 based techniques, with the available radon deficit measurements being too scattered to decisively favour either one, though seeming to be more consistent with the SF_6 measurements.

This presents a problem for models of CO_2 air-sea exchange. We can consider three ways that the discrepancy might be resolved.

1) *K for invasion is greater than K for evasion:* it is axiomatic to the scaling procedure used to convert between the transfer rates derived for the various gases that equation (ii) holds and that the transfer process is reversible, evasion and invasion having the same transfer rate. While this rests on solid theoretical and experimental foundation in the unbroken surface region, in rough seas where bubbles and spray are important it need not be the case. For example, small bubbles which are injected deep enough below the surface can dissolve completely (Thorpe, 1982), a process which clearly contributes to <u>in</u>vasion but not <u>e</u>vasion. The effect of larger bubbles is to increase the effective pressure at which the atmosphere comes to equilibrium with the seawater, and this too can be modelled as an increase of invasion over evasion. Both these effects mean that the zero-flux condition would not be equality of surface ocean and atmospheric pCO_2 but would occur when the sea surface pCO_2 was higher than the atmosphere by a few μatm. The use of SF_6 budgeting, dual tracer or radon deficit techniques, all of which actually measure evasion only, to infer values for the two-way transfer of CO_2 could therefore be a source of error at higher wind speeds.

We can show however, that this cannot by itself account for the discrepancy between Liss-Merlivat results and the ^{14}C average. Splitting equation (ii) into separate invasion and evasion components, we can write

$$F = K'_{ev}.pCO_{2w} - K'_{inv}.pCO_{2a}$$

from which, at zero flux, we obtain

$$pCO_{2w} = [K'_{inv}/K'_{ev}].pCO_{2a}$$

Now if the invasion rate is of order 1.7 times greater than the evasion rate, as would be necessary to bring Liss-Merlivat into agreement with the ^{14}C measurements, then at steady state the partial pressure of surface water would need to be this factor larger than the atmospheric value. We know however, that today global mean pCO_2 differs from the atmospheric value by no more than a few percent, so we would arrive at a net flux of CO_2 into the ocean which was much greater than the global budget would allow if we were to accept this hypothesis. It is worth noting however, that the zero flux condition does not need to be exact equality of pCO_2: surface pCO_2 could be a fraction of a percent higher than the atmospheric value at equilibrium, and this could lead to errors of a few tenths of a gigaton in calculating the net flux into the ocean.

2) *The non-linearity of the gas exchange curve at high wind speeds is underestimated:*in the Liss-Merlivat curve (Fig 2) the slope of the line at high wind speeds was constrained only weakly from a knowledge of wind tunnel experiments, and not at all by field data. Since this relationship was proposed, the dual tracer measurement "storm" determination has apparently confirmed this portion of the curve, but as discussed above this point may be subject to upward revision if the assumed Schmidt number scaling turns out to be wrong in this regime. If this portion of the curve increased much more rapidly with increasing wind, the effect of episodic high winds

might be sufficiently important that the global mean reached the ^{14}C value. Heimann and Monfray found however, that is was necessary to increase the slope of this portion of the curve by a factor of 16 in order to bring the two into agreement. It seems unlikely that the "storm" determination will increase by more than 50% at the most, so this possibility, while it may help to resolve the discrepancy, seems unlikely to be the complete explanation.

3) *There is significant chemical enhancement of CO_2 exchange:*there is known to be a small effect due to the chemical reactivity of CO_2 with seawater which will enhance the transfer of this gas compared to inert tracers. CO_2 reacts quite slowly compared to its residence time in the molecular-diffusion dominated boundary layer, so this effect appears not to be significant except at low wind speeds where the boundary layer is at its thickest (Emerson, 1975; Hoover and Berkshire, 1979). Wanninkhof (1991) makes a numerical study of the significance of the chemical enhancement, and concludes that the effect accounts for only 1-2% of the global mean CO_2 transfer velocity. However at low wind speeds typical of equatorial conditions it might account for up to 20%.

Berger and Libby (1969) suggested that the presence of small quantities of carbonic anydrase enzyme in seawater might very substantially enhance CO_2 exchange. Subsequent studies could not find the very large enhancements suggested, but there is some scattered and anecdotal evidence from controlled experiments which does seem to support the notion that some "unconventional" chemical enhancement of CO_2 fluxes may be taking place.

CONSTRAINTS ON THE GLOBAL CO_2 CYCLE.

Integration of equation (i) over the global ocean to obtain the flux of CO_2 from the atmosphere into the ocean has been carried out by Tans *et al.*(1990) and Etcheto and Merlivat (1988), using the Liss-Merlivat relations. Tans *et al.* also carried out calculations using a higher "empirical" transfer velocity-wind speed relationship; a straight line intercepting the wind axis at 3 m s^{-1} and passing through the bomb ^{14}C point on the K-U diagram. All of the derived fluxes, even those obtained using this artificially high transfer rate, give lower net fluxes than the previously accepted figure of 2 Gt C yr^{-1}. In these calculations there are substantial uncertainties associated with the surface pCO_2 distribution used, which is as yet inadequately defined in many ocean regions and which exhibits substantial seasonal and spatial variability not as yet properly incorporated into the integration.

The way in which we choose to resolve the inconsistencies between the SF_6 and ^{14}C exchange rates is crucial to the calculation of the global CO_2 flux into the ocean. If we accept that the bomb ^{14}C determination of the exchange rates is correct, then it is very probable that the even the Tans *et al.* calculations using the "empirical" relationship underestimate the net flux from air to sea. The reason is that this straight line relation may be expected to be higher at low to moderate wind speeds, but lower at high wind speeds, than the true non-linear K-U curve. However, the regions of the oceans where lower winds predominate (the equator and subtropical gyres) are mostly net sources of CO_2 to the atmosphere or are at equilibrium, whereas the regions of high winds are the strong sinks. The calculation would therefore overestimate the strength of the source areas and underestimate the sinks and the net sink would be an underestimate of the true value. If on the other hand we insist that the ^{14}C points are in error, then our present knowledge of gas exchange would lead us to apply the much lower Liss-Merlivat curve to the CO_2 calculation. Even given our present inadequate knowledge of pCO_2 in the surface ocean, values of the net flux much lower than 2 Gt C yr^{-1} would then be inevitable.

CONCLUSION

A reasonable body of theory exists to explain and understand the processes by which gases exchange from air to sea and vice-versa. The theory and most of the observations are mainly applicable to the "unbroken surface" regime, where whitecap formation is unimportant for gas exchange. The theory is successful in explaining a number of observations, including the quantitative relationship between the rates of transfer of different gases, the transitions from smooth surface to rough surface and the proportionality of gas transfer to friction velocity.

In the real ocean however, bubbles are important in gas transfer and the theory for this situation is far more complex. We are not as yet sure how to relate gas transfer measurements made with tracers such as SF_6 to CO_2 in rough seas. No technique is currently available which measures CO_2 transfer rates directly, except the ^{14}C methods which give only wide-area, long term means. There appears to be a conflict between these ^{14}C measurements and other techniques which is only partially resolved by considerations of the very different times and areas averaged by the different methods. We have no reason to doubt that both the $^{14}CO_2$ methods and the inert gas tracer methods are basically correct, but it is difficult to reconcile the two methods without postulating some as yet unknown mechanism tending to enchance CO_2 transfer rates over those of inert gases. At present, the most logical approach for modelling CO_2 exchange is probably to scale upwards the Liss-Merlivat curve by a factor of 1.7 necessary to bring its global mean into line with the ^{14}C value.

REFERENCES

Berger R, Libby WF (1969) Equilibration of atmospheric carbon dioxide with seawater: possible enzymatic control of the rate. Science 164:1395-1397.

Broecker HC, Petermann J, Siems W (1978) The influence of wind on CO_2 exchange in a wind-wave tunnel, including the effects of monolayers. J. Mar Res 36:595-610.

Broecker, W S, Peng T-H, Mathieu G, Hesslein R, Torgeson T (1980) Gas exchange measurements in natural systems. Radiocarbon 22:676-683.

Broecker WS, Peng T-H, Ostlund G, Stuiver M (1985) The distribution of bomb radiocarbon in the ocean. J Geophys Res 99:6953-6970.

Cember, R (1989) Bomb radiocarbon in the red sea: a medium-scale gas exchange experiment. J Geophys Res, 94:2111- 2123.

Danckwerts, P V (1951) Significance of liquid film coefficients in gas absorption. Ind Engng Chem 43:1460- 1467.

Deacon, E L, (1977) Gas transfer to and across the air-water interface. Tellus 29:363-374.

Erickson, D J, III (1992) A stability-dependent model for air- sea gas exchange. Submitted to J Geophys Res

Etcheto J, Merlivat L (1988) Satellite determination of the carbon dioxide exchange coefficient at the ocean-atmosphere interface: a first step. J Geophys Res 93:15669-15678.

Heimann, M, Monfray P (1990) Spatial and temporal variation of the gas exchange coefficient for CO_2, 1: data analysis and global validation Internal report No 31, Max Planck Institut fur Meteorolgie

Jahne, B, Huber W, Dutzi A, Wais T, Ilmberger J (1984) Wind/wave experiment on the

Schmidt number and wave-field dependence of air/water gas exchange In: Brutsaert W, Jirka GH (eds) *Gas transfer at water surfaces*. Reidel, Dordrecht p 303.

Jahne, B, Munnich KO, Bosinger R, Dutzi A, Huber W, Libner P (1987) On parameters influencing air-water gas exchange. J Geophys Res 92:1937-1949.

Ledwell J R, (1982) "Gas transfer at the air-water interface", PhD thesis, Harvard University.

Liss, PS, Merlivat L (1986) Air-sea gas exchange rates: introduction and synthesis. In: Buat-Menard P (ed)*The role of air-sea exchange in geochemical cycling*. Reidel, Dordrecht, p 113.

Liss, PS (1983) Gas transfer: experiments and geochemical implications. In: Liss PS, Slinn WGN (eds) *Air-sea gas exchange of gases and particles*. Reidel, Dordrecht. p 241.

Memery L, Merlivat L (1985) Modelling of gas flux through bubbles at the air-water interface. Tellus 37B:272-285.

Merlivat L, Memery L (1983) Gas exchange across an air- water interface: experimental results and modelling of bubble contribution to transfer J Geophys Res 88:707-724.

Monahan EC, Muirchaertaich IO (1980) Optimal power law description of oceanic whitecap coverage dependence on wind speed. J Phys Oceanogr 10:2094-2099.

Monahan EC, Spillane MC (1984) The role of oceanic whitecaps. In: air-sea gas exchange In: Brutsaert W, Jirka GH (eds) *Gas transfer at water surfaces*, Reidel, Dordrecht. p495.

Roether W, Kromer B (1984) Optimal application of the radon deficit method to obtain air-sea gas exchange rates. In: *Gas transfer at water surfaces* (W Brutsaert and G H Jirka, eds) Reidel, Dordrecht. pp 447-457.

Sarmiento JL, Orr JC, Siegenthaler U (1991) A perturbation model model of CO_2 uptake in an Ocean General Circulation Model. J Geophys. Res (in press)

Thorpe SA (1982) The role of bubbles produced by breaking waves in supersaturating the near-surface mixed layer with oxygen. Ann. Geophys 2:53-56

Trenburth KE, Large WG, Olson JG (1989) The effective drag coefficient for evaluating wind stress over the oceans. J Climate 2:1507-1516

Upstill-Goddard RC, Watson AJ, Liss PS, Liddicoat MI (1990) Gas transfer velocities in lakes measured with SF_6. Tellus 42B:364-377

Wanninkhof R, Ledwell JR, Broecker WS(1985) Gas exchange-wind speed relation measured with sulphur hexafluoride on a lake. Science 227:1224-1226

Wanninkhof R, Ledwell JR, Broecker WS, Hamilton M (1987) Gas exchange on Mono Lake and Crowley lake, California. J Geophys Res 92:14567-14580

Wanninkhof R, Bliven LF (1989) Relationship between gas exchange, wind speed and radar backscatter. EOS Trans Am Geophys Union 70:297

Wanninkhof, R, (1992) Relationship between gas exchange and wind speed over the ocean Submitted to J Geophys Res

Watson AJ, Upstill-Goddard RC, Liss PS (1991) Air-sea gas exchange in rough and stormy seas measured by a dual-tracer technique. Nature 349:145-147.

Woolf DK, Thorpe SA (1991) Bubbles and the air-sea exchange of gas in near-saturation conditions. J Mar Res 49:435-466.

LECTURE 2: SURFACE OCEAN CO_2

Dr. Charles D. Keeling
Scripps Institution of Oceanography
University of California, San Diego
La Jolla, California 92093-0220
U.S.A.

Introduction. Close to the sea surface, the waters of the ocean are mixed by wave action so that their temperature and chemical properties are almost uniform vertically. Within this mixed layer, which varies in depth from a few meters to several hundred meters, a single measurement of a chemical property is likely to represent that property throughout the layer. Because the water below is colder or saltier and therefore denser, the layer is stable enough to persist for weeks, months, or even years.

At any location, the properties of the mixed layer nevertheless change over time in response to ocean currents and diffusive exchanges of heat and matter with the denser water below and with the atmosphere and sky above. Even the depth of the layer changes with time, as variable wind and wave action stir the water and warming and cooling alter its density. The layer, whenever it deepens, entrains water from below, so that its chemical properties as well as its thickness is altered.

This lecture describes a study of the carbon cycle of this mixed layer over the course of several years at three locations, one in the Atlantic Ocean, two in the Pacific. At each of these locations measurements were made repeatedly of properties that define the relative proportions of CO_2, bicarbonate, and carbonate in the water. One of these properties is the sum of all three of these inorganic chemical components; it will be referred to as dissolved inorganic carbon, or "DIC". Another is the sum of the electric charges of bicarbonate and carbonate, plus several other negative ions in trace quantities, that together chemically neutralize strong acid titrants such as hydrochloric acid. This property is called the "titration alkalinity". From a knowledge of DIC, alkalinity, and also of salinity and temperature, all of the chemical species of inorganic carbon can be computed, including the concentration of CO_2 and the partial pressure, pCO_2, that this dissolved gas exerts at the sea surface.

Another useful parameter of the mixed layer is the $^{13}C/^{12}C$ ratio of DIC, a quantity sensitive to the changes in abundance of carbon species that occur in the mixed layer when CO_2 gas is transported across the air-sea interface, when marine plants in the mixed layer remove CO_2 from the water to create organic compounds during photosynthesis, or when organic matter residing there is remineralized, releasing CO_2. This property was also measured at all three locations.

No measurements were available in this study to characterize directly the air above the water column at the three locations. As a substitute, measurements of atmospheric CO_2, at other locations described in the first lecture, were used to infer gradients in CO_2 and its isotopic ratio at the air-sea interface. Though not ideal, this approach is adequate

NATO ASI Series, Vol. I 15
The Global Carbon Cycle
Edited by M. Heimann

for calculating air-sea fluxes of CO_2 at the selected locations, because uncertainties in the inferred atmospheric data are small compared with the corresponding oceanic variations.

At one sampling location, in the Sargasso Sea near the island of Bermuda, the data are adequate to establish an average annual cycle in properties of the mixed layer. The time rates of change in these properties were established within this annual cycle, and from these rates carbon fluxes were computed for the major pathways of the marine carbon cycle that involve CO_2.

Data are adequate to establish interannual variations in carbon both at this location and in a region in the tropical South Pacific Ocean for which good seasonal coverage is lacking but the data, for a series of transects, extend over a ten year period. The data at both locations show evidence of natural variations and also an oceanic response to increasing atmospheric CO_2. The latter signal is more obvious in the $^{13}C/^{12}C$ isotopic ratio of DIC than in the DIC concentration itself. Gaseous CO_2 produced by fossil fuel combustion, because it has an isotopic ratio similar to that of the plants that originally produced the fuel, is strongly depleted in the rare isotope, ^{13}C. As a consequence of fuel combustion, the $^{13}C/^{12}C$ ratio of atmospheric CO_2 has decreased over the past century, and in response so has the $^{13}C/^{12}C$ ratio of DIC in sea water. This shift, less influenced by other interannual variations than the concentration of CO_2, is the best present evidence that the oceans are responding to changes in atmospheric CO_2 caused by fossil fuel combustion.

At a third location, in the tropical North Pacific Ocean, the record is too short to warrant a detailed discussion. The limited data indicate, however, that the seasonal cycle of carbon in the tropical North Pacific is less pronounced than in the Sargasso Sea.

Data Base. Seasonal and interannual changes in the oceanic carbon cycle from measurements of DIC in sea water were determined by cryogenic extraction, in which CO_2 was evolved from a weighed sample of sea water in a closed vessal attached to a vacuum line and then frozen into a trap cooled by liquid nitrogen. This method has achieved a standard deviation of 0.5 $\mu M\ kg^{-1}$ (micromoles of carbon per kilogram of sea water) for the mean of pairs of analyses of bottled samples of sea water. The accuracy of analysis approaches the precision, because the analyses depend on manometric measurements of the trapped CO_2 using a constant volume mercury manometer calibrated to 1 part in 4000 in connection with atmospheric CO_2 studies described in Lecture 1.

The DIC data are accompanied by measurements of the titration alkalinity of sea water. These were made following largely the methodology of Brewer et al. (1986), but substituting gravimetric for volumetric control of concentrations and a nonlinear least-squares fit of the potentiometric titration data (Dickson, 1981) for the Gran plot analysis used by Brewer et al. A standard deviation has been achieved for the mean of pairs of single analyses of 1.0 $\mu E\ kg^{-1}$ (microequivalents per kilogram of sea water).

Time variations near Bermuda. Since mid-1983 DIC variations have been documented in near-surface water of the Sargasso Sea at hydrographic station 'S', approximately 30 km southeast of Bermuda (at 31° 10′N, 64° 30′W). Since late in 1984, samples have also been analyzed for alkalinity and nutrients.

As shown by Jenkins and Goldman (1985), hydrographic data of sea water density and oxygen concentration at Station 'S' since 1954 exhibit pronounced seasonal variations from the surface to a depth of 200 meters. Such seasonality is characteristic of anticyclonically circulating subtropical ocean waters. Although concentrations of the organic nutrients, phosphate and nitrate, are low at all seasons, there is evidence of organic productivity in the near-surface water, closely associated with deep convective overturning during the coldest months of the year. This convection varies interannually, as noted by Jenkins and Goldman.

Data for water of the mixed layer show a distinct seasonal oscillation in DIC of approximately 30 μM kg^{-1}, but no regular oscillation in alkalinity (Figure 1, first two panels). Measurements of the $^{13}C/^{12}C$ ratio of DIC for the same water samples (Figure 1, third panel) show a similar oscillation in anti-phase with DIC, with an amplitude in $^{13}\delta$ of approximately 0.20‰ peak to peak. As discussed below, this oscillation implies a seasonally varying uptake of inorganic carbon by soft tissue that explains approximately half of the seasonal cycle in DIC, while the remaining variation in DIC is owing mainly to the combination of air-sea exchange and vertical transport. High DIC concentrations and less positive $^{13}\delta$ occur in early spring when surface water temperature is at its yearly minimum temperature (see Figure 1, fourth panel) and overturns with deeper water. Low DIC and more positive $^{13}\delta$ occur in late summer, shortly after the temperature has reached its seasonal maximum. This time lag in minimum DIC suggests that primary production of organic carbon continues to reduce DIC after the water has started to cool. When DIC is lowest, $^{13}\delta$ is close to its most positive value, suggesting that organic productivity mainly determines the time of minimum DIC. In winter, extremes in DIC and $^{13}\delta$ nearly coincide, because overturning with deeper water is responsible for both extremes. The lack of seasonality in alkalinity implies a lack of pronounced steady growth or dissolution of calcareous organisms in surface waters near Bermuda.

Almost without exception the temperature measurements at depths of 2 and 10 meters are in close agreement, as expected for measurements within the mixed layer. The only exceptions are summer measurements when a very shallow, temporary warm layer sometimes formed at the top of the mixed layer. The chemical measurements show more vertical variation, largely owing to experimental error, especially for $^{13}\delta$. Also an appearance of discrepancy in the solid and dashed connecting lines sometimes is an artifact of missing data.

From data for DIC, alkalinity, temperature, and salinity, the CO_2 partial pressure in water (pCO_2) has been calculated for the waters at Station 'S' using the thermodynamic equilibrium quotients of Dickson and Millero (1987), as shown in the upper panel of Figure 2. The seasonal cycle is of much greater amplitude than that of pCO_2 in the air (dashed line) and of approximately opposite phase. To understand the causes of seasonality in pCO_2, those variations to be expected if the seasonality were due solely to the observed temperature changes have been computed (Figure 2, lower panel, solid line), and also those variations solely due to the observed changes in DIC (same panel, dashed line). In each case the other controlling parameters were held constant at the averages of their observed values. Changes in temperature induce a cycle with a peak to peak amplitude of about 105 μatm, while changes in DIC induce a cycle with an amplitude of about 45 μatm. The difference is close to the observed amplitude of about 70 μatm, because the two cycles are nearly opposed.

In view of the large seasonal cycle in DIC, it is difficult to establish an interannual trend, even with seven years of data. As an attempt (Figure 3, first panel), the data have been analyzed using the same algorithm used to separate seasonal and interannual trends in atmospheric CO_2 by Bacastow et al. (1985). The DIC after seasonal adjustment (Figure 3, second panel) appears to have fallen about by 5 μM kg^{-1} during 1984 and 1985

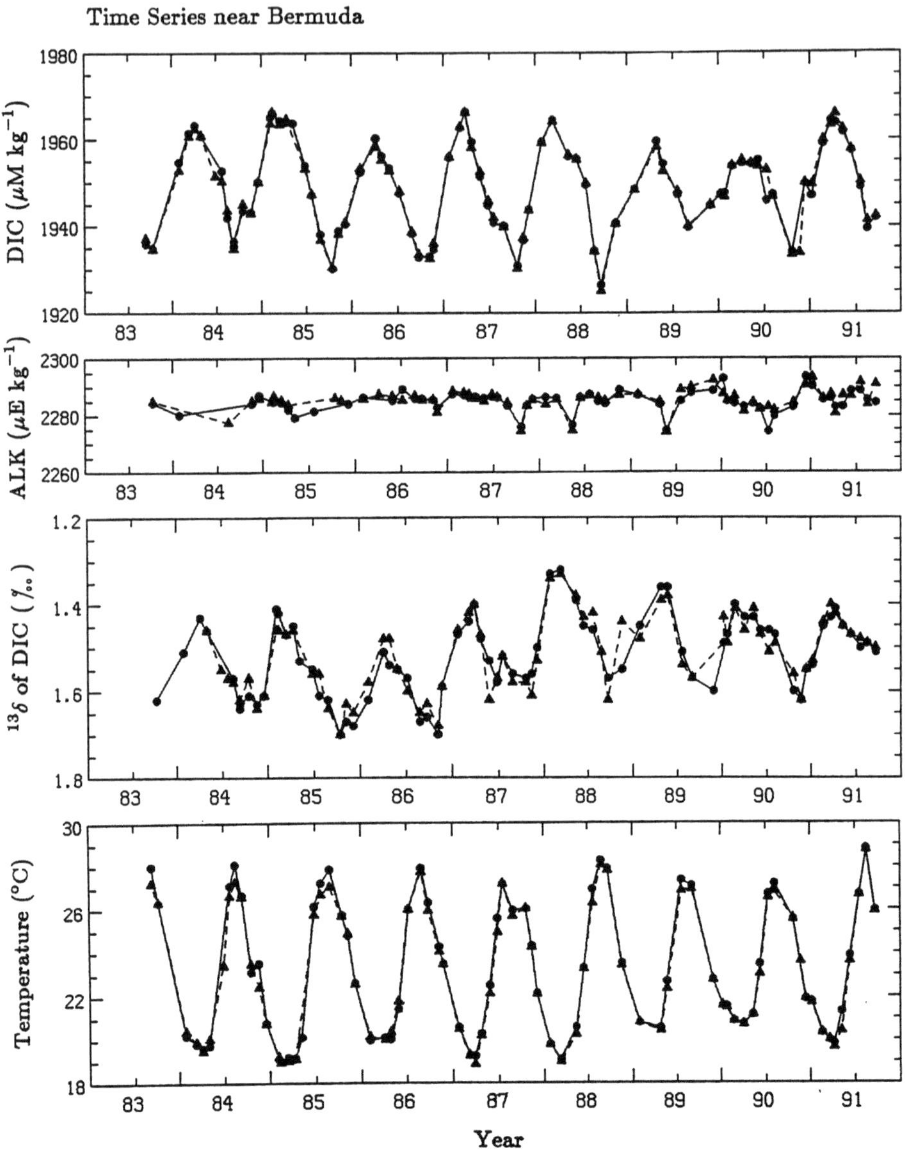

Figure 1. Time series of sea water constituents for Station 'S' near Bermuda at 32°N, 65°W. Solid circles denote samples at 2 m depth, solid squares, samples at 10 m depth. Straight line segments, solid and dashed, connect averages at 2 and 10 m, respectively. Constituents shown are dissolved inorganic carbon (DIC) in $\mu M\ kg^{-1}$, titration alkalinity in $\mu E\ kg^{-1}$, $^{13}\delta$ of DIC in per mil (‰), and temperature in °C. DIC and alkalinity are normalized to a constant salinity of 35 per mil. The vertical scale of $^{13}\delta$ is inverted to facilitate comparison with DIC.

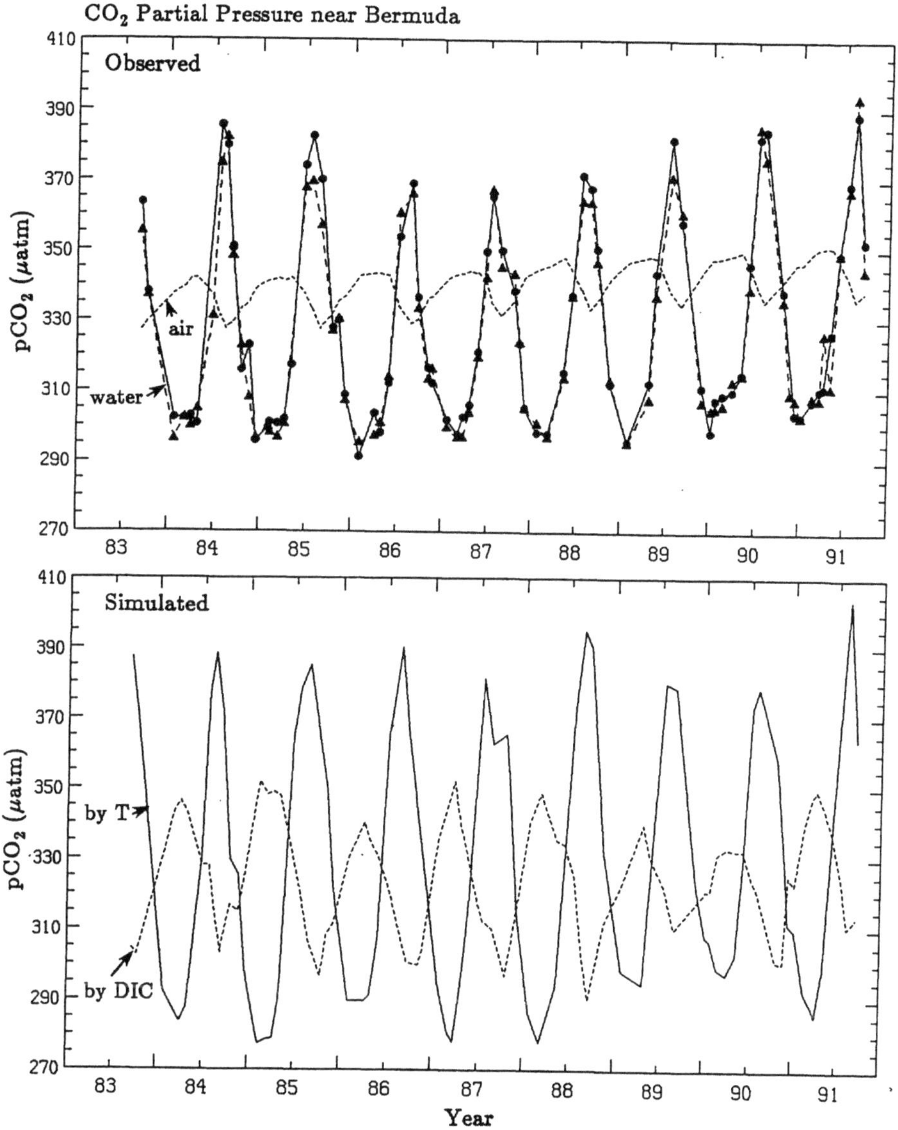

Figure 2.

Upper panel: Time series of pCO_2 in surface water near Bermuda, in μatm, computed from DIC, alkalinity, salinity, and temperature data. Circles, triangles, and solid line segments are as in Figure 1. The dashed line indicates pCO_2 in the atmosphere at La Jolla, California, derived from monthly average concentrations shown as a proxy for the CO_2 pressure in air near Bermuda.

Lower panel: pCO_2 that would occur in surface water near Bermuda if the variation were produced solely by temperature variations (solid line segments) or DIC variations (dashed line segments).

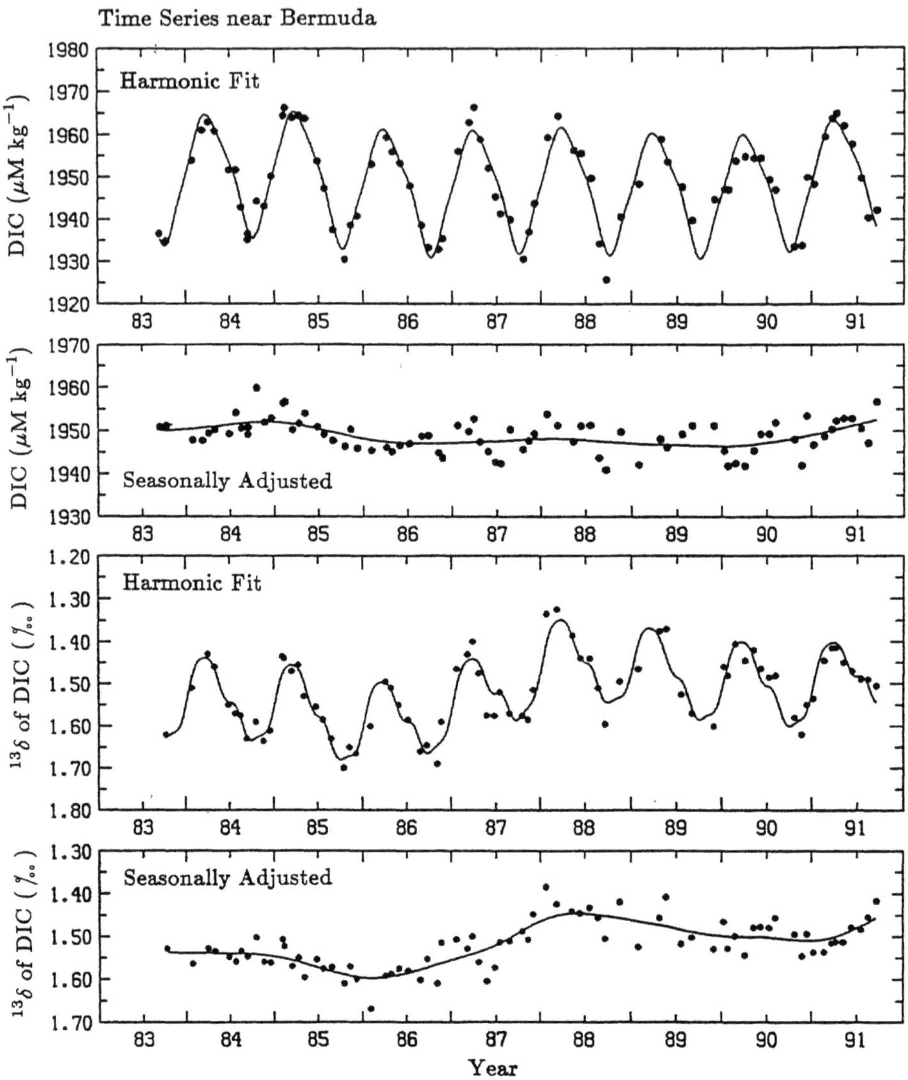

Figure 3.
Time series of DIC and $^{13}\delta$ of DIC at Bermuda, and the same data smoothed and seasonally adjusted by the algorithm used in the analysis of atmospheric CO_2, data described in the first lecture. The annually oscillating curves are fits to a spline function combined with four harmonic terms. The dots denote averages of data at 2 and 10 meters in depth, previously shown in Figure 1. The smoother curves show only the spline function. The associated individual data points are seasonally adjusted.

and afterwards remained approximately constant until a slight rise occurred in 1991. The considerable interannual variability in the amplitude of the seasonal cycle (Figure 1, first panel) suggests that the variable trend may be caused by variable vertical winter-time convective mixing, as described by Jenkins and Goldman (1985). There is no evidence of a build-up of DIC related to fossil fuel combustion. In contrast, the $^{13}\delta$ of DIC, although partially masked by a pronouced seasonal cycle, appears to trend towards more negative values (third and fourth panels). A shorter term interannual variation is also apparent in $^{13}\delta$, with a trough in negative $^{13}\delta$ in 1985 and 1986 and a peak in 1988. This pattern is similar to the trend in $^{13}\delta$ of atmospheric CO_2 of the northern hemisphere as displayed in Figure A.2 of Lecture 1, as though a correlation may exist between DIC variations and El Niño events. Even the DIC trend (second panel) shows a hint of the same pattern.

The average seasonal budget of carbon at Station 'S', has been estimated, as depicted by the model described in Figure 4. The whole record for each constituent was first fit to a harmonic function with respect to time of year, and then, on the basis of this smooth representation of each seasonal cycle, and taking into account sparse additional data on the vertical distribution of each constituent below the mixed layer, fluxes of carbon associated with DIC in the mixed layer were computed. Estimates were made of rates of transport of DIC into the mixed layer from below, both by vertical diffusion and entrainment. Finally, after prescribing isotopic fractionation of carbon for all processes involved, biologically mediated exchange in the mixed layer was estimated. In carrying out these calculations, physical oceanographic data were used to estimate the coefficient of vertical diffusion and the depth of the mixed layer, both of which vary significantly with season. The increase in DIC from air-sea exchange was established from the difference between calculated pCO_2 in the water and atmospheric CO_2 pressure, the latter estimated from data at La Jolla, California. The gas exchange rate was taken to be a function of wind speed according to the formula of Heimann and Monfray (1989). The net effect of withdrawal of DIC by plants, and its subsequent remineralization back to DIC by decay processes, was computed from $^{13}\delta$ of DIC assuming that the isotopic composition of plankton as a function of temperature is as given by Rau et al. (1989).

The exchange of carbon between DIC and organic phases, which will here be called "biological exchange", is the dominant flux, as shown in Figure 5, where it is expressed as an average increase in DIC within the mixed layer, in gC m^{-2} day^{-1}. When this flux is negative, net primary production exceeds respiration. This condition is maximal in March and April during the spring blooming period of phytoplankton. A second, weaker, fall bloom is suggested to occur in October. The air-sea exchange flux is maximal in spring and is generally positive, i.e. CO_2 is usually being added to the mixed layer at this time. The Bermuda area is evidently a strong sink for CO_2. According to the calculations, pCO_2 would have to be raised 34 μatm to halt the net uptake of CO_2 over the year. Upward transport by diffusion is also maximal in spring. Entrainment occurs only in the fall and winter and is calculated to be always small.

At Station 'S', on average, 16 gC m^{-2} of DIC appears to be removed annually from the mixed layer by biological exchange, as a result of net consumption by organic production of 22 gC m^{-2} during shoaling of the mixed layer in spring and summer and net regeneration of 6 gC m^{-2} during deepening. On an annual basis, air-sea exchange adds 21 gC m^{-2} of CO_2 to the mixed layer. Vertical transport by diffusion and entrainment adds a further 18 gC m^{-2} as DIC. The difference between import and export by all four pathways is thus a net gain of 23 gC m^{-2}. A closed annual carbon cycle in the mixed

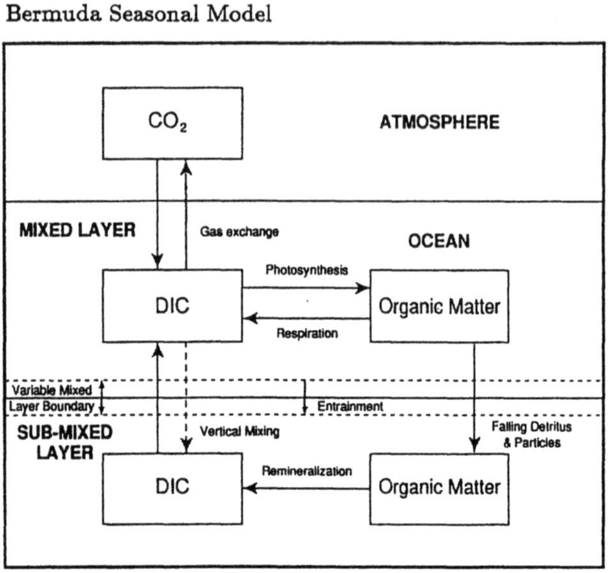

Figure 4. Diagram of the pathways and reservoirs of the carbon cycle in the surface mixed layer near Bermuda, as employed in the seasonal model described in the text.

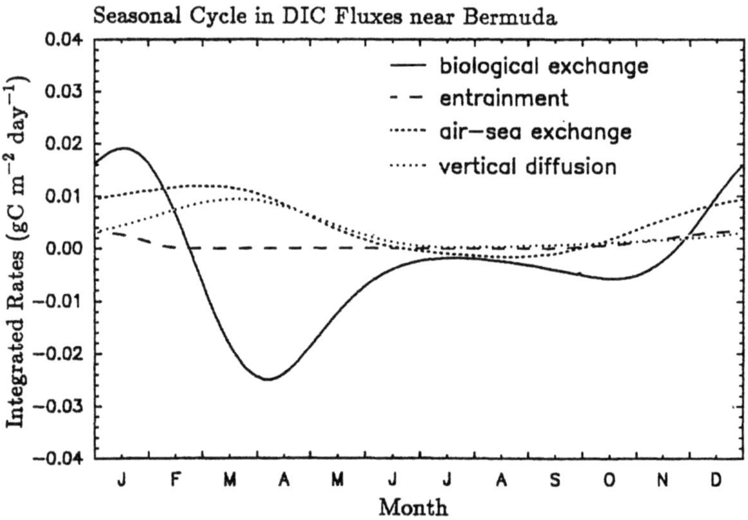

Figure 5. Seasonal variations in the rates of change of DIC, in gC m^{-2} day^{-1}, integrated over the surface mixed layer, as simulated by the seasonal model. Separate rates are shown for biological exchange, entrainment, air sea exchange, and vertical diffusion, as labeled.

layer is achieved, because water incorporated into this layer during deepening has a lower DIC concentration than water left behind when the mixed layer shoals. Numerical sensitivity tests indicate that the deduced seasonal budget of carbon is probably realistic, although it would be considerably better determined if time series of DIC, alkalinity, and isotopic ratios had been routinely obtained throughout the upper ocean column rather than mainly in the mixed layer.

Isotopic data raise new questions about the manner in which net primary production is supported in the surface waters near Bermuda, especially the so-called "new production" which depends directly on the flux of nutrients into the illuminated, euphotic, zone where phytoplankton grow. Jenkins and Goldman (*loc. cit*), on the basis of a mixed-layer-thermocline model, suggested that nutrient injections may be pulse-like, perhaps involving "oases of activity", so short-lived and limited in extent that they are difficult to sample directly. The isotopic data, however, indicate that carbon is gradually withdrawn from near-surface water from March to October as a result of incorporation into soft tissue, without any clear correlation with the observed low phosphate and nitrate concentrations. Perhaps, nutrients, supplied from below, have escaped detection not simply as a result of patchiness, but also because they were in chemical forms not detected by conventional techniques of nutrient analysis.

The above described data for the waters near Bermuda are likely to be valuable generally for improving carbon cycle models of subtropical waters, since there are yet very few locations world-wide with detailed seasonal records of inorganic carbon in sea water. If data continue to be obtained near Bermuda until interannual variations in DIC become clearly evident, the record may also provide a validation of predictions of global change involving the oceanic carbon cycle.

Time series variations near Hawaii. A shore time series has also been obtained near 23° 45′N, 158° 00′W, north of the island of Oahu. Sampling at this station was initiated in 1988 by the University of Hawaii for the JGOFS program (Joint Global Ocean Fluxes Study Program) to study properties of oligotrophic waters of the North Pacific Ocean. Sample collections began in surface water at depths of 5 and 25 meters as soon as the station was established. Data analyzed at La Jolla, California are shown in Figure 6. The concentration of DIC (first panel) shows a seasonal cycle about half as large as near Bermuda, with a hint of an upward trend. Alkalinity (second panel), normalized to constant salinity, shows little variation either with the season or from year to year, as at Bermuda. The $^{13}\delta$ of DIC shows a ragged seasonal cycle with considerable variations between samples at depths of 5 and 25 meters. These variations should be regarded with caution because some of the data may be unsatisfactory. Temperature (fourth panel) varies mainly with season while salinity (not shown) varies erratically, reflecting variable rainfall.

As at Station 'S' near Bermuda, pCO_2 has been calculated from DIC, alkalinity, temperature, and salinity data (Figure 7, upper panel). On average pCO_2 is 7 μatm lower than in the air (dashed line) and of nearly opposite phase. (The mixing ratio of atmospheric CO_2 at Mauna Loa, in parts per million of water-saturated air by volume, is equated with the pCO_2 of the air in μatm at the site, assuming air pressure at the sea surface to be 1 atm.) On average the waters near Hawaii thus appear to be a weak sink for CO_2. As near Bermuda, variations in pCO_2 induced separately by variations in temperature and DIC tend to oppose each other. As with DIC, pCO_2 shows smaller variations with season than near Bermuda, consistent with a smaller oscillation in sea surface

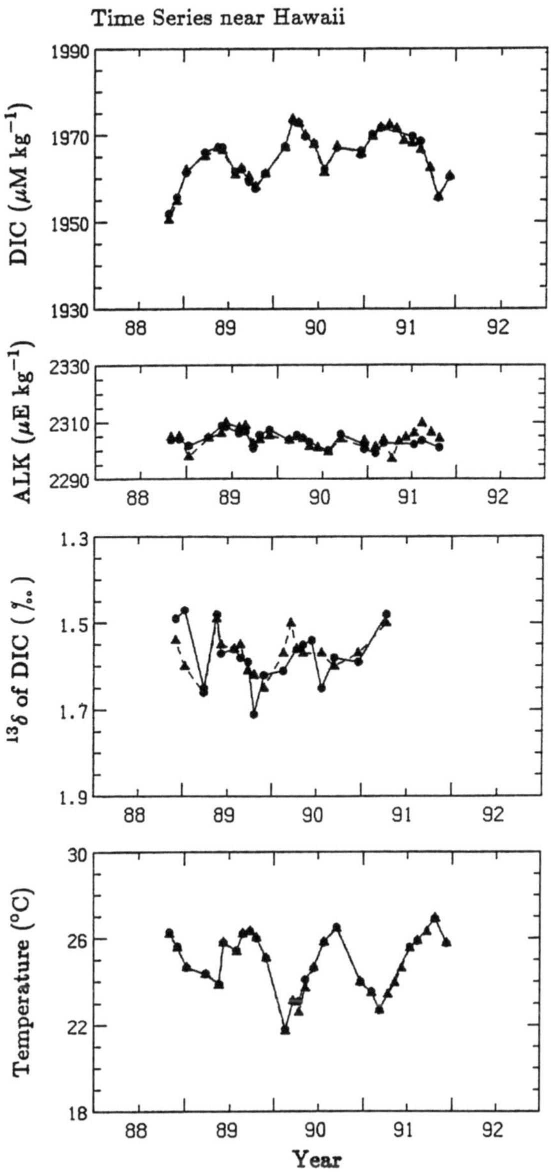

Figure 6. Time series of sea water constituents, as in Figure 1, except that they are for the JGOFS station near Hawaii at 23°N, 158°W. Samples were collected at 5 and 25 m in depth instead of at 2m and 10m, respectively.

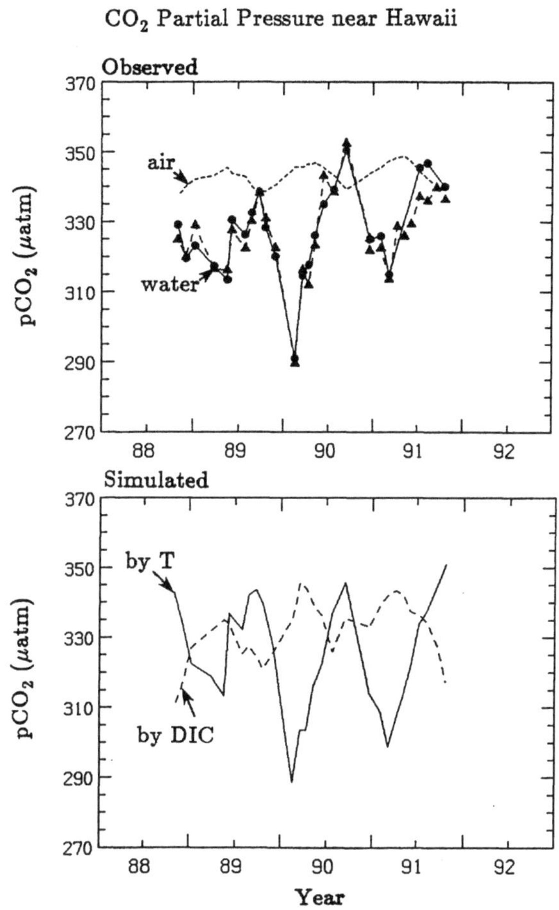

Figure 7.

Upper panel: Time series of pCO_2 in surface water near Hawaii as in Figure 2. The additional dashed line indicates pCO_2 in the atmosphere derived from monthly average CO_2 concentration data at Mauna Loa Observatory, Hawaii converted to the CO_2 pressure in air at the sea surface.

Lower panel: pCO_2 that would occur near Hawaii if the variations were produced solely by temperature or DIC variations, as in Figure 2.

temperature with season. As inferred by Weiss et al. (1982), variations in temperature appear to explain most of the variations in pCO_2 and DIC near Hawaii, taking into account gas exchange at the sea surface. However, a longer record of DIC data involving measurements throughout the upper water column obtained by the University of Hawaii should soon provide a more detailed picture of the carbon cycle at this site.

Time variations near Tahiti. Since 1979, 17 transects north of Tahiti between approximately 8°S and 16°S have been carried out in a collaboration of Scripps Institution of Oceanography with the Centre de l'Office de la Recherche Scientifique et Technique Outre-Mer (ORSTOM). Nine of these occurred near the time of maximum temperature and minimum salinity in March or April, and four at approximately the opposite season. Eleven transects extended across the equator.

To illustrate the data, a series of plots has been prepared of each property versus latitude and afterwards combined into a time series, as illustrated in Figure 8 for the principal variables, temperature, DIC, alkalinity, $^{13}\delta$ of DIC, and pCO_2. To the north of the zone considered in the plots (8-16°S) lies a region of pronounced equatorial upwelling which typically produces low temperature and high DIC in the surface water. This upwelling region has been studied on the transects shown in Figure 9. In 1983 upwelling was practically absent owing to an intense El Niño event (Keeling and Revelle, 1985). As seen in the upper panel of Figure 10, DIC varied across the equator that year mainly in response to temperature. In contrast, upwelling in 1989 was unusually strong, during conditions essentially opposite to those of an El Niño event. Near the equator DIC reached the highest value measured in this study, and the region affected by upwelling was so broad as to be evident on the plot of Figure 8. Upwelling may also affect the data shown in Figure 8 for 1985, 1986, and 1987, although no data exist near the equator to corroborate this. In contrast, alkalinity appears to be unaffected by upwelling, as seen in plots of both Figures 8 and 10. Interannual variability in our DIC data, in addition to that caused by variable equatorial upwelling, results from seasonal variability, which, however, cannot be accurately established because sampling was too infrequent.

As a preliminary assessment of possible long-term changes near Tahiti, linear trends have been estimated in each quantity measured from 1979 through 1989. Because of the greater interference of upwelling in the more northerly data, the assessment is restricted to observations from 13.5° to 17°S. Temperature and salinity show no significant long term trends; temperature decreased at the rate of -0.02 ± 0.05°C yr^{-1}, while salinity rose at 0.01 ± 0.03 ‰ yr^{-1} (± refers to a single standard deviation, positive or negative). The DIC, normalized to constant salinity, increased at 0.3 ± 0.4 $\mu M\ kg^{-1}\ yr^{-1}$, suggestive of a slight trend, but also statistically insignificant. The alkalinity data, after normalization to constant salinity, and excluding doubtful measurements in 1979 and 1980, show no trend (a calculated rise of 0.02 ± 0.3 $\mu E\ kg^{-1}$). In consequence of the small rise in DIC, and no change in alkalinity (after normalization of both quantities), pCO_2 from 1983 to 1989 rose only 0.1 ± 0.8 $\mu atm\ yr^{-1}$, compared to 1.5 $\mu atm\ yr^{-1}$ inferred from the atmospheric data at Mauna Loa Observatory. Closer to a significant shift, $^{13}\delta$ of DIC decreased at -0.012 ± 0.010‰ yr^{-1}, nearly the rate of change of -0.016‰ yr^{-1} predicted by the model of Keeling et al. (1989, pp. 195-199) employing a single deconvolution and variable sea surface temperature.

In conclusion, interannual variability in the carbon cycle near Tahiti restricts our ability to detect secular changes. Lacking more detailed knowledge of the influence of equatorial upwelling and regional circulation of surface waters, we can detect only

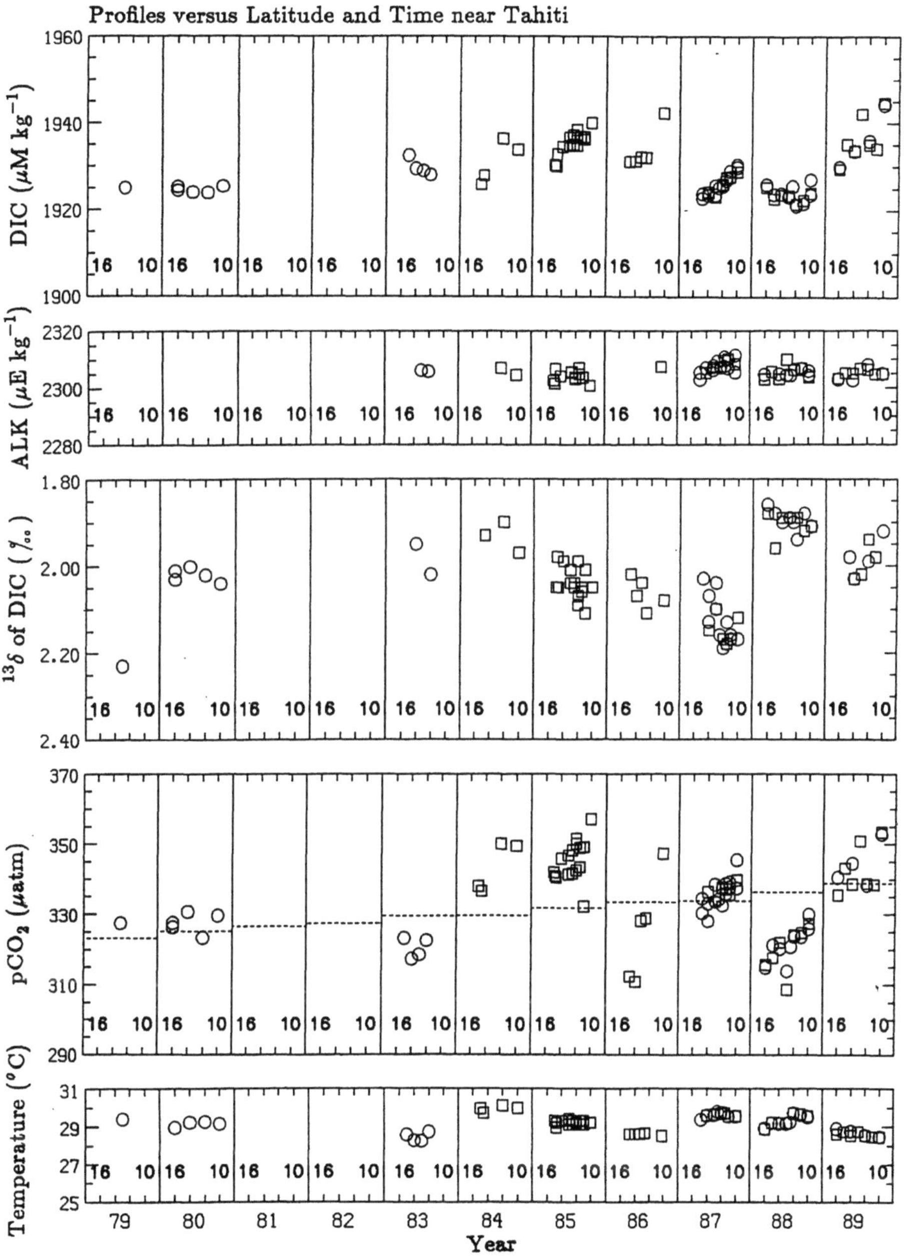

Figure 8. Plots versus latitude of temperature, DIC, alkalinity, $^{13}\delta$ of DIC, pCO_2, and temperature, obtained on transects near Tahiti from 1979 to 1989. The dashed line on the curve for pCO_2 indicates the monthly average CO_2 pressure, pCO_2, in the air derived from data at Mauna Loa Observatory, Hawaii, as in Figure 7. Units are the same as in Figures 1 and 2.

Sampling Locations

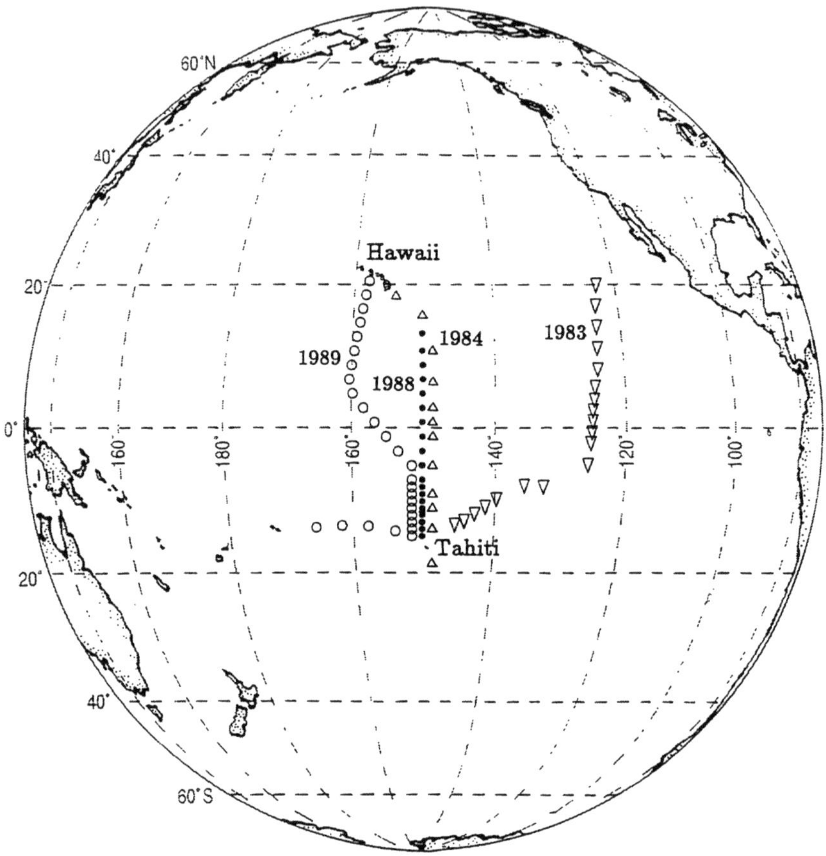

Figure 9. Locations of water sampling between Tahiti and Hawaii for those expeditions shown in Figure 10, below. Inverted triangles, denote the expedition of May, 1983; triangles, that of March, 1984; solid circles, that of March, 1988; open circles, that of March, 1989.

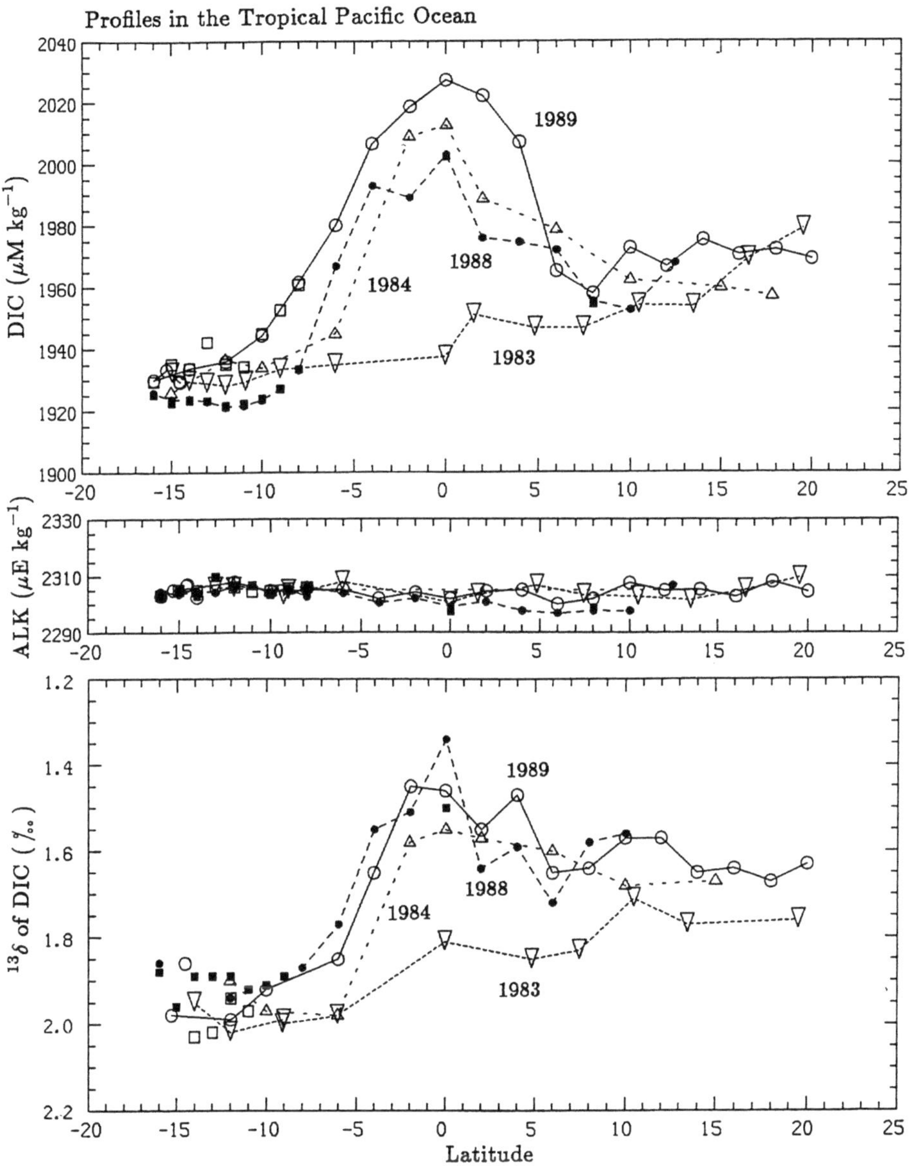

Figure 10. Plots of DIC, alkalinity, and $^{13}\delta$ of DIC between Tahiti and Hawaii, as illustrated in Figure 8 but extended northward to 20°N, to show the influence of equatorial upwelling. Symbols identify each expedition and sampling location, with further details given in Figure 9. Units are the same as in Figure 1.

statistically insignificant changes in bulk carbon chemistry. The data presented above suggest, however, that alkalinity has recently been stationary, while DIC has risen less rapidly than expected globally. In contrast, the isotopic ratio of DIC seems to be responding to the global shift in isotopic composition of the coupled atmospheric-ocean system as expected in response to perturbations caused by human activities. Thus eventually a measureable rise in DIC should appear. Nor can one rule out that alkalinity in the future will show a significant change in concentration in response to human activities.

Epilogue. In contrast to the broad assessment of the carbon cycle that results from a scrutiny of atmospheric CO_2 data obtained from the arctic to the South Pole, measurements carried out in near-surface sea water near Bermuda, Hawaii, and Tahiti serve mainly to indicate how the carbon cycle has functioned on a local scale in three contrasting marine environments. The results of the study for these three regions cannot be generalized to a global assessment of the marine carbon cycle.

The complex local interactions of the annual carbon cycle that occur in the oceanic mixed layer are resolved to a considerable degree when chemical data are taken as often as monthly, as they have been near Bermuda and Hawaii. Seasonal variations are evidently of natural origin, as they are for atmospheric CO_2. Interannual variations are also evident, but are less obvious than for atmospheric CO_2, which shows a pronounced rise in concentration from fossil fuel combustion. The measurements of dissolved inorganic carbon in sea water, indeed, show as yet little evidence of uptake of fossil fuel derived CO_2. In contrast, the isotopic data show a response, although the record is too short to be sure of its magnitude. The isotopic data near Bermuda also shows interannual variations on the time scale of El Niño events. Measurements of carbon will evidently have to be carried out at closely spaced time intervals for several decades, at an array of oceanic stations much larger than the present effort permits, if the extent of human impact on the carbon cycle of the upper oceans against a background of natural variations is to be satisfactorily documented.

Acknowledgments. The research of the Scripps Institution of Oceanography reported here was supported by the Division of Atmospheric Sciences of the U.S. National Science Foundation, Grants ATM88-19398 and ATM91-21986, with co-support from the Chemical Oceanography Program of NSF and the U.S. Environmental Protection Agency. The research was also supported by the Electric Power Research Institute, Contract RP-8011-20. The author is indebted to Nikolas Gruber, Peter Guenther, Timothy Lueker, Stephen Piper, Betsy Stewart, and Timothy Whorf for help in analyzing the data for publication, and for providing many useful suggestions regarding the topics under consideration.

REFERENCES

Bacastow RB, Keeling CD, Whorf TP (1985) Seasonal amplitude increase in atmospheric CO_2 concentration at Mauna Loa, Hawaii, 1959-1982. *J Geophys Res 90*:10529-10540

Brewer PG, Bradshaw AL, Williams RT (1986) Measurements of total carbon dioxide and alkalinity in the North Atlantic Ocean in 1981. In: Trabalka JR Reichle DE (eds) *The Changing Carbon Cycle.* Springer-Verlag, New York, 348-370

Dickson AG (1981) An exact definition of total alkalinity and a procedure for the estimation of alkalinity and total inorganic carbon from titration data. *Deep-Sea Res 28A*:609-623

Dickson AG, Millero FJ (1987) A comparison of the equilibrium constants for the dissociation of carbonic acid in seawater media. *Deep-Sea Res 34*:1733-1743

Heimann M, Monfray P (1989) Spatial and temporal variation of the gas exchange coefficient for CO_2: 1. Data analysis and global validation. Report No. 31, Max-Planck-Institut für Meteorologie, Hamburg, 29 pages

Jenkins WJ, Goldman JC (1985) Seasonal oxygen cycling and primary production in the Sargasso Sea. *J Marine Res 43*:465-491

Keeling CD, Revelle R (1985) Effects of El Niño/Southern Oscillation on the atmospheric content of carbon dioxide. *Meteoritics 20*:437-450

Keeling CD, Bacastow RB, Carter AF, Piper SC, Whorf TP, Heimann M, Mook WG, Roeloffzen H (1989) A three dimensional model of atmospheric CO_2 transport based on observed winds: 1. Analysis of observational data. In: Peterson DH (ed) *Aspects of Climate Variability in the Pacific and the Western Americas.* American Geophysical Union, Washington, DC, 165-236

Rau GH, Takahashi T, Des Marais DJ (1989) Latitudinal variations in plankton $\delta^{13}C$: implications for CO_2 and productivity in past oceans. *Nature 341*:516-518

Weiss RF, Jahnke RA, Keeling CD (1982) Seasonal effects of temperature and salinity on the partial pressure of carbon dioxide in seawater. *Nature 300*:511-513

EFFECTS OF THE MARINE BIOTA ON GLOBAL CARBON CYCLING

Gary Shaffer
Geophysical Institute
University of Copenhagen
Haraldsgade 6
DK-2200 Copenhagen N Denmark

ABSTRACT

Natural carbon cycling in the global ocean and its influence on atmospheric CO_2 concentrations can be examined conveniently in terms of a solubility pump, an organic carbon pump and a calcium carbonate pump (Volk and Hoffert 1985). Here a review is presented of the physical, chemical and biological processes which determine the strength of these pumps. Ocean new (export) production and biogenic calcium carbonate production lie in the ranges 5-10 $GtCyr^{-1}$ and 1-2 $GtCyr^{-1}$ respectively. A new analysis using the HIgh Latitude exchange/ interior Diffusion-Advection (HILDA) model (Shaffer and Sarmiento 1992, Shaffer 1992) shows that the organic carbon pump is about twice as strong as the solubility pump in terms of atmospheric CO_2 drawdown. Model results for the modern, pre-industrial ocean imply that atmospheric transport of CO_2 between high and low-mid latitudes associated with the solubility pump (0.7 $GtCyr^{-1}$ poleward) and the two "biological" pumps (0.8 $GtCyr^{-1}$ equatorward) tended to balance. This indicates that CO_2 outgassing from equatorial upwelling was balanced by CO_2 uptake at mid latitudes. Model results are also used to study the sensitivity of atmospheric CO_2 levels to changes in ocean physics and biology. It is concluded that even very large changes in ocean biology would probably not make a large impact on ocean uptake of CO_2 in the near future. However over century time scales such changes could be important for atmospheric CO_2 evolution.

NATO ASI Series, Vol. I 15
The Global Carbon Cycle
Edited by M. Heimann

1. INTRODUCTION

In coming decades global "greenhouse" warming is anticipated due to increasing concentrations of CO_2 in the atmosphere from fossil fuel burning and changing land use. The future evolution of atmospheric CO_2 concentrations will depend on the rate of fossel fuel burning, the net flux to/from land and the net flux to/from the ocean. According to present best estimates, the fossil fuel emission rate within the last decade was 5.4 ± 0.5 $GtCyr^{-1}$. This may be compared with a yearly net flux of anthropogenic CO_2 to the ocean of 2.0 ± 0.8 $GtCyr^{-1}$ for the same period (Houghton et al 1990).

Ocean flux estimates stem from simple and 3-D models of the perturbation carbon cycle in the atmosphere-ocean system (Siegenthaler and Oeschger 1987, Maier-Reimer and Hasselmann 1987, Sarmiento et al 1992). The (reasonable) assumption made in such abiotic models is that ocean uptake of anthropogenic CO_2 occurs via physical processes alone. The physical exchange rate between the surface and deep ocean is the main factor limiting ocean uptake at present. Marine primary production is assumed to be essentially unaffected directly by present, high atmospheric pCO_2 (partial pressure of CO_2 gas) since nutrients, and not CO_2, appear to limit production.

The potential role of the marine biota for the current atmospheric CO_2 problem should be sought in possible changes in natural carbon cycling, for instance in response to climate change. For a specific setup of ocean and atmosphere physics, ocean biology, ocean surface layer temperature and salinity and ocean-atmosphere inventories of inorganic carbon and alkalinity, there exists a steady-state, baseline level of atmospheric CO_2 such that no net CO_2 exchange occurs between the atmosphere and ocean. (A recent carbon budget estimate indicates a small, net natural flux of CO_2, about 0.6 $GtCyr^{-1}$, from the ocean to the atmosphere required to balance a surplus river input of carbon over carbon burial (Sarmiento and Sundqvist 1992). It is unclear how much of such a natural cycle is taken care of by shelf processes and how much affects the global ocean. In any case such a net flux can be neglected for the purposes of the present paper. Total net CO_2 flux here will refer to what remains after subtracting any small natural cycle of this type). A change in ocean-atmosphere setup would cause the system to drift to a new steady-state with a different baseline pCO_2. During readjustment there would exist a net flux of CO_2 to/from the ocean. An upper bound on this flux (step change in ocean-atmosphere setup) can be estimated from the difference between the final and original baseline pCO_2 divided by an estimate for ocean response time. Then total net CO_2 flux to/from the ocean can be considered to be the sum of an anthropogenic flux, governed by ocean-atmosphere physics and ocean chemistry, and a flux associated with the transition between different ocean-atmosphere setups, governed by ocean-atmosphere physics, ocean biology and ocean chemistry.

The importance of the natural carbon cycle in the ocean for atmospheric CO_2 concentrations is well documented in the historical record. Measurements of pCO_2 in air bubbles from polar ice cores indicate that the baseline pCO_2 level drifted from about 200 ppm to about 280 ppm over the transition from the last ice age to the present interglacial period (Barnola et al 1987). The ocean contains more than 98% of the global reservoir of mobile carbon (Bolin 1986). Even after the buffering capacity of the ocean carbonate system is taken into account (see below), the ocean still has an effective reservoir of about 85%. Thus changes in the physical and/or biological setup of the ocean were probably the main causes for observed glacial-interglacial changes in the chemical composition of the atmosphere (Broecker 1982).

Here I consider the biotic components of the natural carbon cycle in the ocean (Figure 1). First, I present a brief summary of ocean carbonate chemistry. Then I discuss the biological processes leading to carbon flows within the ocean and between the ocean and atmosphere and give present best estimates of the magnitude of these flows. Finally, with the help of a new carbon cycle model of the ocean-atmosphere system I consider the role of ocean physics and biology for controlling atmospheric pCO_2 in the natural carbon cycle.

2. OCEAN CARBONATE CHEMISTRY

The steady state, baseline level of pCO_2 in the atmosphere is coupled tightly over time scales exceeding one year to mean pCO_2 levels in ocean surface layers by gas exchange across the air-sea interface. Concentrations of inorganic carbon species in the ocean are governed by the dissociation relationships for carbolic acid (see Skirrow 1975 or Broecker and Peng 1982 for more details):

Biological and physical CO2 "pumps" remove CO2 from surface waters and thereby provide the main driving forces for the ocean carbon cycle

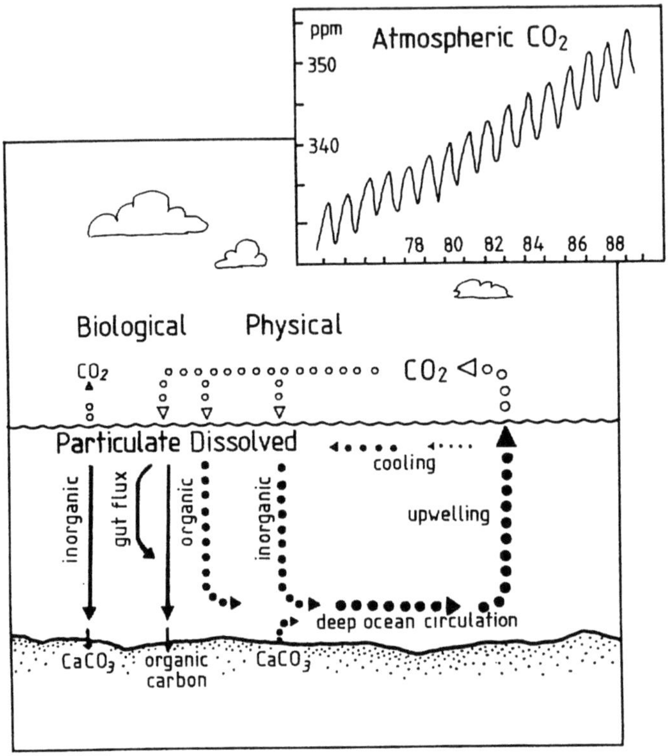

Figure 1: A sketch of natural carbon cycles in the ocean and an inset showing recent changes in atmospheric pCO_2 concentrations. One major question addressed in the text is what do the two have to do with each other. Adapted from the brochure "Oceans, Carbon and Climate Change" of the Joint Global Ocean Flux Study.

$$CO_2 + H_2O \rightleftarrows H^+ + HCO_3^- \qquad (1)$$

$$HCO_3^- \rightleftarrows H^+ + CO_3^{2-} \qquad (2)$$

where CO_2 refers to the sum of dissolved CO_2 and H_2CO_3 (but the concentration of the latter is typically only about 0.2% of that of dissolved CO_2). The chemical equilibrium in equations (1) and (2) is characterized by apparent dissociation constants K_1 and K_2 respectively. Also

$$pCO_2 = \alpha^{-1} [CO_2] \qquad (3)$$

where α is the solubility of CO_2 gas in sea water. The dependencies of K_1, K_2 and α upon temperature (T), salinity (S) and pressure (P) have been determined in laboratory measurements.

From equations (1) and (2) it follows that

$$CO_2 + H_2O + CO_3^{2-} \rightleftarrows 2\,HCO_3^- \qquad (4)$$

and from the above, the partial pressure of CO_2 in sea water may be written as

$$pCO_2 = K_2\,(\alpha\,K_1)^{-1}\,[HCO_3^-]^2\,[CO_3^{2-}]^{-1} \qquad (5)$$

Note that pCO_2 in sea water is a non-linear function of the concentrations of inorganic carbon species.

In any water sample, given knowledge of K_1, K_2 and α from observed T, S and P, two additional measurements are required to determine concentrations of all the above species of carbonate system. Two such observables are pCO_2 and pH ($= -{}^{10}\log[H^+]$). Two other observables commonly used are total inorganic carbon (DIC) and alkalinity (ALK). ALK is defined as the excess of positive charge associated with the cations in a sea water sample over the negative charge associated with cations of strong acids. Since total positive and negative charge must balance, ALK is equivalent to the sum of negative charge in the anions of weak acids which balance the above excess. That is,

$$\text{DIC} = [CO_2] + [HCO_3^-] + [CO_3^{2-}] \quad (6)$$

$$\text{ALK} = [HCO_3^-] + 2\,[CO_3^{2-}] + [B(OH)_4^-] + \ldots \quad (7)$$

Conservation of mass and charge implies that DIC and ALK (both expressed per unit mass) are independent of T and P. Therefore DIC and ALK are useful tracers.

The use of alkalinity (equation 7) to characterize the ocean carbonate system introduces additional unknowns. To a good approximation for oxic conditions, contributions to ALK of weak acid ions other than carbon and boron species can be neglected. For boron,

$$H_3BO_3 + H_2O \rightleftarrows H^+ + B(OH)_4^- \quad (8)$$

characterized by an apparent dissociation constant, K_B, whose T, S and P dependence is known from laboratory measurements. For the carbonate system problem, total boron ($[H_3BO_3]$ + $[B(OH)_4^-]$) can be approximated well as a known linear function of salinity and again the number of equations and unknowns match.

If ALK is approximated by the first two terms only in equation (7) and if the CO_2 term in equation (6) is neglected, equations (5), (6) and (7) imply

$$pCO_2 \sim [\text{DIC}-(\text{ALK}-\text{DIC})]^2\,[\text{ALK}-\text{DIC}]^{-1} \quad (9)$$

This approximate relation will prove useful for the subsequent discussion on how ocean carbon pumps affect atmospheric pCO_2 levels.

For T, S, DIC and ALK in the modern ocean, carbonate chemistry sketched above dictates that $[HCO_3^-]:[CO_3^{2-}]:[CO_2]$ is of order 150:15:1. CO_2 gas entering the ocean hydrates and reacts quickly with the carbonate ion to form bicarbonate ion according to equation (4). Thus pCO_2 and $[CO_3^{2-}]$ are rather tightly coupled. Then a proportional change in pCO_2 leads to a proportional change in $[CO_3^{2-}]$, similar in size but opposite in sign. Since DIC changes by an equal but opposite amount as $[CO_3^{2-}]$ (equations 4 and 6), the proportional change of DIC would be in the same sense as but about an order of magnitude less than that of pCO_2 given above concentration ratios. Such chemical buffering may be expressed more rigorously in terms of a buffer factor, ξ, defined such that

$$\Delta pCO_2\,(pCO_2)^{-1} \quad = \quad \xi\ \Delta DIC\,(DIC)^{-1} \tag{10}$$

which can be derived from the above equations. ξ varies in the range of about 7 to 18 between low latitude and high latitude surface water. Thus a 1% change in DIC leads to a 7 to 18% change in pCO_2 in ocean surface layers. For fixed DIC and ALK at modern levels, pCO_2 increases with T by 4-5% per degree Celcius. Due to carbonate chemistry buffering, the storage capacity of the ocean for atmospheric CO_2 is reduced by a factor $\xi \sim 10$ and the equilibrium time for CO_2 between the atmosphere and ocean is reduced by this factor compared to what one might expect from the ratio DIC:$[CO_2]$ in the surface layer. Still the equilibration time for CO_2, about one year, is about 15 times longer (about the ratio $[CO_3^{2-}]$:$[CO_2]$) than for ordinary gases like O_2.

3. TWO BIOTIC CARBON PUMPS IN THE OCEAN

a) The organic carbon pump

Within the ocean's lighted surface layer, organic matter is formed by photosynthesis in phytoplankton whereby nutrients and CO_2 are taken up and free oxygen is released. Part of the total primary production (the new production) is exported down out of the surface layer via sinking of large particles (Martin et al 1987), mixing and advection of small particles and dissolved organic matter (Najjar 1990) and vertical migration of zooplankton (in their guts, Longhurst and Harrison 1988). In subsurface layers, this organic material is remineralized and oxidized, mainly by the action of bacteria, whereby nutrients and CO_2 are released and free oxygen consumed. Cycling in this organic pump is completed as mixing and advection return nutrients and inorganic carbon to the surface layer and carry free oxygen downward.

The simple equation governing carbon transformations in the organic pump is

$$C^{\circ} + O_2 \underset{\text{production}}{\overset{\text{remineralization}}{\rightleftarrows}} CO_2 \qquad (11)$$

where C° is carbon bound in organic form. Sinks and sources of CO_2 force a repartition within the carbonate system according to equation (4).

New production leads to a sink for CO_2 (equation 11), a decrease in DIC and a draw-down of pCO_2 in the ocean surface layer. Note that total charge in the surface layer, and therefore ALK, remains unchanged (equation 7). The uptake of nitrate and phosphate in new production does however affect ALK. Uptake of these cations increases the positive charge excess thus requiring higher concentrations of weak acid anions (larger ALK) to achieve charge balance. The size of the ALK increase depends upon the relative composition of organic matter in new production. For C:N:P (Redfield) mole ratios in new production of 102:16:1 (Shaffer 1992), the increase in ALK from the nutrient effect would be about 0.18 units for every unit decrease in DIC in new production. An increase in ALK draws down surface layer pCO_2 (equation 9).

The organic carbon pump is strongly controlled by nutrient cycling in the ocean. Over more than 80% of the ocean, surface layer nutrient concentrations approach zero and the rate of nutrient flux into the lighted surface layer from below puts a limit on the amount of new production possible. Uptake experiments (eg Goldman et al 1979) and low N:P ratios (below the standard 16:1 Redfield ratio) in the surface layer outside of high latitudes indicate that inorganic N may be "more" limiting than inorganic P. In as much as availability of nutrients determine new production levels outside of high latitudes, these levels will be directly proportional to the total amount of nutrients in the ocean. The constant of proportionality will however depend upon ocean physics and biology (Shaffer 1989).

The distribution of phosphate in the ocean varies over two distinct time scales. The first is a "long" scale, in excess of 10 kyr, over which mean phosphate inventory is controlled by balances of net sinks and sources like river input, burial and phosphorite formation (Froelich et al 1982). The second is a "short" time scale, less than 1 kyr, over which the spatial structure is controlled by ocean physics and the organic pump (Shaffer 1989). The cycling of nitrogen is more complex. Denitrification and nitrogen fixation can be very large net sinks and sources of inorganic N. Since these processes are coupled to the organic pump there is the potential for both the ocean inventories and the spatial structure of inorganic N to vary over the "short" time scale (ibid). Large changes in ocean inventories of inorganic N over such time scales may lead to climate change via changes in new production and in atmospheric pCO_2 (Shaffer 1990).

Large ambient nutrient levels in the surface layer at high latitudes, and especially in the Southern Ocean, bear witness to a potential for much larger new production there. What then limits primary production in these regions? Three main explanations have been proposed: 1. light limitation due to deep vertical mixing. 2. zooplankton grazing and 3. lack of micronutrients, notably iron (Martin 1990). The latter hypothesis has been used to explain low ice age pCO_2 values (ibid). However, the large new production invoked in such high latitude hypotheses (eg Sarmiento and Toggweiler 1984, Siegenthaler and Wenk 1984, Knox and McElroy 1984) has not found support in the sediment record (eg Mortlock et al 1991). Still there exists a great need to understand the control on new production at high latitudes where probably the greatest potential exists for a change in atmospheric pCO_2 due to the action of ocean biota (see also below).

b) The calcium carbonate pump

Within the ocean surface layer and the upper thermocline, calcium carbonate tests are formed by some species of phytoplankton and zooplankton. Subsequently, these particles sink or are transported to deeper layers in the guts of zooplankton. At depth most of the particles dissolve and release carbonate ions but some accumulate on the ocean bottom. As for the organic pump above, the calcium carbonate pump is closed as mixing and advection carry inorganic carbon back toward the surface layers.

For both types of $CaCO_3$ formed biogenically - calcite (coccoliths, foraminifera) and aragonite (pteropods) - the simple equation describing carbon transformations in the calcium carbonate pump is

$$CaCO_3 \underset{\text{production}}{\overset{\text{dissolution}}{\rightleftarrows}} Ca^{2+} + CO_3^{2-} \qquad (12)$$

Calcium carbonate production near the sea surface leads to a sink of carbonate ion and, therefore, also of DIC. For every unit decrease in DIC there is an accompanying decrease of two units of ALK due to the removal of the calcium ion with a plus 2 charge. The net effect will be to reduce the ALK-DIC difference in the surface layer and to increase pCO_2 (equation 9). This can also be seen in equation 4 (the removal of carbonate ion drives the reaction to the left thereby raising the CO_2 concentration). Thus the two biotic pumps have opposite effects on atmospheric pCO_2.

Whereas nutrient availability controls levels of new production over most of the ocean, there are no such simple paradigms for biogenic $CaCO_3$ production. Clearly ecosystem structure is important. Also some organisms forming carbonate shells like coccoliths appear to prefer warmer temperatures. There is clearly a great need to understand the processes controlling the levels of biogenic $CaCO_3$ production, and thereby the strength of the calcium carbonate pump. As discussed below, observations suggest that in the modern ocean new production of organic tissue exceeds $CaCO_3$ production by about a factor of five in terms of carbon.

Over 300 million years ago planktonic organisms with the ability to form calcium carbonate tests appeared in the ocean. Since then their activities have put a strong constraint on the ocean-atmosphere carbon system. Carbonate ion concentations in the deep ocean have been maintained near their saturation concentrations with $CaCO_3$ such that an approximate balance is maintained between the amount of inorganic carbon entering the ocean with river inflow and the net loss of carbon with planktonic tests collecting on the sea bottom. Changes in circulation and/or organic cycling may temporarily upset this balance in the deep ocean. Subsequently over a time scale of about 10 kyr, dissolution from or accumulation to the large reserves of planktonic tests in deep ocean sediments will restore the equilibrium. In the process, ALK and DIC are added or subtracted to the ocean in the proportion two to one according to above equations and the definition of ALK. In this way ocean-atmosphere inventories of ALK and DIC are tightly coupled. This process also leads to a modification of pCO_2 levels brought on by the initial circulation or organic cycling change.

c) Remineralization and dissolution

To date most research concerned with the two biotic pumps considered here has been directed toward the surface layer processes of primary production and biogenic $CaCO_3$ production. This is natural since these processes drive the pumps. Still, the effect of these pumps on the chemical composition of the ocean and atmosphere will also depend upon the nature of their remineralization/dissolution phases. Deep remineralization/dissolution leads to depressed surface layer concentrations and/or weaker near surface gradients. For instance, outside of high latitudes where new production is limited by the availability of nutrients, levels of new production are lower for deeper remineralization. Also surface layer ALK tends to drop, and atmospheric pCO_2 tends to rise, when dissolution occurs deeper down.

It is clear that ecosystem structure may play a large role in determining where remineralization/dissolution occurs. Surface layer biology will dictate the form of the organic matter exported from the surface layer (large or small particles, DOM). The degree of reworking of this matter by zooplankton in sub-surface layers will affect the depths of remineralization and, perhaps, of dissolution of $CaCO_3$ (Fiadiero 1980). There is a need to include more sophisticated treatments of subsurface remineralization/ dissolution processes in ocean ecosystem models. In this way we would be much better equipped to access the role of the marine ecosystem for regulation the chemical composition of the ocean and the atmosphere.

d) Biogenic carbon flows in the modern ocean

Within recent years global ocean new production has been estimated by a variety of methods including simple and 3-D models (Najjar 1990, Bacastow and Maier-Reimer 1990, Shaffer 1992, see also below), extrapolation of sediment trap data (Martin et al 1987, Berger et al 1987), primary production and f ratios (Eppley and Peterson 1979), and enzyme activity studies (Packard et al 1988). The corresponding new production estimates (in $GtCyr^{-1}$) are 7-13 (N), 4.8 (BM), 5.0 (S), 7.4 (M), 4.3-5.4 (B), 3.4-4.7 (EP) and 21.9 (P).

It has been proposed that as much as 80% of total new production is exported as dissolved organic carbon (DOC, Najjar 1990, Toggweiler 1989). Models (eg. Bacastow and Maier-Reimer 1990, Shaffer 1992) often require remineralization scales deeper than those implied by sediment trap data to be able to fit ocean chemistry data. This is consistent with deeper DOM remineralization. Enzyme activity-based new production estimates include DOC but sediment trap-based estimates do not. While there is as yet no firm basis for accepting a DOC production as high as 80% of new production (Najjar's best fit to ocean chemistry with this proportion is subject to uncertainties stemming from problems with his physical circulation model), it is not unlikely that up to 50% (or even more) of new production may be exported as DOC. With this in mind, total ocean new production probably lies in the range 5 - 10 $GtCyr^{-1}$ and may be larger still.

Estimates of the mean ocean ratio of new production to biogenic $CaCO_3$ production in the ocean surface layer in terms of carbon vary between about 3.5 and 7.5 (eg. Broecker and Peng 1982, Garçon and Minster 1988, this paper). Such global estimates are usually constrained by fitting ocean alkalinity data. Sediment trap data yield complementary, local estimates. These "rain ratios" are thought to be highest at high latitudes where calciferous organisms are less commen. A mean rain ratio of 5 and the above estimates of new production leads to an estimate of 1-2 $GtCyr^{-1}$ for global production of biogenic $CaCO_3$-C.

4. THE ROLE OF MARINE PRODUCTION AND REMINERALIZATION IN THE NATURAL CARBON CYCLE

a) The HILDA model

To gain a better understanding of the interconnections of carbon flows in the natural C cycle and of the relative importance of physical and biological processes in the ocean for determining atmospheric pCO_2, it is useful to formulate and explore a carbon cycle model. Below, I undertake this task with the help of the HIgh Latitude exchange/interior Diffusion-Advection (HILDA) model (Shaffer and Sarmiento 1992, Shaffer 1992) extended to consider the natural carbon cycle of the ocean-atmosphere system.

The HILDA model configuration (Figure 2) consists of a stratified interior and a deep, homogenous high latitude box (HD), each topped by a surface layer (LS, HS) in contact with the atmosphere via exchange coefficients (g_{LS}, g_{HS}). Ocean physics is parameterized by deep upwelling (w), interior mixing (k), vertical exchange between high latitude boxes (u) and lateral exchange between the deep boxes (q). The organic carbon pump is characterized by a mean ocean phosphate concentration, a Redfield ratio (C:P), a high latitude new production, and separate remineralization length scales for phosphate and organic-derived CO_2. New production outside of high latitudes is assumed to be nutrient limited and, thus, its value follows from the choices of the above parameters. The calcium carbonate pump is characterized by a mean ocean alkalinity, biogenic production rates of $CaCO_3$ in the LS and HS layers and a dissolution length scale. The carbon cycle is then closed by either specifying a mean ocean concentration of dissolved inorganic carbon or a value for atmospheric pCO_2. Export of organic carbon from the surface layers (i.e. the new production) is indicated in the figure by f_{OL} and f_{OH}. Export of biogenic $CaCO_3$ is indicated by f_{IL} and f_{IH}.

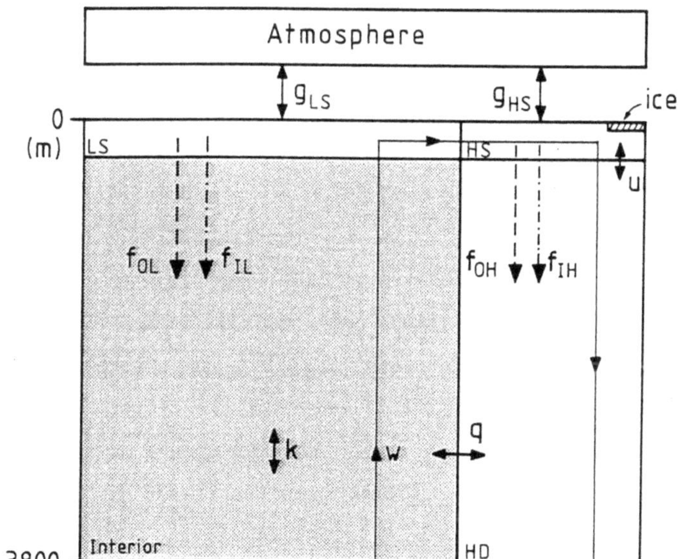

Figure 2: The HIgh Latitude/Diffusion-Advection (HILDA) model for nutrient, oxygen and carbon cycling in the ocean (see text for more details).

The model is calibrated for the "modern ocean" in the following steps: 1. best fit values of ocean physics parameters are found by fitting analytic solutions for temperature and ^{14}C to ocean data (Shaffer and Sarmiento 1992); 2. Given these parameter values, best fit values for organic pump parameters are found by fitting analytical solutions for phosphate and dissolved oxygen to ocean data (Shaffer 1992a); 3. Given all above parameter values, best fit values for $CaCO_3$ pump parameters are found by fitting analytical solutions for alkalinity to salinity-normalized ocean data and 4. Given all above parameters, dissolved inorganic carbon and pCO_2 solutions are found after choosing a mean ocean DIC such that deep interior values match (salinity-normalized) observed values, the assumption being that the deep interior is as yet relatively unaffected by the anthropogenic perturbation. Since carbonate chemistry is non-linear, the last step involves numerical iteration. GEOSECS and TTO data sets were used for chemical species, the Levitus (1982) data set was used for temperature.

With regard to marine production and remineralization, some of the main conclusions of this work are: Total ocean new production is about 5 $GtCyr^{-1}$ of which about 48% takes place at high latitudes (polar North Atlantic and Southern Ocean). The traditional C:P ratio in new production of 106 is consistent with ocean chemistry (the "new" value of 130 proposed by Takahashi et al (1985) is not consistent with ocean chemistry). There is strong vertical fractionation during remineralization of organic matter in the ocean. In the mean, nutrients are released considerably higher up in the water column than CO_2 during this process. Total ocean production of biogenic $CaCO_3$ is about 0.9 $GtCyr^{-1}$ of which about 40% takes place at high latitudes. Thus the overall "rain ratio" is about 5.5. About two-thirds of the biogenic $CaCO_3$ formed near the ocean surface dissolves below 2000 m depth. Thus ocean chemistry is consistent with the conclusion that not all calcium carbonate dissolution takes place on the ocean floor. A significant part of total dissolution appears to take place in the water column (compare with Fiadeiro 1980 or Berner 1977). For the complete HILDA set of best fit physical and biological parameters to characterize the modern, pre-anthropogenic ocean-atmosphere system, atmospheric pCO_2 calculated as sketched above turns out to be 276 ppm, in good agreement with the ice core data.

b) Carbon cycling in abiotic and biotic model oceans

It is useful to consider several, hypothetical ocean states to concentrate on the effects of ocean biota on the natural carbon cycle. In this context I resurrect the "Strangelove" ocean of Broecker and Peng (1982) in a HILDA guise. In this ocean all life has died and the ocean-atmosphere carbon cycle has readjusted to ocean-atmosphere physics and ocean chemistry alone, taken here to be the same as in our "modern ocean" solution. We assume constant, "modern ocean" inventories of inorganic carbon and alkalinity; this is the appropriate assumption for the time scales of decades to centuries considered here. In other words, carbonate compensation is ignored here for the moment.

Consider first an isothermal, isohaline "Strangelove" ocean with T = 18°C (global mean surface temperature) and S = 35 PSU. Lacking processes which can maintain vertical or horizontal differences in DIC or ALK, this ocean is also homogeneous in these properties (see case a in Figure 3. Units are μmol kg^{-1} for DIC and μeq kg^{-1} for ALK. The dots in Figure 3 show mean ocean profiles of salinity-normalized DIC and ALK from the GEOSECS and TTO data). High surface layer DIC and a high atmospheric pCO_2 (643 ppm) result.

Atmospheric pCO_2 is lowered to 460 ppm in a "Strangelove" ocean with modern temperature and salinity distributions in the ocean surface layer ($T_{LS,HS}$ = 20, 2°C, $S_{LS,HS}$ = 35, 34.6 PSU). This is an effect of the solubility pump (Volk and Hoffert 1985, see also Figure 1).

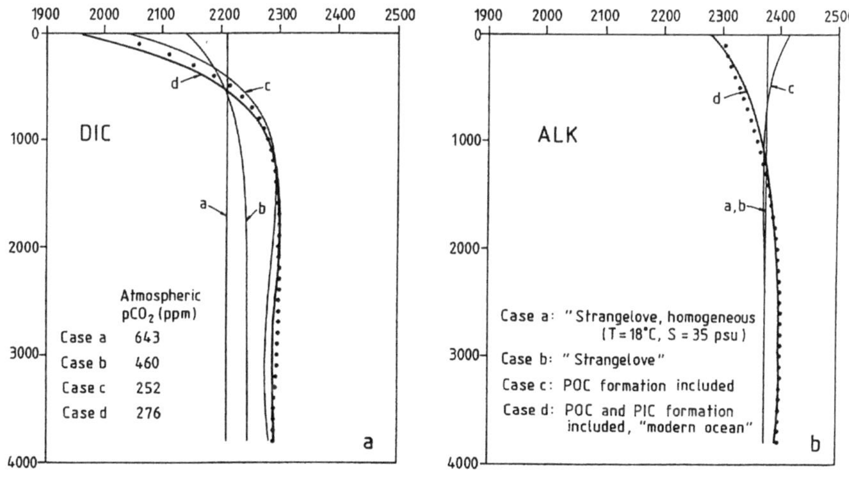

Figure 3: HILDA model profiles for dissolved inorganic carbon (DIC,a) and alkalinity (ALK,b) for four ocean cases all having the same "physics" and inventories of DIC and ALK as the "modern ocean" case. A "Strangelove" ocean is one with the biota removed. Atmospheric pCO_2 values for each case are also listed. POC and PIC refer to particulate organic and inorganic carbon.

Cold water in the high latitude surface layer dissolves more CO_2 than the warmer surface water equatorward. This CO_2 excess is then transported to deeper layers by enhanced physical transport in the polar regions. In this pump CO_2 is drawn out of the low latitude surface layer (resulting in a DIC drawdown there, see case b in Figure 3) and transported via the atmosphere to high latitudes. Alkalinity is unaffected by this pump (Figure 3, case b).

Carbon flows associated with the solubility pump in the "Strangelove" ocean and resulting concentrations of DIC and ALK are shown in Figure 4a. Carbon flows are in $GtCyr^{-1}$ and concentrations are as in Figure 3. The light arrows represent flows in the mean overturning cell (w) whereas the heavy arrows denote "turbulent" or exchange flows (k,u and q within the ocean and $g_{LS,HS}$ for the air-sea exchange). Of course in simple models like HILDA it is not always clear what the physical analog to the real ocean is. For instance u may be interpreted in part as deep polar recirculation and in part as turbulent or convective exchange with shorter time and space scales. The net carbon exchange associated with u is 0.4 $GtCyr^{-1}$ downward as shown in the figure. The net transport of CO_2-C via the atmosphere from low to high latitudes is 0.7 $GtCyr^{-1}$. This amount returns to the low latitude ocean via circulation and ventilation.

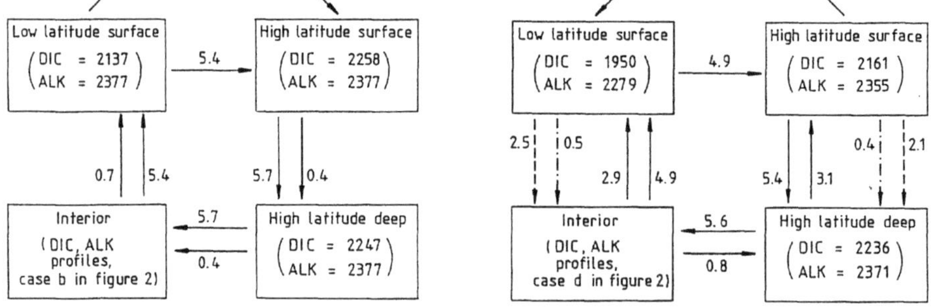

Figure 4: Net carbon flows in $GtCyr^{-1}$ in the "Strangelove" ocean (a) and the "modern" ocean (b). Light and heavy arrows denote advection and diffusive flows respectively. Dashed and dashed-dotted arrows denote export of organic carbon and $CaCO_3$-carbon from the surface layers respectively. Calculated values for DIC and ALK in the well-mixed boxes of the model are also included.

Atmospheric pCO_2 is lowered still more to 252 ppm when plankton production resumes at "modern ocean" levels (still no biogenic $CaCO_3$ production). This large pCO_2 drawdown due to the organic carbon pump can be explained in part by DIC drawdown in the surface layer and in part by an increase there of ALK (and an associated increase in ALK-DIC) due to nitrate uptake in new production (case c in Figure 3).

When the complete "modern ocean" biota is reintroduced the atmospheric pCO_2 rises again to the HILDA pre-industrial level of 276 ppm. Case d in Figure 3 shows that the "modern ocean" $CaCO_3$ pump draws down DIC still more (about as much as the organic pump which circulates about 5 times more carbon! This is a good example of how deep remineralization (dissolution) can have a large effect on surface layer values). Still, the drawdown of alkalinity is about twice as large leading to smaller ALK-DIC and a net effect of higher pCO_2 (equation 9). The degree of atmospheric pCO_2 drawdown by each of the three pumps considered here depends on pump strength and on ambient pCO_2 levels through carbonate chemistry. If the pumps are referenced to modern, pre-industrial pCO_2 levels, the relative strengths among the pumps operating in the natural carbon cycle of the modern ocean (as deduced by calibrating the HILDA model to ocean data) are 1:0.56:-0.16 for the organic pump:solubility pump:$CaCO_3$ pump. In this sense, one might say that biology has been more important than physics in setting pre-industrial concentrations of atmospheric CO_2. Note however that a clear cut separation cannot be made since ocean physics are an integral part in the two biotic pumps described above. Thus these pumping ratios will change for different choices of physical parameters not only due to changes in the solubility pump but also due to concurrent changes in the biotic pumps.

Carbon flows in the HILDA solution for the natural carbon cycle of the modern ocean are shown in Figure 4b. Heavy, dashed arrows represent export of organic carbon from the surface layers while the heavy, dashed-dotted arrows represent the export of biogenic $CaCO_3$. Note that now there is essentially no net exchange of CO_2 between high and low latitudes via the atmosphere. This is consistent with the view that concentrated outgassing of about 2 $GtCyr^{-1}$ in the equatorial region (a figure given by Keeling et al 1989) was approximately compensated by ocean uptake at mid-latitudes in the pre-industrial ocean. High latitudes are not a significant CO_2 sink in this estimate of the natural carbon cycle.

The effect of the biological carbon pumps on natural carbon cycling can be shown more clearly as the "biological perturbation" found by subtracting the "Strangelove" ocean solution from that of the modern ocean (Figure 5). Removal of carbon from the surface layers by new production and $CaCO_3$ production is compensated to a large extent locally by carbon input from below due to physical exchange. Still, the biotic pumps drive a net flow of carbon of 1.3 $GtCyr^{-1}$ "counter-clockwise" around the system. This flow is about double the size of and in the opposite sense to that in the solubility pump. The net carbon flow is associated with enhanced drawdown of DIC in the low latitude surface layer due to the organic carbon and $CaCO_3$ pumps in combination with weak vertical exchange outside of high latitudes.

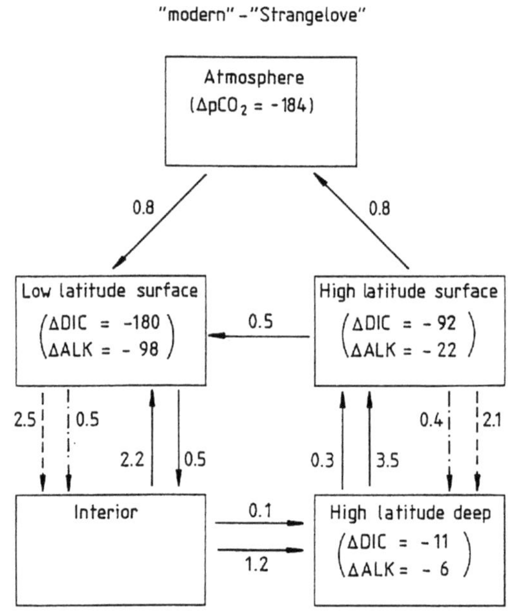

Figure 5: The difference in net carbon flows and DIC, ALK values between the abiotic "Strangelove" ocean and the biotic "modern" ocean. Arrow and unit conventions are as in Figure 4.

c) Sensitivity of atmospheric pCO_2 to changes in ocean physics and biology

To test the sensitivity of atmospheric pCO_2 levels in the natural carbon cycle to changes in physical and biological parameters of the model, new solutions were obtained for doubled values of some of the parameters (Table 1). In the table, model predictions for changes in atmospheric pCO_2 are presented, both for before and for after carbonate compensation. The model deals with carbonate compensation as follows: Over 10 kyr time scales a balance is assumed between river input of inorganic carbon and net loss of carbon through burial of planktonic $CaCO_3$ tests. In the ocean this burial occurs above the lysocline, a sediment horizon below which there is little preservation of the tests. In the model this situation is approximated

Table 1. Sensitivities of atmospheric pCO_2 levels and ocean new production to changes of some physical and biological parameters in the HILDA model of the natural carbon cycle in the modern ocean

Doubling of best fit value for:	ΔpCO_2(2) (ppm)	ΔpCO_2(3) (ppm)	ΔNP(4) (%)
Surface temperature(1)	14.7	11.9	-
Interior mixing (k)	-9.6	-24.8	29
High latitude exchange (u)	34.4	43.7	-
Redfield ratio (C:P)	-116.7	-126.5	100
High latitude new production (f_{OH})	-50.5	-59.1	47
Nutrient remineralization length scale	23.0	20.1	-12
Carbon remineralization length scale	-20.7	-62.1	-
Both remineralization length scales	5.6	-35.1	-12
Low latitude $CaCO_3$ production (f_{IL})	20.3	36.0	-
$CaCO_3$ dissolution scale	-0.9	-3.6	-

(1) 5°C increase outside of high latitudes.

(2) atmospheric pCO_2 change from "best fit", preanthropogenic value (275.1 ppm), before carbonate compensation.

(3) atmospheric pCO_2 change from "best fit", preanthropogenic value, after carbonate compensation.

(4) Percent change of total ocean new production about a standard case value of 5 $GtCyr^{-1}$

in terms of a calcite saturation depth at which the calculated carbonate ion concentration equals the saturation concentration for calcite. For the HILDA best fit solution this depth turned out to be 3580 m. For a new ocean setup, the carbonate ion concentration at this depth is recalculated and compared with the saturation value there. If the new concentration is les (more) than the saturation value, ocean inventories of ALK and DIC are increased (decreased) in the model in the appropriate 2 to 1 proportion to simulate enhanced dissolution (burial) of $CaCO_3$ tests. When the carbonate ion concentration at 3580 m again coincides with the saturation value a new river input/burial balance obtains. Surface layer values of ALK and DIC also become shifted in roughly the same proportion leading to a change in atmospheric pCO_2. Table 1 also shows percent changes in total ocean new production (from the model standard case value of 5 $GtCyr^{-1}$) which follow from the solutions with the prescribed parameter changes.

The results in Table 1 show that a doubling of the Redfield ratio leads to the largest pCO_2 drawdown since the organic carbon pump also strengthens accordingly as reflected in the increase in ocean new production. In a similar way doubled high latitude new production leads to a large pCO_2 drawdown (compare with Sarmiento and Toggweiler 1984, Siegenthaler and Wenk 1984 and Knox and McElroy 1984). Doubled exchange between the high latitude boxes leads to a large pCO_2 increase (see also above references). With no concurrent change in new production at high latitudes, such increased exchange enriches the surface layer with DIC from the deep ocean reservoir. This leads to a pCO_2 increase in the surface layer and in the atmosphere (equation 9).

On the other hand, a doubling of interior mixing leads to only a small increase in atmospheric pCO_2 levels although the oceanic new production increases by almost one third. This can be traced to Redfield coupling of nutrients and carbon in the organic carbon pump. However, this coupling is not absolute due to the additional $CaCO_3$ cycle and the atmospheric connection for carbon but not for nutrients. Thus there is a pCO_2 decrease for increasing mixing and upwelling. Note also that the pCO_2 drop due to this effect is considerably larger after carbonate compensation.

When the remineralization scales are doubled independently, each change leads to about a 20 ppm change in atmospheric pCO_2 but in opposite senses. Deeper nutrient release depresses ocean new production while deeper CO_2 release depresses surface layer DIC values.

Together, the effects tend to cancel. This is another consequence of Redfield coupling. On the other hand, the release of CO_2 at depth makes the deep ocean more acidic, lowers carbonate ion concentration and thereby induces increased dissolution of carbonate tests. As a result, a large, concurrent increase in remineralization scales for nutrients and carbon leads to a rather large decrease in atmospheric pCO_2 concentration after carbonate compensation (compare with Boyle 1988).

A doubling of biogenic $CaCO_3$ production outside of high latitudes has a two-fold effect. First, ALK-DIC decreases in the surface layer leading to increased atmospheric pCO_2 (equation 9). Subsequently, increased ALK-DIC in the deep ocean causes carbonate ion concentrations to rise and more $CaCO_3$ tests to accumulate. As a result ocean inventories of ALK and DIC drop until the river inflow/burial balance is restored. After this carbonate compensation, the initial pCO_2 increase has almost doubled. Large changes in both the $CaCO_3$ dissolution scale and in the biogenic production of $CaCO_3$ at high latitudes (not shown) have relatively small effects on atmospheric pCO_2.

5. DISCUSSION

The results presented above demonstrate that the amount of drawdown of atmospheric pCO_2 due to the action of the marine biota is not simply related to the level of marine new production. The coupling of nutrients and carbon through uptake in photosynthesis (Redfield ratios) puts strong constraints on the ability of the organic pump to sequester atmospheric CO_2 in the ocean. It is changes in total nutrients in the ocean (nutrient inventories) or changes in the efficiency with which marine production uses available nutrients (Redfield ratios, new production at high latitudes) that have the largest relative effect on atmospheric CO_2 levels via the action of the marine biosphere. Processes which redistribute nutrients within the ocean also tend to redistribute dissolved inorganic carbon (vertical mixing, remineralization). Therefore, net effects of such processes via the action of the marine biosphere are relatively small.

The effects by some other aspects of ocean carbon cycling on atmospheric pCO_2 levels are not constrained by this nutrient-carbon coupling. Such aspects are the biogenic production of $CaCO_3$, its dissolution and the carbonate compensation process. The level of $CaCO_3$ production may be controlled by temperature and ecosystem structure. At present, this level may be about a factor of five greater than the flow of inorganic carbon associated with the balance between river input and burial of $CaCO_3$ tests. Processes which redistribute DIC, ALK or temperature in the vertical will change deep ocean concentrations of carbonate ion and lead to new atmospheric pCO_2 levels after carbonate compensation. It is in this way that large changes in vertical mixing and/or carbon remineralization scales eventually lead to large pCO_2 changes in the atmosphere.

The inset in Figure 1 depicts the observed rise in atmospheric pCO_2 in recent years, mainly due to fossil fuel burning. According to present knowledge, the role of the ocean on the evolution of this curve has been to take up part of the anthropogenic emission via air sea exchange and mixing. What impact can the natural carbon cycle, also sketched in Figure 1, have on the future evolution of this curve? As discussed above, it is useful for this assessment to consider CO_2 fluxes to/from the ocean that would occur in association with changes of natural carbon cycling. These can then be compared with the present ocean uptake rate for anthropogenic carbon of about 2 $GtCyr^{-1}$ as calculated from abiotic ocean models.

For a step change in natural carbon cycling and an ocean response time of 100 yr (eg. Maier-Reimer and Hasselmann 1987), it can be shown that the CO_2 flux to/from the ocean during readjustment between initial and final baseline pCO_2 levels in the atmosphere would be about 0.025 $GtCyr^{-1}$ per ppm pCO_2 change. Fluxes would be less for more gradual changes in natural carbon cycling. With the exception of a CO_2 flux to the ocean following a doubling of the Redfield ratio (an event which is unlikely to occur), all upper bound CO_2 fluxes estimated in this way from the ΔpCO_2 results of the doubling scenarios (Table 1, column 1) are less than or much less than the current ocean uptake of anthropogenic CO_2. It appears to be unlikely that "reasonable" future changes of the biological setup of the ocean will make any radical changes in ocean uptake of CO_2 or in the evolution of atmospheric CO_2 concentrations in the near future. Still, fluxes associated with possible changes in the setup of the marine biosphere may be large enough to have a significant impact on atmospheric pCO_2 levels over century time scales.

To be able to predict changes in the strength and structure of ocean biotic pumps, ecosystem models will be needed which consider explicitly biological processes in the ocean. Up until now, progress has been made toward generalized ecosystem models of the ocean surface layer (eg. Fasham et al 1990). For instance, such models can simulate zooplankton control on primary production levels. In future work, attention should also be turned to ecosystem models of sub-surface remineralization. It is hoped that results like those presented in the present paper will provide useful targets and constraints for such future, process-oriented models.

ACKNOWLEDGMENTS

I would like to thank Ola Åkerlund of the Oceanographic Institution, Gothenburg, Sweden for his very competent programming work on the carbon cycle version of the HILDA model. Agneta Malm provided technical assistence with the preparation of this paper. This work has been supported by grants from the geoscience division of the Swedish Natural Science Research Council.

REFERENCES

Bacastow R, Maier-Reimer E (1990) Ocean-circulation model of the carbon cycle. Climate Dynamics 4:95-125

Barnola JM, Raynaud D, Korotkevich YS, Lorius C (1987) Vostok ice core provides 160,000-year record of atmospheric CO_2. Nature 329:408-414

Berner RA (1977) Sedimentation and dissolution of pteropods in the oceans. In: Anderson K, Malahoff A (eds) The Fate of Fossil Fuel CO_2 in the Oceans, Plenum, New York, p 243-260

Berger WH, Fischer K, Lai C, Wu G (1987) Ocean productivity and organic carbon flux. Part 1. Overview and maps of primary production and export production. Univ. of California, San Diego, SIO Reference 87-30

Bolin B (1986) How much CO_2 will remain in the atmosphere? In: Bolin B, Döös BR, Jäger J, Warrick RA (eds) The Greenhouse Effect, Climatic Change and Ecosystems, SCOPE 29, John Wiley, p 93-155

Boyle EA (1988) The role of vertical fractionation in controlling Late Quaternary atmospheric carbon dioxide. J Geophys Res 93: 15701-15714

Broecker WS (1982) Ocean chemistry during glacial time. Geochim Cosmochim Acta 46:1689-1705

Broecker WS, Peng TH (1982) Tracers in the Sea, Eldigio Press, Lamont-Doherty Geological Observatory, Palisades, N.Y.

Eppley RW, Peterson BW (1979) Particulate organic matter flux and planktonic new production in the ocean. Nature 282: 677-680

Fiadeiro M (1980) The alkalinity of the deep Pacific. Earth and Planet Sci Let, 49, 499-505

Fasham MJR, Ducklow HW, McKelvie SM (1990) A nitrogen-based model of plankton dynamics in the oceanic mixed layer. J Marine Res 48: 591-639

Froelich PN, Bender ML, Luedtke NA, Heath GR, DeVries T (1982) The marine phosphorus cycle. Am J Sci 282:474-511

Garçon V, Minster JF (1988) Heat, carbon and water fluxes in a 12-box model of the world ocean. Tellus 40B: 161-177

Goldman JC, McCarthy JJ, Peavey DG (1979) Growth rate influence on the chemical composition of phytoplankton in oceanic waters. Nature 279: 210-215.

Houghton JT, Jenkins GJ, Ephraums JJ (eds) (1990) Climate Change, The IPCC Scientific Assessment, Cambridge University Press, Cambridge

Joos F, Siegenthaler U, Sarmiento JL (1991) Possible effects of iron fertilization in the Southern Ocean on atmospheric CO_2 concentration. Global Biogeochem Cycles 5:135-150

Keeling CD, Piper SC, Heimann M (1989) A three dimensional model of atmospheric CO_2 transport based on observed winds: 4. Mean annual gradients and interannual variations. In: Peterson DH (ed) Aspects of Climate Variability in the Pacific and the Western Americas, Geophys Monogr, 55, AGU, Washington, p 305-363

Knox F, McElroy M (1984) Changes in atmospheric CO_2: influence of the marine biota at high latitudes. J Geophys Res 89: 4629-4637

Levitus S (1982) Climatological Atlas of the World Ocean, NOAA Professional Paper 13 (Government Printing Office), Washington D.C.

Longhurst AR, Harrison WG (1988) Vertical nitrogen flux from the oceanic photic zone by diel migrant zooplankton and nekton. Deep-Sea Res 35: 881-889

Martin JH, Knauer GA, Karl DM, Broenkow WW (1987) VERTEX; Carbon cycling in the northeast Pacific. Deep-Sea Res 34: 267-285

Maier-Reimer E, Hasselmann K (1987) Transport and storage of carbon dioxide in the ocean in an inorganic ocean-circulation carbon cycle model. Climate Dynam 2: 63-90

Mortlock RA, Charles CD, Froelich PN, Zibello MA, Saltzman J, Hays JD, Burckle LH (1991) Evidence for lower productivity in the antarctic ocean during the last glaciation. Nature 351:220

Najjar RG (1990) Simulations of the phosphorus and oxygen cycles in the world ocean using a general circulation model. Dissertation, Princeton University, 190 pp.

Packard TT, Denis M, Rodier M, Garfield P (1988) Deep ocean metabolic CO_2 production: calculations from ETS activity. Deep-Sea Res 35: 371-382

Sarmiento JL, Toggweiler JR (1984) A new model for the role of the oceans in determining atmospheric pCO_2. Nature 308: 621-624

Sarmiento JL, Orr JC, Siegenthaler U (1992). A perturbation simulation of CO_2 uptake in an ocean general circulation model. J Geophys Res 97: 3621-3645

Sarmiento JL, Sundqvist E (1992) Oceanic Uptake of anthropogenic CO_2: a new budget. Nature 356, 589-593

Shaffer G (1989) A model of biogeochemical cycling of phosphorus, nitrogen, oxygen and sulphur in the ocean: one step toward a global climate model. J Geophy Res 94(C2): 1979-2004

Shaffer G (1990) A non-linear climate oscillator controlled by biogeochemical cycling in the ocean: an alternative model of Quaternary ice age cycles. Climate Dynam 4: 127-143

Shaffer G (1992). Biogeochemical cycling in the global ocean 2. New production, Redfield ratios and remineralization scales in the organic pump. Submitted to J Geophys Res

Shaffer G, Sarmiento JR (1992) Biogeochemical cycling in the global ocean 1. A new analytical model with continuous vertical resolution and high latitude dynamics. Submitted to J Geophys Res

Siegenthaler U, Wenk T (1984) Rapid atmospheric CO_2 variations and ocean circulation. Nature 308:624-625

Siegenthaler U, Oeschger H (1987) Biospheric CO_2 emissions during the past 200 years reconstructed by deconvolution of ice core data. Tellus 39b: 140-154

Skirrow G (1975) The dissolved gases - carbon dioxide. In: Riley JP, Skirrow G (eds) Chemical Oceanography, Vol. 2, , Academic Press, London, p 1-192.

Toggweiler JR (1989) Is the downward dissolved organic matter (DOM) flux important in carbon transport? In: Berger WH, Smetacek VS, Wefer G (eds) Productivity of the ocean: Present and Past, J. Wiley and Sons, New York, p 65-85.

Takahashi T, Broecker WS, Langer S (1985) Redfield ratios based on chemical data from isopycnal surfaces. J Geophys Res 90(C4): 6907-6924

Volk T, Hoffert MI (1985) Ocean carbon pumps: analysis of relative strengths and efficiencies in ocean-driven atmospheric pCO_2 changes. In: Sundqvist ET, Broecker WS (eds) The carbon cycle and Atmospheric CO_2, Natural Variations Archaen to Present, AGU Monograph 32, AGU, Washington, D.C.,p 99-110

MODELLING THE MARINE BIOTA

Michael J.R. Fasham
James Rennell Centre for Ocean Circulation
Natural Environment Research Council
Chilworth Research Centre, Chilworth
Southampton, SO1 7NS
United Kingdom

1. Introduction

Marine biota play an important role in the natural carbon cycle of the ocean. Biogenic particles are exported from the euphotic zone to the deep ocean and this 'biological' pump, in conjunction with the 'solubility' pump, maintains the vertical gradient of dissolved inorganic carbon (DIC). Model studies (Bacastow & Maier-Raimer, 1990; Shaffer, this volume) and analysis of GEOSECS data (Volk & Hoffert, 1985) suggest that the biological pump contributes between 60%-83% of the total DIC pump on a world-wide basis. If the marine biota were removed, the atmospheric pCO_2 would increase from its present values of 335 ppmv to 460 ppmv (Shaffer, this volume). However, when modelling the oceanic uptake of anthropogenic CO_2 it is generally assumed that the marine biota play no role in this process (Sarmiento *et al.*,1992). The reason for this assumption is that the high concentration of bicarbonate ions in seawater implies that marine plants should not be limited by CO_2 (Fogg, 1975) and that, therefore, the anthropogenic increase in surface water DIC will not produce any increase in primary production. This situation is in contrast to the terrestrial biosphere where there is evidence for a CO_2 fertilisation effect (Gifford, this volume).

It is, however, possible to envisage ways in which the strength of the biological pump could be altered by either direct human intervention or by anthropogenically induced climatic change. For example, although it is generally assumed that marine primary production is ultimately limited by nitrogen, there are parts of the world ocean (e.g. the Antarctic Ocean and the subarctic Pacific) in which nitrate is not fully utilised and levels remain high throughout the summer. It has been argued (Martin & Fitzwater, 1988) that production in these areas may be limited by a micro-nutrient such as iron. If it was possible to fertilise the upper ocean in these areas with iron, then primary production would rise and the increase in atmospheric CO_2 would

NATO ASI Series, Vol. I 15
The Global Carbon Cycle
Edited by M. Heimann

be reduced (Peng & Broecker, 1991; Joos *et al.*, 1991). Less optimistically, coupled ocean-atmosphere models predict that ocean overturning would be reduced under a doubled CO_2 scenario (Stouffer *et al.*, 1989) which would result in a reduction of both the solubility and biological pumps (Sarmiento *et al.*, 1988; Shaffer, this volume). Shaffer concludes from his modelling study that "it is unlikely that "reasonable" future changes in the set-up of the ocean will make any major changes in the ocean uptake of CO_2....within century time scales". Despite this it would seem only prudent that we should continue to strive to model the role of the biota in the ocean carbon cycle in order both to understand the natural cycle, and to provide an insurance policy in the event of future research revealing that the biota play a larger part in the uptake of anthropogenic CO_2 than at present appears likely.

This paper will begin with a short review of the history of marine ecosystem modelling that might provide a useful reading list for the tyro modeller. There then follows some general observations on ecosystem modelling, after which a simple nitrogen-based model of the biological processes in the euphotic zone will be described. Model validation will be discussed using as an example some attempts to fit the model to observations from Bermuda Station "S" (32° 10'N, 64° 30'W) and Ocean Weather Station (OWS) "India" (59°N 19°W). An application of the model will then be described, namely the problem of predicting the particle flux from the productive surface layer of the ocean. Finally, some comments will be made on the problems still to be solved before global models of the marine biological carbon cycle can be fully developed.

2. A short history of marine ecosystem modelling

The pioneering work in modelling marine ecosystems was carried out by Riley *et al.* (1949) in the USA and Steele (1958) in the UK. It is well worth remembering when reading these papers that computing facilities were non-existent when they were written. Since then details of many and varied models have been published and it is only possible here to refer to a few of these. An early size-structured model was that of Vinogradov *et al.* (1972) who attempted to model the time evolution of a community in the Pacific equatorial upwelling zone. The international programmes investigating coastal upwelling zones in the 1970s also stimulated the development of a number of successful models (Walsh, 1975; Wroblewski, 1977,1980). Another problem that occupied the minds of oceanographers during this period was understanding the processes maintaining the deep chlorophyll maximum in oligotrophic waters and the classic modelling studies on this topic were by Jamart *et al.* (1977, 1979) and Kiefer & Kremer (1981). During the 1980s biological oceanographers became aware of the importance of small organisms, such as picophytoplankton (< 1 µm in size) and bacteria, and the organisms that feed on them, such as protozoa and mucus-net feeders (Pomeroy, 1974). The first model to investigate quantitatively the importance of these ideas was developed by Pace *et al.* (1984). Another important paper of this period was by Evans & Parslow (1985) who emphasised the importance of studying annual cycles, and demonstrated how different cycles of mixed layer depth could give rise to qualitatively different biological cycles. Mention should also be made of the series of papers by Hofmann and her co-workers (Hofmann *et al.*, 1980; Hofmann, 1988; Hofmann & Ambler, 1988) describing models of plankton dynamics on the southeastern US continental shelf.

All the models referred to so far have used deterministic differential equations to describe ecosystem dynamics. Needless to say the real world is highly variable and there have been some papers that have explored stochastic approaches to modelling (Fasham, 1977; Jernigan & Tsokos, 1980; Kremer, 1983). Another novel approach to biological modelling has been developed by Woods & Onken (1982) and Wolf & Woods (1988). They pointed out that light-dependent growth of algae will depend on the historical trajectory of the algae within the turbulent mixed layer and that, therefore, a Lagrangian (following the water motion) solution should be used. They achieved this by seeding their model ocean with an ensemble of particles and solving the algal growth equations for each particle. The plant vertical distribution was then obtained by ensemble averaging at each depth zone.

No all-embracing textbook on marine ecosystem modelling has been published to date but the monographs by Steele (1974), Kremer & Nixon (1977), Platt *et al.* (1981), Nisbet & Gurney (1982), and Walsh (1988) provide useful introductions for the student.

3. Some General Considerations

3.1 Choosing the compartments.

Most ecosystem models are developed on the basis of compartments which may be biotic (e.g. zooplankton, whales) or abiotic (e.g. nitrate, detritus, dissolved organic carbon (DOC)). To build a model it is first necessary to define the compartments and then specify the equations governing the transfer of material between compartments. In choosing the compartments the modeller must attempt to reduce the complexity of nature to manageable proportions, whilst retaining sufficient structure to model the processes that are of interest to him. There are no golden rules for doing this and to a great extent success will be judged by the ability of the model to predict observations and to increase understanding. One useful result from what is called aggregation theory (Iwasa *et al.*, 1987) is that when deciding how to combine groups of organisms into model compartments one should not combine organisms with very different turnover times (turnover time is defined as the organism pool size divided by the flux of biomass through that pool). For example, it would be foolish to define a compartment that combined whales with bacteria. An other less trivial example is the necessity of having more than one DOC or detritus compartment if different components of these pools are being broken down over very different timescales (Jones & Henderson, 1986).

Turnover time is closely coupled to growth rate, which in turn is often found to be dependent on organism size (Peters, 1983). There is therefore a theoretical basis for defining size-related compartments and a number of recent models have been based on size structure (Steele & Frost, 1977; Pace *et al.*, 1984; Moloney *et al.*, 1986; Parsons & Kessler, 1986; Ducklow & Fasham, 1991). However, adding more compartments to a model adds to the dynamic complexity (e.g. Evans, 1988) and can make model interpretation difficult, and so there is much to be said for a stepwise approach beginning with a model containing the minimum number of compartments that might tackle the problem in hand and only adding further compartments as necessity dictates.

3.2 Choice of model currency

The next choice that must be made is the 'currency' of the model, for example carbon, nitrogen, chlorophyll, biomass weight, energy, etc.. Again the choice will be dependent on the problem being studied. Many marine ecosystem models have used mixed currencies and have used element ratios, usually assumed to be constant, to convert from, say, phytoplankton nitrate uptake to carbon assimilation and thence phytoplankton chlorophyll concentration. It is clear that the ideal model of the carbon cycle will need to be multi-currency, and that the biological processes within plants and animals that control their growth may need to be modelled more explicitly in order to achieve this (Anderson, submitted). The model described below uses nitrogen as the currency and where it is necessary to compare results with carbon or chlorophyll equivalents a Redfield C:N ratio of 6.625 and a C:Chlorophyll ratio of 50 was used.

3.3 Modelling the dynamics

There is no biological equivalent of the Navier-Stokes equation that can be used to define how material is transferred between compartments. However, a reasonable consensus has been gradually developing amongst biological modellers on the general form of the equations governing population growth and this has been formalised by Wiegert (1979) using the following equation:

$$\frac{dX_j}{dt} = \sum_{i=1}^{m} e_{ij}\tau_j p_{ij} f_{ij} X_j - (\mu_j + \phi_j + \rho_j) X_j - \sum_{k=1}^{n} \tau_k p_{jk} f_{jk} X_k \ . \qquad (1)$$

The first term on the right side of equation 1 represents the assimilated ingestion or uptake by species j from all the other other modelled species or abiotic sources. The second term represents physiological losses, and the third term the predation on species j by other species. The parameters are defined as follows:

p_{ij} - the preference of species j for the resource i (if predators are dynamically switching between resources then p_{ij} will be a function of the other resources as well:; see section 4.3):

τ_j - the maximum specific ingestion/uptake rate of species j:

f_{ij} - the limitation of ingestion/uptake of species j by resource i. This will normally be a function of X_i:

e_{ij} - assimilation efficiency of species j using resource i:

ϕ_j - specific loss rate due to natural mortality:

μ_j - specific loss rate due to excretion:

ρ_j - specific loss rate due to respiration.

It should be noted that the loss terms in the Wiegert equation are linear in biomass. The mortality of the highest predator in a model can be thought of as a closure term, and Steele & Henderson (1992) have recently pointed out that this closure term might take on other functional forms, such as a quadratic or hyperbolic. They also showed that the choice of the form of this closure can have a profound effect on the system dynamics. It will be shown below that it is necessary to introduce a nonlinear zooplankton mortality in order to model the observations at OWS "India" and, therefore, for full generality the mortality term could be written $\mu M(X_i)$ where $M(X_i)$ is either a constant or a function of X_i.

The equations governing the biological processes must be coupled to a physical model that provides the environmental forcing. This model could range from a simple parameterisation of the seasonal changes in mixed layer depth and incident light at a given position (Evans & Parslow, 1985), to a full three-dimensional general circulation model (Sarmiento *et al.*, 1992). The choice of an appropriate physical model will not be discussed further here, except to remark that there is a considerable benefit to be gained from 'bench-testing' a biological model in a comparatively simple physical model, in order that the biological dynamics can be fully understood before embedding the ecosystem model in more complex physical models.

One issue that must be carefully considered is model initialisation. Often models are run for short periods of time, such as a spring bloom (Fasham *et al.*, 1983; Pace *et al.*, 1984) or an upwelling event (Moloney *et al.*, 1986). In such cases the model evolution may be very dependent on the initial concentrations assigned to the state-variables. Furthermore, the model might give unrealistic results when run beyond the event timescale. Such problems can be avoided if a model is developed to run over a full seasonal cycle of the physical forcing functions. The model should then develop repeatable annual cycles that will be independent of the initial conditions. Care must still be exercised as there may be some starting conditions that will send one of the biological variables to extinction. It should also be noted that the non-linearities of the biological equations can give rise to solutions with multi-year cycles even when the physical forcing has an annual cycle (Evans, 1988).

Most papers describing models tend to emphasise how well, or badly, the model predicts the state-variables when compared to observations. However, it should be stressed that, in order to fully understand the dynamics of a model, it is important to analyse the model inter-compartment flows. If observational data of flow rates (e.g., new production, bacterial production, particle flux) are available then these provide strong constraints on the model. Even without such observations, modelled flows can be used to calculate bacterial or zooplankton growth efficiencies to check that they lie within the known range of experimental observations. The technique of Flow Analysis provides a powerful tool for the analysis of either observed or modelled flow networks (Fasham, 1985; Wulff *et al.*, 1989).

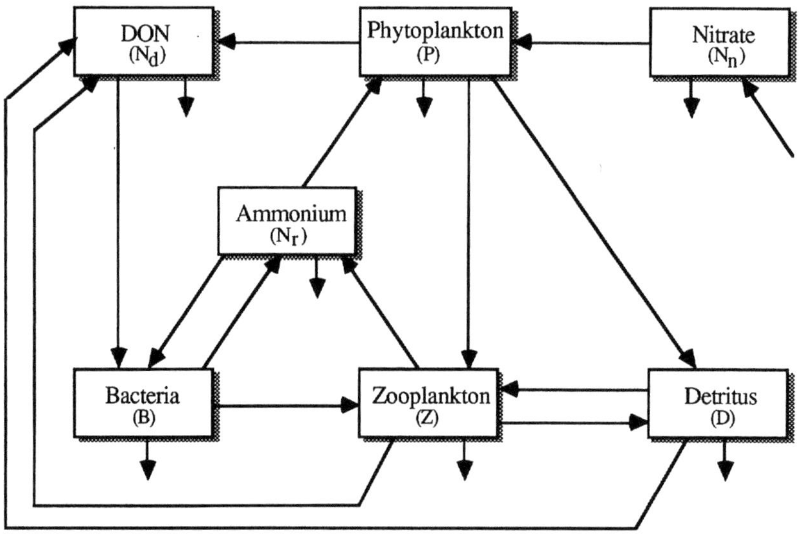

Fig. 1. A schematic diagram showing the compartments and inter-compartmental flows of a nitrogen-based model of the mixed layer ecosystem. The short arrows just below each box represent inputs and outputs between the mixed layer and ocean layers below.

4. A mixed layer biological model

A nitrogen-based model of the mixed layer ecosystem, first described by Fasham, Ducklow & McKelvie (1990; hereinafter referred to as FDM), will be used to highlight a number of the problems that arise when modelling the marine biota. It might appear quixotic to be discussing a nitrogen model in a book devoted to the global carbon cycle. However, as nitrogen is considered to be the nutrient limiting primary production in most of the world ocean then understanding the oceanic nitrogen cycle is a necessary prerequisite to modelling the carbon cycle. Biological oceanographers divide algal primary production into two forms (Dugdale & Goering, 1967; Eppley and Peterson, 1979); "new" production fuelled by nitrate that is introduced into the mixed layer by physical mixing processes, and "regenerated" production, fuelled mainly by ammonium and urea that is recycled within the mixed layer by zooplankton and bacteria. This partitioning can be measured directly by ^{15}N uptake techniques and such data provide extra constraints on any nitrogen model. The ratio of new to total production is a widely-used characteristic of the euphotic zone ecosystem and is called the f-ratio.

FDM considered that the simplest model capable of capturing the essence of the nitrogen cycle within the mixed layer required seven compartments (Fig. 1) viz., phytoplankton, zooplankton, bacteria, nitrate, ammonium, dissolved organic nitrogen (DON), and detritus. The equations governing the model are given in full in the appendix and a full description of the thinking underlying the model is given in the original paper. I will here restrict discussion to a

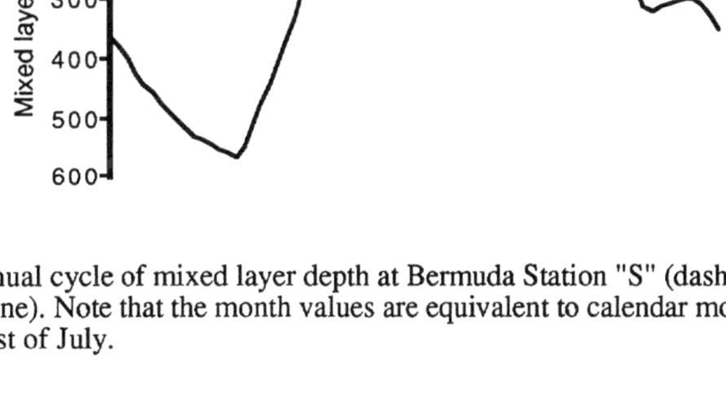

Fig. 2. The annual cycle of mixed layer depth at Bermuda Station "S" (dashed line) and OWS "India" (solid line). Note that the month values are equivalent to calendar months, e.g. the tick mark 7 is the 1st of July.

summary of the ecological processes affecting each compartment and describe, where necessary, changes that have been made since the original description.

4.1 Physical Forcing

The seasonal cycle of mixed layer depth (fig. 2) was specified for the two areas of interest using the climatic monthly averages compiled by Levitus (1982). The deepening of the mixed layer throughout the late autumn and winter entrains nitrate from below the mixed layer

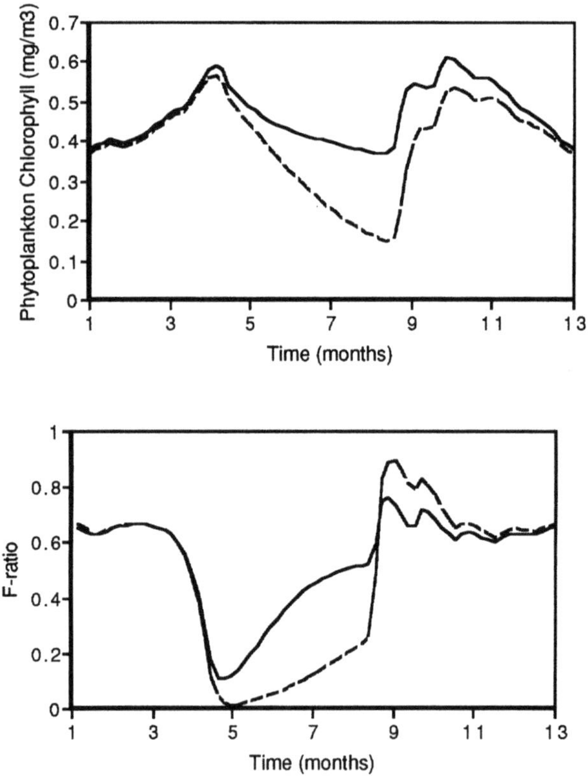

Fig. 3. Top: The annual cycle of phytoplankton at Bermuda for mixing rates of 0.01 m d^{-1} (dashed) and 0.1 m d^{-1} (solid). Bottom: The same for f-ratio. The model sampling interval on this and all subsequent simulation plots is 5 days.

that fuels the new primary production. The amount of nitrate entrained, and therefore the nitrate concentration at the start of the spring bloom, depends on the assumed concentration of nitrate below the mixed layer (N_0 in equation A17) and this quantity is very latitude dependent (Glover & Brewer, 1988). The shallowing of the mixed layer in spring detrains material from the mixed layer but ensures that the phytoplankton left within the mixed layer experience greatly increased average daily light levels, thereby stimulating their growth. Diffusive mixing between the mixed layer and the layer beneath is parameterised simply as the product of a mixing coefficient m and a concentration difference (e.g. equation A17). The choice of a value for m is by no means straightforward. Evans & Parslow (1985) used a value of 3 m d^{-1}, while FDM used a value of 0.1 m d^{-1} and found that this mixing rate meant that cross-thermocline mixing only contributed 15% of the annual flux of nitrate into the mixed layer; the remainder being provided by entrainment in the winter. Many physicists would argue that once the seasonal thermocline is established then mixing across it is very low. Some idea of the sensitivity of the model to lower mixing rates was obtained by running two simulations for Bermuda for mixing rates of 0.01 and 0.1 m d^{-1} . The biggest difference between the two simulations was during the summer; the low mixing simulation has a lower phytoplankton biomass during this period (fig. 3a) which accords more closely with the observations (Menzel & Ryther, 1960), although the values are still too high just after the spring bloom (cf. fig. 7 top). Also the low mixing simulation predicts a low f-ratio throughout the summer (fig. 3 bottom), whereas in the higher mixing simulation the f-ratio gradually increases during the summer as a result of the higher nitrate flux from below the mixed layer.

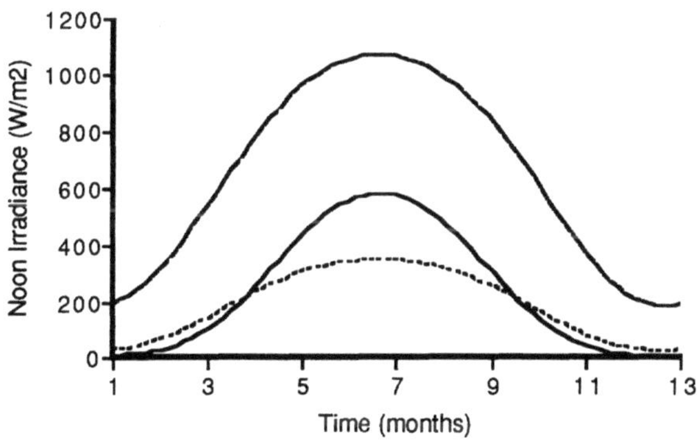

Fig. 4. The annual cycle of irradiance at OWS "India" at the top of the atmosphere (top curve), compared to that at the sea surface for a cloud level of 6 oktas according the the transmittance formulae of Smith & Dobson (1984) (solid line) and Oberhuber (1988) (dashed line).

The other physical forcing function is the solar radiation (equations A5-A7). It is a straightforward matter to calculate the noon solar irradiance for any latitude and time of year assuming a cloud-free sky (Brock, 1981) and data are now available showing the seasonal cycle of cloudiness in certain parts of the ocean (Isemer & Hasse, 1985). However, it is then necessary to calculate the atmospheric transmittance for a given cloudiness and it should be noted that different algorithms (Smith & Dobson, 1984; Oberhuber, 1988) can give markedly different results (fig. 4). It would obviously be desirable to check the calculated values of surface irradiance against observations for a particular area if these are available.

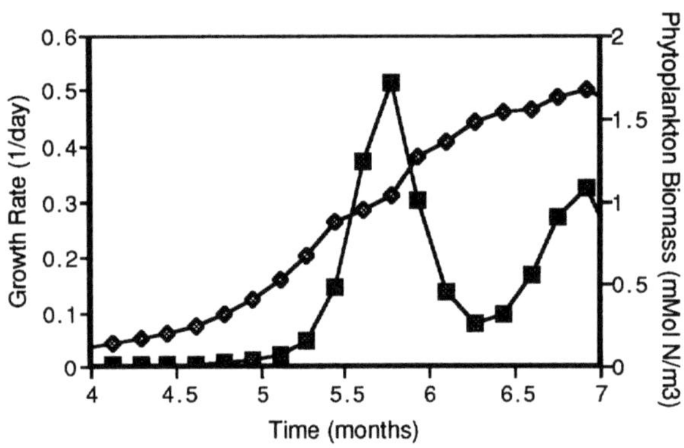

Fig. 5. The variation in modelled phytoplankton growth rate (diamonds) and phytoplankton biomass (squares) at OWS "India". Note the small depression in the growth rate at the height of the bloom.

Once the cloudiness at the surface is known then the light at depth must be calculated using a model of the light transmittance through the water column. FDM used a simple exponential model but showed that the annual primary production was very sensitive to the choice of the attenuation coefficient. It is, therefore, probably worth using more complicated light models (Morel, 1988; Sathyendranath & Platt, 1988) that parameterise the light attenuation as the sum of the attenuation of different wavebands. The number of wavebands chosen will depend on whether good optical data are available. A useful compromise between over-complication and realism is the double exponential (Paulson & Simpson, 1977) which has been shown to give good fits to observed irradiance-depth plots (Fasham *et al.*, 1983; Carr, 1986). It is also necessary to take into account the light absorbed by the phytoplankton itself, which is parameterised by a self-shading coefficient (equation A5). Self-shading is usually only significant during the height of the spring bloom, but in the mixed layer the effect is comparatively small (fig. 5).

4.2 Phytoplankton

The modelling of phytoplankton photosynthesis has probably received more study than any other aspect of quantitative marine ecology, and a vast amount of observational data have been acquired (Falkowski, 1980; Harris, 1986). The first point to emphasise is the difference between gross photosynthesis (total uptake of CO_2) and net photosynthesis (uptake of CO_2 minus CO_2 lost by respiration). Most recent measurements of marine photosynthesis have been made using the ^{14}C uptake method which is generally assumed to measure net photosynthesis although, depending on the incubation time, it probably measures something between net and gross (Dring & Jewson, 1982). The modelling approach adopted in FDM and many other studies (Jamart *et al.*, 1977; Wroblewski, 1977) has been to calculate a light-dependent net photosynthetic rate using one of the many mathematical expressions for the empirically-derived photosynthesis-light (P-I) curve. Such curves are most easily parameterised in terms of the initial slope of the P-I curve α and the maximum photosynthetic rate, P_{max}; a further parameter will be required if photoinhibition is to be taken into account (Platt *et al.*, 1980). There has been much debate about the best equation to be used (Platt *et al.*, 1977; Lederman & Tett, 1981; Fasham & Platt, 1985), however, when modelling annual cycles the choice of the values of α and P_{max} is probably more critical than small differences in the shape of the P-I curve. Once net photosynthesis has been calculated a 'balanced-growth' assumption is made that equates photosynthesis with growth (mathematically this can be achieved by replacing P_{max} in the P-I equation with a maximum growth rate ; see equation A6). The assumption of balanced growth means that processes such as the luxury uptake of nutrients or the storage of carbohydrates by the cell are ignored (for an alternative approach see Tett *et al.*, 1986; Lancelot & Billen, 1986).

Within certain limits, phytoplankton are capable of adapting to varying light levels by increasing the size or number of the photosynthetic units, which implies that α and P_{max} will vary with the ambient light level (Fasham & Platt, 1985; Geider & Platt, 1986). It is more important to take such mechanisms into account when modelling the growth of the deep phytoplankton maximum, rather than phytoplankton in the mixed layer as, in the latter case, the timescale of mixing will probably be greater than the timescale of photoadaptation (Falkowski & Wirick, 1981).

Having calculated the light-dependent growth this value is then multiplied by a function parameterising the effect of nutrient limitation, and the hyperbolic Michaelis-Menten function is the most commonly used (equation A9). However, an alternative approach is to consider that photosynthesis and nutrient uptake are separate processes that provide the substrates carbon and nitrogen for growth. Then using Liebig's law of the minimum, growth is determined only by the limiting substrate (Tett *et al.*, 1986). There is, as yet, no general consensus as to which of these models is best. Recently, a new simple model of balanced phytoplankton growth has been proposed by Sakshaug *et al.* (1989) which expresses growth as a function of irradiance, day-length, cell carbon-chlorophyll ratio, the absorption cross-section of both the chlorophyll molecule and the photosynthetic unit, the maximum photosynthetic yield, and the minimum turnover time of the rate-limiting photosystem. The last four parameters can be estimated experimentally and do not vary greatly, and so, in principle, growth could be estimated from just three parameters. The model appears to have a large measure of generality and explanatory power (Cullen, 1990), but before it could be used in an ecosystem model it will be necessary to predict how the cell carbon:chlorophyll ratio would change with the ambient light (Falkowski & Wirick, 1981) and nutrient level, especially under conditions of nutrient starvation (Cullen, 1990).

We have discussed how nutrient limitation must be taken into account when calculating phytoplankton growth rate. This would be comparatively simple if the nutrient was phosphorus or silica which can exist in only one dissolved inorganic form. However, nitrogen is mainly in the form of nitrate or ammonium (it can also be in the form of urea and amino acids but uptake of these comapounds by oceanic phytoplankton is considered to be small) and a suitable expression for their joint limitation must be devised. If the two forms were of equal utility to the cell then the appropriate limiting function is (O'Neill *et al.*, 1989),

$$\frac{k_2N_n+k_1N_r}{k_1k_2+k_2N_n+k_1N_r}\ , \qquad (2)$$

where N_n, N_r are the concentrations of nitrate and ammonium respectively, and k_1, k_2 are the equivalent half-saturation constants for uptake. However, nitrate and ammonium cannot be considered equivalent since it is energetically more favourable for an algal cell to take up ammonium rather than nitrate, and it appears that nitrate uptake in many algae is inhibited in the presence of fairly low concentrations of ammonium (Dortch, 1990). Wroblewski suggested modelling this inhibition by multiplying a Michaelis-Menten term by a negative exponential function of the ammonium concentration (equation A9). Using data from Walsh & Dugdale (1972) he estimated the value of the inhibition constant ψ as 1.5 (mMol NH4 $m^{-3})^{-1}$, implying a 50% reduction of nitrate uptake by an ammonium concentration of 0.46 mMol m^{-3}, and this value was used by FDM. However, Wheeler & Kokkinakis (1990) have found that, in the subarctic Pacific, nitrate uptake was reduced 50% by ammonium concentrations between 0.05 and 0.19 mMol m^{-3}, implying much higher ψ values of between 3.6 and 14.

I have shown how, given suitable equations for calculating the irradiance at a given depth, describing the P-I curve, and modelling of the effect of nutrient limitation we can, in principle, calculate the phytoplankton growth rate. For a mixed layer model, it is necessary to integrate this growth rate over the depth of the mixed layer in order to calculate the average growth of phytoplankton within this layer. Furthermore, when modelling seasonal cycles it is probably not necessary to consider the details of the diel cycle of photosynthesis, and so the growth rate may also be integrated over the day-length. Evans & Parslow (1985) showed that by a suitable choice of the P-I equation, and by approximating the daily light cycle by a triangular function, an analytical integration could be obtained, thereby saving computer time. Recently, Platt *et al.* (1990) have derived a more general integration that allows for a more realistic parameterisation of the daily light cycle.

All phytoplankton exude a fraction of their net production as dissolved organic matter (DOM), although the amount exuded and the conditions under which the process occurs are still a matter of debate (Fogg, 1983; Lancelot & Billen, 1986). Also, in the light of the recent work on DOM (Susuki, this volume), we need to know not just how much DOM is exuded but the size structure and lability of the exudate; at the moment this information does not exist. Until recently modellers have assumed that a small fraction (5-15%) of carbon production is exuded as DOC (Sharp, 1977). The equivalent fraction for DON exudation is not so well documented although, since the C/N ratio of the exudation products is thought to be high compared to particulate production (Fogg, 1983; Lancelot and Billen, 1986), the fraction is likely to be less than for carbon; FDM assumed a conservative value of 5% (Schell, 1974).

The FDM model assumed that phytoplankton were subject to a natural mortality in addition to the grazing mortality. This natural mortality parameterises ,for example, losses by viral attack (Suttle *et al.*, 1991a) or cell death brought about when photosynthesis is insufficient to meet respiration costs. FDM found that the phytoplankton mortality rate was a very influential model parameter. Furthermore, in a detailed analysis of a large amount of data, Walsh (1983) concluded that phytoplankton mortality and subsequent sinking may be one of the most important loss terms in many marine systems, and yet direct estimates of this rate appear to be nonexistent. FDM decided to use phytoplankton mortality as a free parameter to tune the model to give a required annual net primary production (see section 5).

4.3 Zooplankton

The term zooplankton covers organisms ranging from bacterivorous flagellates through ciliates, copepods, and euphausiids, all of which may be partly herbivorous or carnivorous, to wholly carnivorous chaetognaths. Most biologists would consider lumping all these diverse animal groups into one compartment labelled 'zooplankton' to be a gross over-simplification. Indeed, many modellers have divided zooplankton into two or more size classes (e.g., Moloney *et al.*, 1986; Parsons & Kessler, 1987; Frost, 1987). However, there are some grounds for investigating the behaviour of simpler models. Firstly, one of the main problems in validating the zooplankton compartment is the paucity of suitable seasonal and spatial observations of zooplankton biomass. What data there are has usually been obtained using nets with mesh sizes in the range 200-350 μm (Menzel & Ryther, 1961; Bé *et al.*, 1971; Williams, 1988; Angel, 1988), which sample mainly the mesozooplankton (200-2000 μm) rather than the microzooplankton (20-200 μm). Recent studies (Capriulo *et al.*, 1991; Burkill *et al.*, submitted) suggest that a large fraction of the primary production may be grazed by the microzooplankton. Unfortunately estimating microzooplankton biomass is a time-consuming exercise and so it may be some time before suitable data become available for model testing. Secondly, adding more compartments increases the number of model parameters that must be estimated, increases the dynamic complexity of the system, and adds a computational overhead for models that are to be embedded in 3-D general circulation models (GCMs). However, as will be shown later, the FDM model shows that zooplankton grazing exerts a powerful control on the phytoplankton dynamics. Therefore, although a more complicated zooplankton sub-model will probably eventually be required to produce an ecosystem model that is globally robust, I believe that investigating the behaviour and explanatory power of a one-compartment zooplankton model is an essential first step on the road to these more robust models.

One of the first problems that arises when modelling zooplankton, unless the model consists only of zooplankton grazing on phytoplankton, is dealing with the problem of parameterising grazing on multiple resources. For example, in the FDM model the zooplankton graze on phytoplankton, bacteria and detritus, and it is necessary to specify how much of each resource is grazed at any given time. One approach to this problem (Evans, 1988) is to define a measure of total food, F, as,

$$F = p_1P+p_2B+p_3D \ , \tag{3}$$

where p_1,p_2,p_3 are constants determining the zooplankton preferences for the various food types. The familiar Michaelis-Menten expression can then be used to define the grazing rate on, for example, phytoplankton as

$$G_1 = gZ\frac{p_1P}{k_3+F} \ , \tag{4}$$

where g is the maximum specific grazing rate, and k_3 is the half-saturation constant for grazing. There are, however, certain disadvantages with this method. Firstly, FDM showed that the model results were very sensitive to the values assigned to the preferences. As there are very little quantitative data on zooplankton feeding preferences, or their seasonal variability, this is a matter of some concern. Furthermore, the model zooplankton are a highly aggregated entity incorporating organisms from flagellates to euphausiids. It is therefore difficult, in principle, to assign constant preferences since the mix of zooplankton organisms, and therefore the preferences, will change throughout the year depending on the available food. Such considerations suggests that the assigned preferences should in some way change dynamically as a function of the relative proportion of the food. FDM dealt with this problem by defining weighted preferences p_i

$$p_i = \frac{p'_i}{\sum_{j=1}^{3} p'_j X_j}, \tag{5}$$

where X_j is the concentration of the jth food type and the p'_i are defined as the preferences for the different food types when the concentrations of these foods are equal. Substitution of this expression into equation (4) yields equation (A11) and results in the model zooplankton dynamically selecting the most abundant food organism. One desirable consequence of this parameterisation was that the model was much less sensitive to the choice of the p'_i. For example, altering the preference for phytoplankton from 0.5 to 0.1 only resulted in a 9% change in annual primary production for the Bermuda Station "S" simulation.

Once ingestion has been determined it is then necessary to estimate the loss terms in order to calculate zooplankton growth. These losses are faecal pellet egestion, excretion, and mortality, both natural and that due to unmodelled higher predators. In most models faecal pellet production is usually assumed to be a constant fraction of the ingestion, although this is no doubt an over-simplification of the digestive process (Taghon & Jumars, 1984). FDM assumed that both excretion and mortality were linear functions of zooplankton mortality differing in the fact that excretion resulted only in dissolved products (ammonium and DON), while mortality gave rise to detritus plus dissolved products (arising from the excretion of the unmodelled higher predators). It was pointed out by FDM that the zooplankton mortality represents a model closure term. As discussed in section 3.3, Steele & Henderson (1992) have shown that the mathematical form of this closure term can have a large influence on the dynamics of a model. They studied a three-component system consisting of phytoplankton, zooplankton and a nutrient and considered three forms for the zooplankton specific mortality rate, namely, a constant, a linear function of zooplankton biomass, and a Michaelis-Menten function of zooplankton biomass. They showed that if the constant form was chosen then limit cycle behaviour would occur at high levels of nutrient supply. This model phenomenon had been noticed by Rosenzweig (1971) who termed it the 'paradox of enrichment'. On the other hand, if the linear mortality rate was used then the limit cycle behaviour was avoided and, furthermore, it was found that varying the degree of mortality could produce either a seasonal nutrient cycle with very low summer values (as observed in the North Atlantic) or with comparatively high summer values (as observed in the North Pacific). They conclude that 'the choices (of the form of the closure term) can determine the overall patterns in all the variables and be particularly relevant to the question of nutrient limitation versus grazing control'. They also pointed out that the Michaelis-Menten expression represents a hybrid between the constant and linear forms, being linear at low values of zooplankton biomass and constant in the limit. It will be shown later (section 5) that using a Michaelis-Menten equation for the zooplankton loss term produced a better simulation of the seasonal cycle at OWS "India" and so this formulation has been adopted in the modified FDM model (equation A10).

An important feature of mesozooplankton behaviour is diel vertical migration. Many species migrate from day-time depths below 200 m into the surface at night where they feed on the phytoplankton. Biological oceanographers have speculated for many years on the reasons for this behaviour but it is now generally considered to be a predator-avoidance response (Angel, 1986). This migration, which presumably must have consequences for zooplankton dynamics, has been little modelled (but see Wroblewski, 1975; Andersen & Nival, 1991), although the recent work of Longhurst and his co-workers (Longhurst & Harrison, 1988; Longhurst *et al.*, 1990, 1991) have shown that migrating zooplankton may export primary production from the mixed later by grazing at night at shallow depths and then excreting ammonium at deeper depths during the day.

4.4 Bacteria and Dissolved Organic Matter

Bacteria in the ocean are often classified as either 'attached' or 'free '. The former are attached to particles and obtain the organic material for growth by the exoenzymatic decomposition of the particulate material (Newell & Lindley, 1984), while the latter have a planktonic existence and utilise the DOM in the seawater. Once fluorescent microscopic techniques for counting bacteria became available, it was realised that the majority of oceanic bacteria are free-living (Azam and Hodgson, 1977; Azam *et al.*, 1983). As well as DON, free bacteria can also take up and excrete ammonium. The balance between these two processes determines whether the bacteria act as net utilisers or remineralisers of carbon and nitrogen, which bears on the discussion as to whether bacteria are 'links' or 'sinks' in the marine ecosystem (Ducklow *et al.*, 1986; Sherr *et al.*, 1987; Sherr & Sherr, 1988). If they are large remineralisers this implies a low growth efficiency and only a small fraction of their uptake will be passed on to higher organisms, and so the bacteria will be mainly a sink of carbon and nitrogen. Conversely, if the growth efficiency is high they will be limited remineralisers and will be a strong link of carbon to the higher food chain. Unfortunately, there is currently much debate on whether the growth efficiencies of bacteria are high or low (Bjornsen, 1986; Kirchman *et al.*, 1991)

In a balanced growth steady-state situation the amount of remineralisation depends on the C/N ratio of the DOM being utilised, the C/N ratio of the bacteria, and the gross efficiency of the bacteria (Fenchel & Blackburn, 1979; Goldman *et al.*, 1987). This brings us immediately to the main problem in modelling bacteria and that is determining the character of the substrate used by the bacteria. From radioactive tracer studies it is well-known that bacteria will take up dissolved free amino acids and carbohydates, and that the pool of these low-molecular weight compounds (generally referred to as the "labile" pool) is turned over on timescales of hours to a few days (Suttle *et al.*, 1991). The recent work of Kirchman *et al.* (1991) suggests that, in the euphotic zone, up to 25% of the DOC pool is turned over in time scales of 1-10 days and demonstrates the necessity of including bacteria in models of the carbon cycle.

Susuki (this volume) has postulated that the 'extra' DOC measured using his high temperature catalytic technique is also utilised by bacteria but over slower timescales, and he has coined the term 'semi-labile' for this pool. At the present time we do not know whether the breakdown of semi-labile DOM is mediated by the same bacterial types as utilise the labile fraction. Recent 3-D modelling studies (Najjar, 1990; Bacastow & Maier-Reimer, 1991) have shown that, if it is assumed that a certain fraction of new production is converted into DOM, and this DOM is thereafter broken down to inorganic nutrients over a timescale of *ca* 50 years, then a better agreement can be obtained between the modelled and observed oxygen and phosphate distributions. This is an actively developing field but it is already clear that the modelling of bacterial growth will be an important aspect of carbon cycle models.

One of the critical variables of any model containing bacteria is the fraction of the primary production that cycles through the bacterial compartment. Cole *et al.* (1988) reviewed 70 studies and found that bacterial production averaged 20% of primary production. FDM found that these sort of values could not be obtained from their model unless it was assumed that DON was produced by zooplankton excretion and detrital breakdown as well as phytoplankton exudation. However, more data on the sources of DON in the ocean will be required before these models can be properly tested.

4.5 Detritus

Detritus is defined as non-living organic material that may be comprised of faecal material, exuded mucus, and dead organisms, and it found in a wide variety of forms and sizes (Newell, 1984; Angel, 1984 ; Alldredge & Silver, 1988; Koike *et al.*, 1990). Much of the detrital material sinks out of the mixed layer to the ocean interior where it supplies the food for the deep ocean and benthic ecosystems. However, detritus can also be recycled within the mixed layer by two mechanisms, namely its reingestion by zooplankton (Paffenhofer and Knowles, 1979; Poulet, 1983) or its breakdown into DOM and subsequent uptake by bacteria. In reality the latter process may be mediated by attached bacteria (Newell and Linley, 1984), and some models (e.g. Pace *et al.*, 1984) have included a separate compartment for attached bacteria. FDM considered this to be an unnecessary complication, especially bearing in mind the technical problems in distinguishing attached from free bacteria (Clarke and Joint, 1986).

The detrital sinking rate depends on the size and density of the particles and varies from a few metres to a few hundreds of metres per day (for reviews see Angel (1984), Fowler & Knauer (1986)). However, when modelling detrital sinking it is the average sinking speed for the whole detrital 'community' that is required and, formally, this is given by the ratio of the total downwards detrital sinking flux at a given depth to the average detrital biomass above this depth that is presumed to give rise to this flux. In principle, this average speed can be estimated by carrying out concurrent measurements of, say, the detrital biomass in the top 100 m and the detrital sinking flux at 100 m at different times during the yearly production cycle; the average sinking speed will be the slope of the graph of particle flux versus the depth-averaged detrital mass. In practice it is extremely difficult to separate detrital from living biomass, and so what is usually measured is the total particulate material and the total sinking flux. Alatabet (1989) made such measurements over a period of eighteen months close to Bermuda Station "S", and obtained a significant correlation between the two measurements. By applying a regression analysis to these data I obtained an average sinking speed of 4 m d^{-1} which gives us some idea of the order of magnitude of the sinking speed that should be used in a model. FDM ran their model with two detrital sinking speeds, 1 and 10 m d^{-1} and found that the 10 m d^{-1} value gave the best fit to observations.

A number of biological models have been developed that have divided detritus into a number of different classes each having different sinking rates (Hofmann *et al.*, 1981; Andersen & Nival, 1988). This approach could be taken to its logical conclusion by considering a spectrum of particle sizes and using Stokes' Law to calculate sinking speeds (e.g. Gardner & Walsh, 1990) but, to date, this idea has not been developed. To summarise, no acceptable universal parameterisation of sinking speed has yet been developed and this remains a major problem for carbon cycle modelling.

5. Parameter estimation and model validation

The estimation of the parameter values for a biological model presents many problems. The number of parameters is often large (even the relatively simple model described herein has 28 parameters), observations of some of the parameters show them to be inherently variable (e.g. zooplankton ingestion rate), while others are exceedingly difficult to measure (e.g. phytoplankton natural mortality). The problem is not quite as bad as it seems, because the model may not be very sensitive to some of the parameters and so the value assigned to these parameters will not be too critical. It is, therefore, very important that a parameter sensitivity study should be carried out at an early stage in the model development (e.g. Jamart *et al.*, 1979; FDM). It is also important to identify which aspect of the model is of interest when calculating the sensitivity indices. For example, FDM found that modelled annual primary production was not very sensitive to the values of the bacterial parameters, but this would not have been the case if the bacterial production had been the model output of interest.

Assuming that the most influential parameters have been identified and the literature reviewed for good estimates of these parameters, there will almost certainly be some parameters for which good estimates cannot be obtained. As long as there are not too many of these they can be used as variable tuning parameters to adjust some output of the model to observations at a particular location. For example, FDM varied phytoplankton natural mortality to adjust the modelled annual primary production to equal the observed values at Bermuda Station "S". A more formal approach to this problem has been described by Atkinson (1987) who showed how nonlinear programming techniques could be used to estimate uncertain parameters by maximising the model fit to observations. Bearing in mind the increase in good quality data expected over the next ten years as a result of international programmes such as the Joint Global Ocean Flux Study (JGOFS) (Fasham *et al.*, 1991), such optimising techniques may prove of great use in the future.

Once the parameters have been estimated for a given location a model can, in principle, be validated by simulating the seasonal cycle at the same location for another time period, or for a completely different location, and comparing the model results with the new data. This is easier said than done because biological time-series in oceanic areas are extremely rare. I will now describe an attempt to validate the FDM model by simulating the seasonal cycle at OWS "India" using initially the parameter set determined by fitting the model to the Bermuda Station "S" data (FDM). Data obtained by R.B. Williams and his co-workers at the Plymouth Marine Laboratory (Williams & Robinson, 1973; Williams, 1988 and references therein; also original data kindly provided by R. Williams) at OWS "India" during 1972 were used to test the model.

The initial simulations were made using the parameter set for Bermuda Station "S" described in FDM which used a detrital sinking rate of 10 m d^{-1}. The seasonal cycle of mixed layer depth is shown in fig. 2. Only two parameters were changed from the values quoted in FDM. Firstly, the cross-thermocline mixing rate was set to 0.01 m d^{-1} rather than 0.1 m d^{-1} because as discussed in section 4.1, the former value gave an improved fit to the observations of summer chlorophyll concentrations. Secondly, for the "India" simulation, the maximum phytoplankton growth rate was altered to 1.25 d^{-1}, being the value calculated from equation A8 using a mean annual mixed layer temperature of 10°C. Justifications for the choice of the other parameters was given in FDM and this parameter set will be referred to as model A.

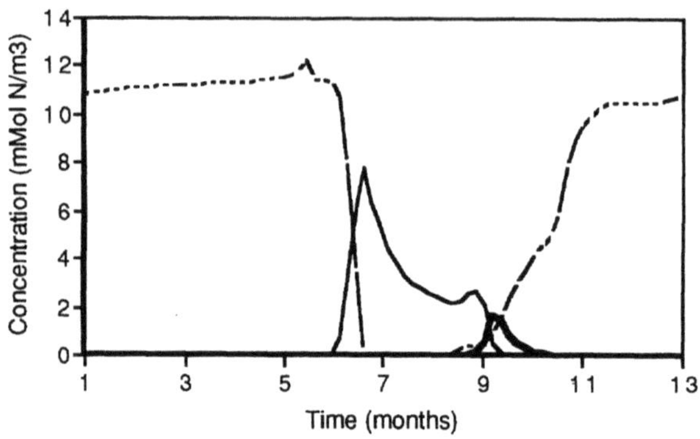

Fig. 6. The modelled annual cycle of phytoplankton (solid line), zooplankton (thick solid line), and nitrate (dashed line) at OWS "India" obtained using model A.

The simulated seasonal cycles of phytoplankton, zooplankton and nitrate for OWS "India" obtained with model A are shown in fig.6. This simulation is a very poor representation of the annual cycle observed at OWS "India". Firstly, the phytoplankton biomass during the spring bloom is far higher than observed (fig. 10) and secondly the summer nutrient levels are far too low; during the period of observations at "India" (1971-1974), the lowest summer nitrate values observed were 0.3 mMol N m^{-3} and more usually were *ca* 1 mMol N m^{-3} (Williams, unpublished manuscript). The reason for this poor fit is that, unlike Bermuda, the winter mixed layer depths are so deep and ambient light levels so low that the modelled winter photosynthesis is virtually zero (fig. 13). This reduces the phytoplankton, and thereby the zooplankton biomass, to extremely low levels, with the consequence that even after growth resumes in the spring it takes until late August before any significant zooplankton biomass is attained. The low spring zooplankton biomass means that the spring phytoplankton bloom can grow unfettered by grazing, until nutrient exhaustion eventually limits photosynthesis in late-July. Thereafter the phytoplankton population gradually declines due to natural mortality until autumn mixing reintroduces nitrate into the mixed layer. Even reducing the zooplankton mortality to zero did not greatly improve the simulation, as there was still a large zooplankton loss due to excretion.

The conclusion is inescapable, model A is not validated by the "India" data and the question now is, how do we improve the model? One possibility is that the model is underestimating the photosynthesis during the winter. The calculation of the daily phytoplankton growth rate (equation A5) assumes that the timescale for mixing throughout the mixed layer is short compared to the phytoplankton growth rate. For mixed layers as deep as 560 m this may not be true. This was tested by repeating the simulation with a minimum winter mixed layer depth of 200 m but it was found to have virtually no effect. Similarly, using a higher maximum phytoplankton growth rate did not improve the simulation. We can conclude, therfore, that the model is robust in predicting very low photosynthesis in winter at the latitude of "India" (59°N) and that the only way to ensure that more zooplankton survive the winter period is to alter the zooplankton sub-model so that losses are minimised during the winter. One way to do this would be to parameterise the total zooplankton loss rate (natural mortality plus excretion) so that zooplankton losses are reduced when biomass levels are very low. The work of Steele & Henderson (1992) discussed above suggested that a Michaelis-Menten function would be the most appropriate (equation A10). As well has having some theoretical justification, the use of a Michaelis-Menten mortality has some biological foundation. Zooplankton mortality consists of natural mortality, which may be due to disease or starvation, and mortality due to higher predators. Both disease and predation are likely to be density dependent processes. The justification for making excretion a density dependent process is less justified, although there is evidence (McClaren, 1963) that zooplankton are capable of reducing their respiratory losses in winter. Similar arguments could be given for assuming a Michaelis--Menten mortality for phytoplankton and so in order to be consistent this was also added to the model (equation A3).

The new parameterisation of zooplankton loss rate requires four parameters, namely: the maximum zooplankton loss rate (μ_2), loss rate half-saturation constant (k_6), and the fractions of the total loss going to ammonium (ε) and DON (δ); any remaining loss is assumed to be exported from the mixed layer. There are now also two parameters required to define the phytoplankton natural mortality. So, how can we estimate these parameters and can we determine a parameter set that will give reasonable simulations for both Bermuda and "India"? It was pointed out by Steele & Henderson (1992) that, despite the critical importance of these loss rate parameters, there are no good observational data that will allow them to be estimated. For the moment, therefore, they must be treated as variable tuning parameters and, as discussed above, methods need to be developed for carrying out this tuning in an objective way. However, in order to test the efficacy of the new model the parameters were derived using an iterative method. Firstly, ε and δ were fixed at 0.7 and 0.2 respectively; these were the annual apportionment obtained by the model A simulation. Secondly, the half-saturation coefficients k_5 and k_6 were arbitrarily fixed at 0.2 mMol N m^{-3}. The parameters μ_1 and μ_2 were then assigned

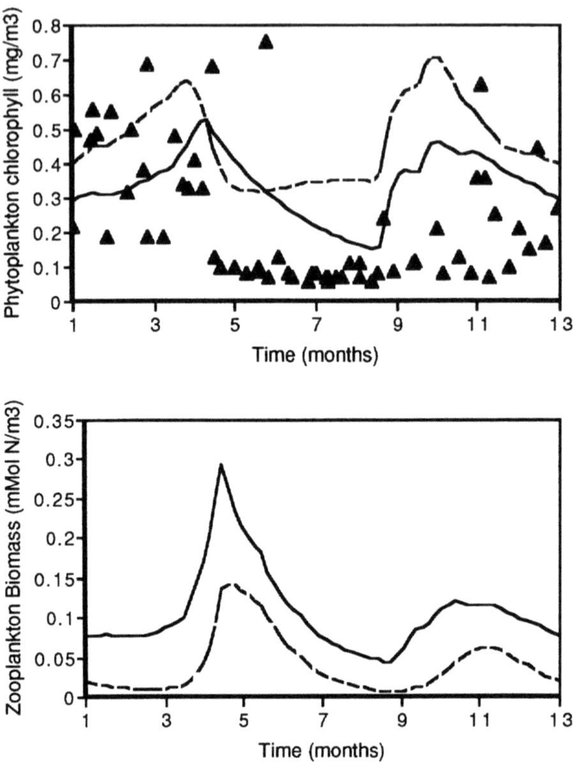

Fig. 7. Top: The modelled annual cycle of phytoplankton chlorophyll at Bermuda Station "S" obtained using the original FDM model (dashed line) and Model B (solid line) compared to observations from Station "S" (triangles). Bottom: The modelled annual cycle of zooplankton at Bermuda Station "S" using the FDM model (dashed) and Model B (solid).

reasonable initial values, and then adjusted using in the following way. Firstly, a Bermuda simulation was carried out and μ_1 was adjusted until the annual net primary production equalled 0.58 Mol N m^{-2} y^{-1}; this follows the procedure used by FDM to estimate μ_1. Using this value for μ_1, an "India" simulation was then carried out in which μ_2 was varied until the summer nitrate value was in the range 0.5-1.0 mMol m^{-3}. This value of μ_2 was then used in a Bermuda simulation and μ_1 readjusted to give the required annual primary production. The process was repeated until convergence, and the final values obtained for μ_1 and μ_2 were 0.05 d^{-1} and 0.325 d^{-1} respectively; this parameter set will be called Model B (see appendix table 1 for a complete parameter list).

The simulated seasonal cycles of phytoplankton (in chlorophyll units) and zooplankton at Bermuda obtained with model B are shown in fig.7, where they are compared with the

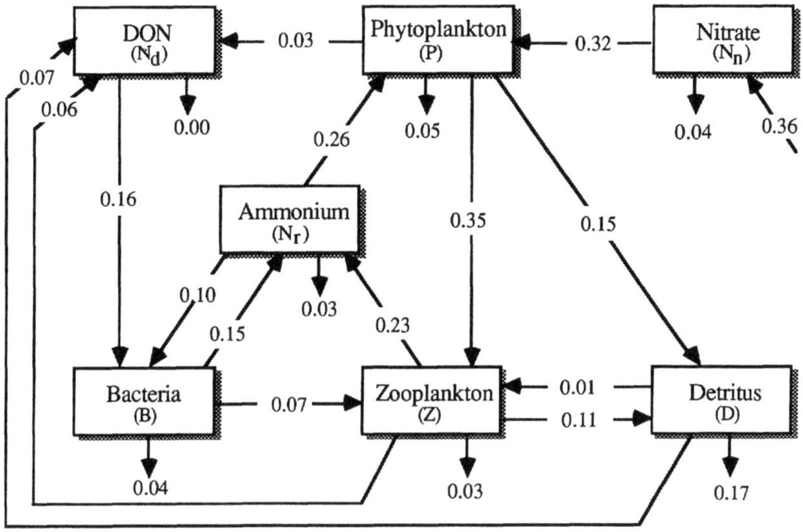

Fig. 8. The annual nitrogen flows (Mol m^{-2} y^{-1}) obtained using the Model B simulation for Bermuda Station "S".

simulated cycles obtained by the FDM model and the phytoplankton chlorophyll data for the period 1958-1960. Neither the FDM model nor Model B gives a very good fit to the chlorophyll data, but it is clear that Model B gives the better fit. The cycles of nitrate and bacteria predicted by the two models were very similar but the Model B zooplankton (and ammonium) concentrations were about twice that of the FDM model, despite the fact that the annual primary production was the same for both models. The reason for this becomes apparent when the annual inter-compartmental flows for model B (fig. 8) are compared with the FDM model (FDM, fig. 17). In Model B 60% of the primary production was grazed by zooplankton with 26% going to detritus as ungrazed dead phytoplankton; the equivalent figures for the FDM model were 22% and 60%. The new mortality formulation has resulted in a complete switch in the fate of the primary production between zooplankton grazing and natural mortality compared with the previous model. As so little data are available on zooplankton biomass and grazing, it is difficult to confirm which of these scenarios are the more realistic, but the Model B scenario accords more with the biological oceanographer's intuitive view of the marine ecosystem. The modelled bacterial production (as a fraction of primary production) varied throughout the year (fig. 11) with an annual average of 19% of the annual primary production, close to the average

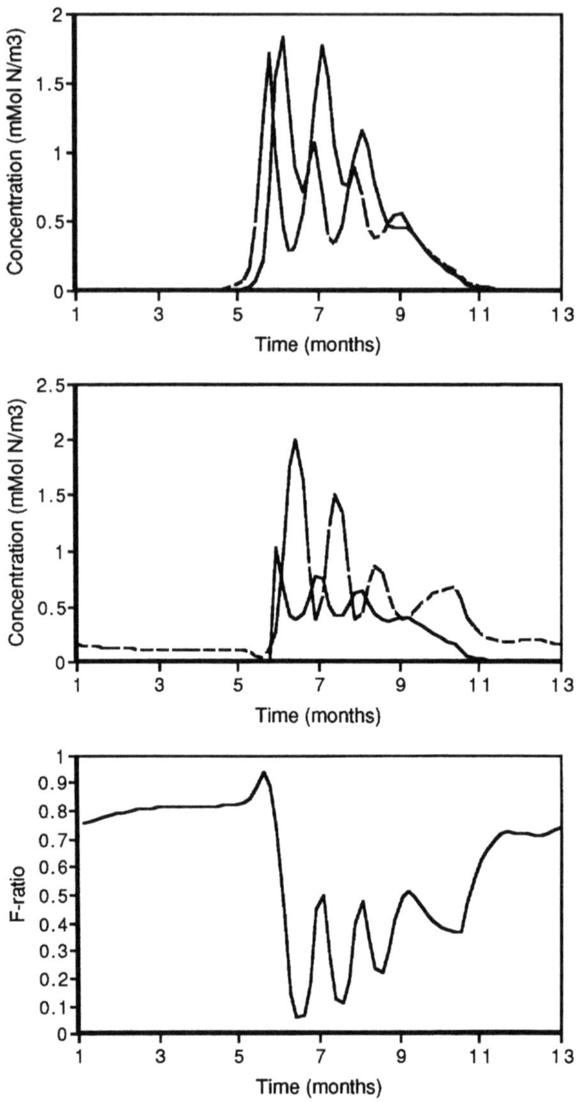

Fig. 9. Top: The modelled annual cycle of phytoplankton (dashed line) and zooplankton (solid line) at OWS "India" obtained using Model B. Midedle: The same for bacteria (solid) and ammonium (dashed). Bottom: The same for the f-ratio.

value of 20% reported by Cole et al. (1988). However, this agreement should not be overstressed bearing in mind that the model value is an annual average whereas the observations used by Cole *et al.* (1988) were from short-term incubation experiments. The modelled particle flux was 0.17 Mol m^{-2} y^{-1} which agrees well with Altabet's (1979) estimate of 0.2 Mol m^{-2} y^{-1} obtained from eighteen months of sediment trap observations at a site close to Station "S".

Another difference between the old and new models is the level of new production. Reducing the cross-thermocline mixing rate to 0.01 m d^{-1} gave a new production of 0.32 Mol m^{-2} y^{-1} compared to 0.40 Mol m^{-2} y^{-1} for the FDM Model. Bearing in mind that both models had virtually the same total production, this means that the regenerated production for Model B was over twice that of the previous model, resulting in a lower f-ratio (0.55) than for the FDM model (0.71). As the latter high f-ratio was considered to be an unsatisfactory result of the FDM Model, the new model can be considered an improvement.

Now let us see how the new model performs at OWS "India". The simulated seasonal cycles of zooplankton, phytoplankton, bacteria, and ammonium are shown in fig. 9. Fig. 10 shows the comparison between modelled phytoplankton chlorophyll and nitrate with observations from OWS "India" in 1972. These results are obviously a vast improvement on Model A. The phytoplankton chlorophyll levels are very close to the observations during the spring bloom (fig. 10 top), although the model bloom starts about 2 weeks later than the observations. This discrepancy is almost certainly due to using monthly climatic values of mixed layer depth in the model rather than actual observations. In 1972 the seasonal thermocline started to form in mid-April, whereas the model mixed layer depths only shallowed to 50m in mid-May (fig. 2).

An interesting feature of the "India" simulation, that was not observed at Bermuda, is the predator-prey cycle between zooplankton and phytoplankton during the summer. This dynamics looks like the limit-cycle behaviour associated with high nutrient levels that was discussed in section 4.3 and may be caused by the fact that the zooplankton loss rate during this period is close to the maximum (fig. 12) and so the closure term will be linear in zooplankton biomass rather than quadratic. However, it is very interesting that the chlorophyll data for 1972 and other years (Williams, 1988) show evidence for a number of peaks and troughs during the summer (fig. 10). More data will obviously be needed before this tantalising correspondence can be taken as support for the model. The limit-cycle behaviour permeates the whole simulation and can be observed in bacteria and ammonium (fig. 9 middle), and also in detritus, DON, and f-ratio (fig. 9 bottom). The cycles in f-ratio are directly related to the cycles of zooplankton. When zooplankton levels are high ammonium excretion is also high leading to higher ammonium levels and ammonium uptake. The higher ammonium levels also inhibit nitrate uptake by phytoplankton and the combined effect of both these processes is to reduce the f-ratio. The reverse process occurs when the zooplankton biomass is low.

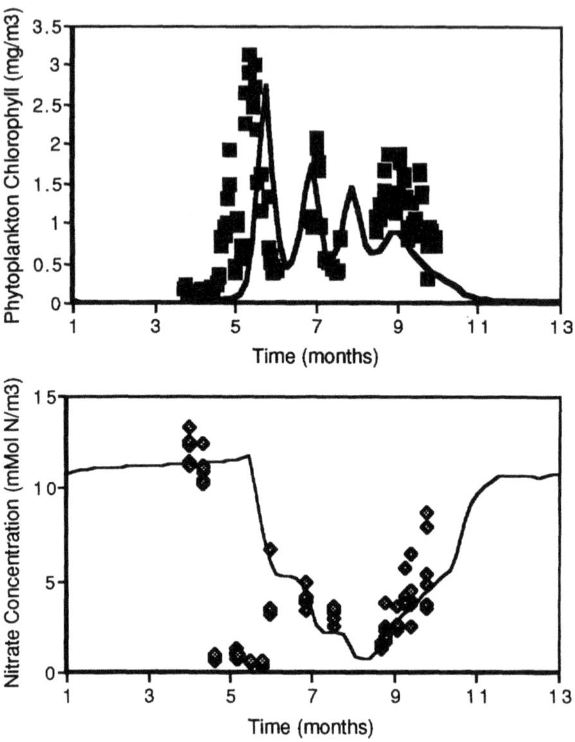

Fig. 10. Top: The modelled change in phytoplankton chlorophyll between March and November compared with the 1972 data from OWS "India". Bottom: The same for nitrate.

It will be remembered that Model B was tuned to give summer nitrate levels between 0.5-1 mMol m^{-3}, which are the observed levels at "India". However, the agreement between model and data for the whole seasonal cycle is far from perfect (fig. 10b). The modelled spring decline in the nitrate values begins late because the modelled bloom begins late, but also the late spring and early summer temporal change is different in the model and data. The data show a fairly rapid decline from pre-bloom values of 10-13 mMol m^{-3} to post-bloom values of *ca* 1 mMol m^{-3}, followed by a rise and then a further decline to values just above 1 mMol m^{-3} before a large rise associated with autumn mixing. In contrast the model shows a series of sharp declines punctuated by periods when there is no change in nitrate, resulting in the minimum nitrate value not being attained until the beginning of August (the periods in which nitrate levels remains constant coincide with the periods of high ammonium concentration which inhibits nitrate uptake). This discrepancy between model and observations suggests that there are still problems to be solved in modelling the obseved nitrogen dynamics.

A caveat must be entered when comparing any model with the OWS "India" data. All the observations were made at approximately the same position and it is known that currents in the area are reasonably high, therefore it is always possible that advective events might introduce biological signals that are atypical of the general area. Of course, such events might equally give a spurious agreement with a model as a disagreement. The only answer to this problem is to obtain longer time series so that unusual events can be averaged out.

The annual primary production predicted by Model B at "India" was 1.26 Mol N m^{-2} y^{-1}, which is in excess of the highest annual value of 0.8 Mol N m^{-2} y^{-1} obtained from ^{14}C observations at OWS "India" during 1974 (Williams, pers. comm.). On an annual basis zooplankton grazing accounted for 80% of the primary production, while bacterial production was 34% of the primary production, both higher values than for the Bermuda

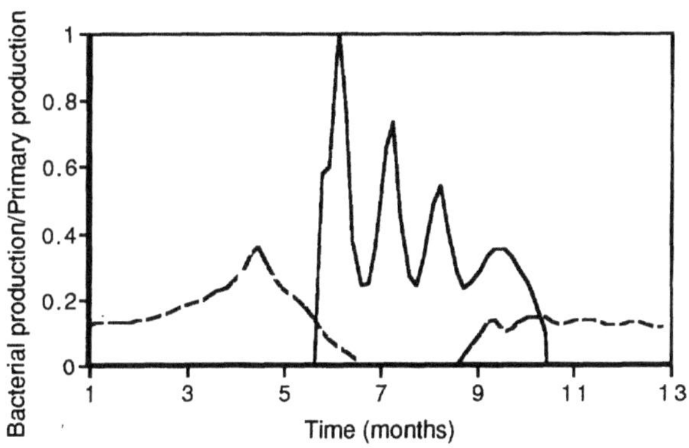

Fig. 11. The modelled annual cycle of the ratio of bacterial to total primary production at Bermuda Station "S" (dashed line) and OWS "India" (solid line). Negative ratios, representing periods when bacterial growth rate was negative, have not been plotted.

simulation. During the spring bloom the modelled bacterial production reached almost 100% of the primary production (Fig. 11).

These results have demonstrated that the introduction of a variable zooplankton loss rate has yielded a greatly improved simulation of the seasonal cycle at OWS "India" and has also improved the simulation for Bermuda compared with the earlier model of FDM. The actual seasonal cycles of zooplankton loss rate for the Bermuda and "India" simulations are shown in fig. 12. The loss rate at Bermuda shows less variation than at "India" and only rises above 0.15 d^{-1} (the constant loss rate in the FDM model) during April and May when zooplankton levels are highest. In contrast, at "India" the loss rate varies by over an order of magnitude throughout the year and is close to the maximum value of 0.3 d^{-1} throughout most of the summer.

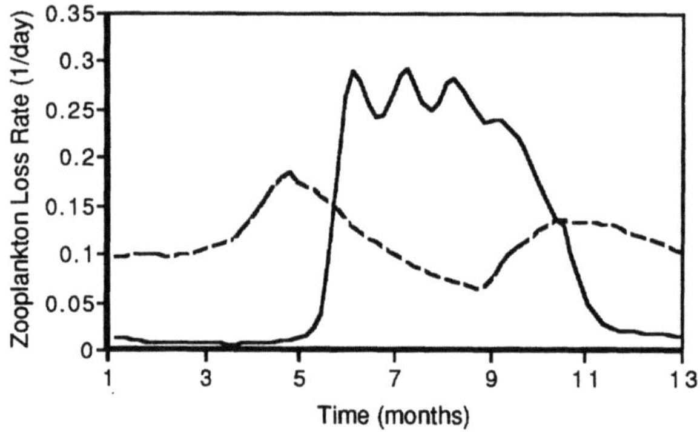

Fig. 12. The modelled annual cycle of zooplankton specific loss rate at Bermuda Station "S" (dashed line) and OWS "India" (solid line).

6. An application: predicting new production and particle flux

The attempt to apply a common ecosystem model to two areas of the North Atlantic with very different seasonal cycles of mixed layer depth, solar irradiance, and annual production cycle has showed that it was necessary to use zooplankton and phytoplankton loss functions that varied with biomass in order to obtain a model that would give reasonable representations of the annual biological cycles at both sites. Further validation of this model will require the use of data at other sites and work is in progress on this task. In the meantime, I will use Model B for illustrative purposes to study a problem of current interest to the analysis of the oceanic carbon cycle, namely how to estimate the export flux from the mixed layer.

I have already discussed how primary production can be divided into new production, fuelled by nitrate, and regenerated production fuelled mainly by ammonium. If we consider a column of water within the euphotic zone and assume that, on an annual basis, the ecosystem is in steady-state then the annually averaged inputs to the column must equal the outputs. It has been argued (Eppley & Peterson, 1979) that the main input to the system is the net vertical and horizontal transport of nitrate, which at steady-state will equal the new primary production, and the main output is the vertical particle flux. If this is so, then obviously the annual new production will be equal to the annual particle flux. As it is the particle flux from the surface layers that drives the biological pump, measuring the strength of the biological pump could, in principle, be achieved by measuring the amount of new production. New production can be measured directly using ^{15}N isotopic techniques. However, such measurements are difficult and time-consuming and so attention has focussed on more indirect methods. For example, methods for estimating the total primary production from remotely-sensed ocean colour observations have been proposed (Platt & Sathyendranath, 1988; Morel & Berthon, 1989) and so if an algorithm relating new to total production could be established that had some universality, then the way would be open for long-term global estimates of new production and thereby particle flux.

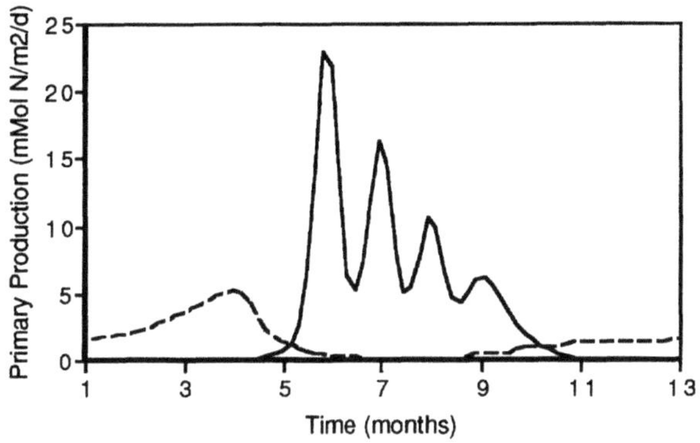

Fig. 13. The modelled annual cycle of total primary production for Bermuda Station "S" (dashed line) and OWS "India" (solid line).

Before investigating methods for estimating new production and particle flux, it should be stressed that the whole concept of 'new production equals particle flux' is a simplification of the true situation. Firstly, there is an export by higher predators, either by long-term migration of these predators or by human fishing activities. The latter loss is usually considered to be trivial in most oceanic areas, except in upwelling areas with large fisheries. Secondly there is an export flux due to zooplankton migration that arises because many zooplankton species migrate into the surface zone at night to feed on phytoplankton and then migrate back to deeper depths during the day where they excrete, egest and respire some of the food eaten at the surface. The carbon and nitrogen export associated with this process has been highlighted in a recent series of papers by Longhurst and his co-workers (Longhurst & Harrison, 1988; Longhurst *et al.*, 1990,1991), who suggested that this contribution may in some case be a significant fraction of the particle flux. Another point to bear in mind is that some exports from the system may not be in a form that will be sampled by a sediment trap situated just below the euphotic zone. For example, part of the zooplankton mortality may be in the form of large carcasses that sink rapidly and are poorly sampled by a trap. Also, material from the mixed layer may be lost to the layers below by detrainment or vertical mixing. It is worth noting that the Bermuda Model B simulation showed that detrital sedimentation only accounted for 53% of the new production, or 29% of the total production (fig. 8); the remaining 47% of new production consisted of detrainment and mixing of ammonium (9%), zooplankton mortality (10%), and detrainment and mixing of phytoplankton and bacteria (28%). Some fraction of the latter two fluxes might well eventually be sampled by a trap, either directly or after having been ingested by midwater zooplankton and then egested. However, taking all these points into consideration, it is more than likely that there will not be a one-to-one correspondence between the real export flux and the flux measured by a sediment trap. Alatabet (1989) provides evidence to support this contention. Using bi-monthly sediment trap samples from a site near Bermuda he estimated the particle flux as 0.20 Mol N m^{-2} y^{-1} (cf. Model B: 0.17 Mol N m^{-2} y^{-1}) and downward mixing of suspended PON as 0.13 Mol N m^{-2} y^{-1} (Model B: 0.12 Mol N m^{-2} y^{-1}). It is also interesting that Murray *et al* (1989) reported that in the equatorial Pacific the f-ratio measured by ^{15}N uptake was consistently 50% higher than the f-ratio estimated from the sediment trap data.

Another possible flux, either into or out of the mixed layer water column, is the advection of semi-labile DOM (see section 4.4). It has been suggested that the advection of DOM from high productivity equatorial areas to low productivity oligotrophic gyres, where it could be remineralised by bacteria, may provide a significant source of nutrients in the latter areas (Bacastow & Maier-Reimer, 1991).

A number of the algorithms that have been suggested for determining new from total production will now be discussed using the Model B simulations for Bermuda and "India", which between them encompass a wide range of primary production values and very different

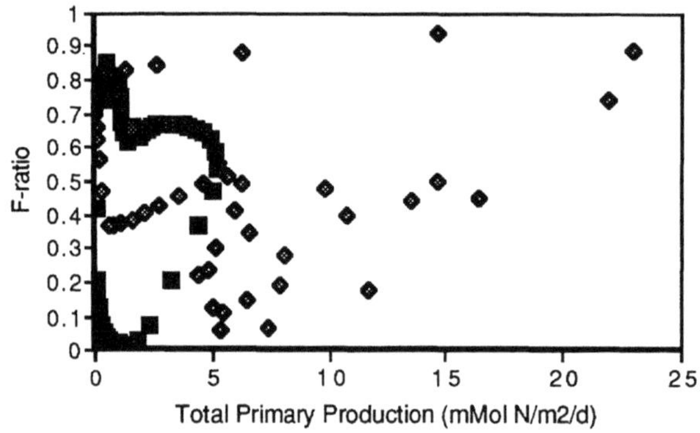

Fig. 14. Plot of modelled 5-day average values of f-ratio against total primary production for Bermuda Station "S" (squares) and OWS "India" (diamonds).

seasonal cycles (fig.13). Eppley & Peterson (1979) suggested that the f-ratio may be a hyperbolic function of total production and this algorithm has been used by a number of workers to calculate new production (Chavez & Berber, 1987) or particle flux (Berger *et al.*, 1989) from estimates of total production. It is, therefore, of interest to see whether this hyperbolic relationship is supported by the model results. 5-day averages of modelled new and total production were used to calculate the f-ratio which was then plotted against total production (fig.14). It is difficult to discern a simple hyperbolic relationship in this plot; the only generalisation that might be made is that the highest primary production values have high f-ratios. If the some of the original plots of f-ratio versus production are studied (Eppley *et al.*, 1979; Harrison, pers. comm.) then a similar conclusion might be reached. Thus neither the model nor, in fact, the observations lend much support to there being a simple relationship between f-ratio and total primary production.

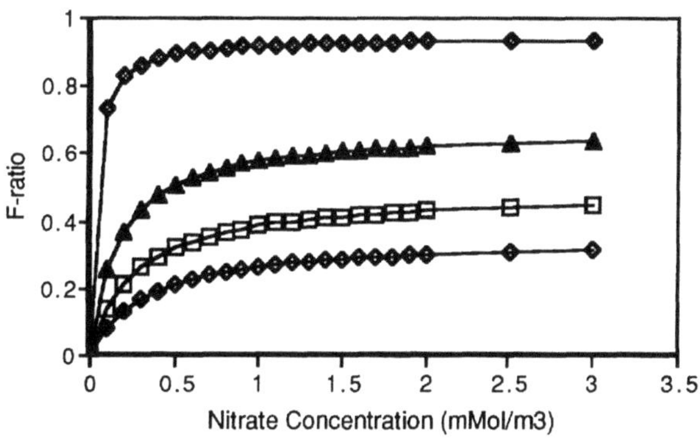

Fig. 15. The theoretical relationship between f-ratio and nitrate concentration for different values of the ammonium concentration derived from eqaution A9..

Another approach has been to try and obtain a relationship between f-ratio and nitrate concentration. For example, Platt and his co-workers (Platt & Harrison, 1985; Harrison *et al.*, 1987) have used an exponential function to fit observations of f-ratio and nitrate from a number of locations in the North Atlantic. In fact, it is possible to derive an analytical relationship between f-ratio and the nitrate and ammonium concentrations from the expressions for nitrate and ammonium uptake (equation A9). For a fixed ammonium concentration there is a hyperbolic relationship between f-ratio and nitrate (fig. 15), but the shape of the curve is very dependent on the ammonium concentration, as has also been pointed out by Harrison *et al.* (1987). The actual relationship between f-ratio and nitrate over a seasonal cycle will therefore depend very much on the associated ammonium concentrations. This is illustrated by comparing the f-ratio vs. nitrate curves for the Bermuda and "India" simulations. The modelled ammonium at Bermuda varied little throughout the year (0.03-0.17 mMol m^{-3}) and so the f-ratio vs. nitrate

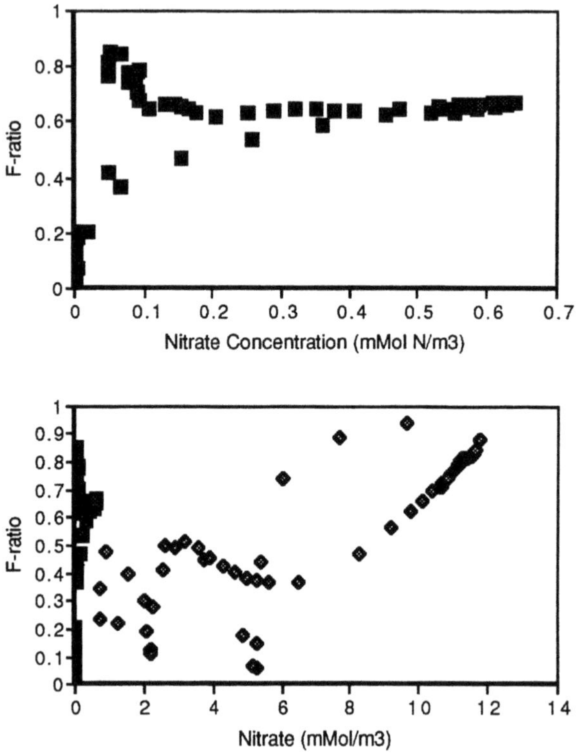

Fig. 16. Plot of modelled 5-day averages of f-ratio against nitrate concentration for Bermuda Station "S" (top) and (bottom) Bermuda Station "S" (squares) plus OWS "India" (diamonds).

curve was not too far removed from a hyperbolic form (fig. 16 top). However, at "India" there were large temporal changes in the modelled ammonium concentration (fig. 9 middle) and so the combined f-ratio vs. nitrate curve is extremely complex and would defy any attempt at curve-fitting (fig. 16 bottom).

These model results have shown that the methods proposed to date for trying to estimate f-ratio would not be very successful if applied to the model data. That does not mean that they will not be successful when applied to real data; the model is not reality! However, these results should make us cautious about these methods and also investigate whether other better methods might be developed. One possibility is to see if new production can be estimated directly from total production without going through the intermediate step of estimating f-ratio. To investigate

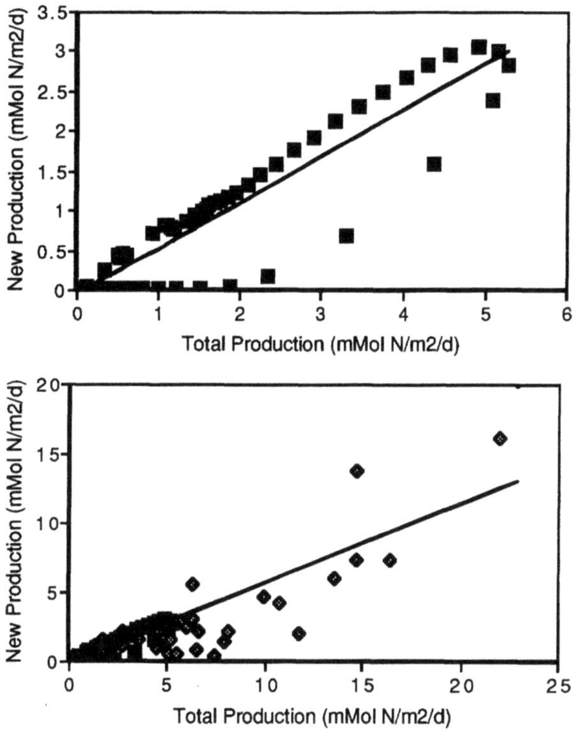

Fig. 17. Plot of modelled 5-day averages of new production against total production for (top) Bermuda Station "S" and (bottom) Bermuda Station "S" (squares) plus OWS "India" (diamonds).

this I have plotted the simulated 5-day new production values against total production for Bermuda together with the fitted regression line (fig. 17 top). The regression is a good fit to the data for most of the values and overall explained 85% of the variance in new production. Furthermore, if one is only interested in the yearly averaged new production then, given a full seasonal data coverage, the regression method will estimate this perfectly because the linear nature of the regression analysis ensures that the regression line passes through the x,y mean of the observations. Fig. 17 (bottom) shows the same plot for the combined Bermuda and "India" results. The slope of the combined regression line was 0.57 ± 0.02 compared to 0.58 ± 0.02 for the Bermuda data alone, and so this method does have the advantage of providing an equation that works both for Bermuda and "India". Again, however we must be cautious and remember

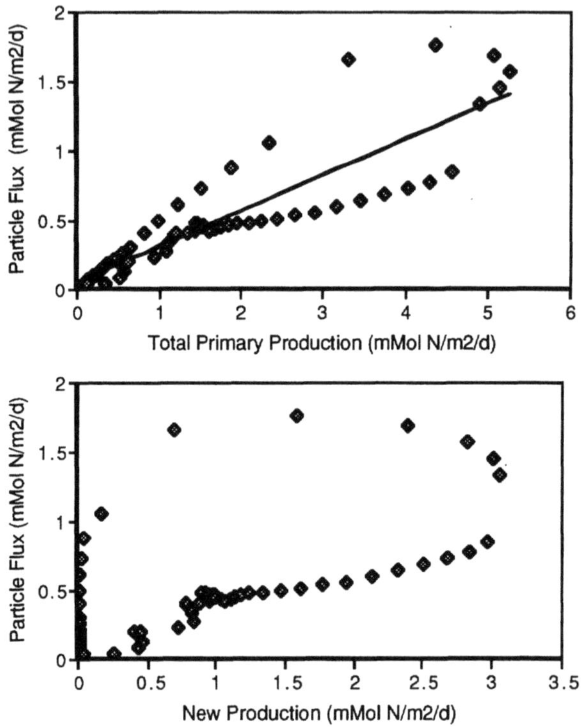

Fig. 18 (a) Plot of modelled 5-day averages of particle flux against total production for Bermuda Station "S" with fitted regression line. (b) The same for particle flux against new production.

that we are dealing with a model with fixed parameters, albeit applied to two areas with very different seasonal forcing. Nevertheless it does suggest that it might be worth analysing the existing data sets of new and total production to see if this method has any utility.

It should be remembered that the main reason for trying to estimate new production was to use it as a proxy for estimating particle flux. I have investigated this using the model by calculating the regressions between the 5-day averages of modelled particle flux and both new and total production. The results are shown in table 1 and the data and regressions for Bermuda are shown in fig. 18. The first point to make is that the regressions on new production were worse than the regressions on total production. This was at first surprising but the reason is that, although on an annual basis new production must correlate with particle flux, there are periods of the year, for example the summer when f-ratios are low, when the correlation is extremely poor (fig. 18 bottom). The second point is that, as was the case for the regressions of new on total production, the slope of the regresssion for the two areas are not significantly different. In this regard it is interesting to note that Pace *et al.* (1987) have calculated the regression between particulate organic nitrogen fluxes measured by traps at the base of the euphotic zone and concurrent primary production measured by ^{14}C incubations. If the Redfield C:N ratio is used to convert their primary production values into nitrogen units then the slope of the regression they obtained was 0.29, very close to the modelled results.

In conclusion, the model results would suggest that the best method for estimating particle flux is by estimating the relationship between this flux and total primary production, rather than attempting the intermediate step of estimating new production. Bearing in mind the problems involved in estimating new production, this may prove a useful result.

Table 1: Regressions of particle flux on new and total production for simulations at Bermuda and OWS "India"

Simulation Area	New Production Regression Slope	Constant	R^2	Total Production Regression Slope	Constant	R^2
Bermuda	0.30 ± 0.02	0.20	44%	0.26 ± 0.02	0.06	79%
OWS "India"	0.25 ± 0.03	0.52	43%	0.23 ± 0.01	0.14	77%

7. The future

I will close this survey of marine biological modelling with a list of the modelling problems I believe need to be solved in order to achieve the long-term goal of predicting the spatial and temporal variations in the oceanic carbon cycle. The problems divide into two classes; those in which there is still considerable uncertainty about how a particular part of the carbon system works and those which, although we think we understand the processes involved, we are uncertain how best to parameterise them. In the first category is the role of DOC in the carbon cycle (see section 4.4), the relative roles of detritus, zooplankton, bacteria, and phytoplankton in creating and decomposing DOC, and, a topic not discussed at all in this paper but of some importance, the biological processes involved in remineralising carbon and nitrogen in the ocean layers below the euphotic zone (Sarmiento *et al.*,1992). Progress here depends on the resources and ingenuity applied to these problems by the biological oceanographic community and it is hoped that international programmes such as JGOFS will provide a timely stimulus. In the second category would be the parameterisation of detrital sinking, the effect of zooplankton migration, and the feeding of zooplankton on multiple resources. Progress will no doubt be made in the future, especially if the parameterisations are based as much as possible on mechanistic models with firm ecological and physiological bases.

Another pressing problem is how best to construct combined carbon and nitrogen models. Most of the existing models that attempt to predict DIC in the euphotic zone do so by assuming fixed 'Redfield' ratios between nitrogen and carbon flows (Peng *et al.*, 1987; Taylor *et al.*, 1991; Bacastow & Maier-Reimer, 1990) and this approach may turn out to be sufficiently accurate for the purposes of modelling the global carbon cycle. However, it is well-known that the C:N ratios of zooplankton, bacteria, sinking detritus, and dissolved organic matter differ from the Redfield ratio. Therefore, the consequences of assuming Redfield ratios needs to be tested using more biologically realistic models. Some progress with coupled C/N models of zooplankton and bacterial metabolism has already been achieved (Anderson, submitted).

Even if all these problems are solved there still remains the problem of whether it is possible to construct a reasonably simple model that can be calibrated by observations and will be geographically robust, in the sense that the same model with fixed parameters will give reasonably accurate predictions of biological production in all the major parts of the world ocean. At the moment we do not know the answer to this question. However we do have some candidate models and we should, as a matter of some urgency, begin to use the biological data sets now being assembled to test their efficacy as global models.

Bearing in mind the large number of problems that still remain we should be very cautious about what claims we make for biological modelling. However, there can be no doubt that the field provides a tremendous intellectual challenge for the coming decade.

Appendix : Model Equations

More detailed derivations of these equations are given in Fasham *et al.* (1990). The state-variable units are mMol Nitrogen m^{-3}. However flows, such as primary production or particulate flux from the mixed layer, are calculated by the model in areal units of mMol N m^{-2} d^{-1}; annual integrated flows are in Mol N m^{-2} y^{-1}. The state-variable symbols are: Phytoplankton (P), Zooplankton (Z), Bacteria (B), Detritus (D), Nitrate (N_n), Ammonium (N_r), and DON (N_d).

1. Mixed layer depth

The mixed layer dynamics are not modelled explicitly, but instead it is assumed that data are available to specify the annual cycle of mixed layer depth, M, as a function of time, t (days). Interpolated daily values of M(t) are used to calculate h(t) as

$$\frac{dM}{dt} = h(t). \tag{A1}$$

Evans and Parslow (1985) pointed out that the effect of the deepening or shallowing of the mixed layer on the concentration of a state-variable is asymmetric for nonmotile biotic, or abiotic, entities such as phytoplankton, bacteria, or detritus. A deepening of the mixed layer will dilute the volumetric concentration, whereas when the mixed layer shallows, material is left behind, or detrained, but the volumetric concentration in the mixed layer will remain unchanged. This can be modelled by defining the variable

$$h^+(t) = \max(h(t),0). \tag{A2}$$

In contrast a motile organism such as a zooplankton might be assumed actively to maintain itself within the mixed layer when the mixed layer depth changes, and so the volumetric concentration of zooplankton will decrease and increase when the mixed layer depth increases and decreases, respectively.

2 Phytoplankton

The equation for phytoplankton is

$$\frac{dP}{dt} = (1-\gamma)\,\sigma(t,M,N_n,N_r)\,P - G_1 - \frac{\mu_1 P^2}{k_5+P} - \frac{(m+h^+(t))P}{M}\,, \tag{A3}$$

where $\sigma(t,M,N_n,N_r)$ is the average daily phytoplankton specific growth rate, G_1 represents the loss of phytoplankton due to zooplankton grazing, μ_1 is the specific natural mortality rate of phytoplankton, k_5 is the half-saturation constant for phytoplankton mortality, γ is the fraction of total net primary production that is exuded by phytoplankton as DON, m is a quantity (units m d^{-1}) that parameterizes the diffusive mixing between the mixed layer and the deep ocean, and $h^+(t)$ has been defined above.

The daily averaged phytoplankton growth rate σ is defined by the equation

$$\sigma = J(t,M)\,Q(N_n,N_r), \tag{A4}$$

where J is the light limited growth rate and Q is a non-dimensional nutrient limiting factor. J is calculated by assuming that the time a particular cell spends at a given depth is long compared to the photosynthesis reaction time, but short compared to the cell division time (Evans and Parslow, 1985). In this case the total daily growth rate averaged over the mixed layer depth can be written as,

$$J(t,M) = \frac{2}{M}\int_0^{\tau}\int_0^{M} F[I_0(t).\exp((-k_W-k_cP)z)]dzdt\ , \qquad (A5)$$

where F[I] is a function describing the phytoplankton photosynthesis- irradiance relationship (the P-I curve), $I_0(t)$ is the PAR immediately below the surface of the water, k_W is the light attenuation coefficient due to water (assumed to be constant with depth), and k_c is the phytoplankton self-shading parameter. Time t is measured in days and is 0 at sunrise and τ at noon. Evans and Parslow (1985) showed that if the following expression was used for F[I],

$$F[I] = \frac{V_p \alpha I}{(V_p^2+\alpha^2 I^2)^{0.5}}\ , \qquad (A6)$$

where V_p is the maximum growth rate and α is the initial slope of the P-I curve, and if the daily variation of $I_0(t)$ with time of day was assumed to be triangular, then equation 5 can be integrated analytically. $I_0(t)$ is defined by

$$I_0(t) = \lambda(1\text{-}a)\ t\ (C)\ S(\theta,t)\ , \qquad (A7)$$

where λ is the ratio of PAR to total solar irradiance, a is the air-sea albedo, t (C) the atmospheric transmittance as a function of cloudiness C, and $S(\theta,t)$ is the solar irradiance at the top of the atmosphere as a function of latitude θ and time t.

The value of the maximum phytoplankton growth rate V_p could be specified using observational data from the area to be modelled. However, Eppley (1972) has collated a large number of growth rate determinations and shown that the maximum growth rates were related to temperature by the equation,

$$V_p = 0.6\ (1.066)^T. \qquad (A8)$$

It is therefore possible to introduce a greater degree of generality into the model by using this equation, especially when modelling basin-wide biological processes (Sarmiento *et al.*, 1992).

The nutrient limitation factor Q was defined as (Wroblewski, 1977),

$$Q(N_n,N_r) = Q_1(N_n,N_r) + Q_2(N_r) = \frac{N_n \exp(-\psi N_r)}{k_1+N_n} + \frac{N_r}{k_2+N_r}, \qquad (A9)$$

where k_1,k_2 are the half-saturation constants for nitrate and ammonium uptake respectively, and is a constant that parameterizes the degree of ammonium inhibition of nitrate uptake.

3. Zooplankton

The zooplankton are assumed to graze on phytoplankton, bacteria and detritus and the full equation is given by,

$$\frac{dZ}{dt} = \beta_1 G_1 + \beta_2 G_2 + \beta_3 G_3 - \frac{\mu_2 Z^2}{k_5+Z} - \frac{h(t)\,Z}{M}, \qquad (A10)$$

where G_1,G_2,G_3 are the grazing rates of zooplankton on phytoplankton, bacteria and detritus respectively, β_1,β_2,β_3 are the equivalent assimilation efficiencies, μ_2 is the zooplankton specific loss rate, and k_6 is the half-saturation constant for the loss rate. The loss rate is assumed to include both excretory losses and losses due to mortality. Fractions ε and δ of the total loss are in the form of ammonium and DON respectively. This leaves a fraction $1-\varepsilon-\delta$ for the mortality loss which, being mainly due to higher predators that excrete large faecal pellets, is considered to be exported directly from the mixed layer.

The expression used for the grazing functions G_1 was

$$G_1 = \frac{g p_1 P^2 Z}{k_3(p_1P+p_2B+p_3D)+p_1P^2+p_2B^2+p_3D^2}, \qquad (A11)$$

where g is the maximum ingestion rate, p_1,p_2,p_3 are repectively measures of the zooplankton preferences for phytoplankton, bacteria, and detritus when their concentrations are equal, and k_3 is the half-saturation constant for grazing. Analogous expressions are used for G_2 and G_3. This grazing formulation was designed to ensure that the zooplankton actively switch their grazing to the most abundant resource (FDM).

4. Bacteria

The equation for bacteria is,

$$\frac{dB}{dt} = U_1 + U_2 - \mu_3 B - \frac{(m+h^+(t))B}{M}, \qquad (A12)$$

where U_1 is the DON uptake, U_2 the ammonium uptake, and μ_3 is the bacterial specific excretion rate. Note that bacteria are assumed to take up ammonium and DON and also excrete ammonium. However, DON and ammonium are not equally interchangeable forms of nitrogen for bacteria. Bacteria obtain their carbon from DON and are thought take up ammonium mainly to obtain sufficient nitrogen to synthesize cell protein. In a balanced growth situation, the ratio of bacterial ammonium uptake to DON uptake should be constant in order to ensure that bacterial biomass of the required C/N ratio is produced from DON with a given C/N ratio . This concept is incorporated into a Michaelis-Menten model of bacterial uptake by first defining a function S as

$$S = \min\,(N_r, \eta N_d)\,, \tag{A13}$$

where η is a function of the bacterial carbon and nitrogen growth efficiencies, and the C/N ratio of bacterial cells and DON (FDM). Then U_1 and U_2 are given by,

$$U_1 = \frac{V_b B N_d}{k_4+S+N_d}\,, \tag{A14}$$

$$U_2 = \frac{V_b B S}{k_4+S+N_d}\,, \tag{A15}$$

where V_b is the maximum bacterial uptake rate and k_4 is the half-saturation coefficient for uptake.

5. Detritus

The detritus equation is given by,

$$\frac{dD}{dt} = (1-\beta_1)G_1 + (1-\beta_2)G_2 - \beta_3 G_3 - \mu_4 D + \frac{\mu_1 P^2}{k_5+P} - \frac{(m+h^+(t)+V)D}{m}\,, \tag{A16}$$

where μ_4 is the specific rate of breakdown of detritus to DON, V is the detrital sinking rate, and the other parameters have already been defined. The first two terms of this equation represent the fecal pellets derived from zooplankton grazing of phytoplankton and bacteria respectively. The third term represents the net loss of detritus by zooplankton grazing, i.e. $((1-\beta_3)\,G_3)-G_3$. The fourth term represents the breakdown of detritus and the fifth term represents the addition of dead phytoplankton to the detrital pool.

6. Nitrate

The nitrate equation is,

$$\frac{dN_n}{dt} = -J(t,M)Q_1(N_n,N_r)P + \frac{m+h^+(t)}{M}(N_0-N_n), \tag{A17}$$

where N_0 is the nitrate concentration below the mixed layer, assumed to be constant. The first term on the right-hand side is the new primary production.

7. Ammonium

The ammonium equation is,

$$\frac{dN_r}{dt} = -J(t,M)Q_2(N_r)P - U_2 + \mu_3 B + \frac{\varepsilon\mu_2 Z^2}{k_6+Z} - \frac{(m+h^+(t))N_r}{M}. \tag{A18}$$

The first two terms represent the uptake of ammonium by phytoplankton (regenerated primary production) and bacteria respectively, the second represents the excretion by bacteria and the third represents the addition from zooplankton excretion. It should be noted that we have assumed that the ammonium concentration below the mixed layer is zero.

8. Dissolved organic nitrogen

The DON equation is

$$\frac{dN_n}{dt} = \gamma J(t,M)\, Q(N_n,N_r)\, P + \mu_4 D + \frac{\delta\mu_2 Z^2}{k_6+Z} - U_1 - \frac{(m+h^+(t))N_d}{M}. \tag{A19}$$

The first three terms represent the production of DON by phytoplankton exudation, detrital breakdown, and zooplankton excretion and messy feeding respectively, while U_1 is the uptake of DON by bacteria. As was the case for ammonium, we have assumed that the labile DON concentration below the mixed layer is zero.

Table A1: Model B parameter set

Parameter	Symbol	Value
Air-water albedo	a	0.05
Ratio of PAR to Total Irradiance	λ	0.43
Cloudiness	C	4 Oktas
Attenuation coefficient of downwelling irradiance	k_w	0.04 m^{-1}

Cross-thermocline mixing rate	m	$0.01\ m\ d^{-1}$
Phytoplankton maximum growth rate (Bermuda)	V_p	$2.9\ d^{-1}$
Phytoplankton maximum growth rate (OWS "India")	V_p	$1.25\ d^{-1}$
Initial slope of P-I curve	α	$0.025\ (W\ m^{-2})^{-1}\ d^{-1}$
Half-saturation constant for nutrient uptake	k_1, k_2	$0.5\ mMol\ m^{-3}$
Phytoplankton maximum specific mortality rate	μ_1	$0.05\ d^{-1}$
Phytoplankton mortality half-saturation constant	k_5	$0.2\ mMol\ m^{-3}$
Phytoplankton self-shading coefficient	k_c	$0.03\ m^2\ (mMol\ N)^{-1}$
Phytoplankton DON exudation fraction	γ	0.05
Nitrate uptake ammonium inhibition parameter	ψ	$1.5\ (mMol\ N\ m^{-3})^{-1}$
Zooplankton maximum ingestion rate	g	$1.0\ d^{-1}$
Zooplankton ingestion half-saturation constant	k_3	$1.0\ mMol\ m^{-3}$
Zooplankton assimilation efficiency	$\beta_1, \beta_2, \beta_3$	0.75
Zooplankton maximum loss rate	μ_2	$0.3\ d^{-1}$
Zooplankton loss rate half-saturation constant	k_6	$0.2\ mMol\ m^{-3}$
Fraction of zooplankton losses going to DON	δ	0.2
Fraction of zooplankton losses going to ammonium	ε	0.7
Zooplankton feeding preferences	p_1, p_2, p_3	0.5,0.25,0.25
Bacterial maximum uptake rate	V_b	$2.0\ d^{-1}$
Bacterial excretion rate	μ_3	$0.05\ d^{-1}$
Bacterial half-saturation constant for uptake	k_4	$0.5\ mMol\ m^{-3}$
Ratio of ammonium:DON uptake for bacteria	η	0.6
Detrital sinking rate	V	$1\ m\ d^{-1}$
Detrital breakdown rate	μ_4	$0.05\ d^{-1}$

References

Alldredge AL, Silver MW (1988) Characteristics dynamics and significance of marine snow. Prog Oceanogr 20:41-82

Altabet MA (1989) Particulate new nitrogen fluxes in the Sargasso Sea. J Geophys Res 94:12771-12779

Andersen V, Nival P (1988) A pelagic ecosystem model simulating production and sedimentation of biogenic particles: role of salps and copepods. Mar Ecol Prog Ser 44:37-50

Andersen V, Nival P (1991) A model of the diel vertical migration of zooplankton based on euphuasiids. J Mar Res 49:153-175

Anderson T (to be published) Modelling the influence of food C:N ratio and respiration on growth and excretion of marine plankton and bacteria. J Plank Res

Angel MV (1984) Detrital organic fluxes through pelagic ecosystems. In: Fasham MJR (ed) Flows of energy and materials in marine ecosystems: theory and practice. Plenum Press New York & London. p 475

Angel MV (1986) Vertical migrations in the oceanic realm: possible causes and probable effects. In: Rankin MA (ed) Migration: mechanisms and adaptive significance. Contr Mar Sci (Suppl) 24:45-70

Angel MV (1988) Vertical profiles of pelagic standing crop in the vicinity of the Azores Front and their implications to deep ocean ecology. Prog Ocean 20:1-46
Atkinson CA (1987) A nonlinear programming approach to the analysis of perturbed marine ecosystems under model parameter uncertainty. Ecol Modell 35:1-28
Azam F, Fenchel T, Field JG, Gray JS, Meyer-Reil LA, Thingstad F (1983) The ecological role of water-column microbes in the sea. Mar Ecol Prog Ser 10:257-263
Azam F, Hodson RE (1977) Size distribution and activity of marine heterotrophs. Limnol Oceanogr 22:492-501
Bacastow R, Maier-Reimer E (1990) Ocean-circulation model of the carbon cycle. Climate Dynamics 4:95-125
Bacastow R, Maier-Reimer E (1991) Dissolved organic carbon in modeling oceanic new production. Global Biogeochemical Cycles 5:71-85
Bé AWH,Forns JM,Roels OA (1971) Plankton abundance in the North Atlantic Ocean. In: Costlow JD (ed.) Fertility of the sea, vol 1. Gordon and Breach Science Publishers. p.17
Berger WH, Smetacek VS, Wefer G (1989) Ocean productivity and paleoproductivity - an overview. In: Berger WH, Smetacek VS, Wefer G (eds.) Productivity of the ocean: present and past . J Wiley & Sons Chichester. p 1
Bjornsen PK (1986) Bacterioplankton growth yield in continuous seawater cultures. Mar Ecol Prog Ser 30:191-96
Brock TD (1981) Calculating solar radiation for ecological studies. Ecol Modell 14:1-19
Burkill PH, Edwards ES, John AWG, Sleigh MA (to be published) Microzooplankton and their herbivorous activity in the north eastern Atlantic Ocean. Deep-Sea Res
Capriulo GM, Sherr EB, Sherr BF (1991) Trophic behaviour and related community feeding activities of heterotrophic marine protists. In: Reid PC, Turley CM, Burkill PH (eds) Protozoa and their role in marine processes. Springer-Verlag Berlin. p 217
Carr MR (1986) Modelling the attenuation of broad band light down a water column.The Statistician 35:325-333
Chavez FP, Barber RT (1987) An estimate of new production in the equatorial Pacific. Deep-Sea Res 34:1229-1243
Clarke KR, Joint IR (1986) Methodology for estimating numbers of free-living and attached bacteria in estuarine water. Appl Env Microbiol 51:1110-1120
Cole JJ, Pace ML, Findlay S (1988) Bacterial production in fresh and saltwater ecosystems: a cross-system overview. Mar Ecol Prog Ser 43:1-10
Cullen JJ (1990) On models of growth and photosynthesis in phytoplankton. Deep-Sea Res 37:667-683
Dortch Q (1990) The interaction between ammonium and nitrate uptake in phytoplankton. Mar Ecol Prog Ser 61:183-201
Dring MJ, Jewson DH (1982) What does ^{14}C uptake by phytoplankton really measure? A theoretical modelling approach. Proc Roy Soc 214B:351-368
Ducklow HW, Fasham MJR (1991) Bacteria in the greenhouse: modeling the role of oceanic plankton in the global carbon cycle. In: Mitchell R (ed) New concepts in enironmental microbiology. Wiley-Liss Inc. New York. p 1
Ducklow HW, Purdie DA,Williams PJleB, Davies JM (1986) Bacterioplankton: a sink for carbon in a coastal marine plankton community. Science 232:865-867
Dugdale RC, Goering JJ (1967) Uptake of new and regenerated forms of nitrogen in primary production. Limnol Oceanogr 12:196-206
Eppley RW (1972) Temperature and phytoplankton growth in the sea. Fish Bull 70:1063-1085
Eppley RW, Peterson BJ (1979) Particulate organic matter flux and planktonic new production in the deep ocean. Nature 282:677-680
Eppley RW, Renger EH, Harrison WG (1979) Nitrate and phytoplankton production in Southern California coastal waters. Limnol Oceanogr 24:483-494
Evans GT (1988) A framework for discussing seasonal succession and coexistence of phytoplankton species. Limnol Oceanogr 33:1037-1036
Evans GT , Parslow JS (1985) A model of annual plankton cycles. Biol Oceanogr 3:327-347
Falkowski PG (ed) (1980) Primary productivity in the sea. Plenum Press New York

Falkowski PG, Wirick CD (1981) A simulation model of the effects of vertical mixing on primary productivity. Mar Biol 65:69-75
Fasham MJR (1977) The application of some stochastic processes to the study of plankton patchiness. In: Steele JH (ed) Spatial pattern in plankton communities. Plenum Press New York and London. p 131
Fasham MJR (1985) Flow analysis of materials in the marine euphotic zone. In: Ulanowicz RE, Platt T (eds) Ecosystem theory for biological oceanography. Can Bull Fish Aquat Sci 213:139-162
Fasham MJR, Denman K, Brewer PG (1991) The Joint Global Ocean Flux Study: goals and objectives. In: Corell RW, Anderson PA (eds) Global Environmental Change. Springer-Verlag Berlin. p 246
Fasham MJR, Ducklow HW, McKelvie SM (1990) A nitrogen-based model of plankton dynamics in the oceanic mixed layer. J Mar Res 48:591-639
Fasham MJR, Holligan PM, Pugh PR (1983) The spatial and temporal development of the spring phytoplankton bloom in the Celtic Sea April 1979. Prog Oceanogr 12:87-145
Fasham MJR, Platt T (1985) Photosynthetic response of phytoplankton to light: a physiological model. Proc R Soc Lond B 219:355-370
Fenchel T , Blackburn TH (1979) Bacteria and mineral cycling. Academic Press New York
Fogg GE (1975) Primary productivity. In: Riley JP, Skirrow J (eds) Chemical Coeanography vol 2. Academic Press London. p 385
Fogg GE (1983) The ecological significance of extracellular products of phytoplankton photosynthesis. Botanica Marina 26:3-14
Fowler SW, Knauer GA (1986) Role of large particles in the transport of elements and organic compounds through the oceanic water column. Prog Ocean 16:147-194
Frost BW (1987) Grazing control of phytoplankton stock in the open subarctic Pacific Ocean: a model assessing the role of mesozooplankton particularly the large calanoid copepods Neocalanus spp. Mar Ecol Prog Ser 39:49-68
Gardner WD, Walsh ID (1990) Distribution of macroaggregates and fine-grained particles across a continental margin and their potential role in fluxes. Deep-Sea Res 37:401-411
Geider RJ, Platt T (1986) A mechanistic model of photadaptation in microalgae. Mar Ecol Prog Ser 30:85-92
Gifford R (1992) Implications of CO2 effects on plants for the global carbon cycle. This volume
Glover DM, Brewer PG (1988) Estimates of winter-time mixed layer nutrient concentrations in the North Atlantic. Deep-Sea Res 35:1525-1546
Goldman JC, Caron DA, Dennett MR (1987) Regulation of gross growth efficiency and ammonium regeneration in bacteria by substrate C:N ratio. Limnol Oceanogr 32:239-1252
Harris GP (1986) Phytoplankton ecology: structure, function and fluctuation. Chapman and Hall London
Harrison WG, Platt T, Lewis MR (1987) F-ratio and its relationship to ambient nitrate concentration in coastal waters. J Plank Res 9:235-248
Hofmann EE, Klinick JM, Paffenhofer G-A (1981) Concentrations and vertical fluxes of zooplankton fecal pellets on a continental shelf. Mar Biol 61:327-335
Hofmann EE (1988) Plankton dynamics on the outer southeastern US continental shelf Part III: A coupled physical-biological model. J mar res 9:919-946
Hofmann EE, Ambler JW (1988) Plankton dynamics on the outer southeastern US continental shelf Part II: a time-dependent model. J mar res 46:883-917
Hofmann EE, Pietrafesa LJ, Klinick JM, Atkinson LP (1980) A time-dependent model of nutrient distribution in continental shelf waters. Ecol Modell 10:193-214
Isemer HJ, Hasse L (1985) The Bunker climate atlas of the North Atlantic Ocean vol 1: Observations. Springer-Verlag Berlin
Iwasa Y, Andreason V, Levin S (1987) Aggregation in model ecosystems I. Perfect aggregation, Ecol Model 37:287-302
Jamart BM, Winter DF, Banse K, Anderson GC, Lam RK (1977) A theoretical study of phytoplankton growth and nutrient distribution in the Pacific Ocean of the northwestern U S coast. Deep-Sea Res 24:753-773

Jamart BM, Winter DF, Banse K (1979) Sensitivity analysis of a mathematical model of phytoplankton growth and nutrient distribution in the Pacific Ocean off the northeastern US coast. J Plank Res 1:267-290
Jernigan RW, Tsokos IP (1980) A linear stochastic model for phytoplankton production in a marine ecosystem. Ecol Model 10:1-12
Jones R, Henderson EW (1986) The dynamics of nutrient regeneration and simulation studies of the nutrient cycle. J Conseil 43:216-236
Joos F, Sarmiento JL, Siegenthaler U (1991) Estimates of the effect of the Southern Ocean iron fertilisation on atmospheric CO2 concentrations. Nature 349:772-775
Kiefer DA, Kremer JN (1981) Origins of vertical patterns of phytoplankton and nutrients in the temperate open ocean: a stratigraphic hypothesis. Deep-Sea Res 28:1087-1106
Koike I, Hara S, Terauchi K, Kogure K (1990) Role of sub-micrometre particles in the ocean. Nature 345:242-244
Kremer JN (1983) Ecological implications of parameter uncertainty in stochastic simulation. Ecol Model 18:187-207
Kremer JN, Nixon SW (1977) A coastal marine ecosystem. Springer-Verlag New York
Lancelot C, Billen G (1986) Carbon-nitrogen relationships in nutrient metabolism of coastal marine ecosystems. Adv Aquatic Microbiol 3:263-321
Lederman TC, Tett P (1981) Problems in modelling the photosynthesislight relationship. Bot Mar 24:125-134
Levitus S (1982) Climatological atlas of the world ocean. NOAA Professional Paper 13, US Govt Printing Office Washington.
Longhurst AR, Harrison WG (1988) Vertical nitrogen flux from the oceanic photic zone by diel migrant zooplankton. Deep-Sea Res 35:881-889
Longhurst AR, Bedo A, Harrison WG, Head EJH, Sameoto DD (1990) Vertical flux of respiratory carbon by oceanic diel migrant biota. Deep-Sea Res 37:685-694
Longhurst AR, Bedo A, Harrison WG, Head EJH, Horne EP, Irwin B, Morales C (1991) NFLUX: a test of vertical nitrogen flux by diel migrant biota. Deep-Sea Res 36:1705-1719
Martin JH, Fitzwater SE (1988) Iron deficiency limits phytoplankton growth in the north-east Pacific subarctic. Nature 331:341-343
McLaren IA (1963) Effects of temperature on the growth of zooplankton and the adaptive value of vertical migration. J Fish Res Bd Can 20:685-727
Menzel DW, Ryther JH (1960) The annual cycle of primary production in the Sargasso Sea off Bermuda. Deep-Sea Res 6:351-367
Menzel DW, Ryther JH (1961) Zooplankton in the Sargasso Sea off Bermuda and its relation to organic production. J Cons 26:250-258
Moloney CL, Bergh MO, Field JG, Newell RC (1986) The effect of sedimentation and microbial regeneration in a plankton community: a simulation investigation. J Plank Res 8:427-445
Morel A (1988) Optical modeling of the upper ocean in relation to its biogenous matter content (case I waters). J Geophys Res 93 (C9):10749-10768
Morel A, Berthon JF (1989) Surface pigments algal biomass and potential production of the euphotic zone: realtionships reinvestigated in view of remote sensing applications. Limnol Oceanogr 34:1547-1564
Murray JW, Downs JN, Strom S, Wei C-L,Jannasch HW (1989) Nutrient assimilation export production and ^{234}Th scavenging in the eastern equatorial Pacific. Deep-Sea Res 36:1471-1489
Najjar RM (1990) Simulations of the phosphorus and oxygen cycles in the world ocean using a general circulation model. PhD dissertation Princeton Univ. Princeton New Jersey
Newell RC (1984) The biological role of detritus in the marine environment In: Fasham MJR (ed) Flows of energy and materials in marine ecosystems: theory and practice. Plenum Press New York & London. p 317
Newell RC, Linley EAS (1984) Significance of microheterotrophs in the decomposition of phytoplankton: estimation of carbon and nitrogen flow based on the biomass of plankton communities. Mar Ecol Prog Ser 16:105-119

Nisbet RM, Gurney WSC (1982) Modelling fluctuating populations. J Wiley & Sons Chichester
Oberhuber JM (1988) An atlas based on the 'COADS' data set: the budgets of heat buoyancy and turbulent kinetic energy at the surface of the global ocean. Max-Panck-Institut für Meteorologie Report No 15 Hamburg
O'Neill RV, Angelis DL, Pastor JJ, Jackson BJ, Post WM (1989) Multiple nutrient limitations in ecological models. Ecol Modell 46:147-163
Pace ML, Knauer GA, Karl DM, Martin JH (1987) Particulate matter fluxes in the ocean: a predictive model. Nature 325:803-804
Pace ML, Glasser JE, Pomeroy LR (1984) A simulation analysis of continental shelf food webs. Mar Biol 82:47-63
Paffenhofer G-A, Knowles SC (1979) Ecological implications of fecal pellet size production and consumption by copepods. J Mar Res 37:35-49
Parsons TP, Kessler TA (1987) An ecosystem model for the assessment of plankton production in relation to the survival of young fish. J Plank Res 9:125-137
Paulson CA, Simpson JJ (1977) Irradiance measurements in the upper ocean. J Phys Oceanogr 7:952-956
Peng T-H,Broecker WS (1991) Dynamical limitations on the Anarctic iron fertilization. Nature 349:227-229
Peng T-H, Takahashi T, Broecker WS, Olafsson J (1987) Seasonal variability of carbon dioxide nutrients and oxygen in the northern North Atlantic surface water: observations and a model. Tellus 39B:439-458
Peters RH (1983) The ecological implications of body size. Cambridge University Press Cambridge
Platt T, Denman KL, Jassby AD (1977) Modeling the productivity of phytoplankton In: Goldberg ED, McCave IN, O'Brien JJ, Steele JH (eds) The Sea vol 6. Wiley-Interscience New York. p 807
Platt T, Gallegos CL, Harrison WG (1980) Photoinhibition of photosynthesis in natural assemblages of marine phytoplankton. J Mar Res 38:687-701
Platt T , Harrison WG (1985) Biogenic fluxes of carbon and oxygen in the ocean. Nature 318:55-58
Platt T, Mann KH, Ulanowicz RE (1981) Mathematical models in biological oceanography. UNESCO Press Paris
Platt T, Sathyendranth S (1988) Ocean primary production: Estimation by remote sensing at local and regional scales. Science 241:1613-1620
Platt T, Sathyendranath S, Ravindran P (1990) Primary production by phytoplankton: analytic solutions for daily rates per unit area of water surface. Proc R Soc Lond B 241:101-111
Pomeroy LR (1974) The ocean's food web, a changing paradigm. Bioscience 24:499-504
Poulet SA (1983) Factors controlling utilization of non-algal diets by particle-grazing copepods: a review. Oceanologica Acta 6:221-234
Riley GA, Stommel H, Bumpus DF (1949) Quantitative ecology of the plankton of the western North Atlantic. Bull Bing Ocean Coll 12(3):1-169
Rosenzweig ML (1971) Paradox of enrichment. Science 171:385-387
Sakshaug E, Kiefer DA, Andresen K (1989) A steady state description of growth and light absorption in the marine planktonic diatom *Skeletonema costatum* . Limnol Oceanogr 34:198-205
Sarmiento JL, Toggweiler JR, Najjar R (1988) Ocean carbon cycle dynamics and atmospheric pCO_2. Phil Trans Roy Soc A:325 3-21
Sarmiento J L, Fasham MJR, Slater R, Toggweiler JR, Ducklow HW (1992) The role of biology in the chemistry of CO_2 on the ocean. In: Farrell M (ed) Chemistry of the greenhouse effect. Lewis Publ. In press
Sarmiento JL, Orr JC, Siegenthaler U (1992) A perturbation simulation of CO_2 uptake in an ocean general circulation model. J Geophys Res In press
Sathyendrananth S, Platt T (1988) The spectral irradiance field at the surface and in the interior of the ocean: a model for applications in oceanography and remote sensing. J Geophys Res 93:9270-9280

Schell DM (1974) Uptake and regeneration of free amino acids in marine waters of Southeast Alaska. Limnol Oceanogr 19:260-270
Shaffer G (1992) Marine Production. This volume
Sharp JH (1977) Excretion of organic matter: do healthy cells do it? Limnol Oceanogr 22:381-399
Sherr EB, Sherr BF, Albright LJ (1987) Bacteria: link or sink? Science 235:88-89
Sherr EB, Sherr BF (1988) Role of microbes in pelagic food webs: a revised concept. Limnol Oceanogr 33:1225-1227
Smith SD, Dobson FW (1984) The heat budget at Ocean Weather Ship Bravo. Atmos-Ocean 22:1-22
Steele JH (1958) Plant production in the northern North Sea. Scottish Home Department Marine Research Report No 7 HMSO Edinburgh
Steele JH (1974) The structure of marine ecosystems. Blackwell Scientific Publications Oxford
Steele JH, Henderson EW (1992) The role of predation in plankton models. J Plank Res 14:157-172
Steele JH, Frost BW (1977) The structure of plankton communities. Phil Trans R Soc B 280:485-534
Stouffer RJ, Manabe S, Bryan K (1989) Interhemispheric asymmetry in climate response to a gradual increase of atmospheric CO_2. Nature 342:660-662
Susuki Y (1992) Dynamic cycle of dissolved organic carbon and marine productivity. This volume
Suttle CA, Chan AM, Cottrell MT (1991a) Infection of phytoplankton by viruses and reduction of primary productivity. Nature 347:467-469
Suttle CA, Chan AM, Fuhrman JA (1991b) Dissolved free amino acids in the Sargasso Sea: uptake and respiration rates, turnover times, and concentrations. Mar Ecol Prog Ser 70:189-199
Taghon GL, Jumars PJ (1984) Variable ingestion rate and its role in optimal foraging behaviour of marine deposit feeders. Ecology 65:549-558
Taylor AH, Watson AJ, Ainsworth M, Robertson JE, Turner DR (1991) A modelling investigation of the role of phytoplankton in the balance of carbon at the surface of the North Atlantic. Global Biogeochemical Cycles 5:151-171
Tett P, Edwards A, Jones K (1986) A model for the growth of shelf-sea phytoplankton in summer. Est Coast Shelf Sci 23:641-672
Vinogradov ME, Menshutkin VV, Shushkina EA (1972) On Mathematical simulation of a pelagic ecosystem in tropical waters of the ocean. Mar Biol 16:261-268
Volk T, Hoffert MI (1985) Ocean carbon pumps: analysis of relative strengths and efficiencies in ocean-driven atmospheric CO_2 changes. In: Sundquist ET, Broecker WS (eds) The carbon cycle and atmospheric CO_2: Natural variations archean to present . American Geophysical Union Washington DC. p 99
Walsh JJ (1975) A spatial simulation model of the Peru upwelling ecosystem. Deep-Sea Res 22:201-236
Walsh JJ (1983) Death in the Sea: Enigmatic phytoplankton losses. Prog Oceanography 12:1-86
Walsh JJ (1988) On the nature of continental shelves. Academic Press San Diego
Walsh JJ , Dugdale RC (1972) Nutrient submodels and simulation models of phytoplankton production in the sea. In: Kramer J, Allen H (eds) Nutrients in natural waters. JWiley & Sons New York. p 171
Wheeler PA, Kokkinakis SA (1990) Ammonium recycling limits nitrate use in the oceanic subarctic Pacific. Limnol Oceanogr 35:267-1278
Wiegert RG (1979) Population models: experimental tools for the analysis of ecosystems. In: Horn DJ, Mitchell R, Stairs GR (eds) Proceedings of colloquium on analysis of ecosystems. Ohio State University Press. p 239
Williams R (1988) Spatial heterogeneity and niche differentiation in oceanic zooplankton. In: Boxshall GA, Schminke HK (eds) Biology of Copepods. Hydrobiologia 167/168:151-159
Williams R, Robinson GA (1973) Primary production at Ocean Weather Ship India (59° 00'N, 19° 00'W) in the North Atlantic. Bull Mar Ecol 8:115-121

Woods JD, Onken R (1982) Diurnal variation and primary production in the ocean - preliminary results of a Lagrangian ensemble model. J Plank Res 4:735-756
Wroblewski J (1975) The verically migrating deep scattering layer - its possible role in the creation of small scale phytoplankton patchiness in the ocean. In: Anderson NR,Zahuranec BJ (eds) Oceanic sound scattering prediction. Plenum Press New York. p 817
Wroblewski J (1977) A model of phytoplankton plume formation during variable Oregon upwelling. J Mar Res 35:357-394
Wroblewski J (1980) A simulation of the distribution of *Acartia clausii* during Oregon upwelling August 1973. J Plank Res 2:43-68
Wolf KU, Woods JD (1988) Lagrangian simulation of primary production in the physical environment - the deep chlorophyll maximum and nutricline In: Rothschild BJ (ed) Towards a theory of biological-physical interaction in the world ocean. Kluwer Dordrecht. p 51
Wulff F, Field JG, Mann KH (eds) (1989) Network analysis in marine ecology .Coastal and Estuarine Studies vol 32 Springer-Verlag

FORMATION AND COMPOSITION OF MARINE PARTICULATES

G. Wefer
Geowissenschaften
Universität Bremen
Postfach 33 04 40
2800 Bremen
Germany

ABSTRACT

The transfer of carbon from the surface ocean to the ocean floor is governed by a number of processes. The particulate flux is related to primary production, which is controlled by physical processes. The release of particles from the surface zone to the ocean's interior is related to biological processes such as formation of larger particles by fecal pellet production and aggregate formation. The most important processes involved in the formation and destruction of particles are described.

INTRODUCTION

After the detection of atmospheric CO2-changes in ice cores (Barnola et al. 1987), several possible mechanisms were proposed to explain these changes. One possibility is to change the amount of carbon leaving the euphotic zone by changing the productivity of the ocean ("biological pumping"; Broecker 1982; see several articles in Sundquist & Broecker 1985). This process, which removes carbon from the surface waters and sequesters it at depth and within the sediments, has been the subject of much discussion since (see e.g. Keir 1988).

The efficiency of the pump depends on the leakage of carbon from the pelagic foodweb, and the associated downward transport of organic matter, both in particulate and in dissolved form. The ratio between organic carbon and carbonate carbon in the downward flux (rain ratio) is of special interest. This ratio plays an important role in atmospheric pCO_2 control. The flux of particulate organic carbon (from tissues) decreases the partial pressure of CO_2 in the surface layers and hence in the atmosphere.

NATO ASI Series, Vol. I 15
The Global Carbon Cycle
Edited by M. Heimann

The removal of carbon from the surface layers by export of carbonate changes the carbonate equilibria and increases CO_2 in the atmosphere. Realistic rain ratio values for the major biogeochemical provinces are necessary for modelling. Part of the atmospheric CO_2 changes of the last 150,000 years, as recorded in ice cores (Barnola et al. 1987) could be explained by changes in the rain ratio. Indications for such changes could be a shift in plankton communities, e.g. from a coccolith (carbonate producer) to a diatom dominated community.

The paper summarizes the knowledge on the regional and seasonal particulate carbon flux in the ocean with special emphasis on the processes of particle formation and disintegration as well as on the ratio between organic carbon and carbonate carbon of the particulate matter in the ocean.

PRODUCTIVITY AND FLUX

The transfer of material to the sea floor is governed by a number of processes which in general are appreciated but not fully understood in detail. Fig. 1 shows a highly simplified sequence for two scenarios: the open ocean with relatively low primary production (PP about 30 gC $m^{-2}y^{-1}$) and the coastal ocean with a primary production of about 120 gC $m^{-2}y^{-1}$.

Particle formation begins with the primary production, which is most commonly measured from carbon fixation using radiocarbon as a tracer. In a compilation of the existing data Berger et al. (1987) produced a map of primary production for the global ocean which is shown in a modified version in Fig. 2. For regions south of 50°S, the values are reduced by a factor of two taking into account new data of the shortness of the period of production in high latitudes (Wefer et al. 1988, 1990; Fischer et al. 1988; Wefer & Fischer 1991). For the southern ocean south of 50°S, Wefer & Fischer (1991) estimated an export production (organic carbon leaving the uppermost 100 m) of 0.17 10^{15} gC (Gigatons) from sediment trap data.

In the low productivity ocean, an export production (flux at 100 m depth, XP(100)) of 10% is expected (Fig.1) compared to about 25% in the more productive coastal ocean. The remaining part is remineralised in the surface water ("regenerated production"). This means that only 3-4% of world export (4 to 5 GtC, Berger et al. 1987) are exported in the Southern Ocean. These figures depend on the initial delivery and the rate at

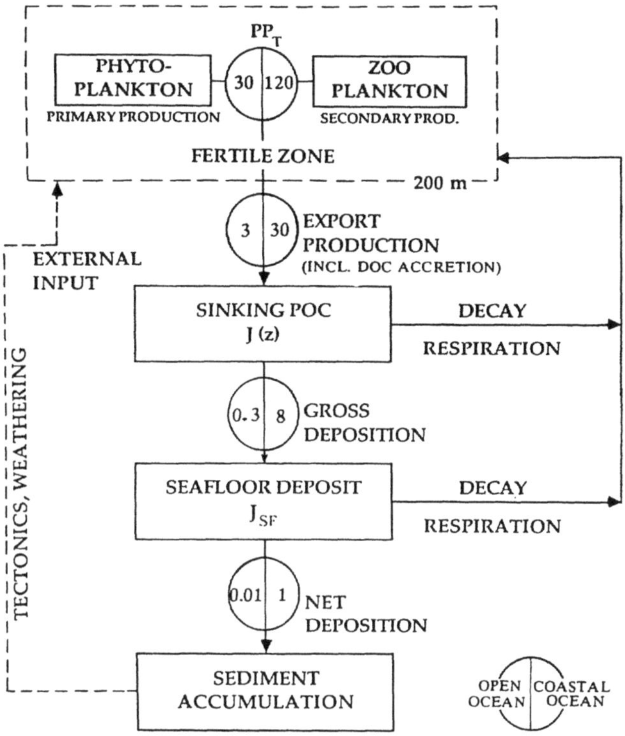

Fig. 1
Sketch of transfer of particulate organic carbon to the sea floor, showing the main elements of the "biological pump". Numbers are fluxes in gC $m^{-2}y^{-1}$; to the left a typical open ocean value; to the right a typical coastal ocean value is given. From Berger et al. (1989).

which the settling material is remineralized. The first is a function of the absolute value of primary production (Eppley & Peterson 1979, Platt & Harrison 1985); the second is a function of consumption and decay rates with depth (Suess 1980).

In Fig. 1 it is assumed that at 100 m depth the fraction of the export (xf) varies between 10% (PP=25 gC $m^{-2}y^{-1}$) and 30% (PP= 225 gC $m^{-2}y^{-1}$) for most of the ocean. Strongly pulsed systems may have a distinctly higher export fraction. Berger & Wefer (1990) defined a "seasonality index" in terms of the number of months which yield one half of the total production (production half-time) and proposed an equation relating the export factor to the level of productivity with a dependency on seasonality. According to

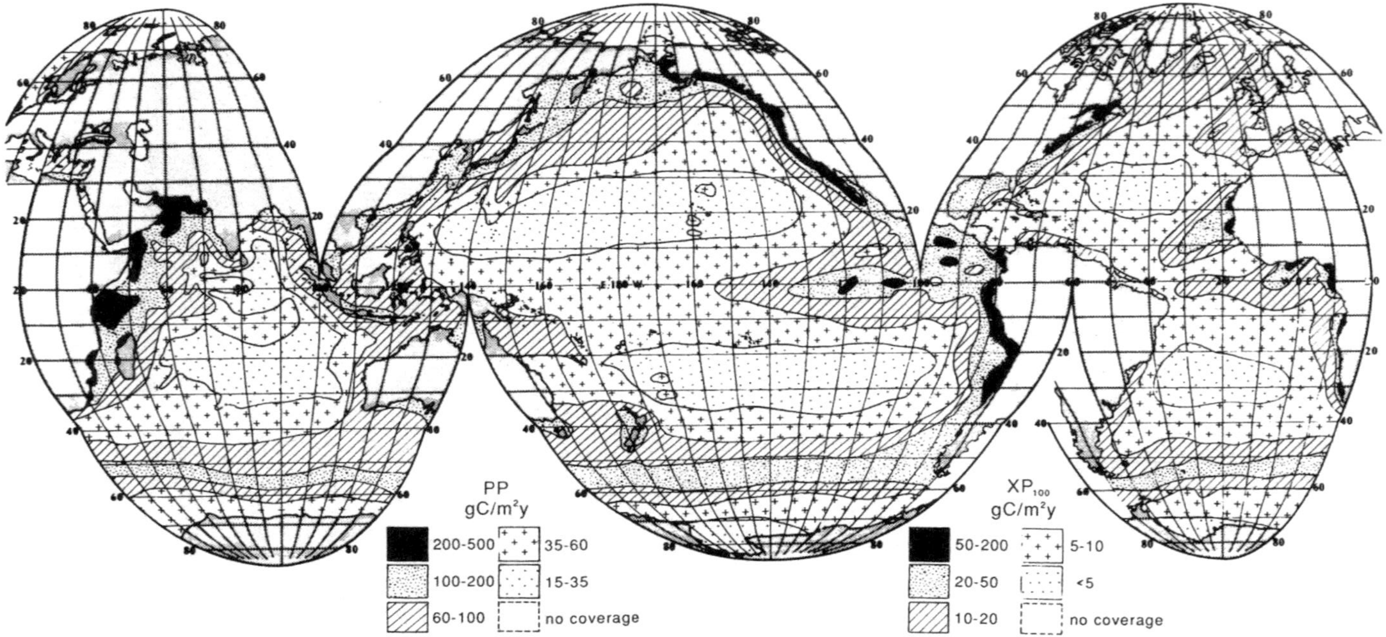

Fig. 2
Global map of ocean productivity, based on maps in Berger et al. (1987), modified for the Southern Ocean (Berger & Wefer 1991). Values in gC $m^{-2}yr^{-1}$. Export production is based on $XP(100) = PP*2\sqrt{PP}/100$. This equation does not include the effects of strong pulsation (Berger & Wefer 1990), which should increase contrasts between high and low production areas even more.

this relationship, strongly pulsed production has a percentage output two times higher than constant production (Fig. 3). This relationship is supported by sediment trap data and other observations.

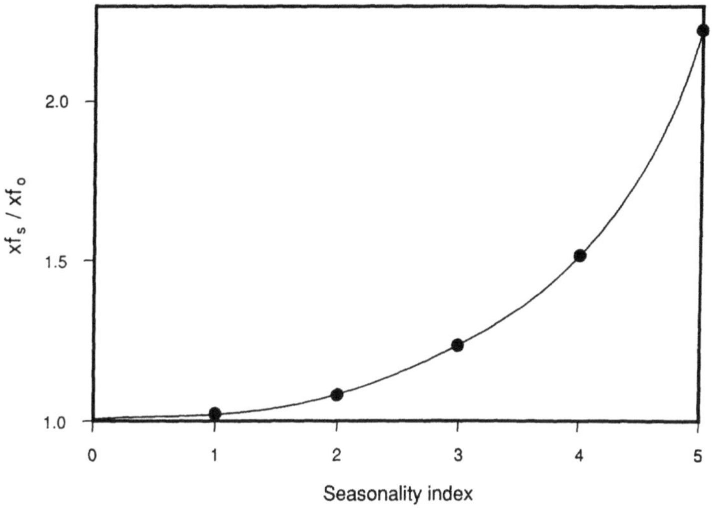

Fig. 3
Factor of increase in export production for constant annual production, depending on seasonality of production (defined as 6 - production half-time in months). XP(0) denotes the case of no seasonality. From Berger & Wefer (1990).

The dependency of export flux on seasonality and intermittency has to be taken into account when estimating carbon transfer to deeper waters during glacial times. Such a process could have led to more efficient downdraw of carbon from upper waters during glacial times when more intense winds would have favored pulsed and sporadic production.

About 10% of the export in the open ocean environment reaches the sea floor; the other 90% of the organic material is remineralised during transfer through the interior of the ocean. At the sea floor about 3% of the arriving flux is buried and the rest is oxidized due to a long residence time at the sea floor. Thus, of the 30 gC $m^{-2}y^{-1}$ produced as primary production, only 0.01 g (0.03%) are incorporated into the sediment. The percentage output is larger in the more productive areas. With a primary production of about

120 gC $m^{-2}y^{-1}$, the remineralisation rate in the surface water and in the water column is probably nearer 75%, respectively, and at the sea floor about 90%. Using these figures, about 1% of the primary production is buried in the sediment. The reason for the differences between the two production systems is most likely to be found in the differences in the community structure and the effectiveness of the food web as well as in the shorter residence time of the material in the water column and at the sea floor. In a more pulsed or episodic system, organisms are less able to recycle the organic material in the surface waters effectively. In the open ocean, the settling speed is presumably lower due to smaller particle sizes. Furthermore, sedimentation rates are lower; the material remains at the sea floor longer and can be decomposed more intensely.

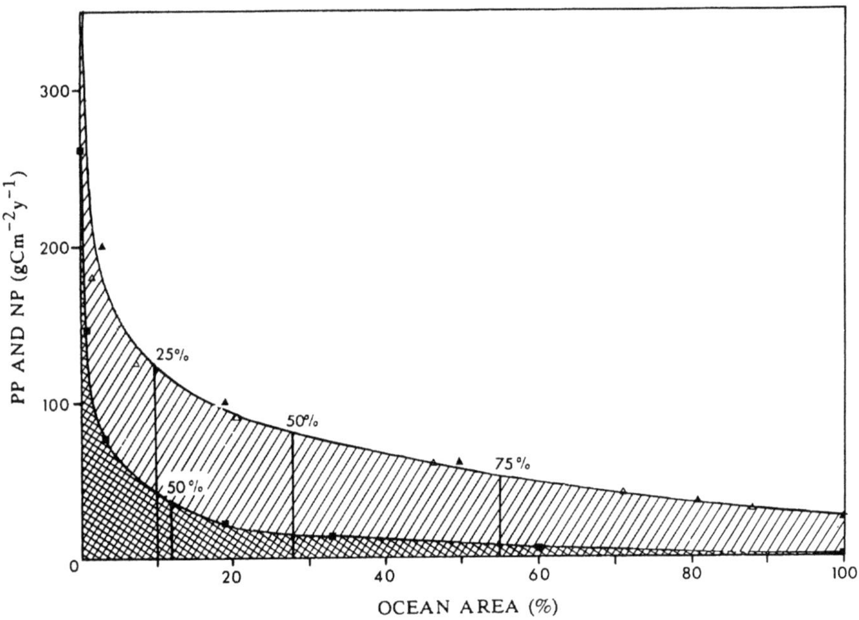

Fig. 4
Histogram of primary production and export production, related to ocean area. Vertical lines: 25% sections for primary production and 50% sections for export production, showing delivery of organic carbon per ocean area. (From Berger et al. 1989).

The differences in primary production between different regions, as depicted in Fig. 2, show the special importance of the high production areas for the global marine carbon cycle. Based on the primary production in the world ocean of Berger et al. (1987), 25% of primary production takes place in less than 10%, and 50% is delivered in less than 30% of the oceans (Berger et al. 1989) (Fig. 4). If we look at the export production, the contribution of the high production areas increases substantially. About 50% of export production is delivered by slightly more than 10% of the ocean (Fig. 4).

Terrigenous organic matter is supplied by air and river transport and adds to the flux of marine organic carbon to the sea floor. Due to its refractory nature, terrestrial material can be enriched during the decomposition of total organic matter (Ittekkot 1988). It is further assumed that a major part of the continental shelf production is exported to the deep sea (Walsh 1989), adding to the organic material decomposed at the sea floor or incorporated into the sediment.

Dissolved organic carbon (DOC)

The concept that the downward flux of organic matter is mostly due to the rain of particulate matter has been challenged by new measurements of elevated concentrations of dissolved organic carbon (DOC) in sea water, especially from the surface ocean (Suzuki, this volume). It is striking that there is a good correlation betweeen apparent oxygen utilization and DOC and that the downward flux of DOC can account for most of the oxygen utilization at depth (Toggweiler 1989).

Today it is not clear how much of the total organic carbon pool at depth is derived from sinking of dissolved organic matter and how much from settling of particles. This ratio must be different for different biogeochemical regions and different mixing systems. From the distribution of dissolved organic carbon in the water column it can be concluded that the uppermost 1000 m of the water column are more affected by dissolved organic matter flux (Bacastow & Maier-Reimer 1991) and that in the deeper part of the ocean particulate transport is more important. When arguing in favor or against DOM flux, we have to take into account that DOM is defined as substances passing through a filter of 0.45 μm pore size. Therefore part of the dissolved organic matter will consist of small particles including small organisms. On the other hand, what is collected in sediment traps as "particulate organic matter" are larger fast settling particles. This means that the flux is underestimated by missing particles of smaller sizes, depending on particle composition and shape. To some degree, therefore, the boundary between dissolved and particulate organic matter is an arbitrary one.

Nevertheless, the measurements do show the need for a better understanding of the coupling of the POM and DOC cycles by bacteria. Large particles might not be a favourable substrate for bacterial growth (see Sharp 1991) and therefore remineralisation of organic matter from large particles during transfer to the seafloor may be minimal.

In models, downward transport of DOM is assumed by advection and diffusion (Toggweiler 1989). It is not taken into consideration that part of the DOM could be transported downward trapped within or attached to fecal pellets. Also vertically migrating zooplankton, feeding in surface waters during night and excreting the organic material during the day in deeper waters, could have an effect on the vertical distribution of DOC in the water. The release of DOC would depend on the gut clearance rate which is not well known (Jahnke 1990).

REGIONAL AND SEASONAL FLUX PATTERNS

Our knowledge on regional and seasonal particle sedimentation has increased substantially through the development of sediment traps, especially time series traps in the eighties. The traps are attached to moorings deployed in different water depths. In general, in deep-sea experiments, the deepest trap is situated well above the seafloor (300-500 m) to avoid the collection of resuspended matter. The shallowest trap is situated between 500 to 1000 m below the surface to avoid the disturbance of vertical particle flux by migrating zooplankton. Different preservatives are commonly used depending on the aim of the study. The different collecting devices, preservation methods, and analytical methods are described and evaluated in Knauer & Asper (1989) and Hurd & Spencer (1991).

Seasonal and interannual sedimentation patterns

In all experiments, large seasonal differences in particle flux are observed. In Fig. 5 three records of particle flux are depicted to show examples from different environments. Seasonality is smallest in tropical and subtropical areas such as the

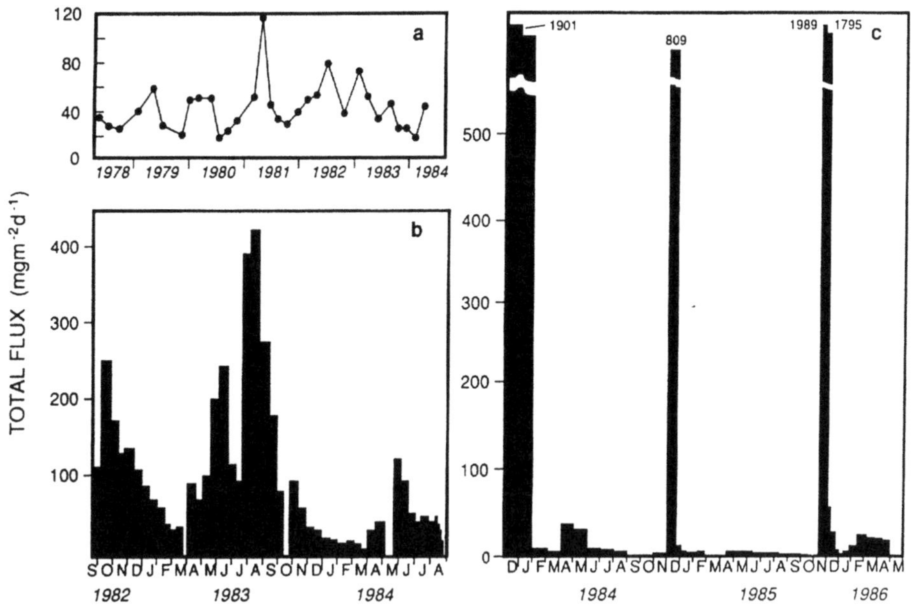

Fig. 5
Records of particle flux. (a) Sargasso Sea (after Deuser 1987), (b) Station P in the Gulf of Alaska (redrawn from Honjo 1984), (c) Bransfield Strait (Antarctica) (after Berger & Wefer 1990).

Sargasso Sea (Fig. 5a), moderate in temporate or subpolar zones such as the Gulf of Alaska (Fig. 5b) and extreme in polar regions such as Bransfield Strait (Antarctica) (Fig. 5c). The small differences in flux rates from the Sargasso Sea (Fig. 5a) are, in part, a result of the two month sampling intervals. The reason for the large difference between flux rates in polar regions is that extremely low sediment fluxes of less than 1 mg m^{-2} day^{-1} were recorded during ice coverage (Fig. 5c).

All experiments showed that seasonal variations of flux are related to biological processes in the surface waters. At several sites, a close relationship between the patterns of estimated primary production and particle sedimentation was observed. Examples are the Sargasso Sea (Deuser et al. 1981) and the Black Sea (Izdar et al. 1987; Honjo et al. 1987a). In the Sargasso Sea, variations of total carbonates, silica, and organic matter flux are tied to the annual cycle of primary production in the surface water, which peaks in early spring and shows lowest values in late fall (Deuser et al. 1981). In the Black Sea, seasonal variations in flux were related to blooms of diatoms, dinoflagellates, and coccolithophorids (Izdar et al. 1987). In the Panama Basin, the particulate flux also directly reflected changes in surface primary production (Honjo 1982). Similar relationships between particle flux patterns and biological processes in the surface waters were obtained at two trap sites in the equatorial Pacific (Dymond & Collier 1988) and three trap sites in the Arabian Sea (Ittekkot et al. 1987). Dymond & Collier (1988) showed that the biogenic particle flux is at least a factor of two lower during periods of intense El Nino Southern Oscillation (ENSO) influence compared to fluxes during other periods. At each site in the Arabian Sea, increased particle flux followed the passage of a monsoon (Ittekkot et al. 1987, Nair et al. 1989). In the Bay of Bengal it was found that the maximum observed flux of particulate matter coincided with river discharge (Ittekkot et al. 1991). The overall flux pattern seems to be controlled by the seasonality of surface water production, which, in the Bay of Bengal, is related to the river inputs. In the Fram Strait seasonal production of biogenic material depends on the behavior of sea ice (Hebbeln & Wefer 1991). Under a closed ice cover the sedimentation is very low, whereas maximum sedimentation is found at the ice margin.

At the Norwegian-Greenland Sea stations, fluxes were not highest during the spring bloom but in the fall between August and October (Honjo et al. 1987b). From plankton investigations in the Norwegian Sea it appears that copepod grazing within the productive layer is responsible for the retention of organic material during the phytoplankton spring bloom (Peinert et al. 1987). In the Bransfield Strait, pulses in the particle flux are not directly related to pulses in the primary production, although more than 90% of the annual sedimentation is concentrated in a surprisingly short period during the austral summer (Fig. 5c). Instead, following phytoplankton growth, vigorous feeding by krill swarms takes place and causes the short sedimentation episodes. During the maximum flux, up to 80-90% of the vertical flux of organic carbon from the surface layer was attributed to krill fecal material (v. Bodungen et al. 1987). Spring thaw processes releasing nutrients to the water column are suggested by Fischer et al. (1988) as being the reason for the variability of particle flux in the northern Weddell Sea where annual biogenic and lithogenic particle fluxes are the lowest yet observed in the ocean. A relationship between particle sedimentation and trade wind/monsoon circulation, inducing upwelling and biological productivity in the surface waters, was found in eastern Atlantic upwelling areas (Wefer & Fischer, ms. subm.) (Fig. 6). The trade wind system and the location of the intertropical convergence zone also control the transport of aeolian dust into the northern hemisphere and equatorial ocean, probably enhancing sedimentation by particle loading.

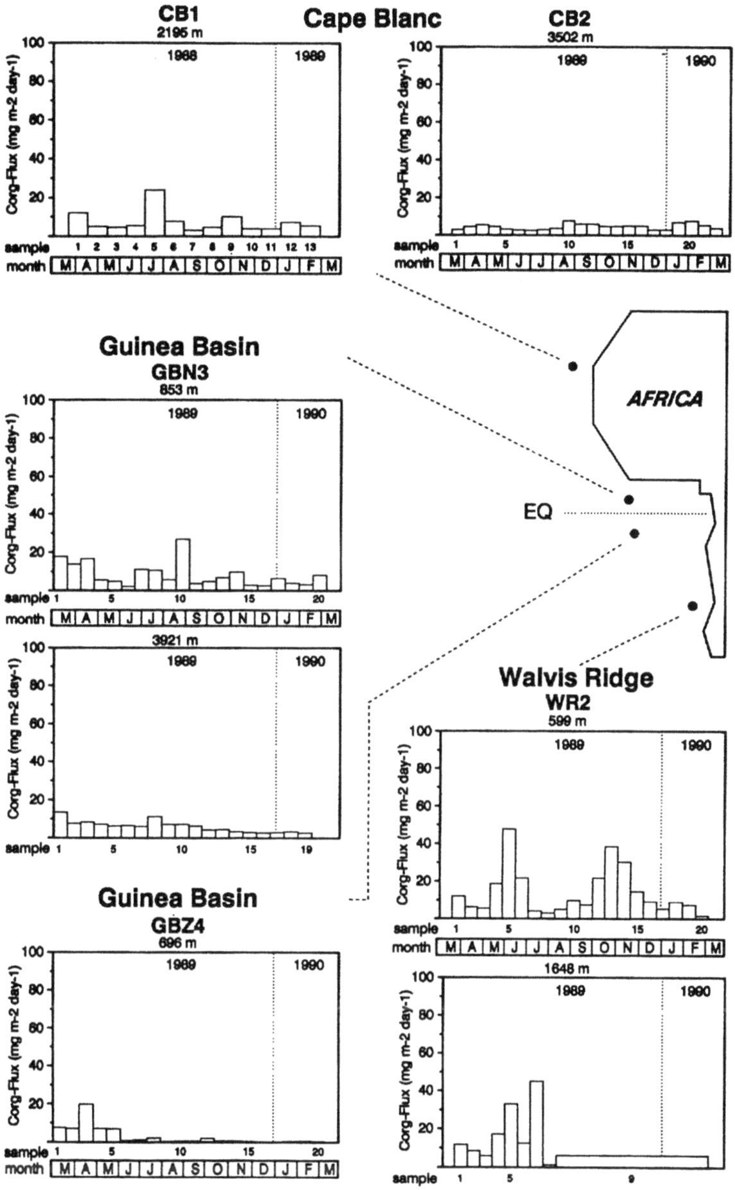

Fig. 6
Seasonal organic carbon flux pattern at several sites in the Atlanic Ocean (from Wefer & Fischer, ms. submitted).

From these observations it can be concluded that the seasonal timing and amount of organic matter production in the surface water is controlled by physical processes such as nutrient supply, light availability, temperature and stability of the surface water but that the release of particles from the productive zone is related to biological processes, that is, formation of larger particles by fecal pellet production and aggregate formation.

Regional variability in composition

Over most of the ocean, total flux in the ocean's interior varies within a range of 7 to 45 g $m^{-2}yr^{-1}$; exceptions are areas of very high productivity and polar regions (for data on seasonal particle sedimentation in deep-sea environments see Wefer 1989).

Also, the values for organic carbon sedimentation vary little over much of the ocean. Typical organic carbon fluxes to water depths of 1.000 to 3.000 m are between 0.5 and 1.0 gC $m^{-2}yr^{-1}$, that is, 1-2% of conventional primary production values.

Flux values that are higher by an order of magnitude occur in high production areas (primary production larger than about 100 gC $m^{-2}yr^{-1}$), such as the Panama Basin and the Bransfield Strait. The reason for the increase is not organic carbon sedimentation, which remains near 2% of the primary production, but that the high total fluxes are derived from a high carbonate (Panama Basin) or silica (Bransfield Strait) contribution as well as large fluxes of lithogenic particles. The relationship between total flux and organic carbon sedimentation reflects the type of surface production. Where carbonate production dominates (e.g. Sargasso Sea, Lofoten Basin), the organic carbon content in carbonates producing organisms is crucial. The same is true for silica production (Bransfield Strait). This is also seen in a compilation of particle flux data by Milliman & Takahashi (ms. subm.). Both silica and carbonate fluxes show a positive correlation with organic carbon sedimentation rates.

Sites dominated by carbonate flux are e.g. the Sargasso Sea, Panama Basin, Black Sea, and the Norwegian Sea. "Dominated" means that carbonate makes up the highest proportion of the major flux components. The main carbonate producers vary between the areas studied. The most important contributers are coccolithophorids in the Panama Basin and Black Sea, and planktonic foraminifera in the Sargasso Sea. In the Norwegian and Greenland Seas, pteropods produce a substantial part of the carbonate flux in addition to coccolithophorids. It is not clear why the predominant biogenic particle flux in the Norwegian/Greenland Sea consists of coccoliths which also grow under the pack-ice (Honjo 1990). Explanations may be sought in processes related to special surface water conditions as a result of mixing between North-Atlantic and Arctic waters or the supply of micronutrients by Siberian rivers discharging into the Arctic Ocean (Honjo 1990).

"Rain ratio" of plankton organic carbon and carbonate ($C_{org}/C_{carbonate}$)

The flux of carbon in the form of organic matter or carbonate is a central element of the carbon cycle of the ocean. The flux of particulate organic carbon (from tissues) decreases the partial pressure of CO_2 in the surface layers and the atmosphere. The removal of carbon from the surface layers incorporated in carbonate skeletons changes the carbonate equilibria and increases CO_2 in the atmosphere. In Fig. 7, sediment trap data are compiled. The sediment trap data from water depths deeper than 2500 m show that most of the values are between 0.4 and 1, irrespective of the area. Also the data from the Norwegian/Greenland Sea are within this range. Higher values are reported

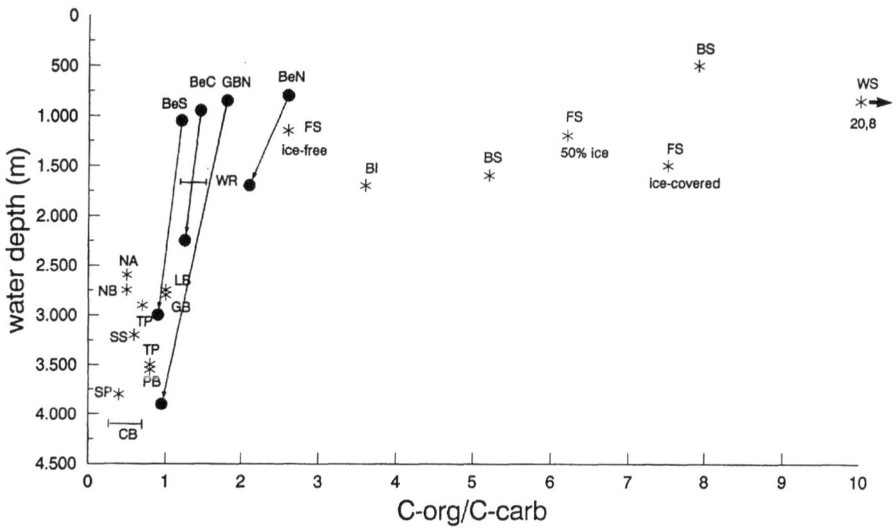

Fig. 7
Corg/Ccarb ratio in relation to water depth in various environments. WS: Weddell Sea; BS: Bransfield Strait; FS: Fram Strait; BI, NA, NB, GB, LB: Sites in the Norwegian/Greenland Sea; SS: Sargasso Sea; SP: Gulf of Alaska; PB: Panama Basin; TP: Tropical Pacific; WR: Walvis Ridge; CB: Cape Blanc; GBN: Guinea Basin; BeS, BeC, BeN: Gulf of Bengal.

from the Southern Ocean and the Fram Strait as well as from shallower depths from the Gulf of Bengal (Ittekkot et al. 1991). The highest value is from the northern Weddell Sea (20.8) where very little sediment is produced.

The ratios decrease with water depth. Examples are the sites from the Gulf of Bengal and the Guinea Basin. If no dissolution of carbonates is assumed, then C_{org} decreases by 50% in the Guinea Basin between 853 and 3921 m (about 3000 m). Similar values were calculated for the Gulf of Bengal stations: about 20% for 1000m and 25% for ca. 2000m. These decomposition rates are in agreement with the $1/z$ relationship for the decomposition of organic matter below the euphotic zone (Suess 1980). A decrease of the ratio with increasing water depth was already shown by Dymond & Lyle (1985) in a flux comparison between sediments and sediment traps in the eastern tropical Pacific and the decrease was attributed to more rapid recycling of organic carbon in comparison to carbonate-carbon.

Berger & Keir (1984) assumed from early sediment trapping results that the rain ratio between organic carbon and carbonate carbon increases with increasing export production. This relationship is also seen in the data of Dymond & Lyle (1985) and in Fig. 7. However, the relationship is probably complex. In the Bay of Bengal maximum river discharge coincides with the maximum observed flux of organic matter showing highest rain ratios (Ittekkot et al. 1991). They point out that this case might be transferred into the past. During deglaciaten fresh-water pulses may have caused similar shifts in ratios resulting in short-term episodes of increased uptake of atmospheric CO_2 by the ocean. A seasonal variability in the C_{org} to $C_{carbonate}$ ratio is also observed in sediment trap data from the Guinea Basin and off Namibia (Wefer & Fischer, ms. subm.). Ratios vary between 2-5 during high organic matter fluxes and were about 1 during low fluxes (Fig. 7).

If the relationship between productivity and rain ratio is applied, a drop in ocean fertility from glacial to postglacial time, which is concluded from sediment core data (for an overview see Sarnthein et al. 1988) can result in a rise in pCO_2 due to a shift from non carbonate secreting (e.g. flagellates and diatoms) to carbonate producing organisms (coccolithophorids and foraminifera). A similar effect is caused by increased carbonate accumulation during ocean transgression over the shelf during postglacial melting of ice caps (Berger 1982, Berger & Keir 1984).

PARTICLE FORMATION AND DESTRUCTION

Organisms not only produce most of the particles, they are also responsible for their rapid transport from the surface water through the ocean interior to the sea floor by enlargement of originally small particles. Applying the Stokesian law, the sinking speed of particles smaller 30 μm is about less than 10 m day^{-1} (Fig.8).

Clay and silt size particles as well as coccolithophorids and small diatoms therefore have to be transported within larger particles, e.g. fecal pellets or aggregates. But also larger particles can be retained in the surface ocean by turbulent mixing (Alldredge et al. 1987).

It is generally thought that biological activity is more important for the packaging/aggregation and fragmentation of the particles in the ocean's interior than purely physical and chemical processes.

First indications of the mode of sedimentation are given by the time lag between plankton blooms and arrival at the traps. In several experiments these time lags are in the order of weeks and are in agreement with calculations of the time needed for larger particles to settle from the surface to the traps (Honjo 1990).

From the delay between production in the surface layer and arrival at the sea floor, Lampitt (1985) estimated sinking speeds of detrital flocs of 100-150 m day^{-1}. Sinking velocities of 74 $\pm$39 m day^{-1} were observed by divers for aggregates ranging from 2.4 to 75 mm length (Alldredge & Gotschalk 1988). Lower sinking speeds (1-36 m day^{-1}) were calculated by Asper (1987) using a camera system in the Panama Basin. On average, larger aggregates (4-5 mm) settled more slowly (1 m day^{-1}) than smaller aggregates (1-2.5 mm) 36 m day^{-1}. Asper (1986) assumed that the larger aggregates are less dense and are resuspended from the nearby continental slope area, while in contrast, smaller aggregates with higher sinking speeds are produced in the surface water.

The relatively short transit times indicate that most of the material caught in the traps settled within fecal pellets or aggregates. Only foraminifera, pteropods and other zooplankton can produce larger particles with sinking speeds that are high enough to reach the deep sea within a few weeks. Typical settling velocities are summarized in

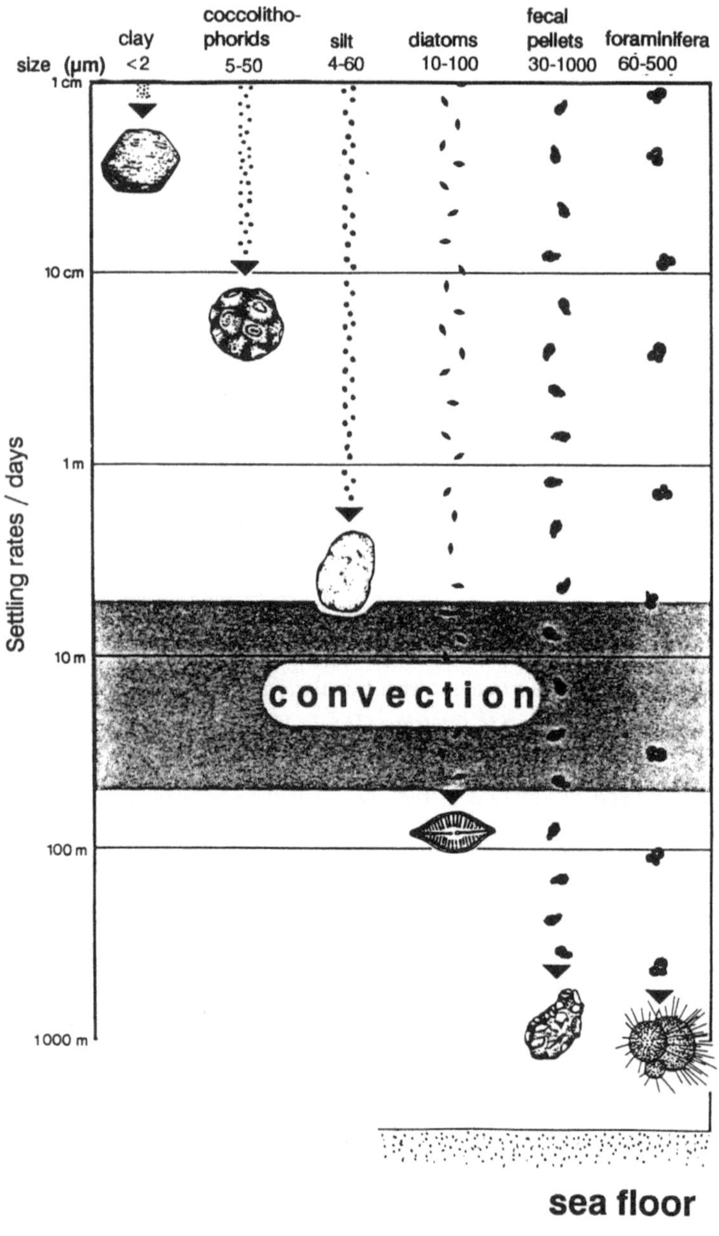

Fig. 8 Average sinking speed of particles with different size and shape in water (after Degens & Ittekkot 1984, modified)

Berger & Wefer (1991), they are from 50-1000 m d^{-1} for krill moults and fecal pellets, 200-2000 m d^{-1} for foraminiferal shells, and 15-500 m d^{-1} for radiolarian skeletons.

The complexity of particle formation and flux is described in Bruland et al. (1989). The most important processes to be considered in the transfer of carbon and associated elements to the sea floor are:

1) Skeletal formation by opal and calcite and aragonite secreting pelagic organisms.

Opal is produced by diatoms (Fig. 9b), radiolaria and silicoflagellates, carbonate is formed by foraminiferans (Fig. 9a), coccolithophorids (Fig. 9c) and pteropods (aragonite). Skeletons of foraminiferans, radiolarians and pteropods and larger diatoms are large enough to settle to the sea floor as single particles.

2) Packaging of fine (living or dead) suspended material (seston) by filtering organisms, into fecal pellets or strings, or onto discarded filters (salps).

These pellets or strings are covered by an organic membrane. The resistance of this membrane against bacterial degradation and mechanical destruction controls the release of the captured finer particles to the water column.

As mentioned, most of the sediments trapped in Bransfield Strait in Antarctica arrive as fecal material. Following phytoplanton growth during austral spring, vigorous feeding by krill swarms takes place and causes a short sedimentation episode. The small phytoplankton particles are incorporated into krill pellets several hundred μm long and about 150 μm wide (Fig. 9e) which have sinking rates of several hundred meters per day. The krill fecal pellets reach the about 2000 m deep Bransfield Strait within days. During settling to the seafloor, the krill feaces were altered due to repackaging in midwater depths by zooplankton coprophagy (Fig. 9f) (v. Bodungen et al. 1987). The feaces are densely packed with detrital material and fine opal from diatom valves. At the Bransfield Strait a large part of the biogenic opal was incorporated into fecal pellets of krill and copepods, shredded to fine diatom debris (1-10 μm in size) (Gersonde & Wefer 1987). Therefore, intact valves found in the sediments must have settled through the water column by means other than krill fecal pellet transport, e.g. as solitary particles or incorporated into or attached to marine snow. Furthermore, there are other feces producing organisms which do not destroy diatom frustules as seen in Fig. 9d.

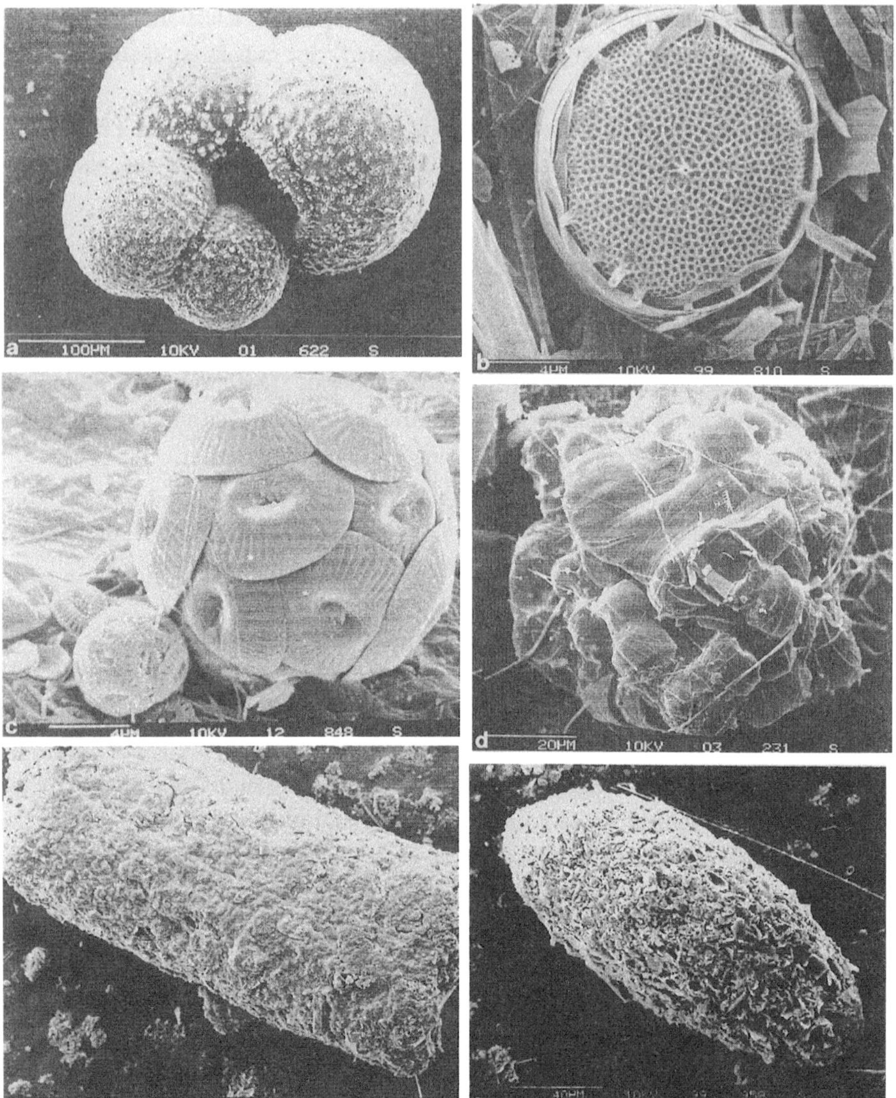

Fig. 9 Scanning Electron Microscope pictures of a) planktonic foraminifera, b) diatom, c) coccolithophorid, d) fecal pellet with diatom valves, e) and f) fecal pellet. SEM pictures C. Samtleben, Kiel.

3) Auto-aggregation of seston by touch or capture (from settling at different speeds or horizontal transport at fronts) to form "marine snow".

Marine snow is regarded as important in plankton interactions and the vertical transfer of organic and lithogenic material from the surface ocean to the sea floor (Shanks & Trent 1980, Silver & Alldredge 1981). There are a number of reports of episodic sedimentation of gelatinous phytoplankton aggregates also containing other materials like coccolithophorids, fecal pellets and clay minerals at the sea floor (Riemann 1989, Thiel et al. 1989). After blooming, coccolith aggregates can settle to the sea floor within days, forming a carpet-like deposition at the sea floor (Honjo 1982).

Due to the fragile nature of the aggregates, difficulties exist in collecting marine snow. Therefore, knowledge on composition and resistance against break-up and sinking speed is limited. In situ observations using time lapse photography are rare. The reasons for aggregate formation are not known; one possibility is the development of a resting stage as a matter of survival strategy (Smetacek 1985, Alldredge & Gotschalk 1989). In a model, Riebesell & Wolf-Gladrow (ms. subm.) argue that particle flux can be generated solely by invoking physical aggregation during phytoplankton blooms.

Other aggregates, composed of various sizes and compositions, may form by encounter of sticky particles during transfer to the sea floor (McCave 1984). The fine particles are attached by secondary metabolic products, essentially polysaccharids. During settling through the water column, aggregates collect finer particles with higher densities like clay minerals, small quartz and feldspar grains and carbonate skeletons from coccoliths. By this process the weight and the settling velocity of an aggregate is increased. The process represents a cleansing of the water column, by acceleration of particle transfer.

4) Refragmentation of fecal material and aggregates by turbulence and the activity of zooplankton (feeding), and return to the pool of small seston.

Aggregates are easily refragmented by weak water movements. Billet et al. (1983) noted that aggregates are easily resuspended in weak currents and disintegrate within 2 days after arrival at the sea floor. Fecal pellets likewise disintegrate during transfer through the water column. In Bransfield Strait krill faeces decreased with depth and finely shredded diatom debris became a more dominant component (v. Bodungen et al. 1987).

5) Injection of fecal matter and aggregates at depth, by migrating zooplankton.

The importance of zooplankton for vertical transport of dissolved and particulate organic matter may have been seriously underestimated due to technical problems in collecting fast swimmers with plankton nets. For many midwater organisms, vertical migration on diurnal or seasonal scales are known (Angel 1986). Diurnal migrations from depths of 600 m (zooplankton) or 1200 m (micronekton) during the day to the surface waters at night are reported. Daily carbon movement into or out of the top 200 m may exceed 50 mg C m^{-2}, a value in the same order as total primary production (Bruland et al. 1989). Feeding is likely to be most intense in the surface waters where food is most plentiful. In deeper waters particulate material is released by defecation and through mortality, and dissolved substances are released by the excretion of metabolic products. Although these processes are poorly understood, they have to be taken into account when estimating vertical fluxes and describing modification processes within the oceans' interior.

In the last few years, observations showed that migrating midwater organisms have a significant modifying effect on the fluxes of organic matter from the productive surface ocean to the deep sea floor. From preliminary data, Angel (1989) estimates that during biological particle modification, 80-90% of the organic carbon flux may pass through in situ bacterial production.

6) Recycling of fecal pellets at depth (repackaging).

In situ repackaging by zooplankton feeding in deeper water (Urrère & Knauer 1981) may lead to a modification of sinking feaces. As an example, evidence for repackaging is given in the Bransfield Strait by the increased flux of densely packed, ellipsoidal fecal pellets collected in traps deeper than about 1000 m (v. Bodungen et al. 1987). Above this depth, up to 80-90% of the particles were string shaped krill feaces (110-180 μm in diameter) of various lengths (120-1600 μm).

7) Dissolution, respiration, and excretion, converting particles to dissolved substances.

Both carbonate and opal skeletons are dissolved during settling through the water column and especially at the sea floor. As a result, large differences exist between the species composition of surface water and surface sediment assemblages (Gersonde & Wefer 1987), indicating that opal skeletons have different solubilities. Milliman &

Takahashi (ms. subm.) estimate that on a global scale about 98% of the opal produced in the surface waters are dissolved in the water column or at the sea floor.

Carbonate is similarly dissolved during settling to the sea floor and especially at the bottom. Dissolution-generated differences between foraminiferal species have been known since decades (for a summary see Vincent & Berger 1981). Due to higher solubility, aragonitic pteropods are only found in sediments shallower than about 3000 m or at particular times, e.g. during deglaciation periods (Sarnthein et al. 1984).

Little is known about disintegration of particles at depth. The consensus is that the transport to the sea floor is in large, fast-sinking particles that are generated from the suspended pool of small particles by aggregation or filtering organisms. Other factors, such as horizontal advection of organic matter at depth from the continental slope (Walsh 1989), have to be considered as well.

SUMMARY

The transfer of carbon as organic carbon or as carbonate carbon out of the productive zone is governed by a number of processes which in general, are appreciated but not fully understood in detail. The amount and composition of material leaving the upper layer of the ocean is related to productivity. In general, seasonal timing, and the amount and composition of material produced are controlled by physical processes such as nutrient supply, light availability, temperature and stability of the surface water. The release of particles from the surface zone of the ocean is related to biological processes such as formation of larger particles by fecal pellet production and aggregate formation. These processes are strongly influenced by seasonality and episodic events.

A number of processes are involved in the formation and destruction of larger fast sinking particles and aggregates. Of special importance for the rapid transfer between surface waters and sea floor are packaging of fine suspended material by filtering organisms, auto-aggregation of seston, injection and recycling of fecal matter by migrating at depth.

REFERENCES

Alldredge AL, Gotschalk CC (1988) In situ settling behavior of marine snow. Limnol. Oceanogr. 33:339-351.

Alldredge AL, Gotschalk CC (1989) Direct observations of the mass flocculation of diatom blooms: characteristics, settling velocities and formation of diatom aggregates. Deep-Sea Res. 36:159-171.

Alldredge AL, Gotschalk CC, MacIntyre S (1987) Evidence for sustained residence of macrocrustacean fecal pellets in surface waters off Southern California. Deep-Sea Res. 34:1641-1653.

Angel MV (1986) Vertical migrations in the oceanic realm: possible causes and probable effects. In: Rankin MA (ed) Migration: Mechanisms and Adaptive Significance. Cont. Mar. Sci. (Suppl.) 24:45-70.

Angel MV (1989) Does Mesopelagic Biology Affect the Vertical Flux? In: Berger WH, Smetacek V, Wefer G (eds) Productivity of the Ocean: Present and Past, Dahlem Workshop Reports, Wiley & Sons, pp 155-173.

Asper V (1986) Accelerated settling of particulate matter by "marine snow" aggregates. Ph.D. Thesis, WHOI/MIT joint educational program, 189 pp.

Asper VL (1987) Measuring the flux and sinking speed of marine snow aggregates. Deep-Sea Res. 34:1-17.

Bacastow R, Maier-Reimer E (1991) Dissolved organic carbon in modeling oceanic new production. Global Biogeochemical Cycles, 5:71-85.

Barnola JM, Raynaud D, Korotkevich YS, Lorius C (1987) Vostok ice core provides 160,000-year record of atmospheric CO_2. Nature 329:408-414.

Berger WH (1982) Increase of carbon dioxide in the atmosphere during deglaciation: the coral reef hypothesis. Naturwissenschaften 69:87-88.

Berger WH, Keir RS (1984) Glacial-Holocene changes in atmospheric CO_2 and the deep-sea record. In: Hansen JE, Takahashi T (eds) Climate Processes and Climate Sensitivity. AGU Geophys. Monogr., 29, 337-351.

Berger, WH, Wefer G (1990) Export production: Seasonality and intermittency, and paleoceanographic implications. Palaeogeography, Palaeoclimatology, Palaeoecology (Global and Planetary Change), 89:245-254.

Berger WH, Wefer G (1991) Flux of biogenous materials to the sea floor: Open questions. In: Hsü K Thiede J (eds) Use and Misuse of Seafloor. Dahlem Workshop Reports, Wiley & Sons (in press).

Berger WH, Fischer K, Lei C, Wu G (1987) Ocean Productivity and Organic Carbon Flux. Part I. Overview and Maps of Primary Production and Export Production. SIO Reference Series, Scripps Institution of Oceanography, Univ. of California, San Diego, SIO Ref. 87-30, 67 pp.

Berger WH, Smetacek V, Wefer G (1989) Ocean productivity and paleoproductivity - An Overview. In: Berger WH, Smetacek V, Wefer G (eds) Productivity of the Ocean: Present and Past. Dahlem Workshop Reports, Wiley & Sons, pp 1-34.

Billet DSM, Lampitt RS, Rice AL, Mantoura RFC (1983) Seasonal sedimentation of phytoplankton to the deep-sea benthos. Nature 302:520-522.

Bodungen v. B, Fischer G, Nöthig E-M, Wefer G (1987) Sedimentation of krill faeces during spring development of phytoplankton in Bransfield Strait, Antarctica. In: Degens ET, Honjo S, Izdale (eds) Particle flux in the ocean. SCOPE/UNEP Sonderbd., Heft 62: 243-257.

Broecker WS (1982) Ocean chemistry during glacial time. Geochim. Cosmochim. Acta 46:1689-1705.

Bruland KW, Bienfang PK, Bishop JKB, Eglington G, Ittekkot VAW, Lampitt R, Sarnthein M, Thiede J, Walsh JJ, Wefer G, 1989. Flux to the seafloor.- In: Berger WH, Smetacek V, Wefer G (eds) Productivity of the Ocean: Present and Past. Dahlem Workshop Reports, Wiley & Sons, pp 193-215.

Degens ET, Ittekkot V (1984) A New Look at Clay-Organic Interactions. Mitt. Geol. Paläont. Inst., Univ. Hamburg, 56:229-248.

Deuser W (1987) Variability of Hydrography and Particle Flux: Transient and Long-Term Relationships. In: Degens ET, Izdar E, Honjo S (eds) Particle Flux in the Ocean. Mitt. Geol. Paläont. Inst., Univ. Hamburg, Heft 62, pp 179-193.

Deuser WG, Ross EH, Anderson RF (1981) Seasonality in the supply of sediment to the deep Sargasso Seas and implications for the rapid transfer of matter to the deep ocean. Deep-Sea Res. 28:495-505.

Dymond J, Collier R (1988) Biogenic particle fluxes in the equatorial Pacific: evidence for both high and low productivity during the 1982-83 EL NINO. Global Biogeochem. Cycles 2:129-137.

Dymond J, Lyle M (1985) Flux comparisons between sediments and sediment traps in the eastern tropical Pacific: Implications for atmospheric CO_2 variations during the Pleistocene. Limnol. Oceanogr. 30:699-712.

Eppley RW, Peterson BJ (1979) Particulate organic matter flux and planktonic new production in the deep ocean. Nature 282:677-680.

Fischer G, Fütterer D, Gersonde R, Honjo S, Ostermann R, Wefer G (1988) Particle fluxes in the Northern Weddell Sea: Seasonality and its relationship to ice-cover. Nature 335:426-428.

Gersonde R, Wefer G (1987) Sedimentation of biogenic siliceous particles in Antarctic waters (Atlantic sector). Marine Micropaleontology 11:311-332.

Hebbeln D, Wefer G (1991) Sedimentation in the Fram Strait: effects of ice coverage and ice-rafted material. Nature 350:409-411.

Honjo S (1982) Seasonality and interaction of biogenic and lithogenic particulate flux at the Panama Basin. Science 218:883-884.

Honjo S (1984) Study of ocean fluxes in time and space by bottom-tethered sediment trap arrays: a recommendation. In: Global Ocean Flux Study, Proceedings of a Workshop, Washington, D.C.: Nat. Acad. Press, pp 305-324.

Honjo S (1990) Particle Fluxes and Modern Sedimentation in the Polar Oceans. Polar Oceanography, Part B: Chemistry, Biology and Geology. Academic Press Inc., pp 687-739.

Honjo S, Hay BJ, Manganini SJ, Degens ET, Kempe S, Ittekkot V, Izdar E, Konuk YT, Benli HA (1987a) Seasonal cyclicity of lithogenic particle fluxes at a Southern Black Sea sediment trap station. In: Degens ET, Izdar E, Honjo S (eds) Particle Flux in the Ocean. Mitt. Geol. Paläont. Inst., Univ. Hamburg, Heft 62, pp 19-39.

Honjo S, Manganini SJ, Karowe A, Woodward BL (1987b) Paricle fluxes, North-Eastern Nordic Seas: 1983-1986. Woods Hole Oceanog. Inst. Techn. Rep. WHOI-87-17.

Hurd DC, Spencer DW (eds) (1991) The analyis and characterization of marine particles. Geophysical Monograph, 63, American Geophysical Union, Washington DC, 472 pp.

Ittekkot V, Manganini SJ, Guptha MVS, Desai BN, Degens ET, Honjo S (1987) Particle flux in the Arabian Sea. Eos 68:1772 (abstract).

Ittekkot V (1988) Global trends in the nature of organic matter in river suspensions. Nature 332:436-438.

Ittekkot V, Nair RR, Honjo S, Ramaswamy V, Bartsch M, Manganini S, Desai BN (1991) Enhanced particle fluxes in Bay of Bengal induced by injection of fresh water. Nature 351:385-387.

Izdar E, Konuk T, Ittekkot V, Kempe S, Degens ET (1987) Particle flux in the Black Sea: nature of the organic matter. In: Degens ET, Izdar E, Honjo S (eds) Particle Flux in the Ocean. Mitt. Geol. Paläont. Inst., Univ. Hamburg, Heft 62, pp 1-18.

Jahnke RA (1990) Early diagenesis and recycling of biogenic debris at the seafloor, Santa Monica Basin, California. Journal of Marine Research 48:413-436.

Kier RS (1988) On the late Pleistocene ocean geochemistry and circulation. Paleoceanography, 3:413-445.

Knauer G, Asper V (1989) Sediment Trap Technology and Sampling. U.S. GOFS Planning Report Number 10, 94 pp.

Lampitt RS (1985) Evidence for the seasonal deposition of detritus to the deep-sea floor and its subsequent resuspension. Deep-Sea Res. 32:885-897.

McCave IN (1984) Size spectra and aggregatgion of suspended particles in the deep ocean. Deep-Sea Res. 31:329-352.

Milliman JD, Takahashi K (ms. subm.) Carbonate and Opal Production and Accumulation in the Ocean. In: Usselman TM, Hay W, Meybeck M (eds) Global Surficial Geofluxes: Modern to Glacial. National Academy Press.

Nair RR, Ittekkot V, Manganini SJ, Ramaswamy V, Haake B, Degens ET, Desai BN, Honjo S (1989) Increased particle flux to the deep ocean related to monsoons. Nature 338:749-751.

Peinert R, Bathmann U, von Bodungen B, Noji T (1987) The impact of grazing on spring phytoplankton growth and sedimentation in the Norwegian Current. In: Degens ET, Izdar E, Honjo S (eds) Particle Flux in the Ocean. Mitt. Geol. Paläont. Inst., Univ. Hamburg, Heft 62, pp 149-164.

Platt T, Harrison WG (1985) Biogenic fluxes of carbon and oxygen in the ocean. Nature 318:55-58.

Riebesell U, Wolf-Gladrow DA (ms. subm.) The relationship between physical aggregation of phytoplankton and particle flux: a numerical model.

Riemann F (1989) Geltinous phytoplankton detritus aggregates on the Atlantic deep-sea bed. Structure and mode of formation. Marine Biology 100:533-539.

Sarnthein M, Thiede J, Pflaumann U, Erlenkeuser H, Fütterer D, Koopmann B, Lange H, Seibold E (1984) Atmospheric and oceanic circulation patterns off Northwest Africa during the past 25 million years. In: U. von Rad et al. (eds) Geology of the Northwest African Continental Margin. Springer-Verlag, Berlin, pp 545-604.

Sarnthein M, Winn K, Duplessy JC, Fontugne MR (1988) Global variations of surface ocean productivity in low and mid latitutdes: Influence on CO_2 reservoirs of the deep ocean and atmosphere during the last 21,000 years. Paleoceanography 3361-399.

Shanks AL, Trent JD (1980) Marine snow: sinking rates and potential role in vertical flux. Deep-Sea Research 27:137-143.

Sharp JH (1991) Review of carbon, nitrogen and phosphorous biogeochemistry. Reviews of Geophysics, Supplement, AGU, Washington, D.C. pp 648-656.

Silver MW, Alldredge AL (1981) Bathypelagic marine snow: deep-sea algal and detrital community. Journal of Marine Research 39:501-530.

Smetacek V (1985) Role of sinking in diatom life-history cycles: ecological, evolutionary and geological significance. Mar.Biol. 84:239-251.

Suess E (1980) Particulate organic carbon flux in the oceans - surface productivity and oxygen utilization. Nature 288:260-263.

Sundquist ET, Broecker WS (eds) (1985) The carbon cycle and atmospheric CO_2: natural variations archean to present. Geophysical Monograph, 32, American Geophysical Union, Washington DC, 627 pp.

Thiel H, Pfannkuche O, Schriever G, Lochte K, Gooday AJ, Hemleben Ch, Mantoura RFG, Turley CM, Patching JW, Rieman F (1989) Phytodetritus on the deep-sea floor in a central oceanic region of the Northeast Atlantic. Biological Oceanography 6:203-239.

Toggweiler JR (1989) Is the downward dissolved organic matter (DOM) flux important in carbon transport? In: Berger WH, Smetacek V, Wefer G (eds) Productivity of the Ocean: Present and Past, Dahlem Workshop Reports, Wiley & Sons, pp 65-83.

Urrère MA, Knauer GA (1981) Zooplankton fecal pellet fluxes and vertical transport of particulate organic material in the pelagic environment. Plankton Res. 3:369-387.

Vincent E, Berger WH (1981) Planktonic Foraminifera and their use in Paleoceanography. In: Emiliani C (ed) The Oceanic Lithosphere, The Sea, Vol. 7. John Wiley and Sons, pp 1025-1119.

Walsh JJ (1989) How much shelf production reaches the deep sea? In: Berger WH, Smetacek V, Wefer G (eds) Productivity of the Ocean: Present and Past, Dahlem Workshop Reports, Wiley & Sons, pp 175-191.

Wefer G (1989) Particle flux in the ocean: Effects of episodic production. In: Berger WH, Smetacek V, Wefer G (eds) Productivity of the Ocean: Present and Past, Dahlem Workshop Reports, Wiley & Sons, pp 139-154.

Wefer G, Fischer G, Fütterer D, Gersonde R (1988) Seasonal particle flux in the Bransfield Strait (Antarctica). Deep-Sea Research 35:1223-1234.

Wefer G, Fischer G, Fütterer D, Gersonde R, Honjo S, Ostermann D (1990) Particle Sedimentation and Productivity in Antarctic Waters of the Atlantic Sector. In: Bleil U, Thiede J (eds) Geologic History of the Polar Oceans - Arctic versus Antarctic . Kluwer Academic Publ., pp 363-379.

Wefer G, Fischer G (1991) Annual primary production and export flux in the Southern Ocean from sediment trap data. Marine Chemistry 35:597-614.

Wefer G, Fischer, G (ms. subm.) Seasonal flux patterns in some eastern Atlantic upwelling areas: relation to seasonal intensified trade wind, ITZC migrations and upwelling events.

DYNAMIC CYCLE OF DISSOLVED ORGANIC CARBON AND MARINE PRODUCTIVITY

Yoshimi Suzuki
Geochimical Laboratory
Meteorological Research Institute
Nagamine 1-1, Tsukuba, Ibaraki 305
Japan

ABSTRACT

Dissolved organic matter (DOM) has an important role in ocean chemistry, in the remineralization of inorganic nutriens, feed web by heterotrophic organisms or as ligand of trace metals. Recent measurements of dissolved organic carbon (DOC) concentration in seawater using a high temperature catalytic oxidation (HTCO) method influenced strongly the biogeochemical ocean community. The HTCO method produced substantially higher values with more possibilities to be related to the consumption of oxygen. The first diel observations of DOC concentrations were done during the bloom study of JGOFS in 1989. DOC concentrations changed from 20 μM C to 50 μM C. This change is primarily due to the physical mixing of water and to biological processes. It was observed that there is a very weak negative correlation between the DOC inventory (0-50 m depth) and primary production. It seems that the rate of decomposition of DOC or DOM affects the rate of primary production. Using a two box model the DOM downward flux was calculated. It is likely that the DOM downward flux of about 5 GtC yr^{-1} plays an important role in the oceanic carbon cycle. The turnover time of DOC is calculated to be 2.9 years in the upper ocean layers and 170 years in the deep waters.

INTRODUCTION

The study of the oceanic carbon and nitrogen cycle is fundamental for understanding the biogeochemical cycles of biologically active elements in the ocean. These cycles are a key to

NATO ASI Series, Vol. I 15
The Global Carbon Cycle
Edited by M. Heimann

the problem of the exchanges of atmospheric carbon dioxide with the ocean. The rate of carbon dioxide uptake by the ocean is controlled by seawater temperature, surface carbonate chemistry, biology and by the mixing and circulation which determine the amount of carbon transported from surface waters to deep water as organic matter (in dissolved and particulate forms) and carbonate.

The role of the ocean in the carbon cycle is characterized as follows: 1) the consumption of carbon in primary production in surface waters; 2) the enrichment of the deep waters in carbon dioxide as the result of the decomposition and dissolution of carbonate and organic matter (dissolved and particulate forms) produced in the surface water; 3) the sinking of water in polar regions, particularly in the North Atlantic, taking CO_2 with it followed by a general bottom water flow toward the equators; 4) the upwelling of some of this water in equatorial regions with a corresponding outgassing of carbon dioxide to the atmosphere and a general poleward flow of surface waters; 5) the general turbulent mixing processes which accompany the meridional circulation .

The first two features taken together are related to the biogeochemical processes in the ocean, the so-called "biological pump" [Revelle, 1990] and the "biodynamic converter" [Suzuki and Tanoue, 1991]. The biological pump is the incorporation of dissolved carbon dioxide in surface waters into living tissue or as carbonate in shells, followed by the downward transport of dead marine organic particles mostly from the euphotic zone. The biodynamic converter is defined as the relation between the flux and the rate constants of production and decomposition of organic matter (in dissolved and particulate forms) within the upper ocean layer.

Dissolved organic carbon (DOC) measured using high temperature catalytic oxidation (HTCO) [Suzuki et al., 1991] has been given much attention in its new role in the carbon cycle. According to the traditional view, it was thought that organic matter in sinking biogenic particles fuels deep sea respiration. HTCO measurements for DOC suggested that DOC concentrations are several times more abundant in surface water and twice as large in deep waters, and that there is a relationship between DOC and oxygen consumption, in contrast to what was previously thought. DOC plays an important role in the biogeochemical cycles of carbon together with POC (sinking particles).

How does DOC play an important role in the oceanic carbon cycle? If the new results of

DOC are confirmed to be correct, the oceanic DOC cycle appears more vivid than previously believed and a heretofore "missing" relation emerges consisting of a dynamic balance between inorganic carbon and oxygen utilization. It appears that DOC measured using HTCO consists not only of a refractory fraction, but also of a fraction which is more active in the oceanic carbon cycle. In order to adequately understand the cycling of organic matter, the essential question is how to fully characterize oceanic organic matter. Unfortunately, it is very difficult to analyze in detail the components of organic compounds with the present technology.

In this paper, I would like to report the significance of the "biodynamic converter" in the oceanic DOC cycle. This question is related to the marine productivity and the balance between organic and inorganic forms of oceanic carbon.

IS THERE A CHANGE OF OCEANIC DOM?

Organic matter present in sea water has various direct and indirect origins although the primary source is the primary production (Figure 1 [De Barr et al., 1989]). It is commonly thought, that DOC consists of various organic compounds with different characteristics which are circulated with different time-scales. The quantitative evaluation of the time-scales and mechanisms of the "biodynamic converter" in the organic carbon-inorganic carbon system represents a key element in the study of the biogeochemical cycling of carbon and nitrogen in the ocean. Is the DOC concentration changing with time? If it is, by how much on daily, weekly, seasonally and yearly time-scale? Up to now there exist only few time-series of DOC measurements.

Although seasonal and diel DOC data from open ocean waters measured using the wet chemical oxidation method are also scarce, Duursma [1961] found differences of 40 to 60 μM between pre-bloom (spring) and post-bloom (late autumn) samples from the upper watercolumn of the North Atlantic. Maximum values were associated with phytoplankton blooms. Holmes et al. [1967] followed the wet DOC concentration of near-shore waters off La Jolla, California, but found no regular seasonal pattern. A similar conclusion was reached by Strikland et al., [1970] who found little evidence of a correlation between the DOC concentration

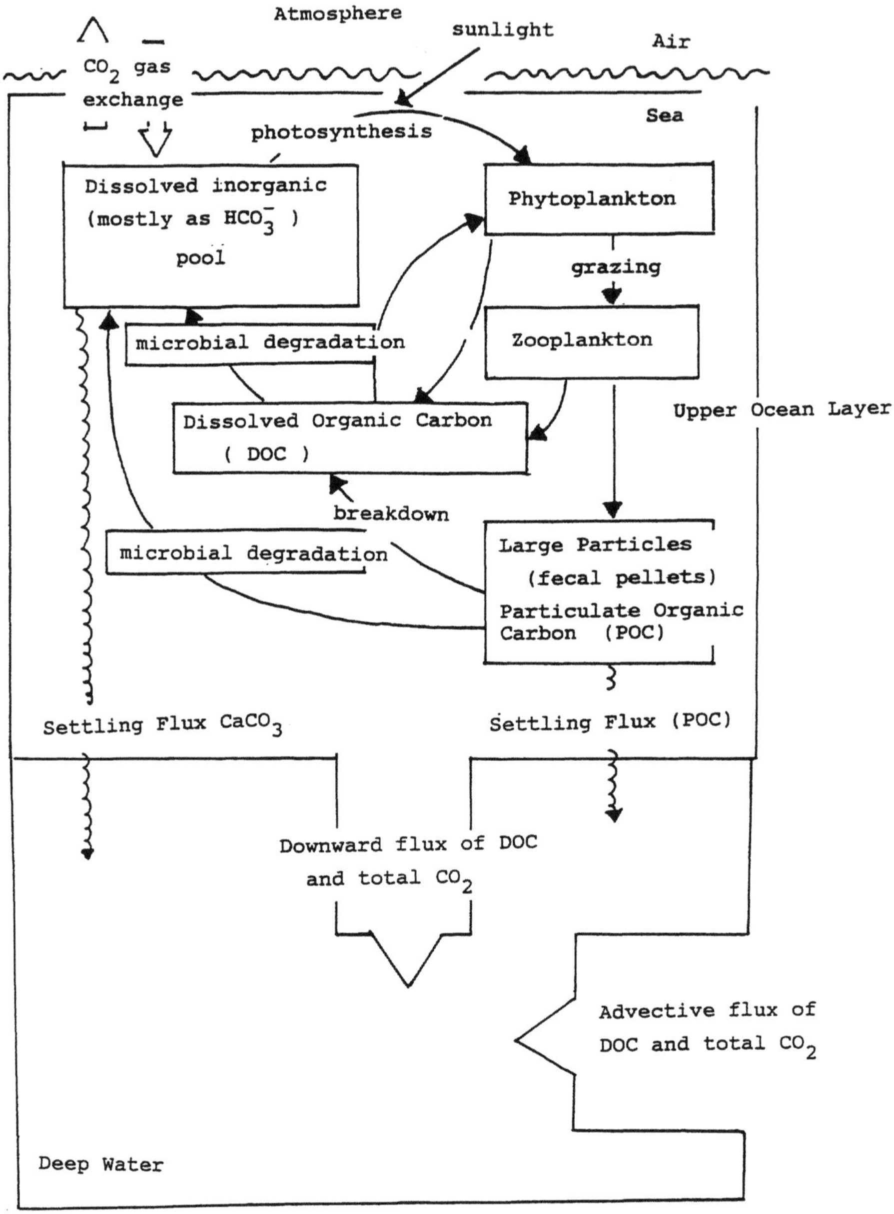

Figure 1. The simplified oceanic carbon cycle in the upper ocean ([De Barr et al., 1989], modified by the author).

and productivity at three coastal stations. But Mel'nikov and Pavlov [1978] reported that DOC concentrated in the upper 100 m of the Arctic Ocean about 2 to 3 weeks after phytoplankton blooms.

There have been a few studies of *in situ* diel dynamics of dissolved organic matter in the open sea. Some temporal patterns emerge, although the effects of physical mixing and biological patchiness cannot be totally quantified in such data. Dissolved carbohydrate, DOC, phytoplankton pigments, and size-fractionated particulated ATP during day and night were measured in the North Atlantic by Burney et al., [1979]. Their data showed that the plankton actively regulates dissolved organic matter in the open sea, and that DOC in the upper 100 m often coincided with accumulations of microzooplankton or phytoplankton. Burney et al. [1981] reported temporal variations from the Sargasso Sea where DOC changes were similar to total carbohydrate, with afternoon-evening increases of 6.7 μM at 80 to 100 m to over 100 μM at 45 m depth. It is uncertain whether these observed diel DOC changes represent biological temporal or physical spatial scale variations without performing a data analysis to separate the two processes. However, even if they are affected by mixing processes on small scales, they must primarily result from underlying biological processes because they are related to plankton communities.

Figure 2 shows the DOC concentration in the upper ocean column at the US-JGOFS NABE (North Atlantic Bloom Study) station at 47°N, 20°W as a function of time (May 18 to May 31, 1989). During the spring bloom diel changes of the DOC concentration were observed. In surface water DOC concentrations range from 170 to 220 μM. The remarkable features of the DOC time series data are patchiness and the occurrence of a sub-surface minimum. The patchiness is one of the important features of the biological contribution with shorter temporal and smaller spatial scales than the water mixing scale. The variations of the DOC concentration are between 20 and 50 μM. The depth of the sub-surface DOC concentration minimum lies between 40 and 70 m, clearly corresponding to the depth of the maximum of the nitrite concentration (Figure 3). The DOC minimum possibly is the result of microbial degradation; it is maintained by small and short scales of biological processes in a "quasi" steady state. Rau et al. [1992] reported time series data of [CO_2 (aq)] and [TCO_2] during the bloom study (April 25 to May 31, 1989). They found a decrease of TCO_2 concentrations in the mixed layer with approximately 50 μM accompanied by an increase of the oxygen concentration (about 30 μM), a decrease in nitrate (about 5 μM), phosphate (about 3 μM) and silicate

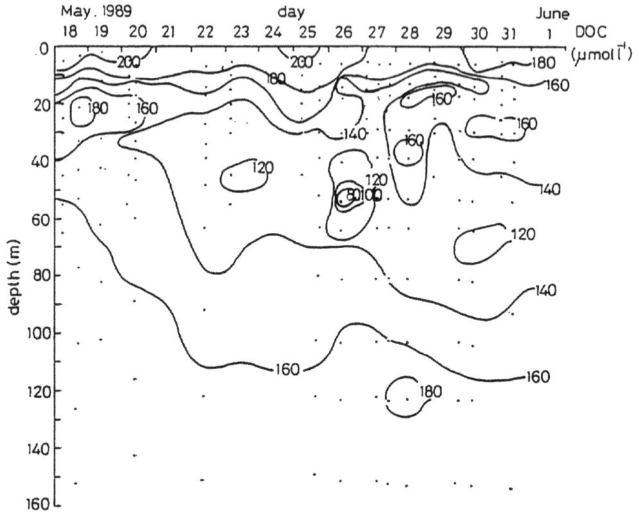

Figure 2. DOC concentrations in the upper ocean during the spring bloom at the US-JGOFS NABE station (47°N, 20°W) in the North Atlantic Ocean, May 18-31, 1989.

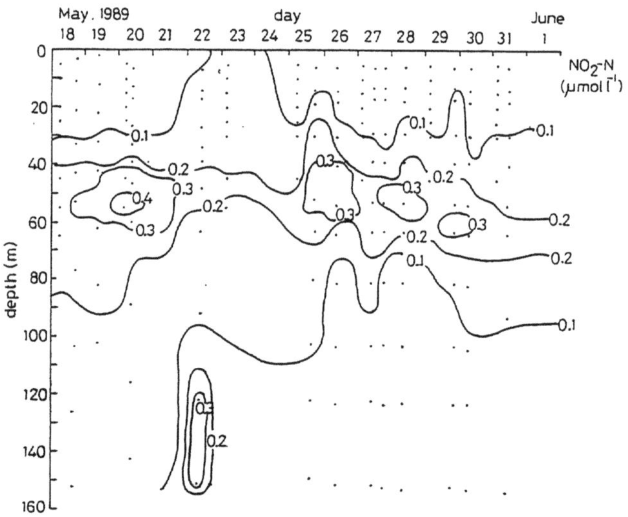

Figure 3. Nitrite (NO_2) concentrations in the upper ocean during the spring bloom at 47°N, 20°W from May 18 to May 31 1989 (measured by N. Williams).

(about 2 μM). They concluded that the decreases in inorganic carbon were due primarily to photosyntate formation. This suggests that the change of 20 to 50 μM in the DOC concentration was caused by biological processes since physical mixing processes hardly affect DOC concentrations in the mixed layer.

Figure 4 shows the diel changes of DOC concentrations at different depths (5 m, 10 m, 20 m and 50 m). Except on May, 19 and May 26-27 DOC concentrations remain within ±20 μM C. The large change around May 26-27 is due primarily to strong convection in the mixed layer (based on other hydrographic data), but not entirely. The temporal patterns of the DOC concentration at each depth are somewhat different which suggests that these were caused by biological processes. Unfortunately, an unambiguous, quantitative separation of the two factors controlling the temporal variations of DOC can not be made at the present time.

Two different clues hint at the mechanism which cause DOC concentration variations in the mixed layer: the magnitude and the steepness or patchiness of the DOC variations. Apparently, it seems that these contradict each other. If changes in DOC were due primarily to mixing by diffusion and advection of sea water with a time scale of a few days then the DOC

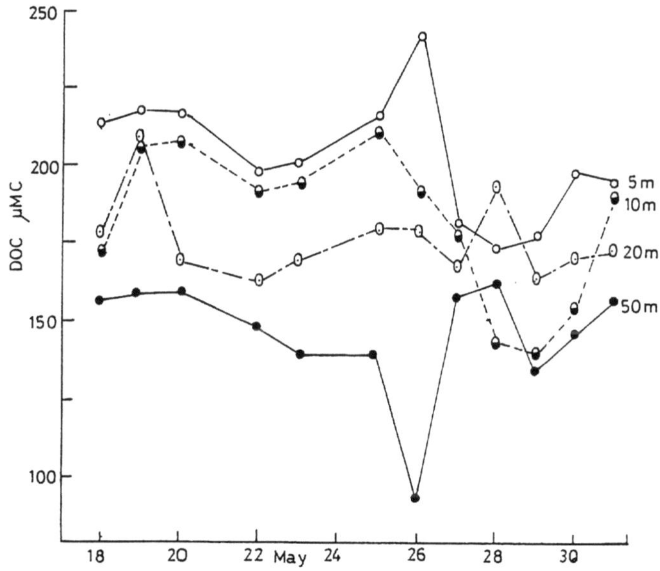

Figure 4. Diel change of DOC concentrations at four different depths (5 m, 10 m, 20 m and 50 m) during the spring bloom at 47°N, 20°W from May 18 to May 31 1989 in the upper ocean of the North Atlantic.

concentration in the 50 m depth mixed layer would be almost homogeneous. This is not the case, thus suggesting that a fraction of DOC varies bioactively by production and decomposition of DOM. Kirchman et al. [1991] conducted experiments to estimate DOC turnover rates in the upper ocean during the spring phytoplankton bloom in the North Atlantic ocean. According to their results, potential turnover rates ranged from 0.025 to 0.363 per day, and were consistent with bacterial biomass production and uptake of dissolved nitrogen (ammonium, nitrate and urea), suggesting even higher rates with roughly 20-40% turning over in days.

In order to determine if an increase in the mixed layer pool of DOC occurred as a consequence of the spring bloom, we have calculated the daily inventory of DOC for the upper most 50 m (Figure 5) [Peltzer et al., 1992]. On May 21, 1989 a large storm passed through the study area. The total organic carbon (TOC = DOC + POC) inventory was 8.4 M C m^{-2} and the DOC inventory was 7.81 M C m^{-2}. The standard deviation of the DOC inventory was about 5%. A change of 5% of the DOC inventory during one day (i.e. 400 mM D m^{-2} day^{-1}) is several times greater than the average daily primary productivity based on C-14 incubations [Martin et al., 1991]. These variations contain contributions by physical and biological processes. Emerson et al. [1991] estimated the net biological oxygen production in the subarctic ocean. He discussed

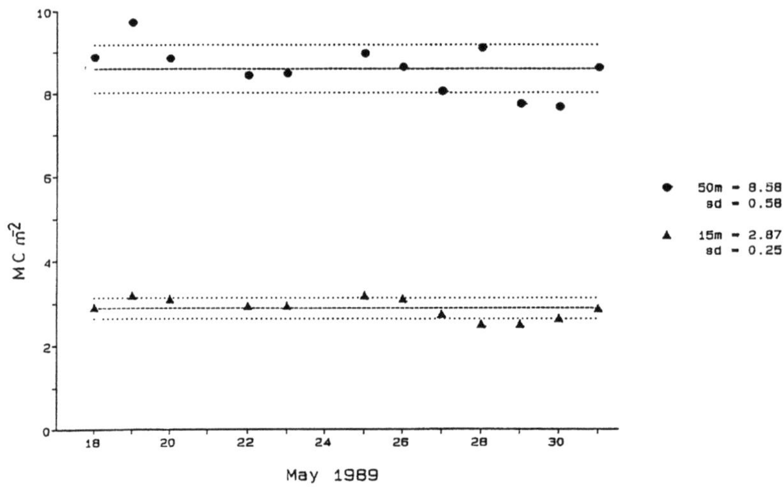

Figure 5. Mixed layer inbentories (0 to 50 m and 0 to 15 m depth) of DOC at 47°N, 20°W as a function of time. Dashed lines represent the means calculated for the time period May 22-31, 1989 [after Peltzer et al., 1992].

the oxygen mass balance which requires a net carbon production in the euphotic zone of 12 mM C m^{-2} day^{-1}, which is 20 to 30% of the primary productivity values determined by the C-14 method. Based on these values the biological contribution to the total DOC budget would only be 0.2%. How can then the apparent imbalance between changes in DOC inventory and primary productivity be explained?

DYNAMIC CYCLE OF DOC AND MARINE PRODUCTIVITY

Dissolved organic carbon (DOC) in seawater constitutes the largest exchangeable carbon reservoir on earth. Peltzer et al. [1992] estimated a global reservoir of DOC measured by HTCO of 1630 GtC, more than double the size of the inventory estimated by Sundquist [1985]. This estimate and that of Martin and Fitzwater [1992] agree very well. A very important question is how much of the biologically labile DOC is in the total DOC pool. Peltzer et al. [1992] estimated the size of the labile DOC pool between 60 and 110 GtC, while the difference betwee the total DOC pool determined by the HTCO method and the inert DOC pool (as determined by the old DOC measuring technique) is 990 GtC. This represents a 9-16 fold increase in the size of the biologically labile DOC pool. It is expected that this biologically labile DOC pool is strongly related to the marine productivity.

Figure 6 shows the plot of TOC (= POC + DOC) and DOC inventories versus net primary production measured by Martin [unpublished data] during the spring bloom study in 1989 in the North Atlantic ocean. A very weak negative or no correlation at all is observed between DOC inventory and net primary production. The DOC inventory on May 20 was 7.61 M C m^{-2} which diminished to 7.03 M C m^{-2} on May 23, while the net primary production increased from 48.3 mM C m^{-2} day^{-1} to 108.4 mM C m^{-2} day^{-1}. The decrease of DOC is 290 mM C m^{-2} day^{-1} corresponding to 43.9 mM N m^{-2} day^{-1} or 0.88 μM N day^{-1} (assuming a Redfield ratio of 6.6). I propose a hypothesis that DOM is recycled with a short turnover time and inorganic nutrients regenerated from DOM are taken up by the assimilation process. This suggests that the rate constant of conversion (the biodynamic converter) from inorganic to organic and from organic to inorganic forms of carbon is a very important factor for determining the magnitude of primary production.

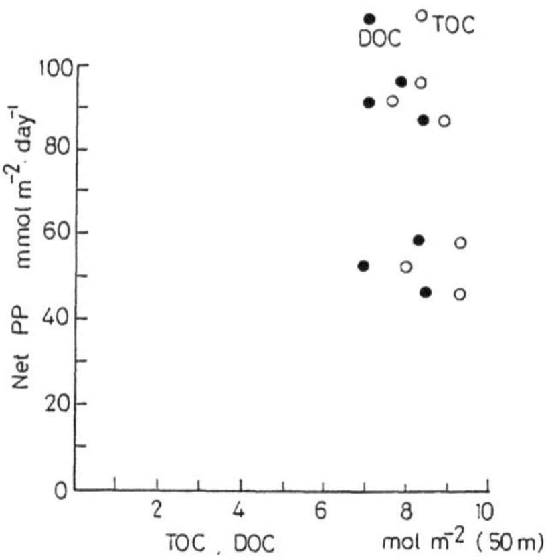

Figure 6. Net primary productivity versus mixed layer inbentories of TOC and DOC during the spring bloom at 47°N, 20°W. (Productivity data measurements from Martin et al. [1991]). Open circles: TOC, closed circles: DOC.

Figure 7 shows the surface distribution of DOC concentrations in the western and central Pacific Ocean. The DOC concentrations range from 118 tp 258 µM with 189 µM as an average value. The highest values of the equatorial Pacific were observed in the South equatorial Pacific (5°S to 13°S), corresponding most likely to an envelope of DOC concentrations in a high productive area. The high values of DOC were also found at around 10°N in the western North equatorial current, where the surface mixed layer is relatively shallow. At the equator, DOC concentrations are relatively low and increase from the east to the west, reflecting the stronger equatorial upwelling in the eastern and central areas compared to the western side. Lower values of DOC are also observed at higher latitudes (north of 40°N) ranging from 118 to 138 µM, with an average value of 126 µM. The highest values were observed around 10°S and 5°S, 150°W with about 270 µM. It seems that the concentrations of DOC are different in the different marine areas, and, if these results are accepted, that the higher concentrations of DOC are observed in the higher primary productivity areas [Berger, 1989]. As discussed in the previous section, DOC concentrations are the result of both, physical mixing and biological processes. Which is more dominant? The contributions of both processes were estimated to clarify the role of the marine biota in controlling DOC.

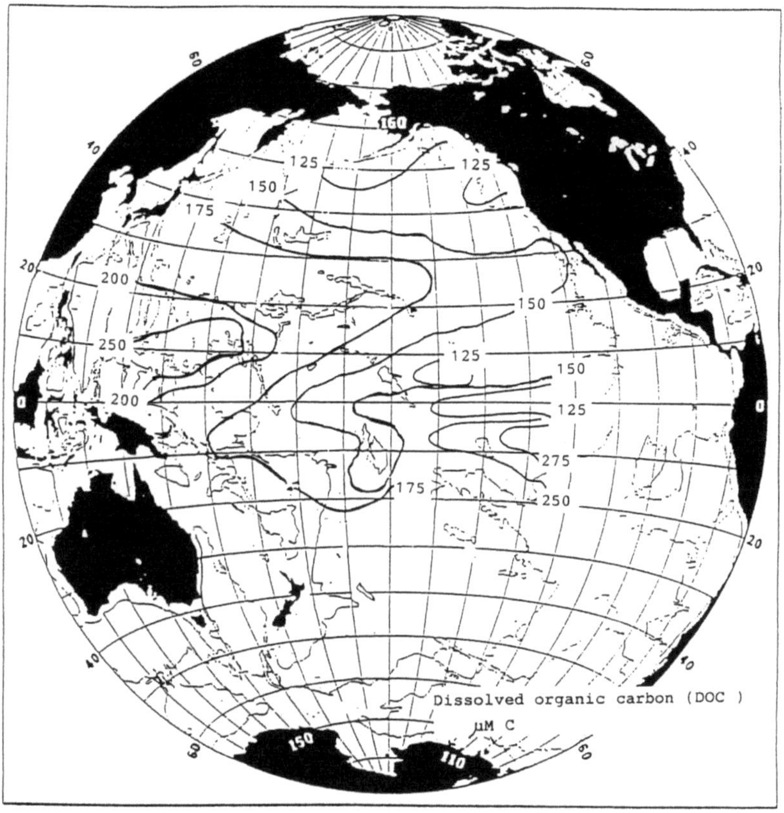

Figure 7. Surface distribution of the DOC concentration in the western and central Pacific Ocean (in µM C).

Vertical profiles of DOC measured using the HTCO method obtained at the equator, 160°W (Japan-TOGA cruise, in January and February of 1989 and 1990), together with the sigma-t profiles are shown in Figure 8. A comparison of the DOC profiles between 1990 and 1989 shows that the station at 160°W experienced remarkable differences in the upper ocean layer. The La Niña event extended into the central and eastern Pacific in January 1989 inducing around the equator a monthly mean surface temperature anomaly of -1 to -2°C. Nutrient concentrations were also higher in 1989 than in 1990, reflecting an enhanced upwelling, particularly at 160°W. DOC concentrations were reduced by about 20 µM C during the La Niña event in 1989. This reduction of DOC in the surface layer is due to the stronger physical mixing and the enhanced biological contribution during the La Niña event.

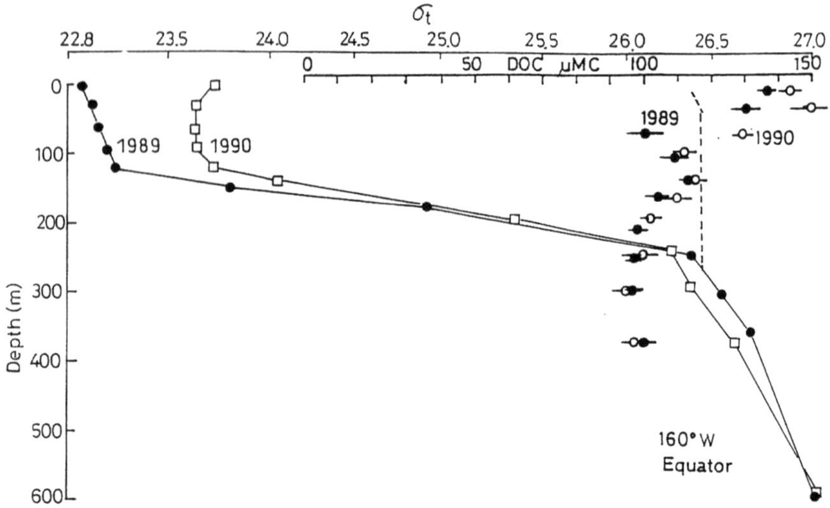

Figure 8. Vertical distributions of DOC and sigma-t at the equator, 160°W in 1990 and 1989. Closed circles: 1989, open circles: 1990. 1989 corresponds to a La Niña event.

Assuming that the DOC, nutrients, dissolved oxygen, and physical parameters (temperature, salinity and sigma-t) profiles of 1990 represent normal patterns, the relative contributions from the physical and biological processes by the la Niña event of 1989 were determined, based on mixing ratios calculated from sigma-t.

The results of the calculation are shown in Table 1. The values of each component

Depth	DOC			NO_3			DO			f_1
m	calc.	obs.	diff.	calc.	obs.	diff.	calc.	obs.	diff.	
0	114	136	+22	6.8	5.5	-1.3	191	199	+8	0.76
29	118	129	+11	6.4	5.2	-1.2	193	197	+4	0.78
57	118	98	-20	6.3	5.7	-0.6	193	195	+2	0.78
86	117	110	-7	6.4	5.8	-0.6	193	194	+1	0.80
114	116	116	0	6.6	6.6	0	192	192	0	0.82

Table 1: Physical and biological contribution to the concentration changes of DOC, NO_3 and dissolved oxygen (DO), during the La Niña event (1989) (in μM C). The calculated values were obtained assuming pure physical mixing with deeper waters. f_1 denotes the fraction of deeper water (150 m) as determined from the observed temperature anomaly. The difference between calculated and observed values reflect the biological contribution.

(DOC, NO_3, DO and sigma-t) of 1990 were used for the computation. About 90% of the variations of the DOCconcentration is effected by the physical mixing of water masses and the net biological contribution is about 10%, ranging from + 20 to - 20 μM C. According to this calculation, DOC possibly is produced in the surface layer and likely is consumed in the subsurface layer. Nutrients are taken up from sea water and dissolved oxygen is produced.

The characteristics of the cycling of DOC in the ocean are dependent on the nature of oceanic DOM, i.e. its source, molecular nature and fate. It is well known, that parts of DOC are labile and are recycled with a very short time scale. The question is the size and the turnover times of the DOC pool. According to the previous estimates, the size of the labile fraction of DOM is about 60-110 GtC [Peltzer et al., 1992], while if the HTCO determined DOC values are correct, it is estimated that the biologically labile DOC pool increases 9-16 fold (to about 990 GtC).

Suzuki and Tanoue [1991] discussed the difference between HTCO-DOC and persulfate oxidation-DOC, called "bioreactive DOC", which is small in the continental margins with high primary productivity. It appears that the cause is the rapid consumption or decomposition of DOM by marine heterotrophic activity and less likely physicochemical removal processes. It appears as if the bioreactive DOC is recycled with much shorter turnover times in shelf waters than in the open ocean. Similar preliminary results have been observed in tropical coral reefs (unpublished data by Suzuki).

Little is known of the chemical and physical nature of bioreactive DOC measured using HTCO. Generally less than 30% of DOC in seawater can be explained by the concentration of carbohydrates, organic acids, proteins and some other identified substances such as humic and fulvic acids [Williams, 1975]. What is the unidentified DOM? If the entire bioreactive DOC constitutes the biologically labile fraction, this unknown fraction may be defined as the "sub-labile" fraction. Possibly this "sub-labile" fraction has a fast (not quick, a fraction with a quick turn-over time is labile DOC) turnover time. Therefore, oceanic DOM tentatively is divided into three main groups: labile, sublabile and refractory. It is important to clarify the molecular nature of "sublabile DOC". This may constitute a new driving force of the oceanic ecosystems.

OCEANIC CARBON BALANCE

At the present time, DOC data measured using the HTCO method are scarce and the exact horizontal and vertical distribution of DOC are not well known. Here, the contemporary oceanic carbon balance was calculated using a two-box model for the Pacific Ocean. The model is illustrated in Figure 9. The continuity of water mass requires

$$k_1W_1 = k_2W_2 \tag{1}$$

but

$$C_1k_1W_1 \neq C_2k_2W_2 \tag{2}$$

The balance of the upper (0-100 m) layer is given by:

$$dC_1/dt = C_2k_2W_2 - C_1k_1W_1 - P + R_1 - D + I - O \tag{3}$$

and the balance of the second layer:

$$dC_2/dt = C_1k_1W_1 - C_2k_2W_2 + d + R_2 - S \tag{4}$$

The symbols in equations (1)-(4) are: kW, the mixing rate of water, W, the amount of water, C, the concentration, P, the primary production, R, the amount of regeneration, I, the gross input of carbon into the ocean from the atmosphere, O, the gross carbon flux from the ocean to the atmosphere, D, the amount of carbon sinking in particles, d, the amount of dissolution, S, the amount of deposition. The index 1 refers to the upper and index 2 to the lower layer.

In the steady state we assume I = O and d = S. The mixing rate of water between the two boxes is one of the key factors. Broecker and Peng [1982] estimated upwelling rates ranging from 5.7 m yr^{-1} in the temperate latitudes and 10 m yr^{-1} in the equatorial region. Assuming that the vertical distribution of only inorganic carbon and inorganic plus organic carbon can be maintained at steady-state by the mixing rate of water and primary productivity, the relationship between mixing rate and primary production was calculated using the two-box model. Figure 10 shows the resulting positive relationship between the mixing rate and net primary production (NPP). Accepted values for the global NPP are between 50 to 60 Gt, which correspond to a mixing rate of 5 m yr^{-1} for the inorganic system and to a mixing rate of 10 m yr^{-1} for the combined inorganic-organic system. Based on this a value of 10 m yr^{-1} is chosen for the

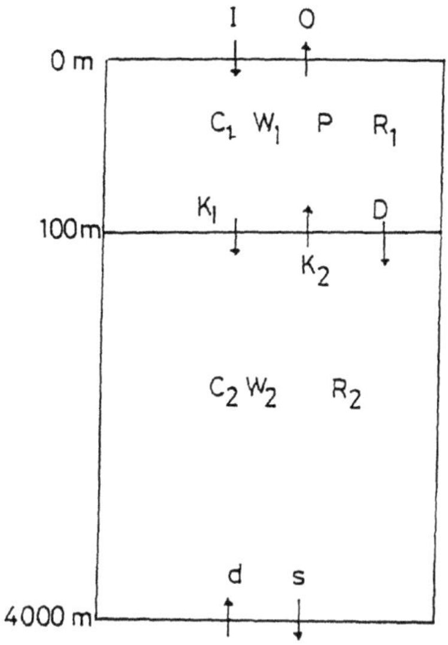

Figure 9. Illustration of the two box model for the calculation of the oceanic carbon balance. The symbols are explained in the text.

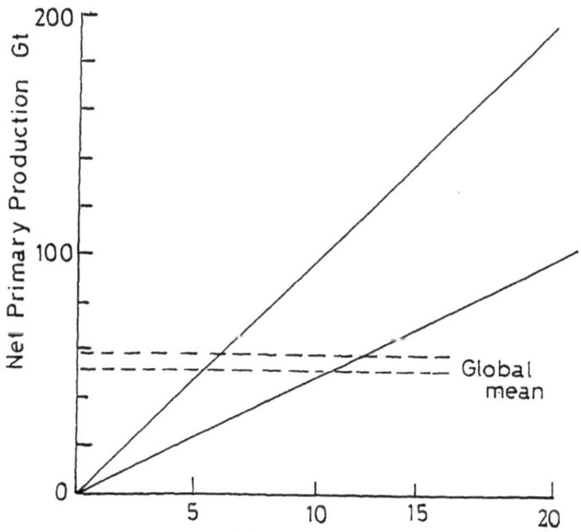

Figure 10. The relationship between net primary production (in GtC) and the mixing rate of water (in m yr^{-1}) for two cases: inorganic carbon only and combined organic and inorganic carbon.

mixing rate in the oceanic carbon two-box model.

Figure 11 illustrates the fundamental principles in the historical view and in the revised view of the oceanic carbon cycle. In the new view DOM emerges as a more important element and the "biodynamic converter" between inorganic and organic forms of carbon consitutes a key element in the oceanic carbon cycle.

The resulting contemporary oceanic carbon balance is given as follows. The amount of net primary production is assumed as 50 GtC yr^{-1}. About 80% of NPP is recycled in the upper ocean. Only 1% (0.5 GtC yr^{-1}) of NPP is transported to the ocean floor in the form of sinking particles. A downward flux of 4.5 GtC yr^{-1} in the form of sinking particles and 5 GtC yr^{-1} by physical downward mixing feeds the DOC pool in the lower box. This net downward flux of DOC by physical processes (i.e. 5 GtC yr^{-1}) is larger than previously thought. The turnover time of DOC is calculated as 2.9 yr in the upper ocean and 173 yr in the deep layer.

Recently, Toggweiler [1989] derived a most realistic simulation of the distribution of phosphate as compared to GEOSECS data when half of the new production is put into a dissolved organic phosphorus pool with a breakdown rate of DOM of 1/200 yr^{-1}. Furthermore, Bacastow and Maier-Reimer [1990] reported that a significantly better agreement between simulated and observed distributions of oceanic phosphate and oxygen is obtained if a large

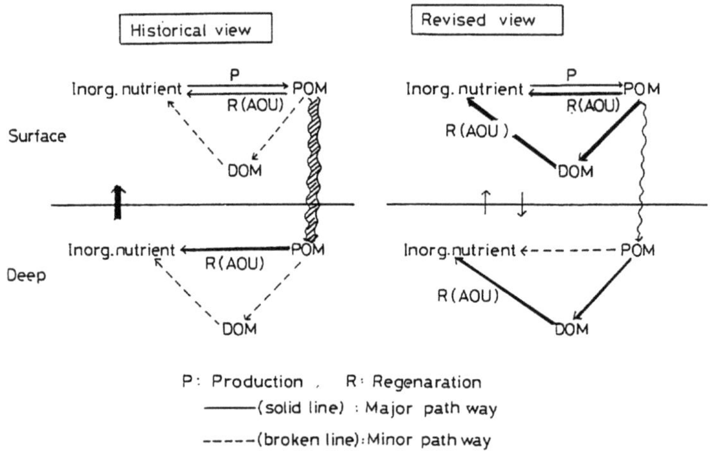

Figure 11. The tradidional and the revised view of the oceanic carbon cycle.

part of the new production flux is carried by dissolved organic carbon in their ocean carbon cycle model. In this latter model study the time constant for the consumption of DOC was taken to be 50 yr for the whole ocean. These calculations agree rather well. It appears that the global average of the turnover time of DOM is less than a few hundred years.

CONCLUSIONS

The oceanic carbon cycle possibly is controlled by the "quasi steady-state" processes of production and consumption and the mixing by the water circulation. The "biological pump" recently has been given much attention in connection with the problem of the "missing" atmospheric carbon dioxide. The strength of the "biological pump" is related to the net primary production. As shown in Figure 10, however, an increase of net primary productivity is limited by the vertical water mixing rate. Is there a possibility that the latter increases with time with a corresponding increase in net primary production? Or can the ocean be driven out of the steady-state?

What has changed during the recent years is the gradient between the atmosphere and the oceans. The concentration of carbon dioxide in the atmosphere has been increasing; certainly it is greater today than 200 years ago. Is there a possibility to inject "excess" atmospheric carbon dioxide into the ocean? Does the "biological pump" give us any satisfactory explanation of the oceanic carbon cycle and how it is related to the air-sea exchange of CO_2?

It is likely that a key to solve these questions lies in the clarification of the "quasi" steady-state on the various temporal and spatial scales maintained by the "biodynamic converter" in upper ocean carbon cycling. In other words, the net downward flux of DOC from the upper ocean to deep water has to be determined accurately. A value of 5 GtC yr^{-1} as calculated with the two-box model is much larger than previously thought. The new aspect of the behaviour of oceanic DOC is its more dynamic nature on various timescales. I think that a quantitative explanation of the enigma of the "biodynamic converter" (i.e. the exchange of DIC to DOC and of DOC to DIC) is required and that this question is deeply related to questions such as role of the marine productivity and the "missing" sink of excess CO_2.

ACKNOWLEDGMENT

I would like to thank Dr. P. G. Brewer (MBARI) for his help and encouragement and also Dr. E. T. Pelzer (WHOI) for his help. I would also like to thank H. Ducklow, L. Lopz, R. Williams, N. Williams (SIO-PACODF), J.H. Martin, S. Fitzwater and C. Hunter (MLML) for their help in obtaining seawater samples.

REFERENCES

Bacastow R, Maier-Reimer (1991) Dissolved organic carbon in modeling oceanic new production. Global Biogeochemical Cycles 5:71-85

Berger WH (1989) Global maps of ocean productivity. In: Berger WH, Smetacek VS, Wefer G (eds) Productivity of the ocean; present and past. Dahlem Workshop, Berlin, John Wiley & Sons, New York. pp 429-455

Broecker WS, Peng T-H (1982) Tracers in the sea. ELDIGIO Press, Columbia University, Palisades, New York

Burney CM, Johnson KM, Lavoie DM, Sieburth JMcN (1979) Dissolved carbohydrate and microbial ATP in the North Atlantic: concentrations and interactions. Deep Sea Res. 26A:1267-1274

Burney CM, Davis PG, Johnson KM, Sieburth JMcN (1981) Dependence of dissolved carbohydrate concentrations upon small scale nanoplankton and bacterioplankton distributions in the western Sargasso Sea. Marine Biology 65:289-292

De Barr HJW, Fransz HG, Ganssen GM, Gieskes WW, Mook WG, Steel JH (1989) Towards a joint global ocean flux study: Rationale and objectives. In: Oceanography 1988, UNAM Press, Mexico DF. pp 11-34

Duursma EK (1961) Dissolved organic carbon, nitrogen and phosphorus in the sea. Netherland J. Sea Reseach 1:1

Emerson S, Quay P, Stump C, Wilbur D, Knox M (1991) O_2, Ar, N_2 and Rn-222 in surface waters of the subarctic ocean: Net biological O_2 production. Global Biogeochemical Cycles 5:46-69

Holmes RW, Williams PM, Eppley RW (1967) Red water in La Jolla Bay, 1964-1966. Limnology and Oceanography 12:503-512

Kirchman DL, Suzuki Y, Garside C, Ducklow HW (1991) High turnover rates of dissolved organic carbon during a spring phytoplankton bloom. Nature 352:612-614

Martin JH, Fitzwater S, Hunter C (1991) Primary productivity data. In: Slagle R, Heimerdinger G (eds) Joint Global Ocean Flux Study, data report North Atlantic Bloom Experiment April-July 1989

Mel'nikov IA, Pavlov GL (1978) Characteristics of organic carbon distribution in the waters and ice of the Arctic Basin. Oceanology 18:163-168

Peltzer ET, Brewer PG, Suzuki Y, Sugimura Y (1992, submitted) Measurement of dissolved organic carbon in seawater by high temperature catalytic oxidation. J. Marine Research

Rau GH, Takahashi T, Des Marais DJ, Repeta DJ, Martin JH (1992) The relationship between $\delta^{13}C$ of organic matter and $[CO_2(aq)]$ in ocean surface water: Data from a JGOFS site in the northeast Atlantic Ocean and a model. Geochimica et Cosmochimica Acta 56:1413-1419

Revelle R (1990) Letter in Forum section.Issues Sience Technology 7(No 2):21-22

Strickland JDH (1970) Introduction. In: Steele JH (ed) Marine food chains. Oliver and Boyd, Edinburgh. pp 3-5

Suzuki Y, Tanoue E (1991) Dissolved organic carbon enigma - Implication for ocean margin. In: Mantoura RFC, Martin JM, Wollast R (eds) Ocean margin processes in global change. Dahlem Workshop, Berlin, John Wiley & Sons, New York. pp 18-23

Suzuki Y, Tanoue E, Ito H (1992) High temperature catalytic oxidation method for dissolved organic carbon determination in seawater - examination and improvement. Deep Sea Research 39:185-198

Toggweiler JR (1989) Is the downward dissolved organic matter (DOM) flux important in carbon transport? In: Berger WH, Smetacek VS, Wefer G (eds) Productivity of the ocean; present and past. Dahlem Workshop, Berlin, John Wiley & Sons New York. pp 65-83

Williams PJLeB (1975) Biological and chemical aspects of dissolved organic material in sea water. In: Riley JP, Skirrow G (eds) Chemical Oceanography vol 2, Academic Press. pp 301-363

INTERHEMISPHERIC TRANSPORT OF CARBON THROUGH THE OCEAN

W. S. Broecker and T.-H. Peng[1]
Lamont-Doherty Geological Observatory
of Columbia University
Palisades, NY 10964

ABSTRACT

A case is made that the Atlantic's conveyor circulation carries roughly 0.6 gigatons of carbon (as ΣCO_2) from the northern to the southern hemisphere. The return transport of this carbon (as CO_2) through the atmosphere would maintain a 1.6 ppm south to north gradient in atmospheric CO_2 content. Such a gradient would go a long way toward explaining the current lower than expected fossil fuel CO_2 supported north to south atmospheric gradient and hence would alleviate the necessity for the large northern hemisphere terrestrial sink proposed by Tans et al., 1990.

During the 1980s, atmospheric CO_2 partial pressures at high latitudes in the northern hemisphere were on the average three μatm greater than that over Antarctica. By contrast, atmospheric circulation models which account for the distribution of fossil fuel CO_2 between air and sea predict a difference of about six μatm (Tans et al., 1990). From verified atmospheric mixing, models show that the current gradient allows a crossequatorial transport of CO_2, an amount approximately equal to that required to produce the observed annual increase in the southern hemisphere atmospheric CO_2 inventory. This requires that uptake of CO_2 by the southern hemisphere combined land and sea surface be close to zero. Tans et al. (1990) interpret this result as an indication that the uptake of fossil fuel CO_2 by the ocean is considerably smaller than that calculated from ocean models. If they are correct, then the amount of excess CO_2 accounted for in the atmosphere-ocean reservoir is only about one half of the combined production through burning of coal, oil and natural gas, and the cutting of

[1]Environmental Sciences Divison, Oak Ridge National Laboratory, Oak Ridge, TN 37830

NATO ASI Series, Vol. I 15
The Global Carbon Cycle
Edited by M. Heimann

forests. As the only other reservoir of significance is the terrestrial-biosphere, this gap between production and inventory points to an enhancement in the growth rate of terrestrial biota driven by CO_2 fertilization. This enhancement must be large enough to produce increases in the mass of the wood and soil humus reservoirs which more than compensates for losses by forest cutting. Furthermore, in order to explain the low magnitude of the observed interhemispheric gradient, this enhancement must take place in the northern hemisphere temperate region. Not only does the magnitude of the growth enhancement coefficient required by Tans et al. (1990) to achieve a balance in the carbon budget appear to stretch biological credibility, but also the reduction in ocean uptake they call upon appears to be incompatible with the distribution of anthropogenic tracers (i.e., bomb ^{14}C and ^{3}H) in the sea.

Heimann and Keeling (1986), Keeling and Heimann (1986) and Keeling et al. (1989) have a quite different explanation for the lower than expected gradient. They call upon a reverse atmospheric gradient prior to the industrial revolution and claim that the release of fossil fuel CO_2 by industrialized nations of the northern hemisphere did not become great enough to balance this natural net northward transport until the mid 1950s. They use, as support for their hypothesis, the evolution of the atmospheric CO_2 content difference between the Mauna Loa, Hawaii and McMurdo, Antarctica. This difference has grown from near zero values in the late 1950s to nearly three ppm in the 1980s. They also point to the interhemispheric carbon isotope composition gradient (in atmospheric CO_2) as being consistent with their interpretation (this conference).

The purpose of this paper is to evaluate the Heimann and Keeling proposal that a south to north interhemispheric transport of about one gigaton of carbon occurred in the preindustrial atmosphere. This northward transport of carbon through the atmosphere would have to have been compensated by a southward transport of carbon through the ocean. For example, were this carbon carried by the 20 Sverdrup flux of North Atlantic Deep Water (NADW) which forms in the northern Atlantic and flows the entire length of the Atlantic Ocean into the Antarctic Ocean, then this water would have to carry 130 μmol/kg more ΣCO_2 than that in the return flow water which balances this export.

A measure of the amount of carbon either gained or lost by any given sea water sample through exchange of CO_2 with the atmosphere can be obtained by subtracting the ΣCO_2 content expected if no interchange across the air-sea interface occurred from the actual ΣCO_2 concentration. Therefore:

$$\Delta\Sigma CO_2 = \Sigma CO_2^{OBS} - \Sigma CO_2^{EXP}$$

The expected total dissolved inorganic carbon content, ΣCO_2^{EXP}, takes into account differences created by the cycling of organic material and $CaCO_3$ within the sea and by the addition and removal of fresh water from the sea surface. It is calculated as follows:

$$\Sigma CO_2^{EXP} = 1900 \frac{S}{35.00} + \frac{1}{2}\left[ALK + NO_3 - 2310 \frac{S}{35.00}\right] + 127\ PO_4$$

where S, ALK, NO_3, and PO_4 are the observed concentrations for the water sample of interest. The three constants are a reference salinity (35.00‰), a reference ΣCO_2 concentration (1900 µmol/kg), and a reference alkalinity (2310 µeq/kg). These constants are chosen as to yield a zero point for the $\Delta\Sigma CO_2$ scale close to the median for surface waters in the ocean. The multiplier for PO_4 (127) is the carbon to phosphorus ratio in marine organic matter (Peng and Broecker, 1987). As shown by Takahashi et al. (1985) and by Broecker et al. (1985), the ratio of O_2 consumption to PO_4 release in the interior of the ocean averages 175 moles/mole. As the N/P ratio in marine organic matter and in sea water averages 16, the amount of oxygen consumed by combusting C and H in the organic matter is 143 moles per mole of PO_4 released (i.e., 175-2x16). Hence the upper limit on the molar C/P ratio is 143 (assumes no bound H in the organic matter). Takahashi et al. (1985) set a lower limit on the molar C/P ratio in organic material of 100 based on $CaCO_3$ corrected regressions of ΣCO_2 versus PO_4 in thermocline waters. These regressions yield only a minimum C/P ratio because of the presence of fossil fuel CO_2; low PO_4 thermocline waters are younger than high PO_4 waters and hence contain more anthropogenic CO_2. The choice of a value within the 100 to 143 range is not well constrained. If the average C/N ratio of 8 reported found by Suzuki (this volume) for the dissolved organic carbon in mid depth water is adopted, a C/P ratio of 128 is obtained. In this case, the oxygen demand by organic bound hydrogen would be 15 molecules per phosphorus atom (i.e., 143-128) and the hydrogen to carbon ratio in the organic material would be 0.47 (i.e., [4x15]/128). Fortunately, the uncertainty in our choice of the C/P ratio does not have an important influence on the conclusions of this study.

The measurements of salinity, NO_3, PO_4, ΣCO_2, and alkalinity used here were made as part of the GEOSECS Atlantic (1972-3), GEOSECS Pacific (1973-4), GEOSECS Indian (1978), TTO North Atlantic (1981), Hudson (1982), TTO Tropical Atlantic (1982-3), and SAVE South Atlantic (1987-9) expeditions. As can be seen from Table 1, where overlap occurs, the agreement among the results from different expeditions is excellent.

Table 1. Comparison of the fossil fuel CO_2 corrected $\Delta\Sigma CO_2$ and $p_{CO_2}^{EXP}$ results obtained on three different expeditions in the North Atlantic averaged by latitude belt. As can be seen, the differences between the expeditions are small. Also no large winter-summer difference is seen at high latitudes.

Latitude Belt	GEOSECS Summer			TTO-NAS Summer			HUDSON Winter		
°N	No. Stns.	$\Delta\Sigma CO_2^*$ µmol/kg	$p_{CO_2}^{EXP}$ µatm	No. Stns.	$\Delta\Sigma CO_2^*$ µmol/kg	$p_{CO_2}^{EXP}$ µatm	No. Stns.	$\Delta\Sigma CO_2^*$ µmol/kg	$p_{CO_2}^{EXP}$ µatm
80-70	0	-	-	11	110	115	41	133	143
70-60	0	-	-	15	94	153	8	102	177
60-50	0	-	-	40	97	157	0	-	-
50-40	0	-	-	23	101	167	0	-	-
40-30	6	33	237	34	48	233	0	-	-
30-20	6	27	260	17	12	265	0	-	-

*Fossil fuel corrected

As can be seen in the plot of $\Delta\Sigma CO_2$, values for surface waters in the southern hemisphere ocean are plotted against water temperature (Figure 1). Both polar and tropical waters have similar values while temperate waters have more positive values. In Figure 2 are shown the $\Delta\Sigma CO_2$ values for surface waters from the northern hemisphere portion of the Atlantic Ocean are plotted against water temperature. Included are both measurements made at high latitude during winter months (Hudson, 1982) and during summer months (Transient Tracers in the Ocean, 1981). Note that no temperate maximum occurs; rather, the ΣCO_2 values increase steadily with decreasing temperature. The most positive $\Delta\Sigma CO_2$ values in the world ocean are found in northern Atlantic. As seen in Figure 3, while fewer measurements are available, the situation in the northern hemisphere portion of the Pacific is similar to that for the southern hemisphere.

It must be kept in mind that the ΣCO_2 content of surface water has increased over the last 100 years because of the buildup of the rise in the atmosphere's CO_2 content. The extent of the buildup depends on the temperature of the water and the year of collection. The thermodynamic buffer factor, R, defined by the following equation:

$$\frac{1}{R} = \frac{\Delta\Sigma CO_2/\Sigma CO_2}{\Delta pCO_2/pCO_2}$$

ranges from about 9 for tropical waters to about 14 for polar waters. Because of this, the extent of buildup is larger for warm than for cold waters. The ΣCO_2 increases for the Atlantic samples expected for equilibrium between atmosphere and surface at the time the water sample was collected are shown in Figure 4. Because of the finite rate of CO_2 exchange between ocean and atmosphere, for any given surface water actual buildup will be somewhat less than that for chemical equilibrium between air and sea. Tracer-calibrated ocean-uptake models show that the degree of equilibration averages about 85% of the equilibrium value. Hence a reasonable estimate of the preindustrial $\Delta\Sigma CO_2$ value can be obtained by subtracting the .85 times equilibrium increase from the observed ΣCO_2.concentration. Corrected ΣCO_2 values obtained in this way for the Atlantic are shown compared with the uncorrected values in Figure 5. As can be seen, this correction does not alter the pattern of $\Delta\Sigma CO_2$ values. A map of the fossil fuel CO_2 corrected surface water values is shown in Figure 6.

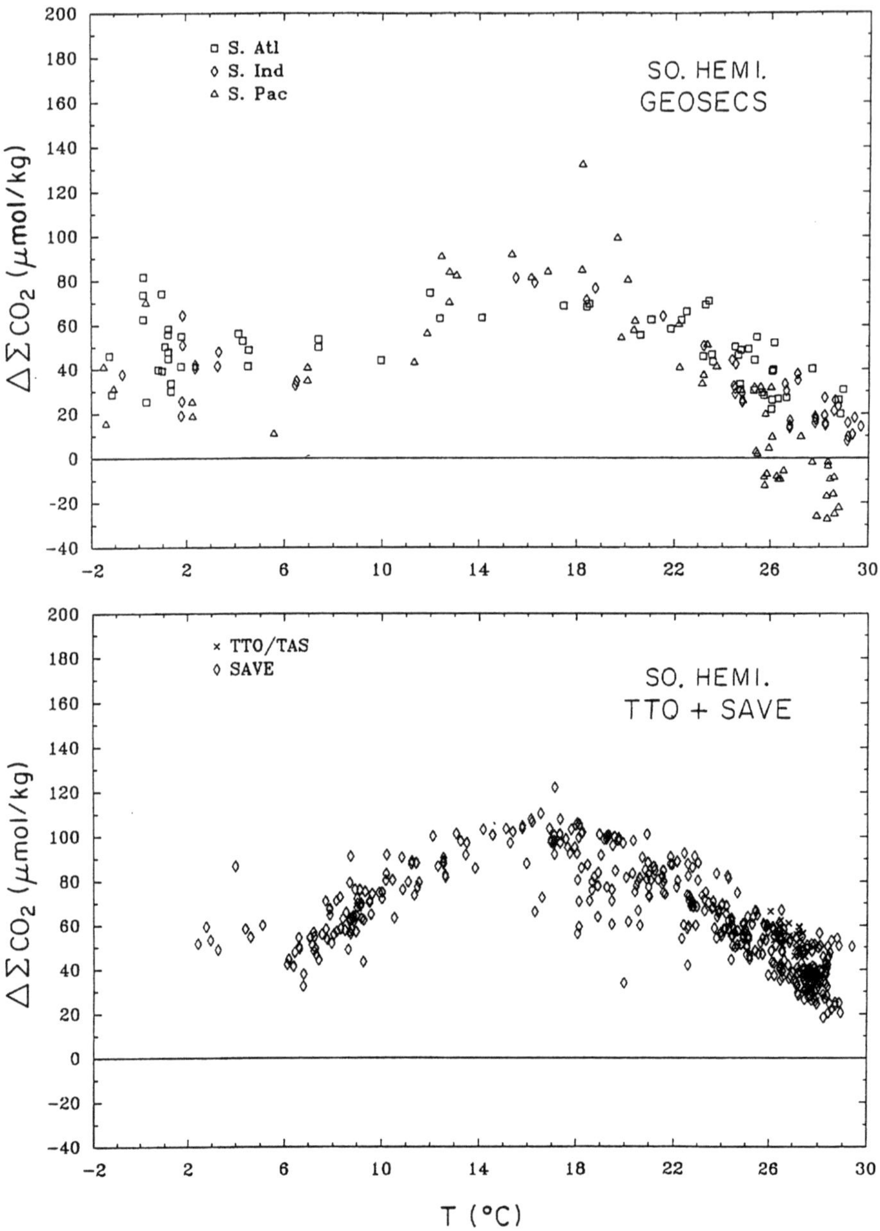

Figure 1. $\Delta\Sigma CO_2$ values for southern hemisphere surface waters. The upper panel is for GEOSECS (1972-3). The lower panel is for TAS (1983) and SAVE (1988).

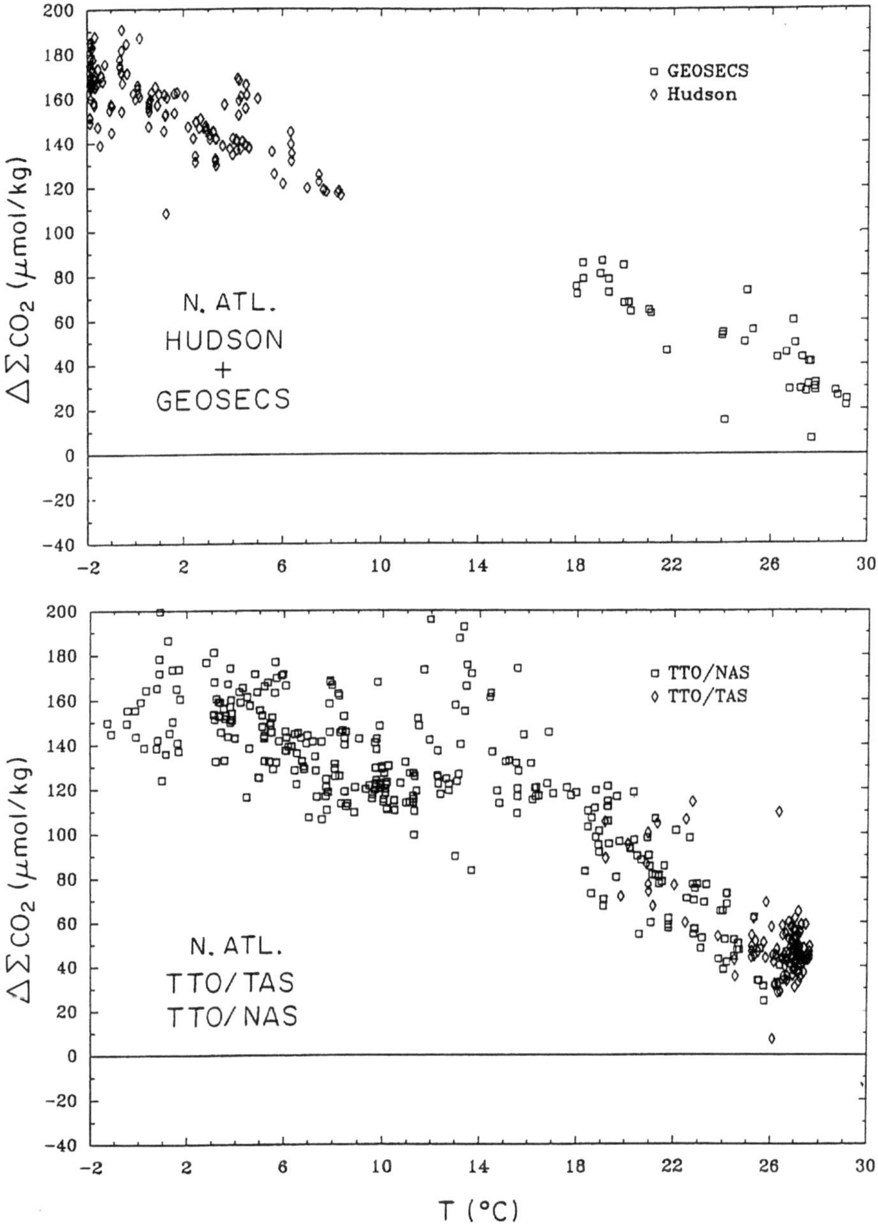

Figure 2. $\Delta\Sigma CO_2$ values for North Atlantic surface waters. In the upper panel are shown results from the GEOSECS (1972) and the Hudson (Winter, 1982) expeditions. The lower panel are shown the results from the TTO northern Atlantic (1981) and the TTO tropical Atlantic expeditions (1983).

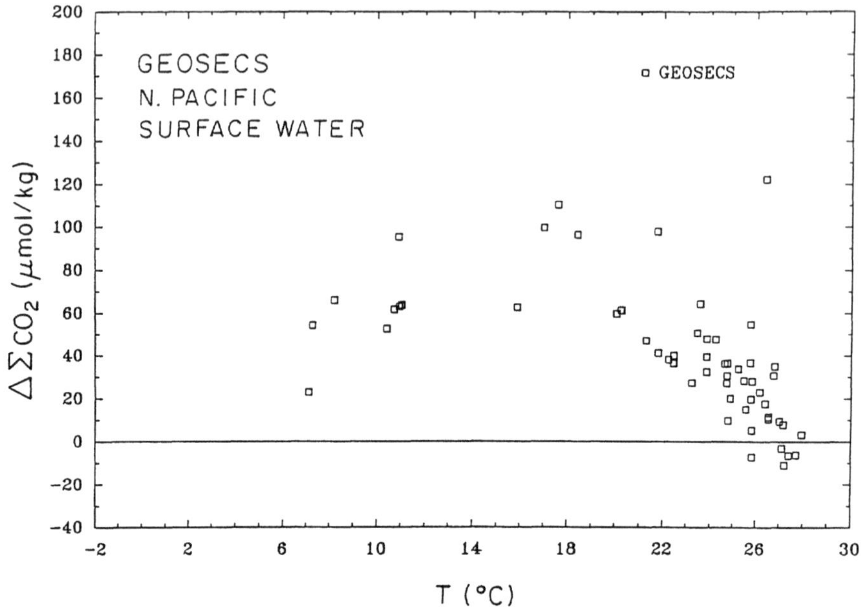

Figure 3. $\Delta\Sigma CO_2$ values for surface waters in the northern hemisphere portion of Pacific Ocean (based on measurements made during the GEOSECS survey).

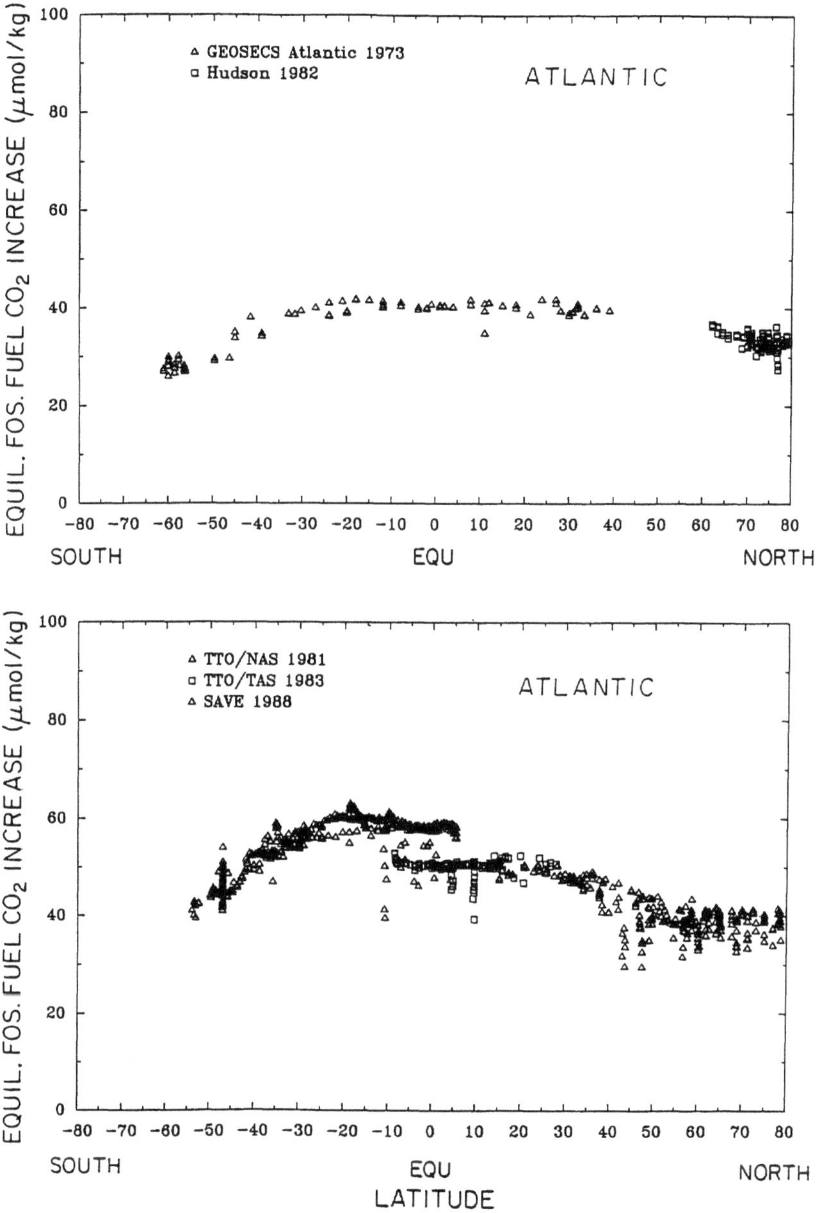

Figure 4. Equilibrium increases in $\Delta\Sigma CO_2$ for surface waters in the Atlantic expected from the increase in atmospheric CO_2 content from a preindustrial value of 280 μatm to the CO_2 content at the time the water sample was obtained. The magnitude of the increase is a function of water temperature.

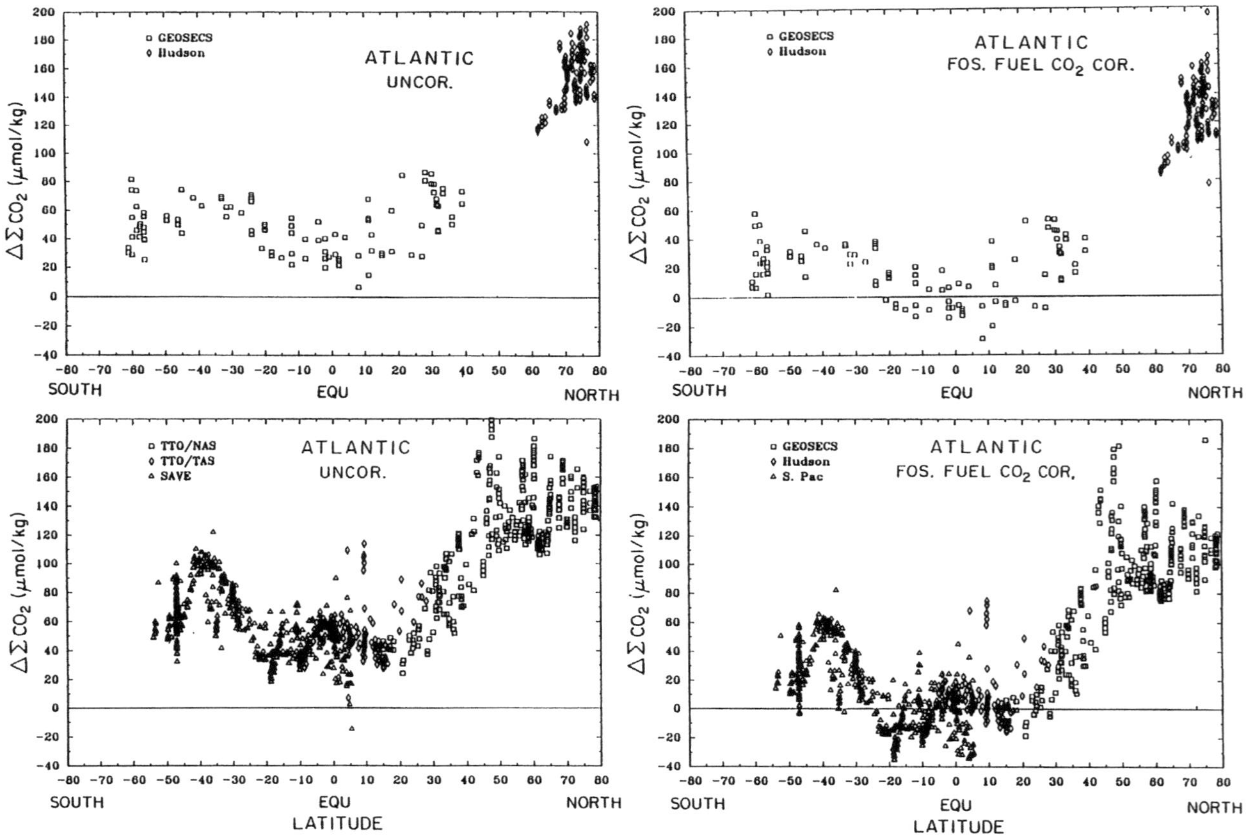

Figure 5. Comparison of $\Delta\Sigma CO_2$ values without (on the left) and with (on the right) the correction for fossil fuel CO_2 uptake.

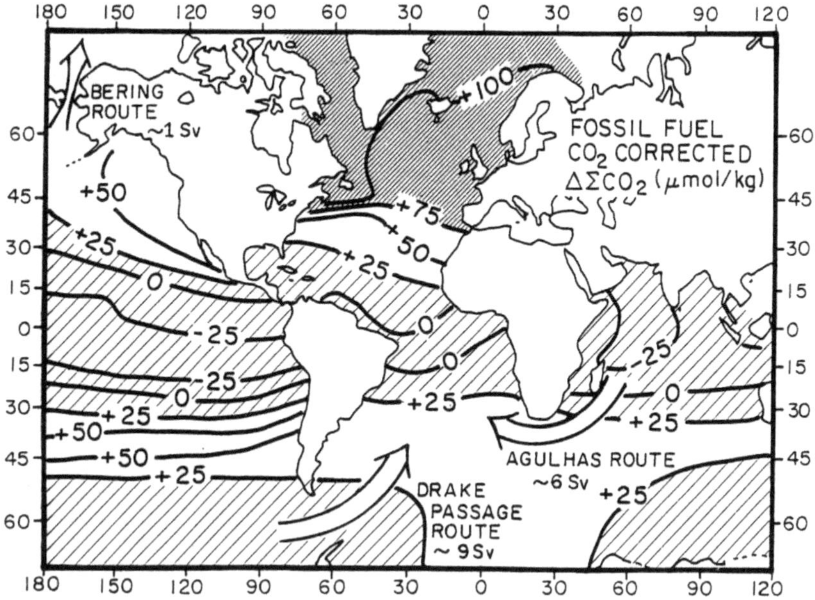

Figure 6. Map showing the geographical distribution of fossil fuel CO_2 corrected $\Delta\Sigma CO_2$ values for surface waters based on measurements of salinity, PO_4, NO_3, ΣCO_2, and alkalinity made during the GEOSECS and TTO expeditions. In the temperate North Atlantic where geographic overlap exists between these two expeditions, the data sets are concordant. It should be noted that winter $\Delta\Sigma CO_2$ values computed from measurements on winter surface waters in the Atlantic north of Iceland during the Hudson 1982 expedition yield values similar to the summer values obtained during the TTO expedition. The return routes for water balancing export from the Atlantic via the conveyor's lower limb are shown by the arrows with rough estimates of the magnitude of their contributions in Sverdrups ($10^6 m^3$/sec). Northward-flowing deep water contributes the remaining four Sverdrups.

Why do waters in the northern Atlantic have high ΣCO_2 values? To answer this question we must consider what we refer to as the expected CO_2 partial pressure for surface waters. It is the CO_2 pressure which would exist were no transfer of CO_2 between the ocean and atmosphere to occur; in other words, it is the CO_2 partial pressure which would exist if the water had the expected ΣCO_2 concentration. Hence,

$$p_{CO_2}^{EXP} = f\left(T,S,ALK,\Sigma CO_2^{EXP}\right)$$

It must be kept in mind that as the choice of coefficients in the ΣCO_2^{EXP} equation was arbitrary, the mean $p_{CO_2}^{EXP}$ value need not equal that for the 1850 atmosphere. Indeed, the average value is somewhat lower. However, as we are interested only in the regional differences in $p_{CO_2}^{EXP}$, this small mismatch is of no consequence. It must also be kept in mind that the $p_{CO_2}^{EXP}$ values have not been influenced by the ongoing atmospheric CO_2 content buildup; rather, they are the partial pressures expected during preindustrial if no CO_2 transport through the atmosphere occurred. As can be seen from Figures 7 and 8, northern Atlantic waters have a lower expected CO_2 partial pressures than any water in the ocean. Hence a tendency exists for CO_2 from other parts of the ocean to move through the atmosphere into the northern Atlantic. As the rate of CO_2 exchange between atmosphere and ocean is rapid with respect to mixing within the sea, most of these expected regional differences in surface ocean pCO_2 are largely eliminated by transport of CO_2 through the atmosphere.

The pattern of expected pCO_2 is influenced by both temperature and nutrient content. For waters of the same phosphate concentration, the higher the temperature the higher the expected CO_2 partial pressure. For waters of the same temperature, the higher the nutrient content the higher the expected CO_2 partial pressure. As cold surface waters have on the average higher nutrient contents than warm surface waters, the two influences tend to counter one another (see Figure 9). The large difference between the expected pCO_2 for surface waters in the Antarctic and those in the northern Atlantic region is a consequence of the difference in nutrient content. The Antarctic waters have on the average 1.6 µmol/kg of PO_4 yielding just the

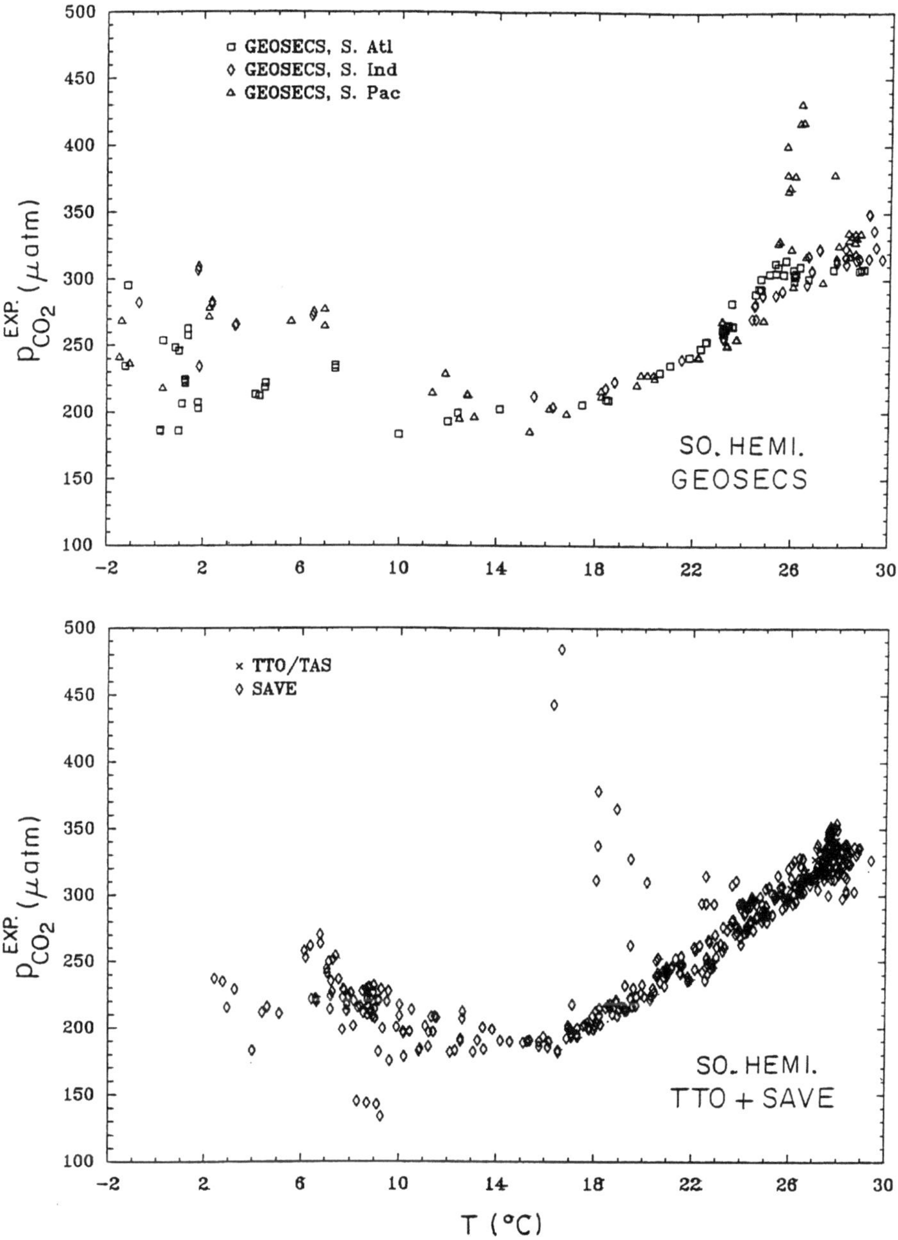

Figure 7. Expected CO_2 partial pressure for southern hemisphere surface waters.

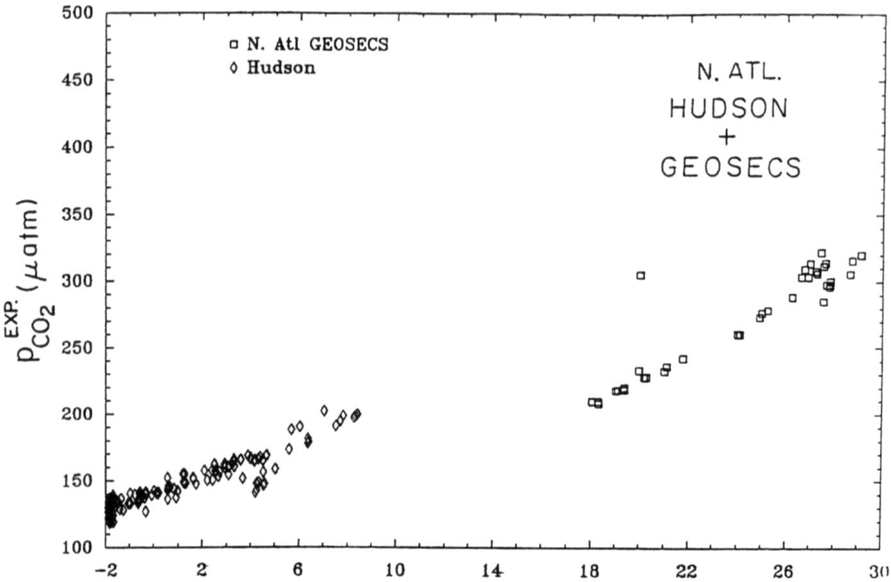

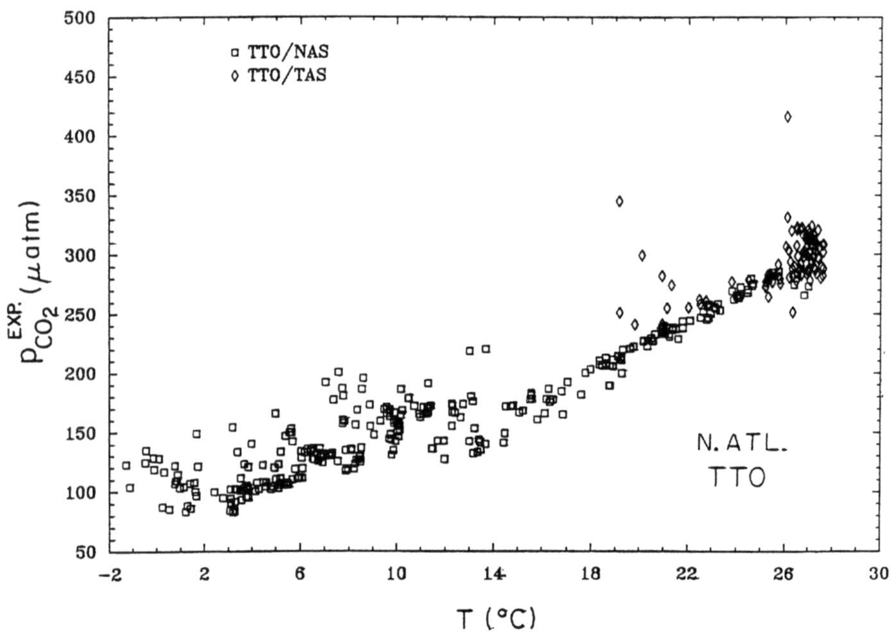

Figure 8. Expected CO_2 partial pressures for northern hemisphere Atlantic surface waters.

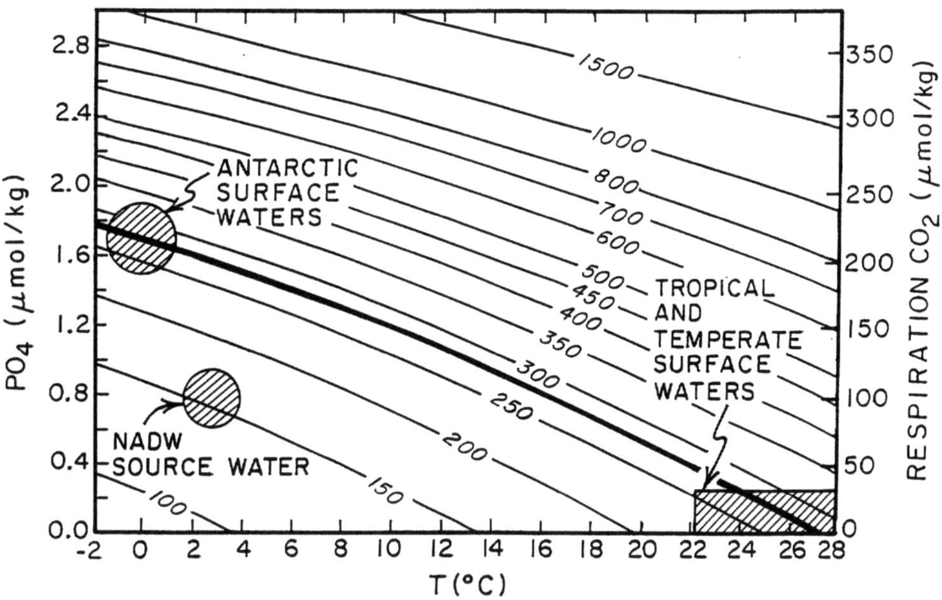

Figure 9. Contours of CO_2 partial pressure (in μatm) for a water sample with a salinity of 35‰, a ΣCO_2 concentration of 1900 μmol/kg, and an alkalinity of 2310 μequiv/kg as a function of water temperature and phosphate content. A C/P ratio of 127 and an N/P ratio of 16 is assumed for the marine organic matter formed by photosynthesis and destroyed by respiration. The heavy contour is that representing the preindustrial atmosphere (280 μatm). Note that in the real ocean fresh water-induced salinity and $CaCO_3$-induced alkalinity differences create additional texture in the surface water CO_2 partial pressure distribution. The important point of this diagram is that the high nutrient content compensates for the low temperature of Antarctic surface waters giving them nearly the same CO_2 partial pressure as warm nutrient-free surface water. By contrast, because of their lower nutrient content, surface waters in the northern Atlantic tend to have much lower CO_2 pressures.

amount of excess ΣCO_2 to compensate for the colder temperature. By contrast, surface waters in the northern Atlantic waters have only 0.9 μmol/kg of PO_4. The amount of excess ΣCO_2 associated with this PO_4 is too small to offset the lower temperature. The difference of 0.7 μmol/kg PO_4 compared to Antarctic surface water (i.e., 1.6-0.9) corresponds to 90 μmol/kg ΣCO_2(0.7x127). With a buffer factor of 14, this yields a 60% reduction in pCO_2! Hence northern Atlantic surface waters have much lower expected pCO_2 values than Antarctic surface waters. Thus the interaction between biological cycling and water transport that dictates low nutrient contents in Atlantic deep waters relative to other deep waters in the world ocean also creates the tendency for transport of CO_2 through the atmosphere to the surface of the northern Atlantic. Of course these transports must be matched by compensating transports through the ocean.

In order to determine the magnitude of the ocean transports, it is necessary to couple the fossil fuel corrected $\Delta\Sigma CO_2$ difference with an ocean mixing model. This is a task for the future. In the meantime, it is only possible to estimate the transport of ΣCO_2 related to the Atlantic's conveyor circulation. Surface waters with high ΣCO_2 values are converted to deep water in the northern Atlantic. This water flows the length of the Atlantic into the Antarctic where it joins the circumpolar current and eventually upwells to the surface. To the extent that this upwelling occurs in the Antarctic, the excess CO_2 carried by the southward flowing deep water is discharged into the southern hemisphere atmosphere. The export of deep water from the Atlantic Ocean is balanced by an import from the south along the surface, at intermediate depths, and along the bottom. The contrast between the $\Delta\Sigma CO_2$ contents of the intermediate and bottom water contributions to this return flow and that in deep water produced in the northern Atlantic are shown by the $\Delta\Sigma CO_2$ versus depth plot shown in Figure 10 for a GEOSECS station from 12°S in the western Atlantic. As none of the water below 500 meters has incorporated a significant amount of fossil fuel CO_2, no correction is needed. Clearly, the southward flowing North Atlantic Deep Water (NADW) carries an excess of ΣCO_2 over the northward flowing overlying Antarctic Intermediate Water (AAIW) and underlying Antarctic Bottom Water (AABW). That this contrast applied to the full range of waters in the deep

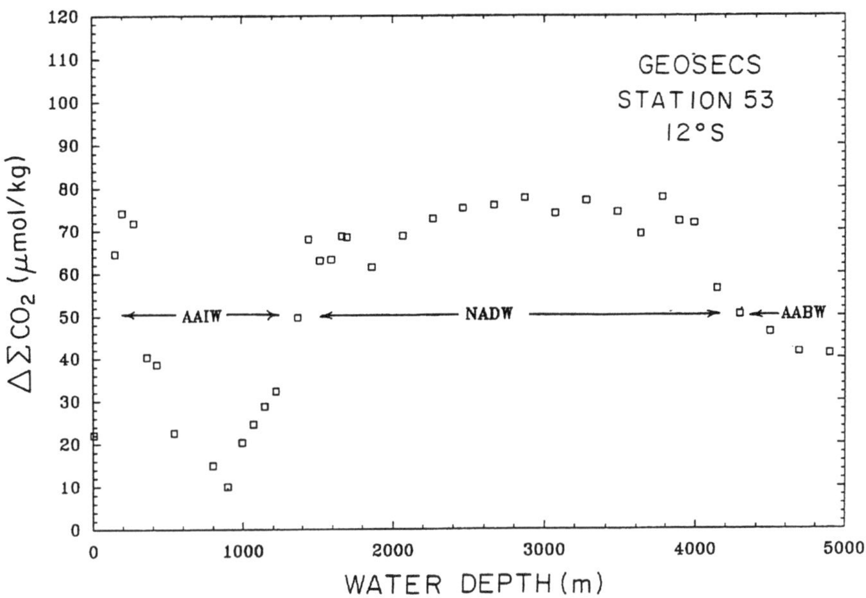

Figure 10. $\Delta\Sigma CO_2$ versus water depth at GEOSECS station 53 in the western Atlantic (12°S). The southward flowing North Atlantic Deep Water (NADW) has higher values than the northward flowing overlying (AAIW) and underlying (AABW) waters of Antarctic origin.

Atlantic demonstrated in Figure 11 where $\Delta\Sigma CO_2$ is plotted against PO_4^*(-1.95+PO_4+O_2/175) for all waters deeper than 1000 meters in the Atlantic Ocean (40°N to 40°S). Deep waters formed in the northern Atlantic have PO_4^* values close to 0.73 μmol/kg. Those formed in the Antarctic have values in the range 1.6 to 1.7 μmol/kg. As PO_4^* is a quasiconservative property, the value for a given deep water provides a measure of the relative contribution of the northern and the southern component. As can be seen, the $\Delta\Sigma CO_2$ values fall along a linear trend. Relatively pure NADW has a $\Delta\Sigma CO_2$ value of about 100 μmol/kg and relatively pure Antarctic component, a $\Delta\Sigma CO_2$ value of 20 μmol/kg. The difference of 80 μmol/kg is similar to the difference between the fossil fuel corrected $\Delta\Sigma CO_2$ values for surface waters in the northern Atlantic and Antarctic (see Figure 6).

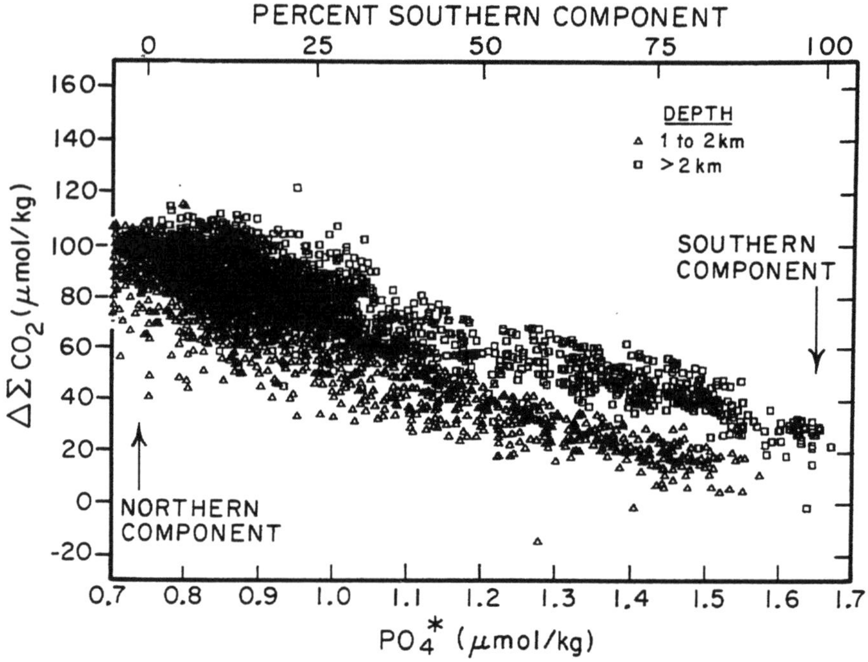

Figure 11. $\Delta\Sigma CO_2$ versus PO_4^* for intermediate waters (triangles) and for deep waters (squares) in the Atlantic Ocean (40°N to 40°S). PO_4^* is a measure of the contribution of Antarctic-derived waters to any given water parcel. The PO_4^* and $\Delta\Sigma CO_2$ values are calculated from measurements made as part of the GEOSECS, TTO/NAS, TTO/TAS, and SAVE expeditions in the Atlantic Ocean.

This is not the whole story for the main contributors to the return flow of water to the Atlantic is thought to be along the surface, one Sverdrup via the Bering Straits, six or so around the tip of Africa via the Agulhas retroflection (Gordon, 1986), and nine or so around the tip of south America via the circumpolar current. These waters all have $\Delta\Sigma CO_2$ values similar to those for Antarctic surface waters (see Figure 6).

Hence the tendency for ocean transport of CO_2 from the northern to the southern hemisphere is surely there. If the difference in $\Delta\Sigma CO_2$ between southward flowing NADW and the northward flowing water is taken to be 80 μmol/kg, then with a transport of 20 Sverdrups (Broecker et al., 1991) about 0.6 gigatons of carbon per year could be transferred from one hemisphere to the other as a result of the Atlantic's conveyor circulation. This is about twice the estimate made by Brewer et al. (1989) for the carbon transport through a section across the Atlantic Ocean at 25°N. However, as Brewer et al. made no attempt to correct for fossil fuel CO_2 being carried northward in the upper waters, their estimate must be considered a minimum.

Of course, this scenario requires that during preindustrial time the southern hemisphere ocean was emitting 0.6 gigatons of carbon per year to the atmosphere while the northern hemisphere portion of the Atlantic was absorbing this amount. Assuming an average CO_2 invasion rate of 18 moles/m^2yr for a pCO_2 of 280 μatm, this requires an average air-sea pCO_2 difference of -13 μatm for the entire North Atlantic (area $0.6x10^{14}m^2$) and of +13 μatm for the Southern Ocean (area $0.6x10^{14}m^2$). However, as has been shown by Sarmiento (in press), a sizable portion of the uptake in the North Atlantic can be accounted for by river run off. CO_2 is being absorbed by plants and soils and is returned to the ocean as HCO_3^- and organic matter. Thus the average pCO_2 deficiency in northern Atlantic surface waters need not have been as large as 30 μatm.

ACKNOWLEDGEMENTS

This research is supported by ongoing grants to Broecker from the Department of Energy CO_2 Program LLNL B130547 and from Exxon Corporation. Lamont-Doherty Geological Observatory Contribution No. 0000.

Peng's research is sponsored by the Carbon Dioxide Research Program, Environmental Sciences Division, Office of Health and Environmental Research, U.S. Department of Energy under contract DE-AC05-84OR21400 with Martin Marietta Energy Systems, Inc. Publication No. 3864, Environmental Sciences Division, Oak Ridge National Laboratory .

REFERENCES

Brewer PG, Goyet C, Dyrssen D (1989) Carbon dioxide transport by ocean currents at 25°N latitude in the Atlantic Ocean. Science 246: 477-479

Broecker WS, Takahashi T (1985) Sources and flow patterns of deep-ocean waters as deduced from potential temperature salinity and initial phosphate concentrations. J. Geophys. Res. 90: 6925-6939

Broecker WS, Virgilio A, Peng T-H (1991) Radiocarbon age of water in the deep Atlantic revisited. Geophys. Res. Lett. 18:1-3

Gordon AL (1986) Interocean exchange of thermocline water. J. Geophys. Res. 91:5037-5046

Heimann M, Keeling CD (1986) Meridional eddy diffusion model of the transport of atmospheric carbon dioxide. 1. Seasonal carbon cycle over the tropical Pacific Ocean. J. Geophys. Res. 91:7765-7781

Keeling CD, Heimann M (1986) Meridional eddy diffusion model of the transport of atmospheric carbon dioxide. 2. Mean annual carbon cycle. J. Geophys. Res. 91: 7782-7796

Keeling CD, Piper SC, Heimann M (1989) A three-dimensional model of atmospheric CO_2 transport based on observed winds: 4. Mean annual gradients and interannual variations. In: Peterson DH (ed) Aspects of climate variability in the Pacific and the Western Americas, Geophysical Monograph 55. American Geophysical Union, Washington, DC. p 305

Peng T-H, Broecker WS (1987) C/P ratios in marine detritus. Global Biogeochemical Cycles 1:155-161

Sarmiento JL, Sundquist ET (1992) Oceanic uptake of anthropogenic CO_2: A new budget, Nature 356:589-593

Tans PP, Fung IY, Takahashi T (1990) Observational constraints on the global atmospheric CO_2 budget, Science, 247, 1431-1438

SUMMARY OF WORKSHOP ON RADIOCARBON DISTRIBUTIONS AS A CONSTRAINT FOR GLOBAL CARBON CYCLE MODELING

Coordinator: U. Siegenthaler
Physics Institute
University of Bern
Sidlerstrasse 5
CH-3012 Bern
Switzerland

Rapporteur: T.-H. Peng
Environmental Sciences Division
Oak Ridge National Laboratory
Oak Ridge TN 37830
U.S.A.

Radiocarbon is the most important tracer used for placing constraints on modeling of the global carbon cycle. Natural ^{14}C is produced in the upper atmosphere and subsequently is oxidized to ^{14}CO and later to $^{14}CO_2$ which is then absorbed into the terrestrial biosphere or transferred into the ocean through sea-air exchange process. The natural ^{14}C serves as a clock (with half life of 5730 years) in the natural system for processes with time constant of 102 to 104 years. The nuclear bomb tests in the late 1950s and early 1960 have produced a great amount of ^{14}C in the atmosphere. This bomb-produced ^{14}C signal spread into the global carbon system and is useful for studying processes with time constants on the order of 10 years. Because the mean age of fossil fuel CO_2 molecules in the atmosphere is on the order of 30 years, the distribution of bomb-^{14}C, with its current mean age of 30 years after its peak level in mid-1960s, offers the optimum analogy for fossil fuel CO_2 distribution in the ocean-atmosphere system.

The distribution of natural ^{14}C in the pre-industrial atmosphere remained approximately constant (with 5% variations caused mainly by the changes in production rates). Direct measurements of atmospheric ^{14}C were made since 1953. The atmospheric ^{14}C levels before this time are obtained by measuring the ^{14}C content in tree rings of known age. Data of ^{14}C variations in the atmosphere for the past 9000 years are thus made available. The natural ^{14}C in the ocean is considered to be at steady state between input from the atmosphere and radioactive decay. However, since the industrial revolution in the 19th century, the atmospheric ^{14}C levels have been reduced by the dilution of ^{14}C-free fossil fuel CO_2 released into the atmosphere (i.e.

NATO ASI Series, Vol. I 15
The Global Carbon Cycle
Edited by M. Heimann

Suess effect). By 1950, the Suess effect had reached 20 , although the increase in atmospheric pCO_2 was from 280 atm to 310 atm which corresponds to a change of 10%. The difference between the expected and the observed decrease has been used extensively to extract information on the operation of global carbon cycle. It can be explained by the fact that the ocean is chemically buffered against the uptake of CO_2, but not against the uptake of an isotopic signal, and by dilution of the isotopic signal by the land biota.

Very few measurements of ^{14}C in seawater were made before 1960s. A large scale oceanic survey of the ^{14}C distribution was made during the GEOSECS program for the Atlantic Ocean in 1973, the Pacific Ocean in 1974, and the Indian Ocean in 1978. By that time, the surface ocean waters had already been invaded by bomb-produced ^{14}C. Determination of the pre-bomb distribution of ^{14}C in the seawater is important for understanding the carbon cycle in the ocean, in its own value as well as in estimating the amount of bomb-produced ^{14}C in the ocean. Inventories of bomb ^{14}C in the global ocean have been estimated using a few available pre-bomb surface water ^{14}C measurements made in the late 1950s and early 1960s, and aided with the penetration depth of bomb ^{3}H measured simultaneously during the GEOSECS program.

Suggestions were discussed during the working group session for reconstructing pre-bomb ^{14}C concentrations in surface waters. In addition to coral growth rings, clam shells have been considered which have growth rings that can be dated to obtain the ^{14}C time series in the water it grew. Collections of these shells in world museums are available for regions such as the higher latitudes where coral reefs do not grow. Shell middens in coastal regions can also provide valuable information on pre-bomb ^{14}C in surface waters. The terrestrial organic components of middens can be used as an index of time, and the coexisting shells can then be dated to reveal the ^{14}C levels, in sea waters.

The sea-air CO_2 exchange rate can be obtained from the distribution of natural or of bomb-produced ^{14}C in the ocean. The flux of ^{14}C transported across the sea-air interface can be expressed as:

$$F = d/dt\ (\text{Ocean } ^{14}\text{C Inventory}) + (^{14}\text{C Decay})$$

Because the natural ^{14}C distribution in the ocean-atmosphere system was at steady state, the change with time of the natural oceanic ^{14}C inventory is equal to zero. We obtain

$$F = (^{14}C\ decay) = A_{oc}\ k_g\ (C_{atm} - C_{surface}),$$

where k_g is the CO_2 exchange coefficient and A_{oc} is the ocean surface area. Based on pre-bomb ^{14}C values, the CO_2 exchange rate is estimated to be 15 to 18 mol m^{-2} yr^{-1} at atmospheric pCO_2 of 280 ppm. Alternatively, the change of bomb ^{14}C inventory in the water column can be used to estimate the CO_2 exchange flux. Such estimates give 15 to 20 mol m^{-2} yr^{-1} at an atmospheric pCO_2 of 280 ppm. These CO_2 exchange fluxes are much higher than the Liss and Merlivat (1986) estimates based on laboratory and field gas exchange experiments which give a value of 8 to 11 mol m^{-2} yr^{-1}. The reason for this discrepancy is presently not understood.

Estimated bomb ^{14}C inventories in the Antarctic Ocean are very low. They are lower than the value expected based on oceanic freon data. Reasons for such low values must be sought because of their relevance to the ocean dynamics in the Southern oceans. Broecker et al. (1985) proposed that upwelling of low ^{14}C waters in this area dilutes the ^{14}C content, and the upwelled water carried the ^{14}C invaded from the atmosphere northward into the Southern temperate regions. Another proposed possibility is that the surface water in the Antarctic Ocean sinks to mid-depth and then is carried along isopycnal surfaces toward the north. Because of sparse data availability in the Antarctic Ocean, the interpretation of ^{14}C in the Antarctic Ocean is difficult. More ^{14}C measurements in this region are required. Currently, some additional ^{14}C data are available in limited regions of the southern oceans measured by German, French, and New Zealand expeditions. A recent SAVE program added more ^{14}C measurements in the Atlantic sector. In the coming years, the WOCE program will provide additional opportunity to take samples in the Pacific and Indian sectors of the Antarctic Ocean. The only limitation is that very few high latitude cruises are planned in the WOCE program. In addition to ^{14}C analyses, measurements of the distribution of CFCs will be a valuable aid for model calibration and validation. A problem is that data collected are usually kept for a long time before becoming available for the public use.

Bomb- ^{14}C in the ocean has been used as an analogue to anthropogenic CO_2. However, there are limits to this analogy. The CO_2 gas reaches chemical equilibrium with the surface water about 10 times faster than ^{14}C atoms. Hence the uptake of CO_2 is less sensitive to the value of the gas exchange rate than the uptake of ^{14}C. Comparison of the input of bomb ^{14}C to various latitude zones of the ocean with the inventory of bomb ^{14}C water columns shows the redistribution of bomb ^{14}C in the ocean caused by the dynamics of water circulation. Similar features are observed for the comparison of model input with model inventory of the excess CO_2 distribution. However, the differences between input and inventory are much larger for CO_2 uptake than for bomb ^{14}C uptake.

Based on consistent estimates of oceanic CO_2 uptake using various types of ocean models calibrated with bomb ^{14}C distribution, we conclude that the current mean oceanic uptake is 2 GtC yr^{-1} with an error of 0.5 GtC yr^{-1}. Taking the mean fossil fuel emission in the 1980s to be 5.5 GtC yr^{-1}, and the amount remaining in the atmosphere to be 3.2 GtC yr^{-1}, the oceanic uptake of 2±0.5 GtC yr^{-1} implies that the terrestrial ecosystem could be in neutral state without net input or output of CO_2 into the atmosphere. The tropical deforestation is likely compensated by enhanced growth in tropical as well as temperate zones. However, to predict the atmospheric CO_2 level in the future, we need to know more precisely the critical processes that control the global carbon cycle.

Distribution and inventories of bomb ^{14}C in land biota are not well known. To estimate the total inventory of bomb ^{14}C in the global carbon reservoirs, we need to know this information. A rough estimate gives about 30% of bomb ^{14}C in the land biota, mostly in the soil organic carbon pool.

The distribution of ^{14}C in the atmosphere since the bomb tests can be used as a constraint for studying atmospheric chemistry, such as the fate of carbon monoxide in the atmosphere. The north-south ^{14}C gradient can also be used as a constraint on atmospheric circulation modeling. However, the most useful constraints based on local atmospheric ^{14}C distribution are in the study of regional mixing of fossil fuel CO_2 with the atmospheric CO_2.

For future measurements, modelers should prepare a strategy for data collection on the basis of their analysis of model results. This strategy should include where and how often observations of ^{14}C and other tracers should be made. Field experimentalists and modelers should work together closely. Some plans for monitoring strategic regions or places in the ocean for tracer distribution, such as the Southern ocean, need to be initiated.

Radon profiles in the atmosphere should be measured so that there will be data for calibrating atmospheric GCMs which can then be used for predicting the distribution and gradient of CO_2 in the global atmosphere. Stable isotope tracers, such as ^{13}C, contain important information regarding sinks and sources of carbon in the biosphere and in the ocean. High precision ^{13}C analyses should be made wherever oceanic ^{14}C measurements are made.

SUMMARY OF WORKSHOP ON INVERSE PROBLEMS

Coordinator: I. G. Enting
CSIRO Division of Atmospheric Research
Private Bag 1
Mordialloc, Vic 3195
Australia

Rapporteur: M. R. Manning
DSIR Physical Sciences
P.O. Box 31312
Lower Hutt
New Zealand

The inverse problem, in the context used here, is the determination of sources and sinks of CO_2 from measurements of concentrations in the atmosphere. The connection between source and sink fluxes (called simply source fluxes below) and atmospheric concentrations is by transport through the atmosphere with associated dispersion and mixing. Atmospheric transport models can be used to define a linear relationship between hypothetical source fluxes and the resulting concentration fields, however the eddy diffusive component of atmospheric transport means that this linear relationship cannot easily be inverted.

The inherent nature of the inverse problem can be considered from different points of view. The diffusive processes involved in atmospheric transport increase entropy and there is an associated, and inescapable, loss of information regarding spatial and temporal structure in going from source fluxes to concentrations. From an algebraic point of view the linear relationship between source fluxes and concentrations is via a badly conditioned matrix so that source fluxes estimated from concentrations are unstable with respect to small variations in the concentration fields. Where the concentrations used to determine source fluxes differ from the "true" representative values, the associated errors in the derived source fluxes are amplified. Similar problems occur in many other areas of geophysics and have been recognized for a number of years and discussed in the literature of mathematics and geophysics.

The basic instability of the inverse problem leads to potentially large uncertainties in corresponding estimates of source fluxes due to a number of factors. Both systematic and random measurement errors in the observations themselves mean that the concentrations used to deter-

NATO ASI Series, Vol. I 15
The Global Carbon Cycle
Edited by M. Heimann

mine source fluxes are not the exact values required by the model. In addition there is often no exact measurement analogue for the values required by the model. For example models usually predict spatially averaged values whereas measurements are at specific sites. Data selection criteria used to remove data subject to local influences usually bias the resulting screened data to specific wind conditions, and this may also be difficult to relate to model predictions.

The problems of correspondence between model predictions and available data are most severe for low dimensional models. For example two dimensional models predict zonally averaged concentrations and would require observations of zonally averaged concentrations to determine zonally averaged inverse source fluxes. Three dimensional models naturally allow a closer correspondence between modelled and observed values but do not completely avoid these problems. In both cases inadequate spatial and temporal distribution of the observations will limit the ability to define source flux fields.

Control of instability in the inverted source fluxes, due to errors in the concentration values used, has been addressed to some extent in the literature and is the subject of ongoing study. However, an additional cause of uncertainty in inverted source fluxes arises from errors in the transport model itself. Enting (this conference) has described a formalism in which the effect of differences between "true" and modelled transport can be considered. At this stage it is not known how significant such errors might be for inverted source fluxes in practical situations.

Despite the difficulties involved in inverting transport models to obtain source fluxes, this remains an important way to determine global scale sources in the carbon cycle. Several techniques are available to ensure that estimated source fluxes are constrained to reduce the instabilities inherent in the inversion process and, provided these are used carefully, the inverse modelling approach should continue to be a useful technique.

Controlled inversion techniques that have been studied include the use of smoothed concentration fields to obtain correspondingly smooth estimates of the source fluxes (Enting and Mansbridge, 1989), and constrained inversion (or Bayesian) approaches in which component terms of the source flux fields are kept close to *a priori* estimates of these (Enting and Pearman, 1987). An alternative approach uses singular value decomposition of the linear relationship between source fluxes and concentrations (Manning this conference). In this case control of instabilities is achieved by treating an overdetermined problem in which there are fewer degrees of freedom in the allowed source fields than in the available concentration fields, and discarding poorly determined components of the general solution.

Most detailed studies of inversion techniques have been done with 1 and 2 dimensional transport models. These suggest that about 30 degrees of freedom can be identified in zonally averaged source fluxes, e.g. 6 or 8 Fourier components in each of 5 or 6 latitude bands. It is important now to apply inversion techniques to 3 dimensional models which will more properly define the limits to temporal and spatial resolution of source fluxes inherent in the present CO_2 measurement network.

An important outcome of this work will be to determine how extension of the monitoring network can improve our ability to define CO_2 source fluxes. In particular we anticipate that measurement sites closer to the large seasonal fluxes from the terrestrial biosphere will be required. Analysis of the inverse problem will also enable an evaluation of the comparative value of aircraft measurements versus high tower measurements over land.

The question of inverting models other than atmospheric transport ones was raised. For example models of the ocean mixed layer might be inverted to identify forcing terms from observations. No specific studies of this type were cited, but it is generally the case that diffusive processes are the limiting factor in inversion procedures. Thus modelled systems without large diffusive components can probably be inverted reasonably well given that appropriate data are available.

Some new ice core measurements of CO_2 and CH_4 concentrations from Greenland were shown and compared with contemporaneous values from Antarctic cores. This raises the issue of inverting temporal histories of source fluxes given records of this type. Some mathematical aspects of this problem have been considered by Enting and Mansbridge (1987).

In view of resolving the discrepancy between inverted source fluxes and estimates of CO_2 uptake by the oceans based on pCO_2 measurements (Tans et al 1989, Enting and Mansbridge 1989, Tans et al 1990) it is important to identify the contributions of other known sources. In this regard the contribution of CO oxidation has been considered (Enting and Mansbridge, 1991) with non-seasonal CO sources, but it was suggested that such a study should be carried out with more realistic seasonal CO sources.

REFERENCES

Enting IG, and Mansbridge JV, (1987), Inversion relations for the deconvolution of CO_2 data from ice cores, Inverse Problems 3:L63-69

Enting IG, and Mansbridge JV, (1989), Seasonal sources and sinks of atmospheric CO_2 direct inversion of filtered data, Tellus 41B:111-126

Enting IG, and Mansbridge JV, (1991), Latitudinal distribution of sources and sinks of CO_2: results of an inversion study, Tellus 43B:156-170

Enting IG, and Pearman GI, (1987), Description of a one-dimensional carbon cycle model calibrated using techniques of constrained inversion, Tellus 39B:459-476

Tans PP, Conway TJ, and Nakazawa T, (1989), Latitudinal distribution of sources and sinks of atmospheric carbon dioxide derived from surface observations and an atmospheric transport model, J Geophys Res, 94:5151-5172

Tans PP, Fung IY, and Takahashi T, (1990), Observational constraints on the global atmospheric CO_2 budget, Science 247:1431-1438

SUMMARY OF WORKSHOP ON INTERANNUAL VARIATIONS IN THE CARBON CYCLE

Coordinator: T. Volk
Dept. of Applied Science
New York University
26-36 Stuyvesant Street
New York, N.Y. 10003
U.S.A.

Rapporteur: R. Keeling
Scripps Institution of Oceanography
University of California, San Diego
La Jolla, California 92093-0220
U.S.A.

Ever since Bacastow [1976] showed that variations in atmospheric CO_2 concentration correlated with the Southern Oscillation phenomenon, there has been interest in understanding the origin of this phenomenon. Measurements of CO_2 made since 1976 continue to exhibit the correlation [Keeling et al., 1989], so there evidently does exist some sort of real connection between atmospheric CO_2 variations and El Niño-Southern Oscillation (ENSO) events.

Some clues to the origin of this phenomenon may be found in the measurements of $^{13}C/^{12}C$ ratios of atmospheric CO_2 which allow CO_2 of terrestrial and marine origin to be distinguished. Measurements of $^{13}C/^{12}C$ ratios have now been made for more than a decade [Keeling et al., 1989, Francey et al., 1990] and now span two ENSO events. The interpretation of the $^{13}C/^{12}C$ data is confused, however, because the two research groups engaged in these measurements disagree on the magnitude of the $^{13}C/^{12}C$ changes associated with ENSO events. Measurements by the Scripps-Groningen collaboration [Keeling et al., 1989] indicate that relatively large (~ 1 ‰) changes in $^{13}C/^{12}C$ratio occur in conjunction with ENSO events, whereas measurements by CSIRO [Francey et al., 1990] indicate that the changes are substantially smaller. The CSIRO isotope data indicate that the ENSO-CO_2 correlation can be explained on the basis of oceanic fluxes of CO_2, while the Scripps-Groningen data indicate that both terrestrial and oceanic fluxes must be involved. Because both groups employ air-sampling stations with a similar geographical distribution, the only apparent way to resolve these contradictory findings is to assume that one or both groups have experienced calibration diffi-

NATO ASI Series, Vol. I 15
The Global Carbon Cycle
Edited by M. Heimann

culties which, remarkably, have correlated with the ENSO phenomenon! Until these discrepancies are resolved, any investigation into the origins of the ENSO-CO_2 relationship will require a subjective decision as to which set of $^{13}C/^{12}C$ measurements is correct.

An understanding of the ENSO-CO_2 correlation might provide clues into the sensitivity of the carbon cycle to climate forcing on the 1 to 4 year time scale. As a sort of boundary condition on carbon cycle models, the ENSO-CO_2 correlation may be useful for improving our understanding of the dynamics of the carbon cycle on other time scales as well. An understanding of these changes in the carbon cycle might also aid our understanding of interannual variability in other biogeochemical cycles, such as the N or P cycles.

Although the variations in atmospheric CO_2 and $^{13}C/^{12}C$ associated with ENSO events are quite small (~ 1 ppm, ≤ 0.1 ‰) the carbon fluxes associated with these events are of the same order of magnitude as fossil fuel burning, i.e. several GtC yr^{-1}. The smallness of the atmospheric signals makes it difficult to exactly quantify the phasing of the CO_2 signals in relationship to the seasonal cycle in CO_2 and in terms of the phasing with latitude. It seems unlikely that the origins of the ENSO-CO_2 correlation will ever be adequately understood using the CO_2 and $^{13}C/^{12}C$ data alone unless there is a substantially larger number of permanent air sampling stations.

A large number of mechanisms are potentially responsible for the ENSO-CO_2 correlation. One mechanism which has been explored is the cessation of upwelling in the eastern tropical Pacific. In non-ENSO years, the upwelling is normally a source for atmospheric CO_2 which presumably is balanced by CO_2 uptake at higher latitudes. During ENSO events, the upwelling ceases. Volk [1989] and Siegenthaler and Wenk [1989] have examined this mechanism and concluded that cessation of upwelling, at best, accounts for a small fraction of the observed atmospheric CO_2 variations. Other oceanic mechanisms worth investigating include ocean temperature variations at higher latitudes, variations in deep-water formation, variations in air-sea exchange rates, and changes in oceanic biological cycling.

If the Scripps-Groningen isotopic data is accepted as correct, the carbon fluxes associated with ENSO events evidently involve a release of several gigatons of carbon from terrestrial ecosystems which is accompanied, and largely offset, by increased oceanic uptake of CO_2 [Keeling et al., 1989, Siegenthaler, 1990]. Kaduk and Heimann [1993], using a two-dimensional global model of the terrestrial biosphere, have examined the role of rainfall and temperature variations on carbon uptake and release and inferred that these effects cannot alone account for the observed ENSO-CO_2 correlation. Other terrestrial mechanisms worth considering include variations in irradiance, human land use, shifts between C3 and C4 plants, fires, temperature and rainfall effects on soil respiration and net primary production. At least on smaller geographical scales, it is clear that these types of processes can produce very large interannual variations in carbon storage.

Because of the large number of mechanisms which are potentially involved, isolating the dominant processes responsible for the ENSO-CO_2 correlation may be very difficult. Substantial progess will probably require better global data sets for such variables as temperature, windspeed, precipitation, surface ocean CO_2 partial pressure, cloud cover, and will certainly require resolving the discrepancies in the $^{13}C/^{12}C$ isotopic data. We have yet another outstanding mystery in the intricate workings of the global carbon cycle.

REFERENCES

Bacastow RB (1976) Modulation of atmospheric carbon dioxide by the Southern Oscillation. Nature 261:116-118

Francey RJ, Robbins FJ, Allison CE, Richards NG (1990) The CSIRO global survey of CO_2 stable isotopes. In: Wilson SR, Ayers GP (eds) Baseline Atmospheric Program (Australia). Bureau of Meteorology/CSIRO Division of Atmospheric Research, pp 16-27

Kaduk J, Heimann M (1993) The climate sensitivity of the Osnabrück biosphere model on the ENSO time scale. In: Müller F, Breckling B (eds) Proceedings of the ISEM's 8th International Conference on the State-of-the-Art in Ecological Modelling and Ecological Engineering. Christian Albrecht University, Projektzentrum Ökosystemforschung, Kiel, Germany.

Keeling CD, Bacastow RB, Carter AF, Piper SC, Whorf TP, Heimann M, Mook WG, Roeloffzen H (1989) A three-dimensional model of atmospheric CO_2 transport based on observed winds: 1. Analysis of observational data. In: Peterson DH (ed) Aspects of climate variability in the Pacific and the Western Americas. American Geophysical Union, Washington, pp 156-236

Siegenthaler U, Wenk T (1989) Modelling the ocean's role for El Niño related CO_2 variations. In: World Meteorological Organization, Environmental Pollution Monitoring and Research Programme, No 59, Extended Abstracts of Papers presented at the Third International Conference on Analysis and Evaluation of Atmospheric CO_2 Data, Present and Past, p 189

Siegenthaler U (1990) El Niño and atmospheric CO_2. Nature 345:295-296

Volk T (1989) Effect of the equatorial Pacific upwelling on atmospheric CO_2 during the 1982-1983 El Niño. Global Biogeochemical Cycles 3:267-279

SUMMARY OF WORKSHOP ON DISSOLVED ORGANIC CARBON IN THE OCEAN

Coordinator: J. R. Toggweiler
U.S. Department of Commerce/NOAA
Geophysical Fluid Dynamics Laboratory
P.O. Box 308, Princeton University
Princeton, New Jersey 08542, U.S.A.

Rapporteur: J. Orr
U.S. Department of Commerce/NOAA
Geophysical Fluid Dynamics Laboratory
P.O. Box 308, Princeton University
Princeton, New Jersey 08542, U.S.A.

Sugimura and Suzuki [*Marine Chem*, 24: 105-131, 1988] reported that a new high temperature catalytic technique for oxidizing dissolved organic carbon (DOC) in seawater recovered up to four times more carbon in surface water and two times more carbon in deep water than older, wet-chemical methods. This working group was charged with a discussion of the implications of these new high estimates of dissolved organic carbon on our understanding of the carbon cycle. The Sugimura and Suzuki results have spawned a great deal of interest and controversy because they suggest that the main sink for the production of new organic matter by marine organisms is not the organic matter in sinking particles, but dissolved organic matter which advects and diffuses into the deep sea. Because questions regarding measurement technique and the absolute accuracy of any of the techniques are still very much unanswered, we devoted much of our discussion to the findings of the recent JGOFS DOC intercomparison workshop in Seattle [P. M. Williams, *U. S. JGOFS News, 3, no. 1*, 1991].

Peter J. le B. Williams attended the Seattle workshop and gave us a synthesis of the workshop findings. Prior to the workshop, several dozen investigators were asked to analyse a set of seawater samples (surface water, deep water, and oxygen minimum water) collected at the JGOFS Hawaii Ocean Time-Series Station. The workshop analysts employed the whole variety of oxidation techniques, including high-temperature catalytic oxyidation (HTCO), persulphate oxidation, and ultra-violet oxidation (the latter two referred to hereafter as wet-chemical oxidation (WCO)). There was no unequivocal evidence from the workshop results that DOC concentration determined by HTCO are elevated in all marine and freshwaters over that measured by WCO techniques.There was some evidence at the workshop that much of the dif-

NATO ASI Series, Vol. I 15
The Global Carbon Cycle
Edited by M. Heimann

ference between techniques and between the various laboratories might be explained by different blank levels. While blanks might be consistent within each laboratory's set of analyses, they might vary greatly between instruments or labs. Uncertainties at this level preclude our making very strong inferences about the role of DOC in the oceanic carbon cycle.

R. Toggweiler reported on a modelling study which was recently published [Najjar et al., *Global Biogeochem. Cycles,* 6:45-76, 1992]. One of the most pertinent results of the study concerned the equatorial Pacific. Najjar et al. examine output from a three-dimensional nutrient cycle model in which the biological uptake of upwelled nutrients is partitioned between a sinking particle phase, in which the organic matter in sinking particles is remineralized immediately below where the particles are produced, and a dissolved organic phase, in which organic matter is remineralized slowly and is advected laterally by uper ocean currents. An important role for the dissolved organic pathway is implied by the fact that the particle-only model preduces an unrealistic subsurface build-up of nutrients in equatorial upwelling regions. Basically, the upwelling system has a hard time getting rid of upwelled nutrients when organic particles, produced as a consequence of the upwelling, are remineralized in subsurface water which immediately upwells again. If the equatorial biota produce significant quantities of dissolved organic matter instead of sinking particles, the dissolved organic nutrient pool can carry upwelled nutrients laterally out of the upwelling zone in the surface flow. Inorganic nitrate and phosphate cannot remove nitrogen and phosphorus in this way because NO_3 and PO_4 concentrations are near zero in the water which flows out of the upwelling region across 10°N and 10°S.

Some of Y. Suzuki's unpublished DOC data from the equatorial and tropical North Pacific were presented in the workshop. A map of the surface distribution showed large-scale variations in DOC. DOC levels were low in recently upwelled water in the eastern equatorial Pacific and were high in water flowing out of the equatorial region to the north. These results are broadly consistent with the Najjar et al. hypothesis if one assumes that nitrogen and phosphorus are incorporated into dissolved organic matter along with carbon. Carbon taken up into DOC and later released during DOC remineralization should have a big effect on pCO_2 levels all over the Pacific. Variations in the production of equatorial DOC during different phases of the ENSO cycle, if large enough, might have an impact on the variation in atmospheric CO_2 associated with the ENSO cycle.

SUMMARY OF WORKSHOP ON THE RELATIVE ROLES OF PHYSICS AND CHEMISTRY IN THE MARINE CARBON CYCLE

Coordinator: G. Evans
JGOFS Büro, Institut für Meereskunde
University of Kiel, Düstenbrooker Weg 20
D-2300 Kiel, Germany

Rapporteur: J. Parslow
CSIRO Division of Fisheries
GPO Box 1538
Hobart, Tasmania 7001, Australia

In his introductory remarks, the coordinator suggested that the working group consider the relative roles of physics and biology under three headings: (i) the natural carbon cycle in the oceans; (ii) the short-term uptake of anthropogenic CO_2 by the oceans; (iii) feedback effects of anthropogenic climate change on the marine carbon cycle.

The coordinator suggested the catchphrase "physics mixes things together, biology separates them" as a useful generalization to shape thinking about the relative roles of physics and biology.

THE NATURAL CARBON CYCLE

There is general agreement that biology plays a significant role (along with physics) in determining regional and seasonal variation in surface pCO_2 and air-sea fluxes. A quantitative understanding of this role is required if predictions of regional and seasonal net fluxes are to be obtained from ocean models. These predictions are needed to provide independent checks of flux estimates based on direct measurements of pCO_2, or on atmospheric inverse models, as the Tans et al. controversy demonstrates. This kind of understanding can also help us to intelligently interpolate among direct observations of key geochemical variables such as surface pCO_2, which can vary rapidly in time and space.

To improve our understanding of the natural carbon cycle, we need better models and improved field and experimental data. There is a growing effort to develop integrated physical,

NATO ASI Series, Vol. I 15
The Global Carbon Cycle
Edited by M. Heimann

chemical and biological models, based on physical circulation models ranging from box-diffusion to 3-D general circulation models. The workshop debated the relative effectiveness of large scale geochemical surveys of variables such as O_2, DIC, delta-^{13}C and inorganic nutrients, versus direct biological rate measurements (primary production, sediment trap fluxes, remineralization and grazing rates), in constraining models of biological processes. It is clear that both kinds of measurements have advantages and disadvantages, and that there is a role for both in a balanced field program. It was noted that process studies making direct biological rate measurements need to make coincident measurements of fluxes due to advection and mixing.

There are clearly serious problems for any field program in adequately resolving time and space scales of variation for both fluxes and pools, especially in the surface layer. There is a danger that failure to resolve variability will lead to biases, particularly where nonlinear dependencies are involved (for example, estimates of air-sea exchange based on average wind fields).

SHORT-TERM UPTAKE OF ANTRHOPOGENIC CO_2

It was generally accepted by the working group that the uptake of anthropogenic CO_2 can be predicted by physical models using the concept of a perturbed pCO_2 field (so-called "$\Delta\Delta pCO_2$"), at least until significant changes in climate and ocean circulation have occurred. (It is worth noting again that we can't measure "$\Delta\Delta pCO_2$", but only a field of "ΔpCO_2" whose seasonal and regional pattern is strongly influenced by biological effects.) There is a need to develop a program of field measurements of DIC and alkalinity, to establish an ocean inventory of inorganic carbon, and directly confirm the predictions of the physical models. The models to date have largely been parameterised and/or validated using other tracers.

FEEDBACK EFFECTS OF ANTHROPOGENIC CLIMATE CHANGE

While the ocean biota play a highly significant role in altering the CO_2 distribution within the ocean, and the partitioning of CO_2 between atmosphere and ocean, there are constraints which limit the potential feedback effects of climate change on this role. These can be summed up by a kind of null hypothesis. In an ocean with depletion of inorganic N or P everywhere at the surface, fixed Redfield ratios in all organic-inorganic exchanges, no biogenic $CaCO_3$ flux, and a fixed DOC pool, there would be no room for changes in biological activity to affect surface pCO_2. The principal point here is that, given fixed Redfield ratios, there is a fixed relationship between DIN and DIC. Where deep water rich in nutrients is upwelled, it brings with it a proportionally increased DIC. New production then serves only to deplete both nutrients and DIC to baseline levels. From this viewpoint, new production, which has formed the cornerstone of biological approaches to carbon cycling, is seen as almost irrelevant to atmosphere-ocean exchanges. The discussion therefore focussed on those ways in which these constraints may be circumvented.

There are of course regions of the ocean (high latitudes and Equatorial upwelling) where surface nutrients are not depleted. Changes in the seasonal and regional pattern of depletion in these regions would produce shifts in surface pCO_2 and ultimately in the partitioning of CO_2 between atmosphere and ocean. There are competing hypotheses about the constraints on nutrient depletion in these regions, involving light limitation, zooplankton grazing and micro-nutrient (iron) limitation. It should be possible to resolve these questions within the next 5 to 10 years, given adequate field experiment and modelling. Current models of grazing control are more robust than early models, but need to be tested by direct measurements of microzoo-plankton abundance and grazing, or indirect measurements (e.g NH_4 concentrations and fluxes). Light limitation should also be better understood through a combination of modelling and existing or newly collected data on seasonal changes in water column structure. Experi-ments are planned to test the iron limitation hypothesis at a variety of scales.

Predicted changes in the extent of nutrient depletion following changes in climate and ocean circulation will obviously depend on resolution of these questions. However, it is worth noting that the predicted effect of a "successful" iron fertilization experiment, in which nutri-ents are depleted throughout the Southern Ocean, is to increase net ocean uptake by about 0.6 Gt C p.a.

The existing and potential variation in Redfield ratios is not well-known. There is evi-dence that organic nitrogen in sinking particulate material may be remineralised at shallower depths than organic carbon. Geoff Evans presented a simple 3-layer model to show that, given some vertical separation of C and N recycling, changes in mixing rate that vary with depth will change the vertical distribution of DIC and consequently surface pCO_2. Similar results have been demonstrated in GCM models by changing the rates of deep-water formation. However, there is a tendency for biological and physical effects to oppose one another in these model experiments. That is, increases in CO_2 uptake due to biological effects tend to be compensated by decreases in CO_2 uptake due to decreased physical mixing/overturn.

The processes controlling the remineralization of C, N and P in the water column are not well understood. They are represented in current models by assuming an empirical depth-scale for remineralization. In reality, this depth-scale is determined by a balance of sinking rates, repackaging by mid-water scavengers and flocculation, and microbial activity. One might expect these to depend in turn on other factors such as temperature, and the composition of the surface community. Better quantitative understanding will be required to model these processes, and predict consequences of change in climate and ocean circulation. It was recommended that existing and new sediment trap data be analysed to try to identify relationships between remineralization rates and other environmental variables, and that process studies pay more attention to remineralization.

Physiological studies suggest that the C:N and C:P ratios in primary production can vary significantly, depending on environmental conditions such as light and nutrient. In view of this potential variability, the apparent constancy in Redfield ratios is somewhat surprising, although this may be due to large scale averaging in time and space in deep water. The critical question is whether significant shifts in these ratios on a regional or global basis could result from climate change. Such changes could result from physiological responses or shifts in community structure. It is clear that we need to better characterize the variability in these ratios in the field now, and understand the factors controlling them, before we can make sensible predictions.

The biogenic $CaCO_3$ flux is dissolved deeper than where organic C is remineralised. This allows the marine biota to act as a carbonate pump, whose magnitude depends on the rate at which ocean circulation recycles nutrients to the sea surface, and the ratio of $CaCO_3$ to organic carbon in particulate matter exported from the surface layer. The latter depends in turn on the community composition: certain phytoplankton, notably coccolithophorids, and microzooplankton such as foraminifera and pteropods, contribute most of the $CaCO_3$ flux. The processes controlling community composition are perhaps the least understood and most difficult to model, although there is some evidence for the importance of simple environmental factors such as temperature and silicate concentration. In order to match observed carbonate profiles,

GCM models must assume that some $CaCO_3$ dissolves above the level of the lysocline. This may occur in zooplankton guts.

The carbonate pump is not subject to the stoichiometric controls on organic carbon fluxes imposed by Redfield ratios, and could in theory lead to quite large increases in surface pCO_2 under the right combination of ocean circulation and community composition.

The DOC pool in the oceans is of comparable size to the atmospheric and terrestrial biosphere C pools. This pool has historically been regarded as being refractory, with a turnover time on the order of hundreds of years. A new measurement technique (high temperature catalytic oxidation) has identified a large previously unmeasured DOC fraction which varies substantially with depth, and apparently has a rapid turnover time. This discovery has a number of implications. First, it may change our understanding of the major fluxes of carbon and nitrogen in the ocean. Perhaps more importantly, a substantial pool of DOC with a relatively short turnover time has the potential to increase or decrease significantly in response to shifts in climate and ocean circulation, and these increases could produce net DIC fluxes which are significant compared with atmospheric shifts. Again, we are not in a position to predict the direction or magnitude of changes in DOC resulting from climate change.

As well as these four primary mechanisms, the workshop considered some other ways in which the distribution of marine carbon could be substantially altered. The above arguments assumed that the total pools of inorganic N and P in the ocean remain fixed. At least in the case of N, there is substantial but poorly quantified turnover associated with N-fixation and denitrification. The latter is probably related to the extent of oxygen depletion in the oxygen minimum layer, raising the possibility of feedback between primary production, remineralisation and nitrogen supply. It is also conceivable that anthropogenic inputs of nitrogen through river runoff and atmospheric deposition will in the future significantly increase C removal from surface waters.

LESSONS FROM LONG-TERM RECORDS

The glacial/interglacial record shows a decrease of 80 ppm in atmospheric CO_2 in glacial periods. Changes in delta-^{13}C and direct calculations suggest a decrease in organic carbon, probably terrestrial, of about 450 Gt C during glacial maxima. This implies a net sequestering of about 600 Gt C in the oceans or marine sediments, accompanied by a drop of 80 ppm in surface ocean pCO_2. This requires a substantial shift in the marine carbon cycle.

EQUILIBRIUM SHIFTS AND NEAR-TERM FLUXES

Assessments of the relative importance of different components of the global carbon cycle depend on both the time scale and the scenarios under consideration. There is an understandable tendency to focus on the next 50 years or so, under a scenario of continued and increasing release of fossil fuel CO_2 from a current level of almost 6 GtC p.a. To be "important" in this context, shifts in natural systems must result in very large net rates of carbon uptake or release, of the order of 1 or 2 Gt C p.a. It is difficult to identify potential shifts in marine biological activity which satisfy this criterion. On the other hand, there are a number of possible shifts in the marine carbon cycle which, over periods of hundreds of years, could produce shifts in equilibrium atmospheric CO_2 levels of the order of 80 ppm. In the long run, these kinds of changes may be seen as important, especially if and when releases of fossil fuel CO_2 are substantially reduced. It is also of course possible that, in a system as complex as the marine carbon cycle, the initial or transient response to climate change may be different in character and even sign to the equilibrium response.

RECOMMENDATIONS FOR FIELD STUDIES

The group recommended that process studies aimed at improving understanding of the marine carbon cycle (eg in JGOFS) be targeted at key hypotheses and processes, and at particular areas, rather than attempting to provide a comprehensive and global measurement and balance of fluxes. Surveys must be designed to provide the spatial and temporal coverage and the variables needed to test large scale integrated models, and to detect change in community structure and function at regional scales. This kind of change could be detected directly (observations of community composition) or indirectly (changes in geochemical variables) but for understanding and model development we will require both kinds of observation.

SUMMARY OF WORKSHOP ON TERRESTRIAL CARBON CYCLING

Coordinator: I. C. Prentice
Department of Plant Ecology
Lund University
Östra Vallgatan 14, S-223 61 Lund
Sweden

Rapporteur: W. Emanuel
Environmental Sciences Division
Oak Ridge National Laboratory
Oak Ridge, Tennessee 37831-6335
U.S.A.

Carbon exchanges betweeen the atmosphere and terrestrial ecosystems influence the reponse of atmospheric CO_2 concentration to fossil fuel emissions. But we do not understand terrestrial carbon cycling well enough to fully explain the CO_2 increases that have occurred during the fossil fuel era or to project future changes. The terrestrial carbon cycling working group assessed the current understanding of the extent and distribution of carbon storage in vegetation and soils and of the responses to these pools to perturbations such as land use, rising CO_2 levels, and climatic change. This report summarizes the major issues addressed by the working group.

Carbon losses due to decomposition and fire approximately balance the large annual uptake from the atmosphere by plant growth. But natural disturbances and human activities both cause significant net fluxes into or out of the atmosphere. Forest harvest and clearing and the stimulation of plant productivity by rising atmospheric CO_2 levels appear to have been the most important perturbations to terrestrial carbon pools during the fossil fuel era.

A widespread decrease in the extent of forests over natural conditions is indisputable. However, estimates of fossil fuel era changes in terrestrial carbon storage derived from forest clearing and harvest data continue to be inconsistent with atmospheric CO_2 measurements and estimates of oceanic uptake. A recent analysis of CO_2 data indicates a substantial net sink in the tropics for carbon from the atmosphere while dramatic forest clearing appears to be causing major carbon releases in that region.

Laboratory experiments in which tree seedlings and saplings are grown under controlled

NATO ASI Series, Vol. I 15
The Global Carbon Cycle
Edited by M. Heimann

environmental conditions suggest that rising atmospheric CO_2 levels increase plant growth rates and sizes. Such a CO_2 fertilization effect may cause a net terrestrial sink that offsets losses, even in regions where forest clearing causes large releases. Casual observations including increased growth on permanent plots, required changes in forestry yield tables, and scattered tree-ring records suggest that a CO_2 fertilization effect may be occurring, but field measurements or experiments have yet to give conclusive evidence.

Global inventories of carbon in vegetation and soils are very uncertain. These are derived by multiplying estimates of mean carbon density per unit area in each major biome by estimates of the areal extents of those biomes. Mean densities ignore the large variability in carbon storage and fluxes across biomes. That variability is complex, and the normal distribution frequently does not describe it well. Furthermore, most global summaries of terrestrial carbon are based on mean densities derived from a small number of samples or from the judgement of researchers with field experience in particular regions. Where data are used, they are frequently from unrepresentative study sites - ecologists tend to collect measurements in unusual or particularly pristine ecosystems. Estimates derived from forestry data tend to be lower than those based on data gathered by ecological research programs such as the International Biological Program conducted in the 1970's. However, forestry data are biased because they emphasize wood products.

Carbon storage in roots and its turnover are very poorly understood. Only a few primary measurements have been made in forests. There is also substantial carbon in microbial biomass, but it is very difficult to assess the magnitude of that pool separate from that of the dead organic matter pool. Laboratory experiments frequently demonstrate that plants grown at higher CO_2 concentration develop larger root structures so that belowground carbon pools are candidates for a significant response to rising CO_2 levels. At this point, such responses are virtually undetectable in field circumstances.

The role of natural disturbance in determining the distribution of terrestrial ecosystems and their properties receives inadequate attention. Many disturbances, such as fire, tend to reduce carbon storage and almost certainly are responsible for maintaining lower stature ecosystems such as grasslands over large regions where climatic conditions would otherwise permit ecosystems with higher carbon standing stocks such as forests. But disturbance events also initiate recovery sequences during which disturbed ecosystems are net sinks for carbon from the atmosphere. A CO_2 fertilization effect can cause increases in average carbon storage in ecosystems undergoing reoccurring cycles of disturbance and recovery because plants grow more quickly through the recovery period between disturbances.

Compartment models that are now used to analyze the carbon cycle lack adequate representations of the processes that are responsible for plant growth and the decomposition of dead organic matter - this forces too much dependence on uncertain carbon inventories for model calibration. Currently available vegetation models that simulate carbon dynamics based on the characterisitics of the plants involved and their interactions treat only small land units, are computationally demanding, and have not been fully perfected for several important ecosystem types. Such models, however, can be used to analyze biome level carbon exchanges by Monte Carlo sampling, and global-scale simulations appear feasible in the foreseeable future. Models need to exploit correlations between carbon storage and environmental and edaphic conditions in order to account for spatial variability, but this requires better global data on temperature, precipitation, ground radiation, topography, and soil characteristics.

The analysis of atmospheric CO_2 data in conjunction with models that simulate carbon exchanges between the atmosphere and oceans gives a great deal of insight about sources and sinks of carbon in terrestrial ecosystems. Terrestrial carbon cycling research needs to take strong guidance from the results of such studies. Projects need to be organized to verify the results of such studies and to explain the causes of sources and sinks in terms of carbon cycling processes and perturbations to terrestrial ecosystems including land use, rising CO_2 levels, and climatic variations.

The addition of measurement stations for CO_2 concentration over land to the existing network of stations in remote locations is likely to provide more information about terrestrial sources and sinks. Measurements of CO_2 exchanges between the atmosphere and ecosystems in the vicinity of meaurement stations can indicate the primary ecosystem responses that are contributing to observed CO_2 variations. Portable flux measurement equipment that can be in place for substantial time periods is now available. Ground measurements can be augmented by data obtained by aircraft in order to understand the spatial pattern of exchanges over larger areas.

Total ecosystem carbon cycling studies are required in order to explain flux and concentration measurements in terms of the processes involved. Such studies are expensive and require large investments of labour; thus, only a limited number of representative ecosystems can be studied extensively, and variables and properties that can be relatively easily measured must be sampled globally in order to extend results. Particular attention to ecosystems, such as tropical forests, for which very little data on carbon cycling is now available is needed.

A good deal of information pertaining to terrestrial carbon cycling can probably be obtained through formal surveys of countries by organizations such as the United Nations and the Food and Agriculture Organization. Countries now report fossil fuel production to the United Nations, and those data are the primay basis for estimating CO_2 emissions from fossil fuels. Similar data on land cover and land use assembled by each country should contribute substiantially toward resolving carbon cycling issues. The Food and Agriculture Organization assembles data on timber yields and agriculture, but reporting is too infrequent and is not comprehensive.

Existing projects are addressing many of the needs in terrestrial carbon cycling research. But there is a serious lack of coordination and integration. One implication is that various measurements are inevitably from different locations, making it extremely difficult to draw strong conclusions about the responses and trends that are affecting CO_2 concentration. There is a great deal of variation in the quality of global terrestrial data. For example there exists a reasonable data base for soil carbon and nitrogen in different types of ecosystems, but there is no similar data set for carbon in vegetation or in litter. Computerized maps of ecosystem distribution that can be used to estimate carbon pools do no use the same ecosystem naming conventions as do these basic data sets. A coordinating team of investigators is badly needed.

SUMMARY OF WORKSHOP ON MEASUREMENT AND MODELLING OF THE TERRESTRIAL NET CARBON FLUX

Coordinator: P. G. Jarvis
Institute of Ecology and Resouce Management
University of Edinburgh, Division of Biological Sciences
Darwin Bldg., Mayfield Road
Edinburgh EH9 3JU
United Kingdom

Rapporteur: R. F. Houghton
The Woods Hole Research Center
P.O. Box 296
Woods Hole, MA 02543
U.S.A.

List of measurements/research topics that address changes in the amount of carbon held in terrestrial ecosystems (not in any order):

1.Change in area of forests
   a.remote sensing
   b.land-use statistics, forestry statistics

2.Measurement of flux
   a.surface boundary layer (eddy covariance method)
   b."planetary" boundary layer (high towers, aircraft)

3.Tree rings

4.Soils/peat

5.Continental CO_2 monitoring stations

6. $^{13}C/^{12}C$ in air

7.Studies of whole ecosystems (Closure of C, N, and water budgets)

8.Paleo-analogues

WHAT IS THE SIZE OF THE TERRESTRIAL FLUX WE ARE LOOKING FOR?

The annual accumulation of carbon on land is estimated with global carbon models to be near zero. Keeling et al. (1989) require a net terrestrial sink of less than 1.0 Pg C; Siegenthaler and Oeschger (1987) based on a deconvolution show a net release of 0.5 Pg C in 1984. If the

NATO ASI Series, Vol. I 15
The Global Carbon Cycle
Edited by M. Heimann

release from deforestation and changes in land use is 1.5 to 3.0 Pg C per year (Houghton, this issue), the uptake required to balance this release must be approximately 1.5 to 3.0 Pg C per year. The analysis of Enting and Mansbridge (1991) suggests a larger sink in the tropics than in the temperate boreal zones. The estimate by Tans et al. (1990) suggests a 2.0 to 3.4 Pg sink in temperate zone systems and another 2.0 or so in tropical systems, depending on the magnitude of the deforestation flux. Thus, a global sink on the order of 2 Pg C per year is sought. It is thought to be largest in the tropics and may be in the temperate zone as well.

Table 1. Magnitudes of terrestrial carbon pools and fluxes and relative changes induced by a global terrestrial sink of 2 Pg C per year.

Stocks	PgC	Relative annual change
Global soil carbon including peat	1500	0.1%
Global vegetation	550	0.4%
Fluxes	**$PgC\ yr^{-1}$**	**Relative change compared to preindustrial size**
Gross exchange (global photosynthesis or respiration)	110	2%
Net primary production	50	4%
NPP of forests (50%)[1]	25	8%[2]
NPP of wood (40%)[1]	10	20%[2]

(1) Half global NPP assumed to occur in forests; 40% of forest NPP assumed to be wood.

(2) The estimates that forest NPP might be 8% higher, and that wood production might be 20% higher than preindustrial rates assumes that the entire 2 Pg C storage occurs in these pools. The estimates are upper bounds (see below).

CHANGES IN AREA OF FOREST

A change in the area of forests is more easily observed than a change in the amount of carbon (in vegetation and/or soil) per unit area, because changes in forest area can be observed directly and because such disturbances affect large changes in carbon. (Changes in the area of forests obviously involve changes in carbon per unit area, but the latter are relatively well documented when initiated by landuse change). Because of the great confidence associated with documenting changes in land use relative to documenting changes in carbon per unit area, it is important to determine the net flux associated with changes in land use even in regions where geophysical evidence suggests a net carbon sink. The flux of carbon from land-use change may show either a net source or sink, but knowing it helps constrain any source of sink independent of land use.

CHANGES IN CARBON PER UNIT AREA

The largest signal of a terrestrial sink is expected to be in tree rings (Table 1). If the entire missing sink of about 2 Pg C per year is in wood, the 20% increase in annual growth above preindustrial growth should be readily seen in tree rings. On the other hand, the lack of an observed increase in ring growth cannot preclude a terrestrial sink because the sink may have been in grassland soil, in moist tropical forests without annual growth rings, or in other forests without annual growth rings, or in other non-forest ecosystems. The lack of annunal rings in tropical moist forest is particularly bothersome because some geophysical as well as physiological evidence suggest the greatest sink should be tropical. Trees in tropical seasonal forests (monsoon forests) may have rings (e.g. teak), but large areas of tropical evergreen forests do not.

The possibility that forests have accumulated carbon in leaves or soils without leaving a record in growth rings seems remote. It seems more likely that if one cannot see an indication of an increased growth in tree rings, that there has been been no sequestering of carbon in these forests (including wood and soil). There is one more reason to look in tree rings for the missing sink. The carbon/nitrogen ratio in wood is many times higher in wood than in soil (300 or 400 as opposed to 10 or 13). If nitrogen limits the accumulation of carbon, more carbon can be stored as wood than as soil carbon, sediments, or any other reservoir, with the exception of peat, which probably has a high C/N as well.

Soils are also a potential sink, although detecting a change in soils will be difficult (Table 1). One of the few ways to measure a change retrospectively appears to depend on analysis of archived pre-bomb ^{14}C soil profiles. Comparison of pre- and post-bomb ^{14}C profiles will provide an estimate of change in the rates of soil carbon turnover.

Another means for measuring an accumulation of carbon in terrestrial ecosystems is the direct measurement of flux. Net fluxes are thought to be measurable with an accuracy of 5-10%. These percentages are somewhat higher than the expected signal of 4% (Table 1), but Table 1 presents a worst case. The calculation in the Table assumes that the annual terrestrial sink is evenly distributed over time and over all ecosytems. On the contrary, some ecosystems are more likely than others to accumulate carbon. Furthermore, annual fluxes of carbon are separated in time into day and night and into seasonal fluxes. Thus, one can measure, separately, photosynthesis and respiration. In the end, one is left with finding a small difference between large fluxes, but the separation of the gross fluxes into time and space should help in this determination as well as in others.

The major argument against measuring flux is that most forests are probably accumulating carbon, not because of changing environmental conditions, but because they are reovering from a fire, blow-down, flood, harvest, or other disturbances. If a net accumulation of carbon is measured, how should it be interpreted? How can the natural accumulation of carbon be distinguished from the perturbational accumulation of carbon?

Measurement of carbon flux above forests is useful for three additional reasons. First, the measurements can serve as a validation for models of forest growth. Second, the measurements provide the basis for scaling up to larger areas. And, finally, continuous or regular measurement of natural systems will provide information on the response of forest carbon balance (photosynthesis and respiration) to short-term (year-to-year) variations in climate/weather.

FIVE DIRECTIONS FOR RESEARCH

Resolution of terrestrial carbon sources and sinks requires five types of research:

1. Continued estimation of the flux of carbon from changes in land use, globally. Expand the program to include use of satellite imagery for determination of deforestation and reforestation rates and fate of deforested lands. Explore the use of ERS-1 satellite to determine biomass directly.

2. Evaluation of the record of tree rings to show a net terrestrial sink of carbon. The analysis of rings widths will have to be different from that used in most previous dendrochronological studies. Most previous analyses have removed the long-term trends and focussed on year-to-year variation. The question here concerns the long-term trends. The cause of the trend is

secondary. The first question is whether trees are growing more rapidly now than 50 years ago. The secondary questions are the factors responsible for any increased growth (e.g. CO_2 fertilization, temperature, soil moisture, increased nitrogen availability). The analyses will have to relate ring widths to carbon accumulation in the tree and, eventually, to carbon storage in the stand and ecosystem, including soils.

3. Comparison of ^{14}C profiles in pre- and post-bomb soils.

4. Extension of theory and measurements that would allow calculation of integrated fluxes of carbon over large areas. Current meaurements address scales of either less than a few square km (eddy correlation method) or hemispheric (inversion techniques based on geographic variation of atmospheric CO_2). An intermediate scale is crucial. A combination of high towers and aircraft is probalby required for this regional scale. The network of CO_2 monitoring stations should be expanded to include continental stations. Stable isotopes of carbon as well as concentrations should be measured.

5. The four approaches above, address the primary question of whether a terrestrial sink can be found, and how large it is. It is also important for predictive purposes to understand the factors responsible for a storage. For addressing this question, we must initiate and carry out several long-term whole ecosystem studies. Measurements should include not only net fluxes of carbon, but fluxes from important processes (e.g. photosynthesis, autotrophic respiration, heterotropic respiration, decomposition, etc.). Measurements should be made to close budgets of carbon, nitrogen, energy, and water. Experiments with elevated CO_2, temperature, and water should be carried out.

MIX
Papier aus verantwortungsvollen Quellen
Paper from responsible sources
FSC® C105338
FSC
www.fsc.org

If you have any concerns about our products,
you can contact us on
ProductSafety@springernature.com

In case Publisher is established outside the EU,
the EU authorized representative is:
Springer Nature Customer Service Center GmbH
Europaplatz 3, 69115 Heidelberg, Germany

Printed by Libri Plureos GmbH
in Hamburg, Germany